THE PORTUGUESE COLUMBUS

THE PORTUGUESE COLUMBUS

Secret Agent of King John II

Mascarenhas Barreto

Translated by
Reginald A. Brown

MACMILLAN

English translation © The Macmillan Press Ltd 1992

All rights reserved. No reproduction, copy or transmission of this publication may be made without written permission.

No paragraph of this publication may be reproduced, copied or transmitted save with written permission or in accordance with the provisions of the Copyright, Designs and Patents Act 1988, or under the terms of any licence permitting limited copying issued by the Copyright Licensing Agency, 90 Tottenham Court Road, London W1P 9HE.

Any person who does any unauthorised act in relation to this publication may be liable to criminal prosecution and civil claims for damages.

First published in Portugal in Portuguese in a longer version by Ediçoes Referendo, Lda 1988 as
O Português Cristóvão Colombo: Agente Secreto do Rei Dom João II
Copyright © Augusto de Mascarenhas Barreto, 1988

English translation first published in Great Britain 1992 by
THE MACMILLAN PRESS LTD
Houndmills, Basingstoke, Hampshire RG21 2XS
and London
Companies and representatives
throughout the world

ISBN 0–333–56315–8

A catalogue record for this book is available from the British Library

Printed in Great Britain by
Billing and Sons Ltd, Worcester

To my master, the late Father Antonio da Silva Rego,
and all the mariners of the Discoveries

Contents

List of Tables — xiii
List of Illustrations, Plates and
 Genealogical Trees — xiv
Preface — xix
Acknowledgements — xxi
Translator's Note — xxii

PART I INTRODUCTION

1 The Cathars — 3
   The Island of St Lawrence — 4
   The Persecution — 6
   The Portuguese Monarchs — 8

2 The Inquisition — 13

3 The Portuguese Jews — 19
   The Compulsory Conversions — 28
   The All Saints' Hospital — 30
   Political 'Estates' — 31

4 The Peninsular Crusades and the Templars — 34
   Crusades against the Muslims — 35
   Crusades against the Saracens and the Turks — 36
   The Extinction of the Templar Order — 38
   The Order of Christ — 42

5 The Discoveries before Columbus — 46
   The Canary Islands — 48
   The 'Sagres School' — 51

viii CONTENTS

	The University of Lisbon	53
	Madeira and the Azores	55
	Sailing West – Brazil	57
	Greenland and Newfoundland	62
6	**The Diplomatic Treaties of King John II**	**66**

PART II THE HERMETICISM OF COLUMBUS

7	**Documentation of Columbus's Origins**	**81**
8	**The Language of Columbus**	**86**
9	**The 'Foreign Columbuses'**	**97**
	The Supposed 'Shipwreck'	100
	A Distorted Miracle	104
	Catalan Navigation	107
	The Island of Genoa	109

PART III EMPIRICAL AND SCIENTIFIC NAVIGATION

10	**Portuguese Nautical Cryptography**	**117**
11	**The Enigma of Columbus's Voyages**	**133**
12	**The Problem of the Legendary Thule**	**143**
13	**More Theses on the Discovery of America**	**149**
14	**The Mystery of the Seven Cities**	**162**
	The Conspiracy	164
	Cypangu	167
	Espionage	167
	The Myth of St Brendan	168

PART IV BIOGRAPHICAL PROBLEMS

15	**Enigmatic Lineage**	**179**
	Fornari's Work	183
	The *Capitulaciones* Diploma	184
	The Origins of Columbus	186

16 The Descent of a Weaver — 189
- Notarial Documents of Genoa and Savonna — 191
- The Real Dates of Columbus's Birth and Arrival in Spain — 195
- Muratori's Document — 197
- Suspect Documents — 198
- The Missing Deed *'Mayoradgo'* — 199
- The Fake Deed *'Mayorazgo'* — 200
- The *Royal Deed* and the *Memorial* to Diogo — 205
- The *Memorando* of Debts — 207
- Conclusion — 212

17 Stratagems and Deceits — 214
- The Lawsuit Against the Crown — 215
- The Suspect Final Paragraph of Fernando's Will — 218
- The Lawsuit of Succession — 219
- The *'Military Codicil'* — 221
- The Letters of Columbus — 223
- The Handwritten Note in the *Imago Mundi* — 227
- The King's Letter — 228
- Conclusion — 230

PART V COLUMBUS REVEALED

18 Dispersing the Mist — 235
- The Origins of Colón-Zarco's Name — 237
- Previous Interpretations of Columbus's Siglum — 241
- Two Seventeenth-century Messages — 244

19 The Prophecy of the Sibyl — 246
- Valentim Fernandes — 246
- The Meaning of the Prophecy — 248

20 Astrology and Alchemy — 253
- The Origins of the Perestrelo Family — 255
- The Alchemists — 257
- Palaeo-Christian and Alchemical Symbols — 257

21 The Labyrinth of the Cabala — 262
- The Western Cabala and its Symbology — 263
- The Rite of Mithra — 271
- The Temurah Method and the Names of 'Babel' and 'Tomar' — 273

x   CONTENTS

      The Gematric Method   273
      The Excellence of the Ternary   274
      The Hexagon and the *Dextratio* Processes   275
      The Mirror Process   278
      The Notariqon Method and a Hebraic Temuric Example   279

22 The Other Side of the Mirror   283
      The Meaning of the 'S'   285
      The Umbilical Centre and the Triangles   286
      Irrelevant Messages   298

23 From Light to Darkness   301
      'Ferens' and Frustrating Messages   307
      The Sibyl's Advice   312
      'Câmara'   314
      The Final Solution   316
      The Mysterious Monogram   319
      The Discoverer Prince   320

PART VI   CHECK AND MATE

24 After the Heat of the Battle   325
      'Terra Rubra'   325
      The Fulling and Weaving Industry of Beja   327
      Bartolomeu Colón at the English and French Courts   328
      A Coincidence of Toponymy   330
      Zarco in Madeira   338

25 A House on Madeira and a Convent in Lisbon   342
      'Esmeraldo'   342
      The Convent of Santos (Saints)   343
      Was Filipa Promised in Marriage to Colón-Zarco?   346

26 The Genealogical Web   349
      The Obscurity of New Christian Surnames and the Inquisition   352
      The Tombs of Isabel and Filipa Moniz   354
      The Navigators Converge on Beja   356
      Juvenile Marriages   357

27 'Terra Rubra' and La Rábida   360
      Vila Ruiva   361
      Alvito   363

	The Friars Marchena	364
	The Coutinhos	370
	The Order of St James and the Convent of Santos	371
28	**Huntsmen and Admirals**	**373**
	The Aguiars	375
	The Eliminated Brothers	378
	Bartolomeu and Diogo Colom	382
	Admiral Kinsmen of Colón-Zarco	384
29	**Stopovers at the Azores, Portugal and Arzila**	**388**
	The 'Nina' at Santa Maria	390
	Albergaria, Castanheira and Rui da Câmara	392
	Columbus Visits his Family	393
	The Letter to Santángel	397
	The Siege of Arzila	399

PART VII ON THE TRAIL OF THE LAST INCOGNITOS

30	**The Protectors of Columbus**	**407**
	Genealogical Relationships	407
	All Columbus's Protectors Were his Relatives	413
31	**Adulterers and Adulterators**	**418**
	Queen Juana of Castile and her Daughter 'the Beltraneja'	419
	Beatriz Henriques and Nuno Pereira of Serpa	421
	Intentional Omissions	421
	The Expedition to Tangier	423
	Governors of Fronteira	423
	Nuno Pereira de la Cerda and his Two Marriages	425
32	**The Damsel of Córdoba**	**427**
	The Notarial Documents	430
	Cousins and Lovers	436
33	**Distortions Corrected**	**439**
	The Name of Don Fernando's Grandfather	439
	The Two Portuguese Grandmothers of Don Fernando	440
	Moliarte, Colón-Zarco's Brother-in-Law	448
34	**The Heraldic Secret**	**456**
	Heraldic Symbology	457
	Colón-Zarco's Coat-of-Arms	465

35 A Question of Physiognomy	470
The Blondish Red-Haired Admiral	471
Portraits of Columbus	473
The Panels of the Holy Prince	474
The Jew and his Emblematic 'Margarita'	476
St Bernard's Mission	479
36 The 'Brothers' and 'Cousins' Gambit	481
The 'Pre-Discoverers' of the Antilles	481
Chronology of Columbus's Life	486
The Probable Identity of Álvaro Damán	492
The Colonnas	498
Gambit	504
The 'Promised Land': America	505
Appendix 1 Portuguese Kings of the First Two Dynasties	508
Appendix 2 Portuguese Vocabulary Used by Columbus	509
Appendix 3 Toponymic Evolution	513
Appendix 4 The Hebrew Alphabet	514
Appendix 5 The Ten Sephiroth	517
Appendix 6 Latin Words that Correspond to the Arrangement of Letters in the Triangles	520
Bibliography	531
Index	553

List of Tables

4.1	Western 'Crusades' in the Iberian peninsula against the Saracens and the Turks	34
4.2	Eastern Crusades in the Middle East against the Muslims	34
4.3	Western 'Crusades' against the Cathars	35
22.1	Deciphering Colón-Zarco's siglum: the three-columned chart	294
22.2	Deciphering Colón-Zarco's siglum: eliminating synonyms and experimenting with verbs	296
22.3	Deciphering Colón-Zarco's siglum: initial combinations	299

List of Illustrations, Plates and Genealogical Trees

Illustrations

1.1	The Capet family tree	9
4.1	The 'anal kiss'	40
5.1	The 'return' from Mina	59
6.1	Columbus in the Canary Islands (1492)	73
6.2	The Treaties of Toledo and Tordesillas	74
6.3	Columbus in Jamaica (1493)	75
6.4	Columbus in Arzila (1502)	76
6.5	Columbus's fourth voyage (1502–4)	77
10.1	The *Tables of the declination of the sun*	121
12.1	Columbus in Iceland and Greenland (1447)	142
13.1	The Corte-Real inscriptions	150
13.2	The Dighton Rock inscriptions	152
13.3	The Nautical Chart of Pizzigano (1474)	154
13.4	Twelfth-century Viking warships	155
13.5	The Runic Ielling Rock	156
13.6	The map of Siurdi	157
14.1	The map of Toscanelli (1474)	165
14.2	The Nautical Chart of Pizzigano (1467)	170
14.3	The *Bianco Atlas* map	173
14.4	The Pareto map (1455)	174
16.1	The Genoese and coeval accounts of Columbus's descent	190
18.1	Colón-Zarco's siglum	238
18.2	Previous interpretations of Colón-Zarco's siglum	242
18.3	Ribeiro's image 'on the other side of the mirror'	242
18.4	Saúl Ferreira's transliteration of Colón-Zarco's siglum	243
19.1	The official signature of Valentim Fernandes	248
19.2	The *Book of Marco Polo* (1497)	249
19.3	Phoenixes and eagles	250
20.1	Alchemy and esoteric alchemy	255

20.2	Sexual dualism	260
21.1	The *Sephirothic* system	264
21.2	The *Sephirothic* decenary	265
21.3	The Three Pillars (or Columns) of the Descent of Power	266
21.4	The *Sephirothic* Ternaries and the Auras	266
21.5	The five- or ten-petalled rose	267
21.6	The pelican	267
21.7	The phoenix instead of the pelican	269
21.8	The rope and the fishing net	270
21.9	The two principles of Persian Zoroastrian dualism	271
21.10	Number and symbol	276
21.11	The sacred lance and the sacrificial lamb	276
21.12	The '*dextratio*'	277
21.13	The Macrocosm and the Microcosm	278
21.14	Using the Temurah to decipher Colón-Zarco's siglum	279
22.1	Colón-Zarco's siglum, showing the 'colon'	284
22.2	The Ss in the siglum	285
22.3	Temuric analysis of the siglum: the ternary and the mirror	286
22.4	Temuric analysis of the siglum: the ternary, the umbilical centre and the mirror	286
22.5	Temuric analysis of the siglum: using *sephirothic* triangles and Greek characters	287
22.6	Order of *sephirothic* triangles	287
23.1	Deciphering Colón-Zarco's siglum: using the Western Cabala	302
23.2	The two Shields of David	302
23.3	The 'roof' of *Xpo*	303
23.4	The mirror method of deciphering *Xpo*	303
23.5	Repetition and the Ss	305
23.6	The lower 'mirror' triangle	305
23.7	The upward- and downward-pointing vertices of the triangles	306
23.8	*Xpo* as a link between God and man	307
23.9	Placing FERENS in the *Sephirothic* system	308
23.10	Using the mirror method for FERENS	310
23.11	FERENS in the two hemispheres of the *Sephira Prima*	310
23.12	The 'candelabrum of Seven Arms' and the 'Table of the Bread of Atonement'	312
23.13	Reading the *Sephirothic* system 'in the tracks of the sun'	313
23.14	The Câmara (Zarco) coat-of-arms	315
23.15	The Marinhos and the Dornelas coat-of-arms	316
23.16	The first colon in the base of Colón-Zarco's siglum	317
23.17	The Altars and the Root	317

xvi   LIST OF ILLUSTRATIONS, PLATES AND GENEALOGICAL TREES

23.18	Deciphering 'Cuba'	317
23.19	Colón-Zarco's monogram and siglum	319
23.20	Deciphering Colón-Zarco's monogram	319
24.1	The heraldic symbol of the tower	332
24.2	The Beja region and the Indies	340
25.1	The most notable descendants of King John I of Portugal	344
26.1	Descendants of Henrique Moniz	355
26.2	Descendants of Isabel Moniz and Bartolomeu Perestrelo	356
28.1	Alão de Morais's *Pedatura*	376
31.1	Henrique Henriques de Miranda as Governor of Fronteira	424
33.1	The monasteries of Arrábida and La Rábida and the churches of Guadaloupe	451
34.1	The mirror of the soul	458
34.2	Heraldic symbology	459
34.3	Coats-of-arms of families genealogically linked to Colón-Zarco	462
34.4	Divisions of the heraldic shield	464
34.5	Colón-Zarco's coat-of-arms	466
34.6	The two coats-of-arms of the Henriques, Colón-Zarco's and Columbus's family's coats-of-arms	468
35.1	Maximilian, The Holy Roman Emperor	474
35.2	The Jew and his 'Margarita'	477

Plates

I	The Sagres School
II	The altars at Tomar and Newport
III	The twelfth-century church at Borgund
IV	A window in the Monastery of Christ at Tomar
V	Henry of Bolingbroke
VI	Sir Thomas More
VII	A sixteenth-century English cannon
VIII	A sixteenth-century Portuguese cannon
IX	St Catherine
X	Colón-Zarco's portrait
XI	The Polyptych of Nuno Gonçalves
XII	Prince Fernando, Duke of Viseu and Beja
XIII	Prince Fernando's eldest son, Dom João
XIV	Prince Fernando's second son, Dom Diogo
XV	An enlarged reproduction of Colón-Zarco's portrait
XVI	The Holy Princess Joana, daughter of King Alphonse V and first cousin to Colón-Zarco

Genealogical Trees

I	Relation: Cabral/ Corte Real/ Camões/Gama/Colón-Zarco	362
II	Vila Ruiva/*Terra Rubra*: ascendants and relatives of Nuno de Mello	365
III	Alvito: ascendants and relatives of Rodrigo da Sylveira	366
IV	Menezes/Henriques/Aguiar/Colón-Zarco	367
V	Who received Colón-Zarco in Portugal?	398
VI	Arzila	402
VII	Relation: Marchena/Cabrera (Moya)/Guzmán (Medina Sidonia)/La Cerda (Medinaceli)/Colón-Zarco	409
VIII	Origin of Henriquez	410
IX	Relation: Noroña/Câmara/Ulloa/Deza/Moya/Menezes	411
X	The two wives of Nuno Pereira	425
XI	The Empress Isabel and Fernando Colón	444
XII	Câmara	445
XIII	Henriques	447
XIV	Teixeira	452
XV	Moniz	452
XVI	Mello	453
XVII	Colón-Zarco's ancestry	454
XVIII	Relation: Aguiar/Vasconcelos/Câmara/Perestrelo	493
XIX	Relation: Góis/Teixeira/Moniz/Câmara	497

Preface

I originally wrote this book in 1971–2. Dr Henrique Barrilaro Ruas was the first person to read and criticise it. He was followed by my mentor, Professor Father Antonio da Silva Rego who encouraged me to publish the work. However, he pointed out the lack of irrefutable evidence as to the real identity of Christopher Columbus, the 'Admiral of the West Indies', and insisted that I add a fifth part that included the decipherment of the cabalistic siglum. Although the book was ready for the printers before Easter 1972, it was imperative to solve the riddle of Columbus's signature. It took me fifteen years.

Convinced that the enigma of Columbus lay in the temuric Cabala, I looked for a beginners' book on the subject and found one in the National Library in Lisbon: *A Treaty on the Cabala* by Rabi Ahib Alvarez de Souto, published in 1708. There was nothing else! It is a modest little work, small and slim, without one single diagram to explain the Sephirothic system and printed in Hebrew. I resorted to the help of a life-long friend, José Eduardo Pisani Burnay who gave me access to his huge library and the sources of information I needed.

And now, with the cryptographic siglum deciphered after 500 years of concealment, all the euphoria of my discovery has evaporated as it becomes clear to me that the solution could not have been arrived at without the work of my predecessors who suspected that Columbus was Portuguese in the same year I was born! They lifted the veil, but they did not have fifteen years to explore the enigma and embroil themselves in day dreams that were undoubtedly enlightened but not completely clear. I must also mention the clear-sightedness of some Spanish and Central American investigators who have helped to refute the thesis that Columbus was Genoese.

To many, this work may seem as a deliberate eulogy to Jewish intelligence and its advanced pre-Renaissance culture, as well as to the tenacity of the astrologers and alchemists of the Templars when they roved into the field of experimental science. This was not my aim, however. I went in search of historical truth, to return to the master that which belonged to the master; to him who gave himself in sacrifice to serve Portugal in the Christian missionary epic.

MASCARENHAS BARRETO

Acknowledgements

My thanks go to the following:

My son Paulo Augusto de Sampayo e Mello Barreto, who gave me a long-standing collaboration. He reproduced the engravings, maps, cartography, coats-of-arms and illustrations.

Gilberto de Abreu, who gave me the most assiduous support.

Filipe de Sousa, for allowing me to consult and study the copy of the *Raccolta* that he brought from Genoa – the only example to which I had access in Portugal.

Carlos Miguel Lima de Abreu d'Araújo, for reading the book before it was printed and making some contributions of genealogical material.

Vasco Sequeira Costa, for pointing out several mistakes in genealogical dates.

Henrique Barrilaro Ruas, who gave me such precious advice and encouragement.

José Eduardo Pisani Burnay, who generously gave me complete freedom to consult his vast library.

José Rebelo Vaz Pinto, whose documentary support and criticism allowed me to correct some imprecise interpretations.

Janina Klawe, Professor of the University of Warsaw, the first non-Portuguese intellectual who publicly declared support for my thesis.

António Champalimaud, always devoted to Portugal's destiny.

Translator's Note

The greatest problem encountered with the translation of this thesis was what to do with the multitude of names that appear in it.

Firstly, the navigator that English-speaking people know as 'Christopher Columbus' takes on a new personality in this work, and as it takes shape the name 'Columbus' undergoes a gradual metamorphosis from its initial Portuguese (Colom) and Castilian (Colón) forms to its newly-discovered Portuguese form.

All other names in the work have been left in their original form, with the exception of kings and the 'household' names of the Discoveries that have already been Anglicised, such as Prince Henry the Navigator, Ferdinand Magellan, Amerigo Vespucci (Vasco de Gama has always been known by his Portuguese name). Bartolomeu Dias retains his Portuguese name as a namesake of his appears in the book, and it would look out of place if both the English and the Portuguese forms appeared side by side.

Portuguese and Spanish noblemen are distinguished by the use of *Dom* and *Don* respectively. I have omitted the feminine *Dona* used by ladies at the time of the Discoveries, as it is a form of address today for anyone of the female sex, while the masculine Dom is still reserved for members of the nobility.

I thank my two colleagues José Manuel de Carvalho Oliveira and Adrian Secchi for reading the translation and making very valuable suggestions for its improvement.

REGINALD A. BROWN

Part I
Introduction

1 The Cathars

In order to understand the personality of the navigator who became known in English as 'Christopher Columbus' because his name was wrongly translated, and to understand his role as a secret agent of the Order of Christ in the service of King John II of Portugal (1481–95) it is first necessary to analyse some of the remote origins of the spirit of the Templars that exercised a commanding influence over the first Portuguese dynasty (see Appendix 1, p. 508), and continued to do so until the reign of King Manuel I (1495–1521). It is also necessary to understand the Catharism of Languedoc which spread north to the boundaries of Burgundy and, above all, the cult of the Holy Ghost that has always pervaded the faith of the Portuguese. We must also take into account the crucial role played by Jewish science in the phenomenon of the maritime Discoveries.

From the end of the eighth to the end of the ninth century, a mystic movement called *Catharism*, which had its origin in Constantinople, developed in Eastern Europe. The Cathars maintained that they were Christian, but *dualists*, admitting to two equally eternal principles – Good and Evil: souls were therefore creatures of Good that were liable to succumb to the temptation of Evil. Jesus Christ was an angel that the principle of Good, God – represented by the Holy Ghost – had sent to earth to reveal His true nature to man. The name 'Cathar' had already been given to Purity and the ideas of the perfect and good found in Platonic doctrines.

Moving through the Mediterranean, this religion spread from the Black Sea to Western Europe, invading Greece and taking deep root in the Macedonian Balkans before stretching across to the Pyrenees. From there, it spread into the Kingdom of Aragon to the south and into a crescent-shaped area to the north. It expanded along the banks of the rivers Aude and Garonne to the north-west, through Narbonne, Carcassonne, Toulouse and Agen, and north-east along the River Hérault by way of Montpellier and Nîmes in the centre of Languedoc. From there, it stretched its tentacles along the Rhône to Bourg St Andeol and Tornon and up the River Saône to Mâcon and Lyon in Burgundy.

The eastern Cathars called themselves 'Bogomils', which in Slavic dialect means 'God's Beloved People' (in the same way that Jews considered themselves 'God's Chosen People'). During the Middle Ages, the Byzantine, Greek, Hungarian and Bulgarian propagators of their religion, who spread their word while trading, were known in the west of Europe as Bulgars or Bugres which simply derived from the word 'Bulgarian'. Wherever they settled they became known by different names – 'Patarines' in Italy, 'Ketzers' in Germanic regions, 'Occitanians' or 'Albigenses' in the south of France.[1] They were also called the 'Poor People' of some towns, or 'weavers', as that was the trade of most of their number.

The word 'Albigense' comes from the Languedoc town of Albi, which stands on the River Trane. The name may have been chosen because *alba* means 'white' or 'pure', or more probably because the leader of the defensive struggle of the Cathars against the Roman Catholics was Raymond-Roger, Viscount of Albi. Around the year 1000, the city of Toulouse was already looked upon as the 'Rome of the Cathars', the geographic centre of western Catharism.

The Cathar movement in time split up into various sects that professed the doctrines of Manichaeism, Paulicism, Bogomilism or other doctrines that the Roman Catholics generally designated as 'heretic'; some of these sects did indeed deny, in open opposition to the Church of Rome, the verity of the Incarnation, the Crucifixion and the resurrection of Jesus Christ. The Cathars referred to themselves as *'prud'hommes'* (good men) or *parfaits*, and St Bernard[2] himself considered them as 'charitable and virtuous', leading their daily lives according to the moral principles of true evangelism; this explains the tolerance that he showed towards the 'faithful of the Paraclete' (Holy Ghost).

THE ISLAND OF ST LAWRENCE

The Cathars of the South of France spread into Catalonia and Aragon. Many of them had already settled along the Catalan coast during the rule of the Count of Barcelona, Raymond of Beranger IV. In 1147 – the year that the first king of Portugal, Alphonse I (Afonso Henriques) conquered Lisbon from the Moors – the Cathars, based on the River Ebro and the Balearic Islands, attacked and sacked the Mediterranean ports of Catalonia and the Genoese shipping that traded with Barcelona. In an attempt to put an end to this piracy, Raymond IV proposed an alliance with the Genoese, using his army and the Genoese to conquer Tortosa (which stood on the Ebro) and the Balearic Islands. The Abbot

Ferdinand Ughelli,[3] of the Cistercian Order, relates that Raymond IV conceded various privileges to the Genoese that helped in the conquest of Tortosa in 1148. He donated two-thirds of an island which is in the middle of the estuary of the River Ebro, to the Church of St Lawrence of Genoa, donating the rest of the island to them two years later. Only in 1158 were the Canons of St Lawrence able to take possession of the island, however, as they had to buy the rights that a Genoese citizen called Bonvassallo (and some of his fellow citizens that he represented) possessed over it. For more than half a century, this island was known as the 'Island Parish of the Genoese', and was inhabited by a group of Patarine *'prud'hommes'*. By the sixteenth century, however, this name had disappeared and only the name of the river remained; the region had, however, become one of the centres of contact between Italian and Albigensian Cathars.

The cult of the Albigensian Cathars dominated the whole region of Occitania around the year 1200 – Toulouse at this time was not only more populous than Paris but, in certain respects, more civilised. The population of the region did not only defend the heretics among them because they were *'prud'hommes'*. The first French sovereigns of the Capetian dynasty[4] were lords of only some territories of the *Langue d'oïl*: they controlled others, but only as suzerains. Their power was so reduced in the regions of the *Pays d'oïl* that they were kings of barely half of France.[5] Pope Gregory VII, alerted to the Albigense movement, had many doubts regarding the danger that the Capet agents attributed to it and was convinced that intensive missionary action would recover his errant flock. But his successor, Alexander III, foreseeing that a new religion in Europe would weaken Christian unity, which in turn would increase the danger of a resurgence of Islamic power (since the Muslims still controlled a great part of the Iberian Peninsula) and being swayed by information that came from those of his dioceses that were strongly influenced by the intentions of the Capets to extend their power, authorised a persecution of the Provençal Cathars in 1179.[6]

In 1181, the papal legate, Henry d'Albano, gathered an army recruited from the *d'oïl* language-speaking counties and attacked Occitania. The attackers committed such atrocities that both the nobles and people, who were mainly Roman Catholics, revolted; some considered the attack as papal interference in the government of their territories and others, who were really heretics, transformed the invasion into a vehicle of propaganda against the Roman Church. Although Albigensian Catharism was no more than a doctrine followed by the common citizens, artists and merchants, it became the standard of rebellion of the Occitanians against the Capet invaders, and by 1206 it could be considered to be a politico–religious force.

THE PERSECUTION

At this time, a papal legate, Pierre de Castelnau, had formed a league with several lords of *Langue d'oïl* area who were friendly vassals of the Capets and who in 1207 had denounced the Duke of Narbonne and the Count of Toulouse, Raymond VI, to Pope Innocent III as being too tolerant towards the heretics, requesting their excommunication. The Pope, who had recommended his legates not to be too harsh in their attitudes, had the good sense to ask Raymond VI to intervene in order to 'put an end to the violence against the Catholics'. The nature of the intrigues woven by Castelnau and the League of the Capets's sectarians can be seen in the terms of this request.

As the Count of Toulouse denied the charges of violence against the Catholics and refused to punish his own people by force of arms, he was excommunicated in the same year:

> In the name of the Holy Father, I sentence you to total excommunication. You are therefore bereft of all civil rights. Whosoever shall deprive you shall do good. Whosoever shall kill you shall be blessed. We shall oblige your neighbours to consider you as an enemy of Jesus and a persecutor of the Church and they shall be permitted to take possession of all your lands that they may be able to occupy.

In this era of deep religious feeling, excommunication brought far-reaching consequences: people went in fear of eternal condemnation, churches were closed, nobody could attend Mass or receive the Holy Communion, the dead could not be buried in Holy ground. Raymond VI preferred publicly to humiliate himself and accepted flagellation at the hands of the new papal legate, Arnaud Amaury, while lying on the floor of the church of St Giles. The legate then entered Béziers with a *langue d'oïl*-speaking army and massacred the whole population of 25 000 people; not one man, woman or child was spared. The German monk, Cesar of Heisterbach, relates that in reply to the population's request for holy sanctuary in the churches, Amaury ordered his army: 'Kill them all. The Lord will know those that belong to Him!'[7] Following this, Amaury fell on Carcassonne. Despite the pleas of the suzerain of the city (Peter II of Aragon) for Amaury to spare the Catholic population, the latter took the city with the utmost cruelty. The Viscount of Trencavel fell in the defence. Raymond decided to appear before the Councils of Saint Giles (1210) and Arles (1211) in order to try and justify his actions; such were the conditions demanded, however, that he was forced to take up arms. The great feudal lords of the north of France, who were neither allied to or materially dependent on the Capets, were not interested in going to war against Raymond. But a minor lord, Simon de Montfort, appeared before the Count of Toulouse with a powerful contingent of Crusaders and defeated Raymond at Muret in 1213. Father Langlois describes Montfort's actions after his victory:

> He ordered his troops to burn the cornfields, uproot the vines, cut down the trees, demolish the houses, obliterate paths and transform one of the most beautiful countries in the world into a barren desert.

Innocent III then invested Simon de Montfort with the government of the County of Toulouse; but such was his brutality that the people rose up, giving Raymond VII (son of Raymond VI) the chance to take over the reins of power. The new Count, however, was killed in 1218. Philip II, who had ascended to the throne of France in 1180, managed to persuade Innocent III officially to request him to fight the Cathars, and took the opportunity to conquer all of Languedoc. Raymond VII's daughter, Branca de Castela, continued the war – now against Louis VIII – until the Treaty of Paris of 1229 brought it to an end. Branca married Alphonse of Poitiers and a period of peace followed, during which the extermination of the Albigensian Cathars who had escaped the sword continued by fire. One more revolt broke out, led by Trencavel's son, who managed to reconquer the region of Carcassonne in 1240 with the aid of some Catalan volunteers. Jean de Belmont, a general of King Louis IX of France (known as 'St Louis') marched against him with an army of Crusaders. After the fall of the great Albigensian fortress of Peypertuse, a truce was signed at Montréal.

The Treaty of Lorris in 1243 finally brought peace. From then on until 1328, the extermination continued in the torture chambers and the fires of the Inquisition, thus 'cleansing' the world of the Albigensian dualists, the followers of the Gospel of St John and worshippers of the Holy Ghost, those who had simplified the rites and reduced the Catholic sacraments to only one: the Baptism of the Holy Ghost or *Consolamentum* (consolation) in the 'Secret Scene' of this new church called the Paraclete.

In the same way that the Jews in Portugal adopted names of animals, metals, trees and plants as surnames when they were Christianised, the bourgeois Cathars chose surnames connected with the dove: *Pigeon*, *Pigeonneau*, *Colombe*, etc. among the Albigenses; *Taube*, etc. among the Ketzers; *Gôlúbi*, etc. among the Slavs; *Palomo*, *Palombo*, *Colombo*, *Columbrete*, etc. among the Navarrese, Aragonese and Catalans; *Columbo*, *Colombo*, *Columbano*, *Collumbella*, *Columbati*, *Piccione*, etc. among the Italian Patarines. Some European Jews also adopted surnames derived from 'dove'. (There are still some 'Palumos' and 'Palumbos' in England, for example, as a result of the Jewish exodus from the Iberian Peninsula.) Certain Spanish merchants altered their Patarine or Hebrew name, 'Piccione', to 'Pizone' or 'Pizon', as this could give them the chance of claiming a pseudo-descendency from the Roman high nobility, but derived from the word for 'mortar' and not that for 'little pigeon'. (There are also many towns and villages all over the Iberian Peninsula with this etymological origin: *Pizón* in Spain and *Pisão* and *Pisões* in Portugal, for example.)

The disguising of these names related to 'dove' happened only after the persecutions of the Cathars, as worshippers of the Holy Ghost, occurred all over Europe. (Later, there took place the persecution of the Jews, for whom – as we shall see – the identification of Jesus Christ as a Biblical angel and not as God incarnate in man, forming one sole person with the Paraclete, made it easier for them to accept the enforced Christianisation through the sacrament of Baptism, which they considered as a mere 'pro-forma', simply receiving this sacrament of the Holy Ghost intrinsically, in the same way that Jesus Christ had been baptised by John the Baptist.)

THE PORTUGUESE MONARCHS

Count Henry of Burgundy, knight of the Crusades, came to the Iberian Peninsula some time before 1095, and the King of León gave Henry the hand of his daughter Teresa and the Portucalense county (see below), a name that may come from (the Holy) *Calice* (Chalice), one of the palaeo-Christian symbols that were particularly worshipped by the Cathars and the Templar Knights. According to some theses *Portus Calixis* (the same as Porto Grail) = Portugal. Nothing indicates that Count Henry had embraced the Cathar doctrine, but as a Burgundian he would certainly have defended the reasons for the struggle of Languedoc against the Capet forces and condemned the repressive measures taken by the envoys of the Roman Church. He had been a Crusader, and in Jerusalem had had contacts with the Templars, to whom the principles of the dualist religion had been revealed (with its remote origins in the Zoroastrian doctrine), and who may possibly have secretly defended them. The Kings of France had tried to impose vassalage on the Dukes of Burgundy, who were only theoretically subject to the crown and enjoyed an irritating independence. With the denunciation of the Languedoc Catharism that was already spreading over Provence, Burgundy and Aragon, the Burgundians found themselves involved in the Albigensian campaigns. The lasting rivalry between the Burgundian and Provençal nobility and the royal house of Capet also divided the people of Languedoc, making government of the dukedom difficult.

THE CATHARS 9

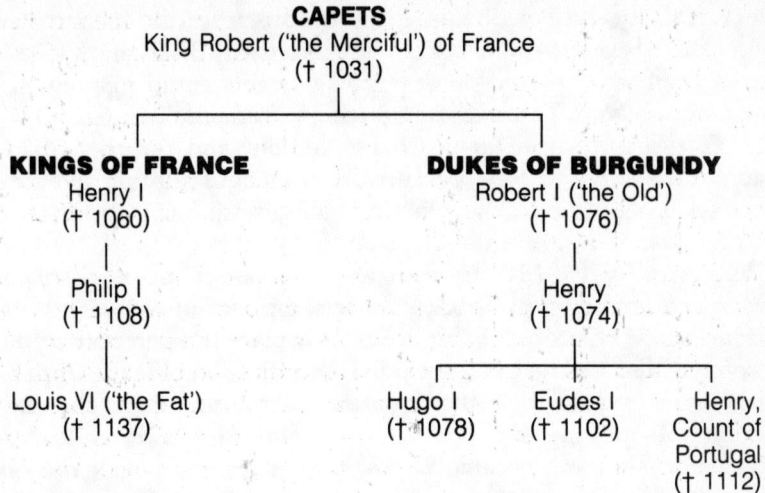

1.1 The Capet family tree.

Count Henry was the brother of Dukes Hugo and Eudes of Burgundy (see Figure 1.1), and nephew of the Abbot St Hugo of Cluny and Queen Constance, wife of King Alphonse VI of León and Castile. The name of Count Henry is mentioned for the first time in 1072;[8] ten years later, he was still considered to be a *puer*, so he must still have been under fourteen years of age. By February 1095, he was married to Teresa, the daughter of the King of León. Where had he been during those twenty-three years between 1072 and 1095? We know only that at this time young men took up arms before they were seventeen.

Count Henry wore the cloak of the Crusaders, which has led some authors to suppose that he had been in the Holy Land during that period; others believe that he had won the insignia of the Cross in the battles fought against the Almoravids in the Iberian Peninsula. He was made Lord of the whole of the Portucalense county, which in 1097 already included the River Tagus, except for the region of Santarém and the coastal area of Lisbon. On the death of his father-in-law, he called himself '*Henricus Dei gratia comes et totius Portucalensis [sic] dominus*'.[9] He died in 1112 and was buried in Braga. The Knights, scholars and artists of Henry's court were naturally Burgundian and Provençal, and during the whole of the Alphonsine dynasty started by his son Alphonse I, the culture that predominated was designated as 'Provençal', although it was influenced by the Troubadors of Toulouse.

The cultural influence of the search for the Holy Grail, which took root in Portugal during the Alphonsine dynasty and continued up to the reign of King Manuel I, was spread by the friars of Cister and by the

Templars. It was inspired by the heterodox doctrines of Joachim of Flora (born in Calabria c. 1435) that divided the History of Humanity into the 'Age of the Father' (based on the Old Testament), the 'Age of the Son' (based on the New Testament) and the 'Age of the Holy Ghost' (up to the defeat of the Anti-Christ). This gave rise to the Eternal Gospel, a fusion of the Old and New Testaments, which would lead to an era of universal brotherhood in which Christians, Jews and Muslims would live in peace in an atmosphere of spiritual fraternity.[10]

The *Search for the Holy Grail* texts (see p. 86) were translated into Portuguese in the reign of King Dennis, under the aegis of the Holy Queen Isabel. So well did the Portuguese people assimilate its heterodoxy and disguise it that the Church of Rome, unable to suppress it, finished up by tolerating the situation.[11] The heterodox elements of Joachim of Flora's ideology, transformed into an ecumenical doctrine of religious proselytism, became the basis of the concept of expansionism of the Christian Empire that was manifested by Prince Pedro (brother of Prince Henry, known as 'the Navigator') in the fifteenth century. It is clear that Rome disapproved of this deviation from Catholic orthodoxy on the part of the Portuguese Kings, as it always tended to ignore their pleas when the interests of Castile were at stake (with very few exceptions, as in the cases of the Portuguese Pope John XXI and Pope Martin V, son of Agapito Colonna, Bishop of Lisbon). According to Alexandre Herculano, there were occasions during the first dynasty when, on the orders of the Kings, some of the Pope's envoys were insulted while collecting the tribute due to His Holiness.

While the Templars followed the Gospel of St John, they were not Cathars. The doctrine the Cathars observed was designated as the Religion of St John, while the Order of the Temple was not a religion. The Portuguese Cathars, therefore, did not belong to the Order of the Temple, although some of them may have entered the Order as foreigners coming from Languedoc and Aragon. (This hypothesis, however, is unconfirmed.) It is however, certain, that Dom Afonso of Portugal, the bastard son of King Alphonse I, was the Grand Master not of the Order of the Temple, but of the 'Religion of St John', as the inscription on his tombstone in the Church of St John of Alporão (today the Archaeological Museum of Santarém) bears witness. It is legitimate to conclude, therefore, that he had been a Cathar,[12] and that the mastership of the 'Religion' had been transferred to Portugal following the defeat of the Albigenses.

Pope John XXII (who was born at Cahors, not far from Albi, and was Pope from 1316 to 1334) founded an 'Order of Christ' in Rome,[13] as an institution of the Holy See. At the same time, King Dennis of Portugal (1279–1325) ventured to propose the creation of another 'Order of Christ'. While being closely linked to the Roman 'Order of Christ', the

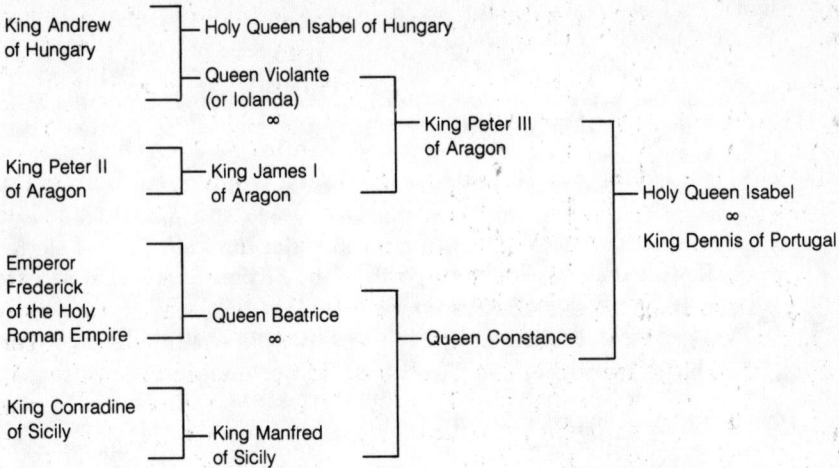

1.2 The ancestry of the Holy Queen Isabel.

Portuguese institution was clearly a metamorphosis of the Order of the Temple, which had very opportunely converted some qualified Jews, now 'New Christians', into its ranks. King Dennis married Isabel, a Princess of Aragon – the centre of Spanish Catharism. Although she was regarded as a saint by the Portuguese, she was canonised by Pope Urban VIII only 300 years after her death – unusually late, according the common rule in the Middle Ages.

It was actually King Dennis and Queen Isabel who were the driving force behind the resurgence of the cult of the Holy Ghost in Portugal.[14] The Holy Queen Isabel, daughter of King Peter III of Aragon, was born in Saragossa in 1271 and died at Estremoz (Alentejo) in 1336. She was the great-granddaughter of King Andrew of Hungary and of the Emperor of the Holy Roman Empire, Frederick II. She was also the grand-niece of the Holy Queen Isabel of Hungary (see Figure 1.2).

Columbus professed a particular devotion to the Holy Ghost, but did not adhere to Catharism; it will be shown that he belonged to the Order of Christ, the successor of the Order of the Temple: he was practically a 'New Christian' (as was the navigator João Gonçalves Zarco, the discoverer of Madeira, who was Columbus's grandfather).

Notes and References

1. During the Middle Ages, Occitania comprised the territories where *'langue d'oc'* was spoken. Its linguistic boundaries stretched from the northern

limits of Poitu, Limousin, Marche, Auvergne and Dauphin – beyond Poitiers and Grenoble therefore – to the Pyrenees, to include Guyenne, Gascony, Languedoc and Provence. The *d'oc* language resisted the *d'oïl* language that was spoken to the north of that frontier. The nomenclature of this linguistic difference comes from the way in which the word *'oui'* (yes) was pronounced.

2. St Bernard, who was born in Burgundy, entered the Cistercian Order in 1114 and founded the celebrated Monastery of Clairvaux the following year. It was due to his powerful influence that Innocent III was elected Pope. Regarded as the eloquent apostle of the Second Crusade, St Bernard was canonised by Pope Alexander III in 1173.
3. Ughelli (1717–22); Abbot Ughelli was born in Florence in 1595 and died in Rome in 1670.
4. The Capets were descendants of Hughes Capet, Duke of France and Burgundy, who became King in 996, the last year of his life. His successors lost practically all of Burgundy and managed to reconquer it only after the Albigensian Wars.
5. Manteutter (1970); Dondaine (1972); Fournaison (1964); Durban (1968); Niel (1955).
6. 1179 was the last year of the reign of Louis VII (known as 'the Youth').
7. Written 60 years after the event, this may simply be the reproduction of a legend.
8. *Charles de Cluny*, no. 3516.
9. *'dominante a flumine mineo usque in Tagum'*, it is called *'comes (Conde) Portugalensis . . . omnis Portugalensis provincia'*, see *Documentos Medievais Portugueses*; *Documentos Régios*, I, vol. I, Doc. part III; Ribeiro, J. Pinto (1633); Merêa (1933).
10. Brunetti (1974); Broomfield (1977).
11. Saraiva, Book I (1953).
12. According to this inscription on his tomb, Dom Afonso had renounced the mastership of the Religion of St John and died at Santarém in 1207. In the same church that is devoted to St John the Evangelist and the Holy Ghost there are two other tombs, both from the fifteenth century, of Martin de Océm (Ambassador of King John I and Governor of the House of Prince Duarte) and his nephew João de Océm (Chancellor of King Alphonse V) both Counsellors of King Duarte and King Alphonse.
13. Dinis (1931).
14. For an understanding of this Portuguese historico-religious phenomenon one must read the seminal work of Quadros (1986–7), vol. 1, 3rd part, pp. 157ff.

2 The Inquisition

The Inquisition was initially instituted to help the Church in the struggle against the expansion of the Cathars, who, especially the Albigenses, were considered heretics. It was established in France in 1229, following the Council of Toulouse, and the Pope granted the functions of the inquisitors to the members of the Order of Preachers of St Dominic. However, it was at the Council of Béziers in 1245 that the main rules of the inquisitorial laws were defined.[1]

The activity of the 'Holy Office of the Inquisition' later became more intensive in the north of Italy, especially in Milan and Genoa, where the numbers of Patarines[2] had proliferated. It was also established in the Kingdom of Aragon and (to a lesser extent) Catalonia, but it never became very generalised in León, Castile or Portugal. It also functioned in England, up to the time that Henry VIII (1509–47) proclaimed himself spiritual leader of the English Church, denied the authority of Rome and took the inquisitorial system into his own hands, placing it at the temporal service of the State.

In the thirteenth century, it was the Dominicans who preached against the Cathar heresy. According to Friar Luis de Sousa,[3] the Order of Preachers settled in Portugal in 1217, their first monastery being founded by Friar Soeiro Gomes on the Montejunto Mountain Range, near Alenquer. A second monastery was founded at Figueira-Velha, near Coimbra, ten years later. Such was the repression that Friar Soeiro Gomes exercised on people (through the imposition of exaggerated fines, terrible corporal punishments and confiscation of land and assets) that King Alphonse II was forced to intervene. In fact, it seems that there were no Cathars in Portugal that needed correcting, or at least they did not show themselves; there were many Moors already converted to the Christian faith and even more non-converted Jews. However, the only ones that could be considered heretics were those that had professed the Roman Catholic religion and then repudiated it to follow another cult.

It is illogical to try and appreciate the behaviour of the men of the

Middle Ages by comparing this period with today. Society was very different, still barbarous in many ways. Many Catholics traditionally practised rites of a primitive origin or even of Roman mythology (like that, for example, of taking food and money to their grave[4] so that they could 'cross the river'). The idea that there is a boatman, Charon, ferrying the dead between life and Purgatory is presented in Gil Vicente's play *Auto das Barcas* (Gil Vicente was a New Christian whose daughter was examined by the Inquisition because she was suspected of being a Jewess) and still exists today among some of the inhabitants of the Trás-os-Montes and the Beiras, in spite of the fact that they are confirmed Catholics.

The ignorance of the people who lived far away from the urban and civilised centres led them to carry out irregularities in their Christian rites; although these acts were repugnant to good Catholics, they did not raise any presumption of heresy. Such people lived in a world of extreme violence and material and spiritual fear; their latent barbarism, however, led them to challenge the agents of evil itself (whether they were at war, or on the gallows, in the face of devastating epidemics or meteorological disaster, or even imagining the terrors and fears of Hell). While they feared the effects of war, it did not stop them turning themselves into mercenaries at the first signs of spring; the fear of the hangmen did not turn them from the path of crime, and they robbed the victims of plagues, earthquakes and fires.

The most repugnant aspect of the Inquisition was the method used for eliciting a confession of crime: interrogation, prison, torture, fire. But all this had been a part of the temporal laws of States since ancient times, and continued to be so in many States in more recent times: it is the evil of Cathar dualism which is always present in the human soul. What is strange is that this inquisitorial violence was perpetrated under the auspices of the men of the Church, although it was always established in the service of men of the State. But was this really the case? Let us consider the case of Portugal.

At the beginning of the sixteenth century, when King Manuel I (in compliance with his marriage contract, see below) yielded to the conditions imposed by the Catholic Monarchs of Spain (his future in-laws) concerning the expulsion of the Jews, he decided to opt for mass forced conversions. When excesses were committed, he set up a committee of notable clergymen to report on the situation. Alexandre Herculano summarised this ecclesiastical report (of which only the first sentence is transcribed here, as this alone is sufficiently illuminating):

> Recalling the words and works of Christ, the apostles and the first priests, the meekness with which Christianity should be instilled, the respect that should be felt for the free will of mankind in

adopting a new creed and the indulgence that was exercised in relation to the fragility and deviations of the new converts that came spontaneously and without coercion to be baptised under the auspices of the Bull of 7th April, the brutal tyranny with which the conversion of the Jews was carried out in Portugal is in stark contrast with this wonderful picture of tolerance and moderation of the first years of the Church.

In *De Rebus Emmanuelis Gestis*, Dom Jerónimo Osório (1506–80), who was Bishop of Silves and nicknamed the 'Portuguese Cicero' because of his eloquence, considered the expulsion of the Jews as 'iniquitous and unjust'. His treatise, written in Latin, was widely read throughout Europe but was translated into Portuguese only two hundred years later (by Francisco Manuel do Nascimento, who was persecuted by the Holy Office in 1778; he fled to France, where he died in 1819). But the Bishop of Silves was not the only prelate to speak out in favour of the Jews. Another was the great Jesuit preacher and writer, Father António Vieira (1608–97), the precursor of the abolition of slavery and a defender of the Jewish cause before Pope Clement X, saying that the Jews were 'unjustly, tyrannically and barbarically' persecuted by the Inquisition.[5] He was also persecuted, and escaped burning at the stake after two years of imprisonment to which he was sentenced by the Court of the Holy Office of Coimbra.

Portugal was the last country to institute the Court of the Holy Office and did so only half a century after Spain;[6] papal authority to exercise its activities was requested only in 1531 (i.e., more than three centuries after its establishment in France, Aragon and Italy). The establishment of the Inquisition in Castile and Portugal is rather surprising, as the contact between the Iberian Christians and Jews and Muslims was already long-standing and it had created a certain tolerance among the different beliefs. This religious coexistence had not harmed Catholicism, and had even contributed to making the country richer through Jewish participation in the fields of economics, finance and science. The most powerful financiers of the Crown and the nobility were undoubtedly the Jews. They were not only bankers, merchants and tax collectors, but also (as in the reigns of Kings Dennis and Ferdinand in Portugal) Ministers for Finance and scholars, protected by the Crown and by the Order of the Templars (see Chapter 3). When this Order was extinguished in the rest of Europe, the Jews continued to be accepted by its Portuguese successor, the Order of Christ.

While the Spaniards were expelling the Jews in the fifteenth century and systematically confiscating their assets, the Portuguese were doing everything possible to keep them in the country. They also received those expelled by Spain, although they always tried to convert them to

Christianity. In his book *The Portuguese Inquisition*, António José Saraiva wrote, clearly and objectively:

> The Catholic Monarchs didn't want Portugal to reap the material benefits from the expulsion of the Jews and so one of the conditions that they imposed on King Manuel on his marriage to their daughter was the expulsion of the Jews from Portugal.
>
> The King ceded to this demand, but tried to lessen the impact of the results: in fact, he promulgated a law that obliged the Jews to either leave the country or be converted (1496). But he tried everything within his means to force them to stay. One of the methods was forced baptism of children under 14 years of age, who were forcibly separated from their families. Another method was to confront the families that preferred to leave with difficulties of every kind. Even so, about twenty thousand people gathered at the Estaus Palace in Lisbon with the object of departing, but almost all of these were baptised.
>
> By using such a machiavellian process, the King of Portugal managed a mass conversion that was merely formal and apparent. This was enough for him to achieve his aim of keeping the Jews in Portugal and, at the same time, not break his promise to the Catholic Monarchs.
>
> Nobody had any doubts about the value of these pro-forma convertions and so it was decreed, on 30th May 1497, that the New Christians could not be hindered in their religious practices for a period of twenty years and that, after this period, accusations regarding the practice of Hebraism could only be tried judicially within such strict provisos that it would be almost impossible to bring any charge . . . Various facts show that this legislation was very effectively applied. Besides this, the Portuguese authorities firmly resisted any attempts by the Spanish Inquisition to persecute the 'New Christians' who had taken refuge in Portugal. The law that forbade people of Hebrew origin and their assets from leaving the country, the only discriminatory law that was still in force in relation to converted Jews, was eventually abolished. It was obviously thought that such lax legislation was the only way to avoid clandestine emigration.
>
> The policy of forced assimilation followed by King Manuel I completely altered the traditional behaviour of the Jewish minority. In fact, once discrimination was abolished, the ex-Jews started marrying 'old Christians', and before very long even the most important families of the Portuguese nobility had Jews among their ancestors.

At the beginning of the 17th century, the Jesuit priest, Diogo de Aredo, remarked that 'the New-Christians have already merged with the old-Christians in such a way that there is hardly any respectable family that doesn't include many men and women of Hebrew blood'.

The inquisitors of the time, in an official reply to an enquiry from the king, calculated the number of Jewish residents in the kingdom at 'two hundred thousand families, each of many people. They are part of the nobility that are in your Majesty's books and the knights of the Military Orders, as well as secular knights, including the greatest of the land; and those in the Treasury become more prosperous, as they are the only ones that have money, contracts, goods and the greatest power in the kingdom'.

Two hundred thousand 'numerous' families must add up to at least 1 million people in a country that must have had 2 million at the most. It was, therefore, popular belief in Portugal at the beginning of the 17th century that much more than a third of the population of the country was linked to the ex-Jews by ties of blood.

What had the Jews done in the Iberian Peninsula, and more particularly in Portugal, from the foundation of the Kingdom to the reign of King Manuel I, to have enjoyed such protection and privilege? The reply to this question that is set out in the following chapters will justify the reasons that led the Portuguese navigator, João Gonçalves Zarco (whose ancestors had prayed in the Jewish synagogue in the Rua Nova in Tomar – the headquarters of the Order of the Temple – and who was the grandfather of Christopher Columbus) to become a Christian and a member of the Order of Christ.

Notes and References

1. At the Council of Verona in 1184, the bishops had been ordered to find out for themselves, or through informers, about the behaviour of the people suspected of heresy. There were various levels of suspects: convicted; apostates, people who had denied their religion and given themselves up to heresy; penitents, who had confessed their repentence; and relapsed, who had returned to heretical practices and were condemned without appeal.
2. 'Patarino' was the name given to Italian Cathars and corresponded to the name of a coin of the lowest denomination in the Italian states. This is why the 'Poor People of Christ' themselves of Milan, Naples, Venice and Genoa, chose the name.

3. His secular name was Manuel de Sousa Coutinho, a Portuguese nobleman born in 1555. He entered the Order of Malta while still very young. He was taken prisoner by a Moorish galley in the Mediterranean and taken to Algeria, where he spent almost a year of imprisonment (1576). There he met the great Castilian writer, Miguel de Cervantes, author of *Don Quixote de la Mancha*, who referred to him in his work *Trabajos de Pericles y Segismundo*. As Philip II of Spain, after the death of King Sebastian in the North of Africa, had usurped the throne of Portugal, Sousa Coutinho went into exile in Panamá. He entered the Order of the Dominicans in 1614 and took the name of Friar Luis de Sousa. He wrote several historical and religious works, such as *The History of St Dominic, The Life of Friar Dom Bartolomeu of the Martyrs, of the Order of Preachers* and the *Annals of King John III*. He died in the Monastery of Benfica (Lisbon) in 1631.
4. The present-day funeral of some Jews still involves the throwing of some coins into the grave, and always an umbilical kiss.
5. Inéditos de Vieira, Bibliographic Archive, Coimbra University.
6. At the end of the fifteenth century – when the Inquisition was almost stagnant in the rest of Europe – Hebraism arose as a new threat for the Church of Rome in the imagination of the Catholic Monarchs of Spain. In order to fight it, the Inquisition was established in the kingdom in 1481, at the same time that it raised its head again in Italy (mainly in Genoa, Milan, Venice and Naples). If it was dangerous to be a Patarine in Italy, it was no less fatal to be Jewish in Castile.

3   The Portuguese Jews

According to the annals, the Council of Toledo of 633 (on the proposal of St Isidor) dealt with the situation of the Iberian Jews, and proclaimed:

> In the future no Jews should be obliged to embrace Christianity: those that were forced to do so in the reign of Sisebut and received the sacraments must remain Christian (so that they would not be heretics).

It is strange that, as the Jews considered themselves the 'chosen people', chosen by God (thus creating a conception of racism that forbade their sons to marry women that were not of Hebrew blood), the Jews tried to persuade people of other races to follow their religion. It is even more illogical that they tried to do so among the less fortunate – slaves and those who were their dependants through the contraction of debts. As they were accused of this practice, however, the Council of Orleans of 538 ordered that 'If a Jew forces one of this slaves to apostate, he shall lose that slave'.

This order was confirmed one hundred years later at the Council of Toledo and the following was added: 'The slaves of Jews shall be free from this day forth'.

This subject was obsessively dealt with at subsequent Councils[1] until, at the Synod of Palencía (Castile) it was decided to separate the Jews from the rest of the population; the Synod declared unequivocally that: 'The Jews shall have separate quarters in the cities where they live. Hence we have the *Ghetto* in Italy, the *Carrière* in France, the *Judenviertel* in Germany and the *Judiaria* in Portugal, etc.

With the mass Christianisation (whether voluntary or coercive) of the Portuguese Jews, and with the Jews' family links with Christians and their subsequent dissemination throughout the towns and cities where they lived, the Lisbon *Judiaria* was no longer exclusively Jewish by the reign of King John III, some of its inhabitants being Catholics of non-

Jewish origin. This process, in fact, had been under way since the end of the fourteenth century, when King John I had donated the goods of Jews that had emigrated to Castile to Christian knights, as will be explained later. The existence of Jewish quarters in Portugal, therefore, was of a much shorter duration than in other countries (where some of them survive today).

The Jews enjoyed a privileged situation in Portugal, as they had chosen their own quarters. In Lisbon, they had occupied the manorial section of the city, on the slopes below St George's Castle (Alfama), where the old palace-like villas of the defeated Moors stood. The *Judiaria* stretched from the beginning of Terreiro do Trigo and the Castle to the Escolas Gerais, the Igreja de Santo Estevão (St Stephen's Church) and the Largo do Chafariz-de-Dentro (present-day names) and their shops stretched along the banks of the River Tagus to the present-day Terreiro do Paço. They squeezed the Moors round to the other side of the Castle to a more modest quarter – the *Mouraria* (Moorish quarter). It wasn't long before the Jews had fortified the area, thus creating a small Jewish State in the middle of the city.

In Strasbourg, by contrast, 2000 Jews were burned to death in their own homes, those that tried to escape the flames having their throats cut. In Mainz, 12 000, including children, were massacred. Popes John XXII and Clement VI published vigorous protests and registered their repugnance of such acts and Innocent III, Alexander II, Gregory IX, Clement V and Benedict XIV all took a similar stand.[2] In the fifth and sixth centuries, Popes Gelasius and (St) Gregory I ('the Great') had manifested their indignation against whosoever had coerced the Jews by obliging them to receive the baptism, destroying their synagogues and profaning their cemeteries; Alexander II wrote: 'it is the duty of good Christians to fight the Saracens that persecute the faithful [of Christ] and spare the Jews, who are pacific and harmless'. Martin V[3] reiterated this in 1417.

Except in the Iberian nations, anti-semitism broke out all over Europe, and at the beginning of the thirteenth century a monk by the name of Radulf began to incite the faithful to exterminate the Jews. Not quite so harshly, Abbot Pierre of Cluny appealed to King Louis VIII (known as 'the Lion') of France in 1225: 'I'm not asking you to kill them, but at least punish them in proportion to their perfidy'; he added: *'Reservatur eis vita, auferantur ab eis pecunia'*.[4]

Not all the members of the Church thought the same way. St Bernard of Clairvaux wrote two notable epistles, one to friar Radulf, and another to the bishops and people of France. Although he exhorted the Christians to take up arms against the infidel in defence of the Eastern Church, he opposed the persecutors of the Jews, although he recommended: *'Non sunt persequendi judaei, non sunt trucidandi, sed nec efferendi quidem'*.[5]

Christian–Jewish relations in the Iberian Peninsula, approaching the ideal of St Bernard's preaching, differed greatly from those of the rest of Europe. The Jews were so well treated that Pope Alexander III wrote to the Iberian bishops in 1166:

> We were very pleased to hear . . . that you had saved the Jews that live among you from being put to death by the Crusaders [foreigners] that are in Hispania to fight the Saracens . . . The cause of the Jews is different from that of the Muslims. The struggle against the latter, who persecute the Christians and expel them from their cities and rightful abodes, is just. The former, however, are willing to serve everywhere.

In Portugal, when King Alphonse I conquered Santarém from the Moors,[6] the Jewish associations were maintained and they were allowed to build their first synagogue, and even their own settlements, in that city.[7] Alphonse I decreed that: 'The Jews may and must present their complaints to the King's officials, alcaides and governors, so that they may redress their grievances'. Alphonse's son, King Sancho I, went even further in the charter he conceded to Almada in 1190, extending this protection to the Portuguese Muslims: 'the complaints of the offended Moors or Jews shall be sent to the alcaide and or the governors, as they were in the reign of my father'.[8]

To reward the services that Jahia-aben-Jaísch (one of the most noble of the Portuguese Jews) had rendered during his conquests, Alphonse ceded him some villages and allowed him to use a noble coat-of-arms. Sancho I appointed the Jew's son, Joseph-aben-Jahia, Royal Treasurer and gave him authorisation to found the first synagogue in Lisbon 'in a beautiful and magnificent situation'.[9] In the following reign of Sancho II, however, it was found necessary to forbid Jews to have servants, so that they could not Judaise them, (in fact, a nonsensical accusation); to prevent this apostasy, it was decreed that:

> after a Jew or Moor has embraced the Christian faith, he may not return to the one he professed before. And if he does, he will lose his head if he does not wish to repent after being admonished.

The apostate, therefore, was given the chance of contrition, being punished only after a recreant relapse. Numerous advantages were offered to the Jews to attract them to the Christian faith and if anyone molested the converts, even with words (by calling them 'turncoats', for example) the accuser would be decapitated.

There were, however, manifest differences between the Christian and Hebrew customs. The laws of the Talmud, for example, permitted and even regulated usury and as the indebted Christians attempted to evade payments that bore extremely high interest, the Jews would wait for them when they went to Mass and shame them by trying to collect their debts there.[10]

Dom Soeiro, Bishop of Lisbon, who opposed the policies of King Sancho II, denounced the King to Pope Gregory IX, complaining that: 'public positions are given, by preference, to Jews in the diocese of Lisbon . . . to the opprobium of the Christians and to the indignation of many people'. Jews certainly held the highest positions – such as Royal Standard Bearer, Admiral, Lord Chamberlain, etc. Dom Soeiro denounced Admiral Fuas Roupinho (natural brother of Alphonse I) and the King's Comptroller and preceptor, Dom Egas Moniz, one of the famous figures of Portuguese history, as being Jews, although already Christianised. (The attitude of the Bishop of Lisboa may be attributed to a grudge that he bore, since the position of Royal Standard Bearer had belonged to his relatives, Dom Soeiro Raimundo and then Dom Lourenço Soeiro, who lost the post to Dom Martim João in 1218; João had Jewish ancestors, and in turn was succeeded by his son, Dom Pedro João in 1221.)[11] No known chronicle or document states that either Fuas Roupinho or Egas Moniz had Jewish blood. If Fuas Roupinho was of Jewish origin, it would obviously imply that Count Henry had had relations with a woman of that race. Chronicles of a much later period omitted his maternal name and in the later genealogical books, care was always taken to omit any Jewish ancestry. The only thing that can be said in favour of this supposition is his name, as the Moors and Jews always wore a short waistcoat called a *'roupinha'*. It was also not because of any Jewish blood that Dom Fuas Roupinho was appointed Admiral (although others of Jewish blood had held the post): the post was given to him because he was the half-brother of the King and a notable warrior and sailor. We know that he commanded three expeditions against the Saracens; in the first he destroyed a powerful Muslim fleet off Cape Espichel in 1180; in the second he made for Ceuta and attacked any ships that left the port; in the third, in 1184, he lost his life in a naval combat, as narrated by the Arab chronicler Ibn-Cadune.

It is thought that Dom Egas Moniz was the son of Moninho Moniz, companion of Count Henry in the war against the Saracens and the Head of the Count's Household. Moniz came from a Suevian Visigothic family; he was preceptor to the Household during King Alphonse I's childhood, so would have been too old to occupy the position of Royal Standard Bearer in the King's reign. But many sons took their father's name at this time, and there were many Egas Monizes. Could the one Dom Soeiro denounced have been the son of Moninho Moniz and a Jewess? Dom Soeiro – who had fought at the conquest of Alcácer do Sal and was an illustrious prelate – could well have known of such a relationship: it is difficult to think that he would have dared to invent such a thing for the ears of the Pope and King Sancho II, especially as the facts were still fresh in everyone's memory and could have easily been denied. Documents that exist (and there are many that no one has yet studied) affirm only that up to the time of King Sancho II, the Steward of the Royal Household, Filipe Daniel, was a son of Jews. As far as the post of Comptroller is concerned, Dom Luis Caetano de Lima[12] appointed its holders only from the time of King Ferdinand, although we know that the post existed before this.

Based on the accusations of the Bishop of Lisbon, Pope Gregory censured King Sancho II; the King, however, limited himself to imposing certain special imposts on the Jews, one of which was to oblige them to equip every ship built by the crown with 'an anchor and a hawser'. In the reign of Sancho II's successor and brother, Alphonse III, the Jews ceased to occupy the posts of inquisitors and attorneys in Christian prosecutions against Hebrews, but they maintained their equal guarantees and rights in lawsuits and were allowed to present (as were Christian plaintiffs) thirty witnesses, on condition that each witness swore on the Torah (the five books of Moses) in the synagogue and before the Rabbi, so as to prove that they would not lie.[13] In the following reign, King Dennis wrote a letter to the judges of Bragança, one of the largest Jewish centres in Portugal, in which he ordered: 'let no-one do harm or injury to those Jews, otherwise they shall turn me against you for this reason'.

King Dennis called on a Jew, a Chief Rabbi by the name of Judâh,[14] to manage the finances of the kingdom. In 1303 he was succeeded in his post by his son, Dom Ghedéliâ-aben-Judâh, who was also Chief Rabbi. The Jews, as a whole, were very useful to the kingdom, but some began to abuse their royal privileges; they persecuted those that were converted to Christianity (whether they were Jews or Muslims), daring to confiscate their goods and even enslaving them due to their indebtedness, as was the custom of the time.[15] They entered the churches to drag out reluctant debtors. They even dared to threaten the bishops themselves with death if the latter opposed them, and there were cases in which they cut off the ears of some of the servants of prelates that had expelled usurers from the Christian churches.

The bishops presented forty complaints to Pope Nicholas IV in 1288, but the king's attorneys declared them to be false. On one count, however, they were guilty:

> Such pride did the Jews take in their long dishevelled hair, that it curled, stood in obscene tufts, and gave them an effeminate appearance.

It is interesting to note that in this period in which the crown's financial backer, the Order of the Temple, was extinguished (due to several serious accusations which were never proved in Portugal, thus allowing the King to transform it into the Order of Christ), the Jews and the converts of Jewish origin (who were also wealthy and financed the state) were accused of similar acts against nature, namely sodomy, especially in relation to minors. (The Templars were accused of adopting such practices during the time they mixed with Semites, Hebrews and Arabs in Jerusalem; particular cases became generalised out of all proportion.)

The real motive for the envy and hatred of the Jews was not caused by their cult or their blood, but by the pride that came from the wealth and the public offices they had acquired; this ostentation vexed the Christians that were less well off, or had been dismissed from office in order to make way for them. The Jews had mainly obtained these posts through their own merit, but some had resorted to bribery in order to achieve positions of power and privilege in commerce. The Moorish, Venetian and Genoese merchants did the same thing but when the whole nation boiled over with the hatred of the rich merchants, the Jews were the most hated because they were also usurers. King Alphonse IV (known as 'the Brave'), King Dennis's son, decided to have the Jews' and Moors' long hair cut off, as well as the hair of any Christian that was worn long or with the forelock in a high wave: 'The Moors' hair shall be cut with a knife if they are dishonest, the Jews' shall be cut with scissors so that they should not be effeminate; the Portuguese' hair shall be cut so that they will feel more manly and virile'. The King also decreed the payment of fines for those that did not pay their taxes; he forbade usury – whosoever lent money with high interest would lose that interest, and the amount lent. (The Jews had, in fact, exploited their debtors, 'at times demanding interest of 33 per cent on the amount lent.) During the plague of 1350, the story was put about that the Jews had poisoned the waters, which led to the killing of large numbers of Jews, 'without the Christian princes being able to put a stop to the slaughter'.[16] King Alphonse also determined that no Christian woman should walk in the company of a Jew, except to buy food, and even then accompanied by two Christian men. (This was to prevent the woman from being tempted by gold and prostituting herself.)

While the Jews were not allowed to acquire landed property in Castile, they were allowed to become landlords in Portugal; the Portuguese Jews became fabulously rich from their agricultural land and usury, and many of their coreligionists emigrated from Spain. One of these was Rabbi Moyseh from Navarra, who King Peter I appointed as Royal Treasurer, allowing him to use the title 'Dom' and to perpetuate the name of Navarra for his heirs. King Peter (a central figure in the tragic story of Inês de Castro), on learning that two of his squires had assassinated a Jew who was selling spices, called them into his presence and, although in tears because of the esteem he felt for them, ordered that their throats be cut, and it was done'.[17]

The Jews of Coimbra also vexed the prior and the priests of the Church of St James by driving them from the Jewish quarter while they were begging for alms for their Christian servants. This led the 'attornies' at the Cortes of Évora in 1361 to launch a campaign for the moral education of the Jews, complaining about the harm that their usury did to the people. King Peter therefore decided to make it compulsory to write contracts that would prevent this usury, and at the same time oblige the Christians – who used any pretext to evade payment of contracted debts – to honour their contracts.

Some Jews were alchemists, others practised what was considered 'witchcraft'; there was at least one case in which a Jew bribed a sacristan to supply him with 'sacred particulars', which he used to perform 'magic'. Denounced as a 'defiler of the Host', King Peter ordered him to be burnt at the stake.[18]

In the following reign, King Ferdinand married Leonor Telles (a woman who was separated from her husband) and gave the title of the Count of Andeiro to her former lover, which caused great consternation among the Portuguese. Andeiro later became the Count of Ourém due to the Queen's influence and, as the King of Castile had married King Ferdinand's daughter, Beatriz and coveted the Portuguese crown, he became the champion of the Castilian monarch's ambitions. When King Ferdinand died in 1383, the Castilians invaded Portugal. An illegitimate son of King Peter I, John (Master of Avis and future King John I) killed Andeiro. In three memorable battles at Atoleiros, Aljubarrota and Valverde, the Portuguese defeated the Castilians against overwhelming odds. During the first invasion, however, the Castilian troops entered Lisbon and burnt the Jewish quarter. Two notable Portuguese Jews, José and Ghedaliah, considering the defeat of the Portuguese inevitable, went into exile in Castile.

The Royal Treasurer in Portugal at that time was Dom Judâh-aben-Mosséh Navarro, who had succeeded his father. He was followed in 1369 by Dom Mossêh Chavirel; in Oporto, the Royal Land Purchaser was Dom Jusaf-aben-Abasis.[19] The Jews and their coreligionists took the side of the power which seemed to be the strongest, especially as the Castilians had burned down the Jewish quarter of Lisbon. But they were wrong and they became the target of the pent-up wrath of the populace, who had already thrown the Bishop of Lisbon (who was Castilian) from the tower of the Cathedral. After the victory of the Portuguese, therefore, the one idea of the Jews, now denied the protection of the Crown and the object of the people's hatred, was to flee.

Despite the fact that the Jews had supported the party of the Queen and the King of Castile, the Master of Avis, Dom John, on learning that some of the people intended to kill two Jews Dom Judas Jadeu and Dom David Negro, decided to give them his protection and, with the aid of a group of knights, drove the populace from the *Judiaria*. Coming across Antão Vasques, the 'Justice of the Peace' of the city, he ordered him, in the name of the Queen, to proclaim 'that nobody do harm or offend the Jews' under the threat of severe punishment. This order was carried out, but in the name of the Master of Avis, as Antão Vasques refused to do so in the name of the Queen.

After the Queen had been banished and the Master of Avis had ascended to the throne as King John I, the insecurity that reigned among the Jewish community led many of them to choose exile in Castile. This exodus enabled John I to reward some of his supporters, distributing among them the goods of Dom Judas Jadeu and Dom David Negro,[20] of two banker brothers, Dom Judas and Dom Moysés Nahum, and of many others.

The Portuguese Jews had chosen a bad time for their exile. In Seville, the archdeacon Ecija Hernando Martinez incited the population against the new arrivals, giving rise to the greatest massacre of Jews that the Iberian Peninsula had ever witnessed: 4000 men, women and children were killed in one day.[21] The slaughter spread to Toledo, Burgos, Valencia and Córdoba, where all the Jewish quarters were burnt and sacked and their inhabitants put to death. It is calculated that about 15 000 Jews perished and there would have been more if many had not managed to escape to the swamps and the mountains. Thousands of Jews took refuge in Portugal; they were being persecuted all over Europe, and had been cruelly expelled from England by King Edward I in 1290.

The aversion to the Jews in Portugal then began to die down, and in 1391 the Chief Rabbi and physician to King John I Mosséh-aben-Navarro, requested that the Bull of Pope Boniface IX (and those of previous pontiffs) that manifested the intention of the Church of Rome to protect the Jews be invoked. The persecutions continued in Aragon and Castile and the Portuguese Jews, looking on John I as their protector, begged him not to authorise forcible conversions. The Portuguese did not interfere with the religious or consuetudinary laws of the Jews (such as, for example, the obligation to celebrate the fourteenth day of Nisan – the first month of the Holy Year – and sacrifice the paschal lamb at nightfall according to the traditional rite), as long as they were not detrimental to the Christian community.

King Duarte (known as 'the Eloquent') one of the most brilliant figures of his time, succeeded his father, John I in 1433. The study of the stars was still closely linked to the predictions of man's destiny and almost all

Christian Kings had a Court Astrologer, very often a Jew; King Duarte's astrologer was Master Guedáliah, who correctly predicted several calamitous events, such as the disaster of the expedition to Tangier in 1437 (see pp. 423–36). During the following reign, of Alphonse V (nicknamed 'the African' because of his conquests in North Africa), some Christian youths offended certain Jews in Lisbon (1449); the Jews complained to the justiciary, João Alpoim, and the youths were found guilty and publicly whipped. Infuriated, the people tried to set fire to the Jewish quarter, and in the ensuing riot, some Jews were killed and many houses were sacked. In order to restore the peace, the Count of Monsanto led a group of officials against the mob, killed a few and arrested many more. The King was in Évora at the time, but on learning of the riot, he punished the prisoners harshly. The executions ceased in 1450 and a period of apparent calm between the people and the Jewish bourgeoisie followed. Meanwhile, certain Jews – in a class of their own in the medieval social stratification in Portugal – had been raised to the nobility, being entailed lands, receiving the highest titles and being allowed to wear the sword of the knight. Authorised – only in Portugal – to acquire land, they became the owners of big farms, estates and manors, by means of transactions that were unfavourable to the indebted Christian nobility. The situation of the rich Jews became not far short of that of the rich Christian nobles: if they were converted to Christianity, Jews could marry their daughters into Christian families, thus accumulating gold and acquiring nobility. The long-drawn-out social evolution that took place turned the Jewish bourgeoisie into the lower Christian nobility.

At the Cortes of Évora in 1482, during the reign of King John II, the representatives of the people complained:

> We are told everywhere of a censurable dissolution among the Jews, the Moors and the Christians, whether it be in their way of living and of dressing and in their conversations. We see Jews mounted on horses and mules with rich trappings, dressed in mantles with fine hoods and silk doublets; they carry swords, and they wear richly decorated cowls that cover their faces so that they are unrecognisable. They then enter the churches to mock the Holy Sacrament and mix with the Christian women, thus committing a grave sin against the Holy Catholic Faith.[23]

Besides protecting the synagogues and severely punishing whoever offended the Catholic church, King John II also acted against pecuniary abuses, thus protecting the poor from the predations of the rich. The representatives of the people also complained that the Jews, by wearing expensive jewels and clothes of fine and rare fabrics, corrupted the people of the less privileged classes who, being unable to afford such ostentation, contracted unpayable debts with the Jews so that they could display an ephemeral and fictitious wealth:

> if before their sin was envy, it is now vanity and momentary glory which could lead them to crime, the men to deceit, robbery and murder, the women to prostitution and other repugnant vices.[24]

On the other hand, the Rabbis complained that their faithful did not attend the synagogue: each had become so rich and powerful that he became estranged from the religion of his ancestors and disdained the protection and strength that he could derive from the unity of his own community.

The Kings of Portugal had protected the Jews for 300 years, allowing them to accumulate wealth and raising many of their number to the nobility, and although the embittered populace raised its voice against corruption among both Jews and Christians, the Kings of Portugal held back from expelling the Jews, wishing to keep them within the frontiers of the country and concede them privileges that would lead to their eventual conversion to Christianity.

THE COMPULSORY CONVERSIONS

There were excellent doctors among the Jews – like Master António – Chief Physician to the king (who acted as António's godfather when he embraced Christianity) but also famous scientists in the fields of mathematics, astronomy and geography. Afonso de Paiva and Pedro da Covilhã, who were sent overland to explore the East in 1497, were both New Christians and it was two other Jews, Rabbi Abraham of Beja and Master José de Lamego who were sent to meet them to collect secret information (see p. 167). Jácomo de Majorca, in the service of Prince Henry, who supplied maps for this expedition, was also a Jew. Masters José and Rodrigo Zacuto, doctors of King Alphonse V were both illustrious mathematicians. In the following century, Pedro Nunes[25] was the most celebrated mathematician of his time. It was the Jews who introduced the printing press into Portugal – in Lisbon and Leiria. When the Jewish quarter was invaded by the populace in 1482, one of the many houses sacked was that of the learned bibliophile, Isahac Abravanel,[26] and a number of priceless books were destroyed.

Coinciding with the expulsion of the Jews from Castile, an epidemic of bubonic plague swept across the Iberian Peninsula and the blame for this fell on the Jews' shoulders. The Lisbon Town Council published an edict, *Preventive Measures for Health*, which ordered the expulsion of the

Jews from the City; Oporto followed suit. King John II opposed these measures, censuring the Council for driving Master José out of the city.[27] It must nevertheless be said that the immigration of Jewish exiles from Spain had been a terrible burden for the Portuguese; their villages had neither lodging nor food for so many refugees. The Portuguese Jews themselves, terrified by what they saw in both the neighbouring country and in Portugal, offered no help whatsoever to their unfortunate coreligionists, keeping whatever they possessed for a possibly adverse future and giving aid only to those who least needed it and whoever wielded the most influence. The archipelagos of Azores, Madeira and Cape Verde already contained numerous converted Jews. The King thought of colonising the fertile but still deserted Island of São Tomé, but he had no wish to do so with Jews, some of whom were accused of all types of social and moral vices, and of not even practising their own religion. Even though he knew that the Jews that had fled to North Africa had been dispossessed of all their goods and the old and the sick that could not be enslaved had been slaughtered, the King understood that he could not convert them against their wishes. So he decided to start his colonisation in a way that history had never before witnessed – with the offspring of Jews forcibly wrested from their parents' arms. In vain did Dom Fernando Coutinho (the Chief Justice and later Bishop of Silves) try and persuade John to change his mind, and in the following year he was similarly deaf to the ideas of Abraão Zacuto (the physician of King Manuel I, besides being an experienced mathematician and astronomer).[28] Nonetheless the period between 1507 and Manuel I's death in 1521 can be considered peaceful; the Castilian Inquisition made every effort to get the fires lit in Portugal, but the King repelled them, alleging that he could not break his word to the Jews or betray the hospitality he had offered them. But in 1504, on Whit Sunday, while two converted Jews were walking along the Rua Nova, they were surrounded and insulted by a crowd of youths; one of the insulted parties drew his sword and wounded one of the offenders, and a riot ensued. As had happened in 1449, proceedings were instituted and the two converts were found innocent. The arrested youths were sentenced to be publicly whipped and to perpetual banishment to the Island of São Tomé. However, after hearing the pleas of the Queen, daughter of the Catholic Monarchs of Castile, King Manuel quashed the sentence of deportation and the guilty parties were imprisoned in Lisbon.

In 1506, during a plague in which about 120 of Lisbon's inhabitants died every day, a glittering light appeared behind the altar of the Church of St Dominic. Some of the faithful interpreted this as a miracle, and when a New Christian voiced his doubts and attributed the light to some other accidental phenomenon, he was immediately put to death and burnt in the churchyard for his supposed blasphemy. Two Dominican friars started roaming the streets, inciting the mob with cries of 'Burn them!', and 2000 New Christians were slaughtered in two days, although many escaped by taking refuge in the houses of Old Christians who were relatives or friends. On learning of the occurence, King Manuel had over a hundred people arrested and 47 of the worst culprits summarily tried. The two Dominican friars who had incited the mob were condemned to death; the rest of the Dominican friars were expelled from Lisbon and their monastery handed over to secular priests. Everyone who was thought to have taken part in the riot had their assets confiscated. In the provinces, however, Dominican friars continued to incite the local populations to extirpate the Jews; the Catholic Monarchs waved Manuel's marriage contract at him and censured him for his leniency. Not wishing to institute the Inquisition and loathing to use his predecessor's method of forcibly separating children from their parents, King Manuel I could do only one thing – banish the Jews. But at the same time, he did not want to see the Kingdom deprived of its best brains; coercive conversion thus seemed to be the most satisfactory solution, and this was the decision he came to.

If at the beginning of the seventeenth century more than a third of the population of Portugal was indeed of Jewish descent, it was a result of the mass voluntary and compulsory conversions that were carried out in the reigns of John II and Manuel I. Among such a large number of coercively baptised families, there were many that would not reconcile themselves to abjuring their faith, but trying to help it survive among their coreligionists (or even trying to spread it) was considered subversive. Some opted for the practice of a mixture of the two religions, and as the occult sciences were widely practised by the Jews, especially those connected with astronomy, chemistry and medicine, some devoted themselves to experiments that were considered to be heretical, and therefore condemned by the Catholic Church.

THE ALL SAINTS' HOSPITAL

Illness was thought to be a stroke of fate in the Middle Ages and any of the sick that were thought to have diseases that could endanger the community were driven away: one became sick through the 'will of God'. A unique system of medical assistance was, however, sent up in Portugal. Three big leper colonies were opened in Coimbra, Santarém and Lisbon and Queen Leonor (wife of King John II) founded the 'Brotherhood of Our Lady of Mercy' (*Misericórdia*) in 1498 for the protection of the poor and the curing of the sick. To organise this work she appointed a friar (Miguel Contreiras) and a bookseller and a candlemaker who were both New Christians. She also founded the first hospital since classical antiquity to offer thermal treatment at Caldas da Rainha, besides setting up the 'Roda' to look after orphans and abandoned children (whose names were kept secret), and protecting over forty already existing hospitals.

King John II himself founded the 'All Saints' Hospital' in Lisbon, which was the biggest and most up-to-date of its time; the one that his cousin, brother-in-law and successor, King Manuel I built in Goa, the 'Royal Hospital of the Holy Ghost', was even bigger. In view of the knowledge of new tropical diseases, a campaign of 'Public Health' was launched in Goa, with one doctor for every 2000 inhabitants, which was unique in the sixteenth century. ('St Paul's College' already operated as school of medicine at that time in Goa.) In the Royal Hospital, most of the medicine and chemistry professors, as well as the surgeons and pharmacists, were New Christians. Garcia de Orta studied cholera; Pedro Nunes, the most notable mathematician of his time, wrote the first book on experimental science (*A Treaty on the Sphere*); Tomé Pires, in his *Epistolas*, described the 'drugs' (medicines) of the Orient for yellow fever and tropical fevers, discovering quinine among the many curative products.

POLITICAL 'ESTATES'

From the first quarter of the thirteenth century, Portugal showed itself to be politically and socially different from the rest of the European nations. Only in England were there demonstrated certain similarities, but even there the 'Magna Carta' signed in 1215 did not concede any rights and was the result of political necessity. Four years before, at the Cortes of Coimbra in 1211, the Portuguese had considered this change as a natural evolution. With the Cortes of Leiria in 1254 (during the reign of Alphonse III) the people began to have a place in these assemblies, their representatives being elected among the '*prud'hommes*' (remember this designation among the Cathars) of the different municipalities, in order to put their problems and protests before the king. The sovereign was the 'judge' in the discussions among the three 'estates' (the social classes) and he often took the side of the people so as to thwart the demands and abuses of the nobility and impede the unwarranted interference of the clergy in matters of temporal politics.

This revolution in medieval politics thus took place in Portugal forty-one years before the same measure was adopted in England, when the representatives of the bourgeoisie took their place in the 'Magnum Concilium' of 1295, in the reign of Edward I.

Were the Kings of Portugal influenced by their Jewish counsellors or by the Templars? It is certain, particularly after the crowning of John I, that the representatives of the people already had a considerable voice in the affairs of the nation and exercised a powerful influence on royal decisions.

Notes and References

1. The Councils of Albi (1254); Montallieu (1258); Bourges (1276); Pont-Andemer (1279); Offen-Hungary (1279); Anse (1300); Treves (1310); Bologne (1330); Valladolid (1322); Avignon (1337–47); Prague (1349); Apt (1365); Lavaur (1368); Palencía (1388); Salzburg (1418).
2. Other Pontiffs later came to their defence, such as Pope Pius IX (and, more recently, Pope Pius XII, who saved many Jews from the Nazis).
3. Son of Dom Agapito Colona, Archbishop of Lisbon (see p. 499).
4. 'Save their lives, but take advantage of their money'.
5. 'Do not persecute the Jews, do not kill them, but do not exalt them too much, either' (Migne, 1868).
6. See pp. 21 and 36.
7. Sousa, João de (1830).
8. *Portugalliae Monumenta Historica*, 'Leges et Consuetudines'.
9. Los Rios (1876).
10. Faria (1963); Ribeiro, J. Pinto (1633); Isaac (n.d.).
11. There had been one Christianised Moor, by the name of Fernão Cativo, who had occupied such a high post at the time of Count Henry. As far as is known, it was only much later that a man of Jewish ancestry became Standard Bearer to King Ferdinand. This was Ayres Gomes da Silva, who died in the conquest of Ceuta, and was the grandson of Gonçalo Ayres de Azevedo or Gonçalo Vasques, who had occupied the same post in the reign of Alphonse IV. However, there could have been many more that the chronicles did not mention, or about whom documents had been purposely destroyed.
12. *Geografia Histórica de Todos os Estados Soberanos da Europa*, vol. 1.
13. *Inéditos da História Portuguesa*, Appendix Chap. XXII.
14. This Jew was so rich that he was able to lend Dom Raimundo de Cardena enough for him to buy all of Vila de Mourão (a country town).
15. Slavery through indebtedness continued in England, for instance, until the nineteenth century, the debtors being deported to the colonies.
16. Brandão (1683).
17. Pina (1790).
18. *Guia Histórico do Viajante em Coimbra*, p. 77.
19. *Chancelaria de Dom Fernando*, Archives of Torre do Tombo; *Pergaminhos*, Book II, Archives of the Oporto Town Hall.
20. He went to Castile to become the Chief Rabbi (see p. 212).
21. *Anales de Sevilla, año de 1391*; Los Rios (1876).
22. The cemetery of the Portuguese Jews in Amsterdam has one of the most

beautiful collections of sixteenth-, and seventeenth-century tumular monuments, a result of the Jewish exodus from Portugal to Holland. The gravestones and stelae, which bear epitaphs in Portuguese and coats-of-arms of the nobility, are artistically sculptured. The city boasts the most sumptuous synagogue of the Western world and the ceremonies, even today, are held in Portuguese. The author owes his visit to this cemetery to his learned friend in Brussels, Gerrit Van Lent of the European Community Council, a scholar of medieval itineraries to Santiago de Compostela.

23. Santarém (1975), (1928), vol. IX; Remédios (1911).
24. Santarém (1928), vol. II, p. 268.
25. A Jew (or, more probably, a New Christian). In the seventeenth century, his descendants were persecuted by the Court of the Holy Office, suspected of heresy.
26. A counsellor to Alphonse V; he died in Italy.
27. Oliveira, Eduardo Freire de (1815), Book I.
28. This surgeon treated the Holy Princess Joana, daughter of Alphonse V and sister of John II, in the Convent of Aveiro in 1480, on the insistence of the latter King. Zacuto knew that his efforts would be in vain on examining the patient, declaring her incurable: 'a malignant and invisible acarid that is eating away her breast'. We may deduce that today it would be cancer, but at the end of the fifteenth century, the prediction of the existence of micro-organism was an extraordinary hypothesis in medical science. In fact, it was only about 50 years later that the Italian, Frascato, suggested the same thing in a poem: he created a character – a shepherd called Syphilis – who, because he tried to destroy the altar of King Sun, was condemned to have his body covered with '*bubas*' (syphilitic boils). In another work, the same author (who was also a doctor and alchemist) wrote that this disease was caused by 'a little animal that cannot be seen'. Through the publication of his work, the diseases then called *morbus galicus* because generically known as '*Syphilis*', some 300 years before the definition of 'microbes' by Pasteur. With the discovery of America by Columbus, the Spanish mariners passed on the disease to the natives, who had no resistance to this new malady. On their return to Castile, the disease had become exacerbated among them and spread all over Europe and was designated as the 'Spanish disease'.

4 The Peninsular Crusades and the Templars

In order to understand this chapter more clearly, the three phases of the Crusades, from the eleventh to the sixteenth century need to be appreciated. Table 4.1–4.3 set these out in chronological order.

Table 4.1 Western 'Crusades' in the Iberian peninsula against the Saracens and the Turks

1st	(1085)	Conquest of Toledo by the Kings of Aragon and Castile and León
2nd	(1147)	Conquest of Lisbon and other towns by King Alphonse I of Portugal
3rd	(1249)	King Alphonse III overcomes the last Islamic resistance in Portugal
4th	(1492)	The Catholic Monarchs overcome the last Islamic resistance in Spain (in the same year that Columbus reached the Antilles)
5th	(1578)	Disastrous expedition of King Sebastian of Portugal and his death in the Battle of Alcazar-Kibir; end of Crusades

Table 4.2 Eastern Crusades in the Middle East against the Muslims

1st	(1095)	Liberation of Jerusalem and the conquest of the Holy Land
2nd	(1144)	Called after unsuccessful attempt to reconquer Edessa (1144) and the conquest of Damascus; loss of Acre (1187)
3rd	(1189)	Muslim reconquest of Acre (1191) and the occupation of the Mediterranean coast of Palestine; four other undertakings all failed; loss of the Holy Land

Table 4.3 Western 'Crusades' against the Cathars

1st	(1209)	War against Raymond VI, Duke of Narbonne and Count of Toulouse
2nd	(1218–1224)	Fifteen years of war against Raymond, his son, Raymond VII and Trencavel
3rd	(1229)	War against Duchess Branca de Castela and the final conquest of the *pays d'oc* by the Capets

CRUSADES AGAINST THE MUSLIMS

When the Muslim wave had overrun Persia in the east and the Iberian Peninsula in the west in the eighth century, the Christians realised that their religion was in danger. This became even more evident when the first Arabs crossed the Pyrenees in 713. Eight years later, the Count of Toulouse succeeded in halting the Moor Zama at the gates of the city. In 731, a powerful army led by Adalramaman drove the Christians across the Dordogne. The defeated Christians took refuge with the Frankish Duke, Charles Martel, who saved France when he defeated the invaders at the Battle of Poitiers in 732. The Saracen retreat was slow, and it was only in 759 that King Pepino (known as 'the Taciturn' and the son of Charles Martel), reconquered Septimania with his victory south of Narbonne.

But it was not only Islam that threatened the survival of Christianity. The Paulician Cathars joined the Arabs and threatened Byzantium. Before dying in battle, their leader, Carbeas, reached the banks of the Black Sea; his nephew, Chrysocheir (who was an ex-Byzantine officer of Greek origin and a convert to Paulicism, conquered Ephesus and, in 870, defeated Basil's army at Tepheric. Two years later, he was defeated by Basil's son Christopher ('the bearer of Christ').

The victories of Islam in the east led Pope Gregory I to plan a military undertaking to recover the conquered countries. However, it was Pope Urban II that started things moving, at the Council of Clermont in 1095, launching a campaign to free the Holy Land. Hordes of Western peasants and knights set out to liberate Jerusalem: it was the start of the Crusades, and in all of them (including those against the Cathars) every type of ruffian imaginable was enlisted: adventurers, criminals, false prophets and fanatics, all out to get what they could from the great adventure.

During the 2nd Crusade, the Christians fell out among themselves and this enabled the Muslims to conquer Edessa in 1144. St Bernard used this disaster as a motive to revive religious fervour and advocate the 3rd Crusade; however, the Christian Kingdom of Jerusalem was lost with the fall of Acre (1191). In the persecutions that followed the fall of Jerusalem, bands of former Crusaders were transformed into bandits who fell on the pilgrims to the Holy Land. In order to defend the Christians from both the attacks of the Turks and the Christian bandits, the Military Orders were established. Their members were men of arms and, although they did not belong to the clergy, they took vows of chastity, poverty and obedience. Some of them rendered such good service that they were given a large house (in the place where the Temple of Solomon was said to have existed) as their headquarters in 1127. Although they called themselves the 'Poor Knights of Christ', they thus became known as the Templars. Some others, of the Order of the Hospital (a place of shelter for pilgrims) of St John the Baptist of Jerusalem, were called the Hospitallers. The Templars and the Hospitallers finally fell out and started fighting each other; the Christians abandoned the Middle East, but the black and white standard of the Templars would now be raised in Europe.

CRUSADES AGAINST THE SARACENS AND THE TURKS

The Christian reaction to the Muslim invasion of the Iberian Peninsula in 711, which was commanded by Tarik-ibn-Zeiad, happened almost immediately. The Cantabrian Prince, Pelagius, obtained the first victory over the Arabs in the Battle of Covadonga (Cangas de Onis) in 718. However, the true Iberian Crusade began only after the capture of Toledo by the Kings of Aragon and Castile in 1085 and the definition of the boundaries of the Portucalense County by Count Henry of Burgundy, as we saw in pp. 8-9. Henry went to the Holy Land in a rescue operation and so distinguished himself on the battlefield that he was granted the right to 'wear the cross'. His son, Alphonse Henriques (Alphonse I) chose this symbol as his coat-of-arms when crowning himself King of Portugal.[1] After Count Henry's death, his widow, in a charter of 1128,[2] granted the Order of the Temple the castles of Soure and Alpreade, as well as 'the unpopulated land between Coimbra and Leiria'.[3] When King Alphonse I continued with the Iberian Crusade to drive the Saracens from the Iberian Peninsula and consolidate his kingdom, he was often accompanied by Templar Knights and their Master, Gualdim Pais.

In 1139 Alphonse I made a deep incursion into enemy territory and met and defeated five Moorish leaders at Ourique (far south of the River Tagus). He took Leiria in 1145 and Santarém, Lisbon, Sintra and Almada two years later. After conquering Santarém in 1147, he granted the city an ecclesiastical charter (Lisbon already had a Bishop at this time).[4] In

1153, Alphonse founded the Monastery of Alcobaça for the Cistercian Order, which he had brought to Portugal with the support of St Bernard himself. One of his companions in arms, Dom Pedro Henriques[5] (natural son of Count Henry and therefore half-brother to Alphonse I) was present at the ceremony. The King conquered Alcácer do Sal five years later and in 1160 the Templars took Tomar on the River Nabão. The King had given the Order the Castle of Cera in the previous year and he now gave it the newly conquered town; it was here that Gualdim Pais would build the castle that became the headquarters of the Order.[6] He also granted the Order a third of all territory that it conquered south of the River Tagus 'on the condition that all the income from the lands would be spent in the service of the King and his successors as long as the war against the Saracen lasted'.[7]

King Alphonse I took Beja in 1162 and Évora three years later. He made an incursion to Trujillo, to the east of Portuguese territory, following this with another to Caceres and Juromeña. After this, he took Serpa. His epic had reached its close. Besides the Monastery of Alcobaça, he had ordered the building of four monasteries and many churches and chapels; St Theotonius, Prior of the Holy Cross, the first Portuguese saint, was canonised during his reign.

Following the death of Alphonse I, his son, Sancho I – who had already led an attack on Seville while still heir to the throne – conquered Silves in the Algarve, in the extreme south of the country. Some Crusaders helped him take the town, but as we saw on page 21, they committed such atrocities against the population that the King expelled them from the Kingdom. Two years later Jacob-ibn-Josef, taking advantage of the plague that was decimating the Portuguese, raised a powerful army in North Africa and crossed the Straits of Gibraltar; he reconquered Silves and the rest of the Algarve. Marching north, he took Alcácer do Sal, Palmela and Almada, facing Lisbon across the River Tagus, and earned the nickname of 'Almansor' ('the Victorious'). He crossed the River Tagus farther north and besieged Santarém, but was forced to lift the siege and withdraw. He died in Seville in 1191 of wounds received in battle.

The Templars helped King Alphonse II reconquer Alcácer do Sal in 1217 and in 1220 they completed the construction of the polygonal chapel of the first Church in the Castle of Tomar, which is today considered to be the finest example of Byzantine art in Europe. King Sancho II conquered Elvas, only to lose it again immediately; but he retook it three years later and followed this by seizing Moura and Serpa in 1232, Aljustrel in 1234 and Mertola, Tavira and Cacela in 1238. He was succeeded by his brother, Alphonse III, who ended the Crusade in Portugal by occupying the whole of the Algarve in 1249. (It was during Alphonse III's reign, in 1276, that the Portuguese Cardinal Pedro Julião, a professor of philosophy and medicine, became Pope John XXI.)

Despite the constant warfare, the Portuguese took advantage of the disputes among the Muslim leaders and frequently made alliances with them. During the periods of truce, ambassadors were exchanged and the knights of both sides took part in tournaments and bullfights, particularly between the tenth and twelfth centuries. From the twelfth century, the populations living in the territories conquered by the Portuguese rarely insisted on maintaining their Muslim religion and were gradually evangelised, a task in which the Mozarabs (Christians) played an important role.

The Castilian Crusade was a much slower process than its Portuguese counterpart and it was only in 1492 (the year that Columbus reached the Antilles) that the Catholic Monarchs managed to overpower the Kingdom of Granada, the last bastion of Islamic power in the Peninsula. By this time, the Portuguese were expanding overseas, reconquering territories in North Africa that had been Christian before the Arab invasion. The attempt to conquer Alcazar-Kibir (the Great Fortress), in which King Sebastian lost his life together with 800 of his knights and 7000 of his soldiers, was considered by Edgar Prestage to have been the 'Last Crusade'.[8] The combined army of Portugal and its Moorish allies amounted to 1500 knights, 15 000 infantry and 36 pieces of ordnance, while the force facing it had 4100 knights, 29 000 infantry and 84 cannon; all the wounded of the defeated army were put to the sword on the battlefield. Although the army had had a long march through exhausting desert territory, the terrain where the two forces faced each other was unfavourable to the Portuguese, and the Christians and their allies were overwhelmingly outnumbered, the King refused to retreat; the spirit of the Templars, whose motto was 'We never refuse to fight' is clearly mirrored in this attitude.

THE EXTINCTION OF THE TEMPLAR ORDER

The Order of the Temple had been founded by Hugues de Payen in 1127 and supported by St Bernard. The Order's statutes became known as the 'Book of the Soldiers of Christ'. The General Chapters of the Order were secret; the main articles obliged members to attend Mass every day, unless they were in battle, in which case they had to recite certain prayers. They had to fast four days a week and could not drink milk or eat eggs on Fridays. They were allowed to hunt only wild animals. They were obliged to shave their heads, but could grow a beard. The Order was made up in four ranks: knights, who had to be the eldest sons of noblemen, squires, chaplains and lay brothers. Each Knight Templar had the right to own three horses and have one squire in his service.

In the middle of the thirteenth century, the Order of the Temple owned about 9000 fortifications, temples and castles in Europe.[9] The Templars' headquarters in France was in Paris, where they settled in 1150; the Order was so rich and powerful that it worried the King himself. After the fall of Acre, the bulk of the Order's treasure was transferred to Cyprus and from there on to Paris. As the coffers of the State were empty because of the Crusades, the King was able to refill them only by imposing excessively heavy taxes. The people began to manifest a certain hatred for the monarch and a subservient admiration for the Order of the Temple, whose purchasing power enriched craftsmen and merchants alike, whether Christian or Jew, and whose fertile lands employed and gave bread to many families. As in Portugal, the sovereign was finally obliged to go cap in hand to the Order and contract enormous loans from it, thus making it the 'Banker of the Crown'.

In 1294 King Philip IV of France antagonised Pope Boniface VIII, who excommunicated the King. The King convened the Estates General[10] in 1302 and it was decreed that the Crown of France would not recognise the authority of the pontiff. The Templars had supported Boniface VIII and his successor Benedict XI, and since the due date for the settlement of the King's debt was approaching, he decided to dissolve the Order. The French Templars were arrested, tortured and exterminated. Wide-ranging accusations were brought: that in the initiation ceremony, the new member was obliged to deny Christ as a false prophet, crucified for his crimes and not for the Redemption of Man's sins; that he had to spit on the crucifix and then tread on it, after which he would be subjected to an 'impure kiss'; that the knights took part in orgies and defiled the Cross during Holy Week; that they worshipped a four-legged idol during secret Chapters-General; that they possessed a lead or silver image of a bearded figure, presumably the Devil; that they were authorised to practice sodomy amongst themselves; that they worshipped only the Holy Ghost and not the Holy Trinity.

To lend weight to these accusations, it was averred that the Templars had spent a long time in the Middle East, in close contact with both Muslims and Jews; that other members came from the *pays d'oc*, where the Albigensian Cathar doctrine had its roots; that they followed the dualist credo and did not believe that Christ was crucified for the redemption of Man; that their stay in the Middle East had introduced them to the practice of sodomy, which was very common among the Arabs, and that this was why they always travelled in pairs, as the insignia of the Order suggested. As far as the 'impure kiss' is concerned (see Figure 4.1), the accusers deduced this from the Templars' habit of prostating themselves one behind the other during prayers.[11] They were also accused of practising the 'umbilical kiss' instead of kissing the shoulder, the vestments or the cheek during the initiation ceremony.

4.1 The 'anal kiss'.

> Seventeenth-century engraving depicting the 'anal kiss' practised on the night of the Sabbath ceremonies. In Hebrew, the 'Sabbath' simply means 'rest', and corresponds to 'Saturday'. However, it also designated an assembly that sorcerers celebrated at midnight on Saturday under the direction of the Devil, who appeared in the form of a satyr. The initiation of new followers of diabolism was then held, in which the initiates had to eat repulsive things such as frogs and the flesh of executed murderers or their victims. After this, everyone present took part in orgies of lechery and obscene songs until the crowing of the cock announced dawn.

(The 'umbilical kiss' is not impure. It is a Jewish practice, their last farewell to a dead person. It is still practised today (on a disinfected sheet) in a rite in which the navel signifies *alpha*, the Beginning (birth) and the deceased himself *omega*, the End; it is a rite that was probably practised by the Hebrews of the primitive Christian Church.

The 'anal kiss', which originated in the 'black mass' of the Sabbath in which the Spirit of Evil was worshipped, is impure and obscene. It was practised in the meetings of witchcraft in which incest and homosexuality were encouraged and which were beneficially repressed by the Inquisition. The Cathars had been accused of practising the 'obscene kiss' with a cat, but it is highly improbable that either group had adopted such ceremonies; the 'umbilical kiss' would have no symbolic meaning with a living person.

Subjected to the worst tortures, very few Templars confessed to the crimes of which they were accused, later denying their confessions. The Grand Master of the Order, Jacques de Molay, requested that his hands be left untied when he was burnt at the stake so that he could pray.

One plausible accusation was based on the fact that many Cathars had been involved in the Order of the Temple – it is, in fact, possible that the persecuted Albigenses sought refuge among the Templars. There was one historic case, but it involved the Hospitallers. The Cathar Navarro Sancho Espada, who had helped to defend the besieged Albigensian castle of Montferrand, managed to find refuge with the Order of the Hospital, later becoming prior of their House in Toulouse. Similar cases may have happened with the Order of the Temple, especially as the Templars refused to persecute the Cathars and even gave them shelter. This led Pope Clement IV to write a letter to the Grand-Inspector of the Temple: 'Let not the Templars try my patience, so that the Church may not be obliged to examine more closely a certain reprehensible state of things, borne until today with an excess of indulgence, because later there shall be no remission'.

Guillaume de Nogaret, the most tenacious persecutor of the Order of the Temple in the service of Philip IV (known as 'the Fair') was present at all the interrogations and tortures. He not only managed to confiscate the immense wealth of the Order for the Crown of France, but also succeeded in having the Order extinguished in the whole of Christendom (or almost, as we shall see, as in Portugal things happened a little differently).

After both the fires and the Order had been extinguished, the guilt of the Templars remained uncertain, as the coercive influence of the torture chambers may have warped the truth. If there had been, in fact, some ignominious cases, they must certainly have been isolated and unknown to the leaders of the Order. The Order of the Temple was the royal banker of King Dennis of Portugal and, like Philip IV of France, he owed them huge sums of money. He had bestowed the towns of Salvaterra-do-Extremo and Idanha-a-Nova on the Order, the income from which was to pay off the debts. The Bull *Regnans in Coelis* of Pope Clement V in 1311 decreed that: 'all the assets of the Order of the Temple should revert to the Order of the Hospitallers'; King Dennis complied with it to the letter, having a process against the Templars prepared and expropriating the castles of Soure, Ega, Pombal, Redinha and Castro-Marim and the properties of Salvaterra-do-Extremo and Idanha-a-Nova. In 1312, he agreed with King Ferdinand IV of Castile and León that it would be better if they deferred, *sine die*, the specific determination of the Bull in relation to the Hospitallers; he came to a similar agreement with King James II of Aragon the following year. The two other Iberian Kings dissolved the Order of the Temple in their respective countries and transferred its wealth to the Crown. King Dennis, however, thought that the Templars had fought bravely in the Crusade in Portuguese territory, without ever having set foot in the Holy Land. He therefore came to the conclusion that they could not have acquired eastern cults and vices and, as they had all been found innocent of the charges brought against them, he decided to request from the new Pope, John XXII, authorisation to found a new Order in Portugal, to occupy the place and the functions of the recently extinguished Templar Order.

THE ORDER OF CHRIST

The new pontiff had been the Chancellor of King Robert of Naples and had always shown himself to be fair towards the Italian Patarines, forbidding them to practise their cult but never manifesting hatred against them. He had been born at Cahors, in Languedoc, one of the cities that had suffered the persecution of the Albigenses, and his father had been a shoemaker (i.e., of the social class from which most of the Cathars had come). John had been elected at the age of seventy-two after prolonged debates in the Sacred College, because many of the Cardinals disapproved of the fact that he had shown too much leniency towards the Albigenses and Patarines and had also protested against the persecution of the Jews. In fact, Pope John had never manifested the slightest sympathy for either Hebraism or Catharism, he simply considered their followers as God's creatures and susceptible to conversion: the Jews had not yet found the true path, Cathars had gone astray. John therefore gave his blessing to the creation of the 'Knights of Christ' in the place of the 'poor' Templars, and in 1319 the Bull *Ad ea ex quibus* transformed the Portuguese Order of the Temple into the Order of Christ.

Other Orders besides those of the Temple and the Hospital had already been established in Portugal. One was that of St James (known as St James 'of the Sword' because this was its emblem), whose members at first depended on the Masters of Castile; the Portuguese branch of the Order separated from its Castilian parent in 1228, its first Master being Dom João Fernandes. Another was the Order of Avis, or Calatrava – the oldest Order in the Iberian Peninsula, created in 1162 by a Bull of Pope Alexander III, its first statutes being based on the reform of the Cistercian Order, which had already been established in Portugal by St Bernard himself.

After the Order of the Temple had been dissolved, most of the Portuguese Templars entered the Order of Christ; their first headquarters and monastery were at Vila de Castro-Marim, and its first Master, or 'Apostolic Administrator' was Friar Dom Gil Martins. The first Statutes of the Order were copied from those of the Order of Avis and the reformed Order of Cister, and for this reason it was subject to the 'Visit' of the Abbots of Alcobaça (see p. 37).

During the reign of King Alphonse V, the headquarters of the Order was moved to Tomar, and from 1443 only Portuguese princes held the post of 'Master': Prince Henry 'the Navigator', his nephew Dom Fernando, Duke of Beja, the latter's sons, João and Diogo, and also King Manuel I. Under Prince Henry's Administration, the Order functioned in symbiosis with the fleets of the Discoveries, its wealth helping in the discovery of new lands, the temporal administration of the conquest being the domain of the King and the spiritual being that of the Order.

With the institution of the Knights of Christ, the black and white standard of the Templars disappeared. The standard that Guillaume de Nogaret had denounced as the dualist symbol of 'the Kingdom of Light' and 'the Kingdom of Darkness' – God and the Devil – was replaced by a red cross, with a smaller white cross superimposed on it, under which the Portuguese would discover new worlds for the world.

Experts, navigators and discoverers like João Gonçalves Zarco and many others belonged to the orbit of the Order of Christ, but the Order was not open only to knights and mariners. Mathematicians, geographers, cartographers and physicians were connected with it and some of them were of Jewish descent. They were allowed to be members as long as they had became a 'gentleman', the fundamental condition to be a knight.

The 'Alphonsine Ordinances' that King Alphonse V had compiled defined the first degree of nobility needed to enter the Order of Knighthood: 'this Nobility can be earned in three ways: the first by lineage; the second through knowledge; the third through Kindness, Good Habits and Ability'. Knowledge and ability led many converted Jews to the nobility and to the Knighthood of the Sea. The Templars had generally sworn to maintain absolute secrecy regarding what they saw and heard within their Order; the Knights of Christ took an oath that went even further – it also kept the secrets of the King himself when he confided in them.

When the Portuguese caravels sailed down the coast of Africa, looking for the way to India, even in the time of Prince Henry, foreigners kept a close watch on everything they did. So measures were taken to prevent this persecution, and the calculation of the distances adopted by the Portuguese seamen was different from those used by the rest of Europe. 'Anyone who dared to sail beyond the limits set by the Portuguese was punished in various ways and would be excommunicated by pontifical documents'.[12] Popes Martin V, Eugene IV and Nicholas V decreed excommunication for all those that ventured into the seas of Guinea without the consent of the Kings of Portugal. If they were caught red-handed, they were thrown into the sea or hanged 'without trial'. In 1475, the future King John II forbade the writing of routes 'by leagues or degrees and the drawing of maps beyond a hundred leagues south of the Island of Príncipe [on the Equator]'. This policy of secrecy became paramount: the prows of the ships ploughed their path through the oceans and the rudders wiped out any trace of them. If before the Portuguese 'the sea was a secret', with them 'nautical gnosis' was as well, so that the 'magpies of the sea' (foreign spies) that followed in the wake of the Portuguese naus would not be able to take advantage of their discoveries.

Notes and References

1. In the 'Hall of Kings' in the Monastery of Alcobaça, there is a sculptural group that represents St Bernard and Pope Eugene III crowning King Alphonse I.
2. Pimenta (1934), doc. 10.
3. Matoso (1939).
4. In *Lisboa Antiga*, vol. II (1936) the illustrious Ullysipographer (from the ancient name of Lisbon, 'Ulyssipon'), Júlio de Castilho denied the existence of a Bishop of Lisbon, due to a mistranslation of the Latin text. In this text, however, written by Osborn, a Norman priest who accompanied the Crusaders who came to Lisbon to help King Alphonse I conquer the city, there is a clear reference to the Christian emissaries that met with their Saracen counterparts: '*Archiepiscopus Bracarensis* (Braga) *et Episcopus Portugalensis*'; on mentioning that they were on the walls of the castle, it says: '*Dato utriusque signo, ipso civitatis alcaide murum cum Episcopo et Primiciis* (for *Primiceriis*) *civiatis stantibus*', i.e. the Bishop and the main figures of the city. See Oliveira, José Augusto de (1936); Schwartz (1953); and also Chapter 5, n. 1 below.
5. He fought alongside King Alphonse I at the Battle of Ourique, was appointed a Peer of the Kingdom of France by King Louis VII and, on returning to Portugal, was Master of the Order of Avis. However, he exchanged this post for the monk's cowl of the Order of Cister. This Saint, who maintained close relations with Dom Pedro in France, wrote a letter to King Alphonse I: 'Your Highness's brother, Pedro, deserving of all honours, told me of all the things you charged him to do and, after victoriously fighting all over France, is now carrying out your wishes in the State of Lorraine so that he may shortly became a Soldier of the Lord of the Armies'.
6. The Synagogue of the Arch, which gave its name to the ancestors of Christopher Columbus, was built in Tomar.
7. Areia (1947).
8. Prestage (n.d.).
9. The Templars had buildings in Portugal, Castile, Aragon, France, Auvergne (an independent Duchy), Rome, Messina (with Sicily and Apulia), England, Ireland, Germany, Dalmatia and Hungary.

10. The first assembly in which the French people were represented took place in 1301, 47 years after this practice had been instituted in Portugal.
11. See the Panel of the Navigators in the Polyptych of Nuno Gonçalves in the Lisbon Art Museum (p. 476 and Plate 11, 2nd panel).
12. Rego (1967).

5 The Discoveries before Columbus

The construction of the Kingdom of Portugal had been a true crusade for the Alphonsine dynasty. The Avis dynasty that followed continued this Crusade beyond the Iberian Peninsula, in the reconquest of North Africa, which had been Christian before it was overrun by the Islamic invaders. With the conquest of the Algarve, King Alphonse III practically defined the frontiers of Portugal, making it, as a nation, the oldest country in Europe. After 535 years of Muslim civilisation in Portugal, indelible marks were left everywhere south of the River Mondego, especially south of the River Tagus. The conversion of the defeated populations was a slow process, in which missionary activity proved to be more effective than military might, but the Christian Mozarabs, who thanks to the tolerance of the Muslims had conserved their faith,[1] practised a *sui generis* religion, impregnated with pagan and Oriental superstitions. The defeated Muslims occupied about a half of the country and they willingly accepted the religion of the victors, as the latter, not wishing to see their country depopulated, granted them many benefits. Portugal had no more than 1 000 000 inhabitants in the twelfth century,[2] so the Portuguese never thought of exterminating or exiling the vanquished populations, as deserted land has nobody to cultivate it or to defend it.

Except for a period of Almoravid hegemony, during which Christianity was repressed, the Moors allowed the Christian cult to be practised in the territories under their control. When this situation was reversed the Muslims were allowed to maintain their religion, but everything was done to coax them into accepting Christianity, either through encouraging mixed marriages, making sure that the Muslims were converted and that the Christians did not apostatise, or by offering grants of land to those who were baptised.

When Lisbon, in the centre of the country and standing on a wide river, was chosen as the capital of the kingdom, it was the first step towards having a base for the Discoveries. Portuguese ships docked in

Brussels in 1184 and four years later the cloth trade with Flanders was regulated, this being followed by agreements with England, Normandy and Brittany. In 1254, the King decided that ships bringing iron and other metals from France and La Rochelle could dock at Oporto and Gaia in equal numbers. King Alphonse I had already charged his half-brother, Dom Roupinho, the first Admiral of the Realm, to prepare the galleys necessary to drive off the Moorish ships that were playing havoc with sea communications in and out of Lisbon. Several naval battles were fought[3] and an undated 'regulation' attributed to either Sancho I or Sancho II shows that the King firmly protected his mariners: *'et defendo quod nullus sit qui malefacere meis marinariis'*.[4] King Dennis's decree of 1298 exempted sailors from serving on the battlefield and from working on land, with the exception of the construction of docks, walls and street giving access to ports; he also had pinewoods planted to protect the arable land near the coast. The trees not only prevented the sand dunes from encroaching inland, but could also be used to expand the new 'oarless' navy (a navy of ships with sails). Two different types of pines were planted – the pinaster, or wild pine (*pinus solis*), which was planted in an area that stretched from Leiria to the coast and was used for making masts, hulls and decks, and the umbrella pine, which was planted in The King's Pinewood at Costa da Caparica and, as the sea wind bent the trees as they grew, was used to make the curved pieces for the stem-post and ribs: figuratively speaking, it may be said that in each tree planted there was a 'Templar finger', in each piece of wood rigged, a hand of a 'Knight of Christ'. This was why King Dennis opposed the definite extinction of the Order of the Temple, and reformed it. There were already shipyards, and the keels of caravels were already being laid on the bank of the Tagus in 1315.

King Alphonse IV extended protectionist measures to foreign mariners who wished to settle in Portugal. He brought Florentine merchants of the 'Company of the Bards' to Portuguese ports, and granted enticing privileges to Nicolo Bertaldi and Beringel Ombeste that even allowed them to trade with the Moors of North Africa, as long as they did not sell embargoed goods such as arms, iron, pitch and cereals.

The Italian Cathars, persecuted in Florence, Naples, Milan, Venice and Genoa, who had already started to emigrate to Portugal in King Dennis's reign – their religious creed not being approved of, but at the same time being ignored by the authorities – matched the Jews as tradesmen and merchants. There was already a Street of the Genoese in Lisbon in Peter I's reign; many of its inhabitants, ex-Patarines, were now the owners of shops and warehouses. They traded with their places of origin, but were afraid to go back because of their fear of the Inquisition. As they did not dare try to spread their doctrine and because they had ceased to be 'the poor people of God' in order to become wealthy citizens and merchants of the Lisbon market, their Catharism gradually lost its fervour; very few were detected by the Court of the Holy Office in the sixteenth century, and the institution must have been fully occupied in rooting them out, as in the reigns of John II and Manuel I they could be distinguished from the Portuguese only by their names, and many of these had been changed to their Portuguese equivalents.

THE CANARY ISLANDS

During the reign of Alphonse IV two important events took place that influenced future relations between Portugal and Castile: the 'finding' of the Canaries, and the Battle of Salado.

The Atlantic islands of the Canaries, which are off Cape Juby on the African coast, may be considered as an insular extension of the Atlas chain of mountains – or, more precisely, the oceanic sequence of the western limit of the Anti-Atlas massif, in a crescent-shaped volcanic line. After being defeated by General Metello, Sertorius (one of the heroes of the Lusitanian campaigns against the Roman occupation of the Iberian Peninsula and the successor of the great chief Viriato)[5] tried to take refuge on the Fortunate Isles[6] of which, according to tradition, the Portuguese knew. The Carthaginians gave the name of 'Atlantides' to these islands, while the Greeks called them 'Western Hesperides' or 'Gorgons' (as they were just off the coastal region occupied by the Gorgon tribe) or even Purpurines (the mollusc, *Murex*, from which purple dye is extracted, is found there in abundance). The islands were known to the Arabs around 1016.[7] The name 'Canaries' probably comes from the Canarii, a tribe that lived in the Atlas Mountains and then settled in the archipelago.

The Portuguese had been to the islands in the fourteenth century,[8] and when (in 1345) Pope Clement VI informed the Kings of Portugal, Castile and Aragon, Sicily and France that he had decided to make the Canaries a principality and donate it to Luís de La Cerda,[9] Alphonse IV of Portugal protested and declared that the Portuguese had been sending their ships there for a long time and had brought back trophies; he had not been able to occupy the islands because first he had been at war with his son-in-law, Alphonse XI of Castile (from 1336 to 1338), and then because he had gone to the aid of Alphonse XI in his fight against the Spanish Saracens at the Battle of Salado. In 1336 the King of Portugal had certainly sent an expedition to the Canaries, commanded by Angiolino Del Teggia of Florence and Nicolo da Recho of Genoa; another Genoese, Lanzarote Malacello, also took part.

After five hundred years of Muslim occupation of the Peninsula, the Moors continued to control the coastal waters. When the Portuguese had settled the conquered territories, they thought it would be to their benefit to hire seamen that were experienced in the constant warfare against the corsairs in the Mediterranean and they chose those that had often defended the ports of the Italian States like Genoa, Florence and Venice. This explains why there was a Passagno as the Admiral of the Realm and Nicolo da Recho as leader of the expedition to the Canaries. Passagno turned out to be a deep disappointment: as a corsair, he was repeatedly defeated, losing galleys and lives; as a navigator, 'he brought nothing new to the Portuguese in the way of seamanship'.[10] His descendants, however, were trained in Portuguese seamanship and Arab-Hebraic cosmography, and Atlantic navigation demanded greater cosmographic knowledge than its counterpart in the 'closed' Mediterranean.

With the so-called 'Sagres School' of Prince Henry, the Portuguese caravels sailed the oceans, their pilots standing head and shoulders above their foreign rivals. It was then that the Italians and Flemings asked the Portuguese Kings if they might sail in their service.

King Alphonse IV, in an effort to gain the goodwill of Pope Clement VI, promised to help Clement's protégé, Luis de la Cerda, and in 1370 La Cerda's son Fernando donated two of the Canary islands to his admiral, Lançarote de France, Lanzarote Malacello's son.

With his second series of 'Sesmarias Laws' in 1375, King Ferdinand encouraged the 'Company of Naus' and decreed extraordinary measures of protection for navigation. At this time, such was the 'forest of masts' before Lisbon that the people of Almada on the southern bank of the river had to cross the river at Santos if they wanted to get to the city. According to the chronicler Fernão Lopes, there were at times as many as 500 ships anchored in the waters off Lisbon. Not having the Canaries, which could have been used as a base to victual and protect their ships on their mission of discovery along the West African coast, the Portuguese had to concentrate their efforts in the fifteenth century on the conquest of North Africa. They conquered Ceuta in 1415 and other cities and towns followed later. It fell to one of John I's sons – one of the incomparable 'Magnificent Generation' – Prince Henry 'the Navigator', as the Apostolic Administrator of the Order of Christ, to forge the destiny of the oceanic and multiracial Portuguese nation. Henry's much-travelled brother, Pedro, brought a wealth of geographic information and a *mappa mundi* from his many trips, the certainty of the existence of an Oriental empire in India and its approximate location. Through the Templar tradition of the Order of Christ, Henry had heard of the explorations of the Francisco João de Montecorvino,[11] and he had read of the journeys of Marco Polo, whose book had been presented to Prince Pedro by the Doge of Venice.[12] He was therefore aware of the approximate geographic situation of India, where a supposed Christian

King (Prester John) reigned and from where spices were sold in Portugal at very high prices. The importance of spices was at that time fundamental in the struggle against hunger, as spices were used to conserve food. Once animals were slaughtered, fish caught, they had to be smoked or salted, methods that did not halt the process of putrefaction for very long.

The main spices came from the India trade. They were shipped through the Persian Gulf or the Red Sea, transported by caravans through Mesopotamia from Iraq, Syria or the tongue of land of Sinai and then shipped to the ports of Italy. Overburdened with the taxes levied by the countries through which they passed and by the profits made by the innumerable merchants that handled the goods, they were so expensive when they reached Portugal that few people could afford them. To cut out the middlemen would be a great economic victory. The writer Henri Vignaud rejects the theory that Prince Henry had conceived a 'plan' to reach India, basing his argument on the fact that at the Prince's time the Hindustan Peninsula was confused with Ethiopia. A letter of 10 January 1502,[13] sent to Vasco da Gama to be handed to the Samorim of Calicut, states that: 'my uncle, Prince Henry, having started the discovery of the lands of Guinea in the year fourteen hundred and forty three, with the intention and desire that by sailing the coast of Guinea [meaning Africa] he would discover the way to India'. Duarte Pacheco Pereira[14] states 'and it was thus that this marvellous mystery was revealed to the virtuous Prince by God, hidden to all other generations of Christianity. He wished it thus that by the hand of . . . Pope Eugene, with his blessings and letters, the conquest and trade of these regions to the ends of all India . . . be bestowed on him and, with this justification, he set about his task'. João de Barros (1496–1570)[15] confirms: 'with these realms and dominions, [King Manuel] also inherited the continuation of an undertaking that was as great as that of his predecessors [Prince Henry and King John II] which was the discovery of the Orient by sea: so much effort, so much work and so much money those seventy five years had cost'. The caravels of the Prince and John II, therefore, had sailed along the African coast with the purpose of rounding it; and as King Manuel I had ascended the throne in 1495, the seventy-five years mentioned take us back to 1420, during the Prince's lifetime.

The Bulls *Romano Pontifex* (1454) of Pope Nicholas V and *Inter Cetera* (1456) of Pope Calixtus III are not very explicit; the latter grants the Order of Christ (of Prince Henry) the jurisdiction 'over the islands, towns, ports, lands and places, from [Cape] Bojador and from [Cape] Non to all of Guinea [meaning Africa] and beyond the southern [western] beach to the Indies'. The Bulls were thus worded owing to the confusion that reigned between Ethiopia and the Hindustan Peninsula –

it was supposed that these two regions were separated by the River Nile, as St Isidor believed: 'there are two Ethiopias: one is near the rising sun [East Africa] and the other is where the sun sets, near Mauritania [West Africa]'. The medieval maps of Hereford (thirteenth century), the ones of Ricobaldo Ferrare, Ranulpho Hidgen and Friar Paolino de Minorita (fourteenth century), plus the one of the Franciscan Friar, João de Carpino (fifteenth century), all mistook the African coast of the India Ocean for India; all of them placed Asia in the wrong position, even though de Carpino had been sent there by the Pope as ambassador in the East;[16] as was Friar Lourentino de Portugal.

When the Portuguese found that the coast curved eastwards at the Ivory Coast, they thought they had opened the door to the Indian Ocean. As has already been mentioned, the Prince knew of the kingdoms in the east, as the chronicler Damião de Gois confirms:[17] '[The Prince] decided to send ships along the coast of Africa with the intention of achieving his purposes, which were to discover, starting out from these western parts, the sea route to India in the east'. It was already supposed, therefore, that India could be reached only by circumnavigating Africa.

A Catalan map of 1433 (now in the Florence Library) bears the following caption: 'In this region, there reigns a great emperor, Prester [presbyter] John, Lord of the Indians, who are black by nature'. But Brito Rebelo[8] demonstrates that the 'India of Prester John' and 'Prester John of the Indies' are two different things and that the Portuguese never used the former expression, which led Henri Vignaud to the conclusion that the Portuguese confused Africa with India.

The Prince sent secret emissaries to Thebes (Egypt) to gather information about Prester John of India, as did John II later when he sent Afonso Paiva, who died in Abyssinia, and Pero da Covilhã, who reached the Hindustan Peninsula. Only then did the confusion come to an end. One thing is certain: Prince Henry's 'Plan of the Indias' had a clearly defined aim.

THE 'SAGRES SCHOOL'

As has already been mentioned, the spice trade was a commercial goal of considerable importance. There was also a desire to learn about the rest of the globe, the nature of its products and the customs of its peoples. But it would be stupid to think – as some writers have done – that only trade and curiosity led the Portuguese Kings to try and reach the Orient. If it had not been for the desire to spread Christianity, the world would not have been discovered so quickly.[19] We can see in Prince Henry's policy of expansion a spirit of the Templars in which the

Order of Christ was steeped: fight the Saracens and the Turks, the enemies of Christianity; evangelise the heathens and the infidels; acquire new knowledge of eastern science and enrich the country through new trade routes: Crusade, Knowledge and Power. The caravels began to explore the oceans (bearing the Cross of Christ on their sails) but along secret routes. Like the Templars, the Prince's men swore an oath of secrecy. Forming an 'academy' around the Prince and accompanying him wherever he went – Lisbon, Lagos, Sagres – they devoted themselves to the study of astronomy, geography, cartography and navigation. The academics of the so-called 'Sagres School' were not only Portuguese; there were foreigners like Father Egídio of Bologna (brought from Genoa by the Bishop of Viseu); Patrício Conti (probably the son of another Conti who had been to India overland) and the notable Jácome de Majorca, Prince Henry's closest collaborator.

Before becoming a Christian in 1391, Jácome de Majorca[20] was called Jafuda Cresques. He was the son of Abraham Cresques (to whom is attributed the famous *Catalan Atlas* of 1375 and who was nicknamed 'Joen Boxoler, the Jew of the Needle' because he made compasses). Following his death in 1410, his son chose another name, Jaime Ribas, and he entered the Prince's service between 1420 and 1427.

The Portuguese and foreigner experts at Sagres secretly devoted themselves to studying not only the nautical sciences, but also the construction of scientific equipment – astrolabes, compasses, and navigation charts – the famous square charts invented by Prince Henry, which defined the position of lands according to the 'calculations of the Sun and the Polar Star'. The 'Sagres School' was thus not a nautical academy in the real sense of the word, but its academics certainly placed theoretical elements of cosmography at the disposal of the Portuguese navigators, which the latter, from the bridges of their vessels, became adept at perfecting during their long voyages.

In his work *Roteiro*, published in 1509, Jerónimo Munzer, Doctor of medicine and also known as Hierónimo Monetarius, writes: 'Prince Henry died in the new town at Cape St. Vincent on the 13th November 1460'. An official letter of 19 September 1460 also states that he was at Sagres. In *Esmeraldo de Situ Orbis* (book I, p. 22), the chronicler Duarte Pereira Pacheco clearly states that: 'He [the Prince] established on Cape St. Vincent, which was formerly called the Sacred Promontory, the town of Terça Naval, situated on the bay of Sagres and which is there today'. The chronicler João de Barros, in Década I (Book I) says: 'he was in a town that he had founded in the Kingdom of Algarve, on the bay of Sagres, to which he gave the name of Terça Naval and is now called the Town of the Infante'. Prince Henry therefore clearly lived in the area between Cape St Vincent and Sagres Point and used it as a base for a naval undertaking. In any case, the name of the 'Sagres School' was not invented by a Portuguese, but by an English historian Samuel Purches in 1625.

It is obvious that the Portuguese navigators would not have been able individually to have improvised the knowledge of astronomy that they proved to possess merely by learning to sail the seas, otherwise they would not have advanced enough to gain the upper hand over their competitors on the ocean routes. Whether in Sagres or in Lagos, Sines or Lisbon, they were able, under the protection of the Prince and the Order of Christ, to exchange experimental observations and the scientific notions that were acquired in their studies or at sea. It is not enough to suggest that 'the benches of the Sagres School were merely the planks of the caravels', in a vain attempt to deny the existence of the Prince's 'School'. Another attempt to decry Prince Henry's knowledge is based on the fact that he did not leave a scientific library. However, the Italian navigator, Cadamosto (pp. 143–4) praised the Prince: 'for his studies of the movement of the stars and Astrology [meaning astronomy]'. The chronicler João de Barros wrote that the Prince, 'studies the humanities [scientific writings] a lot, especially the science of cosmography, the fruit of which has now given this kingdom the seigniory of Guinea'; and the historian Friar Luís de Sousa (1555–1632)[21] describes him 'exercising all his sound scientific knowledge, especially in the fields of Cosmography and Geography', and refers to 'a book that [the Prince] had published regarding the success of these discoveries . . ., which he sent to the King of Naples and which we saw in the city of Valencia in Aragon . . . in the ante-chamber of the Duke of Calabria'.

THE UNIVERSITY OF LISBON

Prince Henry's intellect is also confirmed by other chroniclers, such as Pedro Nunes. Corroborating what has already been mentioned, the mathematician Professor Francisco de Castro Freire[22] informs us that it was the Prince who had suggested to Jácome de Maiorca 'the idea of making plane charts so that the rumbs were reduced to straight lines, thus making things easier for the seamen, who could then pass over to the new reduced charts of Mercator and Wright'.

King Dennis (grandson and spiritual heir of the Astrologer King, Alphonse X, known as 'the Wise') founded the University of Lisbon (then called the 'General Studies') in the Campo de Pedreira de Alfama in 1290. Prince Henry distinguished himself as 'the benefector of studies in the city of Lisbon', settling the University in new premises in the Freguesia de São Tomé in 1431 and reforming the education provided. Henry created the 'Seven Liberal Arts', by adding the *'Quadrivium'* (Arithmetic, Geometry, Astronomy and Music) to the *'Trivium'* (Grammar,

Logic and Rhetoric) that already existed. In addition, he introduced the teaching of Medicine, Theology, Canonic Law and Philosophy and Moral Studies including the 'free philosophy' of the mathematician and astrologer Raymond Lully) and also Law. In the room for the teaching of Law there was a painting depicting Aristotle, flanked by two other anonymous figures in the robes of a Pope and an Emperor, representing the vast science of philosophy and the spiritual and temporal powers. In another room used for formal ceremonies, doctorates were bestowed, followed by a Mass of the 'Holy Ghost'.

Books were a rarity in Prince Henry's time, and were generally the product of laborious copying by monks. The reproduction of texts in 'xylography' (with the characters engraved in inverted relief on a wooden plaque) was limited to thirty or forty copies due to the deterioration of the wood; they were hardly ever scientific works, but religious, or light romances of chivalry or comedies. We know that the Prince possessed the *Tratatus Spherae* by the English astrologer Johannes Sacrobosco, and the *Book of Marco Polo*. According to Dom Fernando Colón (Columbus's son) the Prince also had a copy of *Summulae Logicales* of Pedro Julião (Pope John XXI) and the *Speculum Regium* of Dom Álvaro Pais, Bishop of Silves (1344).

Fernando Colón himself possessed the *Libro español de mano, llamado secreto de los secretos de astrologia compuesto por Infante D. Enrique de Portugal*. Unfortunately, this work, better known as *Secreto de los Secretos de Astrologia* and written by Prince Henry, has been lost, but its existence is mentioned in the Sevillian library of Fernando Colón, Columbus's chronicler. Why was Fernando Colón so interested in Prince Henry's books? What was the relationship between the two? For the moment we will say only that the Prince's place in the University was taken over by his nephew and adopted son, Prince Fernando, Duke of Viseu and Beja and his successor in the Order of Christ.

It was in the second decade of the fifteenth century that the question of the Canaries again reared its head. Practically abandoned, the only people that set foot on them were Biscayans, Normans and Frenchmen, who robbed the natives and made slaves of the men and women. In 1418, about seven of the islands were Castilian and the rest French; the donee of the Castilian island, Guilherme de Las Casas, invaded the French possessions, which were governed by Manchiote (or Maciot), who turned to Lisbon and complained to Prince Henry, who in turn tried to placate the two contestants. The King of Castile, however, protested to Pope Eugene IV, who did nothing except publish a Bull, *Dudum ad nos*, which left the affair to be settled personally between Kings Duarte of Portugal and John II of Castile.

In 1448, Prince Henry signed a secret, but documented, contract with Manchiote, who was then exiled in Lisbon, in which he rented the Island of Lancerote. The Portuguese occupied two other islands two years later and according to Cadamosto, who was in the service of Portugal, the population of the three Portuguese islands had been Christianised by 1445, which had not happened on the two islands that belonged to Castile. After having belonged to Dom Pedro de Menezes, Count of Vila Real, the two islands that the Portuguese had occupied passed into the possession of Prince Fernando, the son of King Alphonse V and the adopted son of Prince Henry. Meanwhile, and still during Prince Henry's lifetime, Portuguese navigators had rediscovered the two islands of the Madeira Archipelago and the nine of the Archipelago of the Azores. The ten islands of the Archipelago of Cape Verde were also discovered, and the African coast down to Cape Ledo in Sierra Leone was surveyed. The Portuguese coast was rich in fish at that time and there were whales off the beaches and in the wide estuaries of the Rivers Tagus and Sado. No fisherman would have dared to sail far from land and face the 'Terrors' of the oceans, unless he had been turned into a sailor in compliance with the mission of the Templars.

The rule of secrecy inherited by the Order of Christ from the Order of the Temple that reigned over the Portuguese nautical science and sea voyages resulted in the almost total disappearance of the sea charts and logbooks of Prince Henry's time, either because they were so well hidden or because they were destroyed by wear and tear. Only leaks of this secret information could have given European cartographers the possibility of charting the Portuguese discoveries on their maps, not only the name of the place discovered but often the name of the discoverer as well. One instance of this is the 'Land of João Coelho', a pre-Columbus discovery of the American continent which Portuguese policy kept secret, as it did other explorations (see Chapter 5). The pillage of the historical libraries and archives of Portugal during domination of Kings Philip II–IV from 1580 to 1640 resulted in the loss of many valuable documents. Earthquakes and fires were also the cause of the disappearance of many books and maps. The Inquisition also burned everything it found in the possession of suspects who devoted themselves to mago-scientific studies – and almost all the astronomers and cartographers were either New Christians or non-converted Jews. Since the policy of secret navigation that was implemented in Prince Henry's time was intensified in the reign of King John II, when secrecy regarding the Discoveries became the overriding aim of the state,[23] we may safely say today that the greater part of the history of the Portuguese Discoveries has still to be discovered.

MADEIRA AND THE AZORES

The rediscovery of Madeira took place in 1419, but it is thought that the island was known to the sailors of Ancient Antiquity. 'The Legend of

Machim' also became common knowledge – the story that an English nobleman came across the island while eloping from England to Spain with his forbidden bride, Ana Harfet. The lovers, it is said, later reached the coast of Morocco. In spite of the many contradictions of this romance, which supposedly took place during the reign of Edward III (the great-grandfather of Prince Henry) somebody must have been there before 1370, the year in which the *Medicis Atlas* shows Madeira to the north of the Canaries, possibly based on Portuguese information, as Italian seamen did then not venture into those areas.

The Jew Abraham Cresques had also traced it into his *Catalan Atlas*, together with Porto Santo farther east. However, he gave the islands Portuguese names 'Ilha da Lenha' (Island of Timber) and 'Ilha de Este' (Island of the East).[24] The Portuguese had thus certainly already been to Madeira.

In 1413, the planisphere of Meciá de Viladestes indicates the existence of the same islands, as does the Venetian Giacomo Giroldi in 1423, who called them 'P. Santo, the Island of Madeira, the Wild Islands and the Desert Islands'. It seems that a Castilian pilot, Juan de Amores, was a prisoner of the Moors in Morocco and while there met Machim and his beloved: was it he who spread the story of the legend, or had he invented it on the basis of other information? Even the least experienced sailor knows whether he is sailing east or west – it is difficult to believe, therefore, that a ship leaving England for Spain could sail so far off course and so far south. Besides this, the 'three-day journey' of the legend would not even have taken them to the Azores, let alone Madeira. Perhaps in recognition of the poetic legend of Machim, the Portuguese later gave the name of Machico to one of their settlements.

João Gonçalves Zarco rediscovered the Archipelago in 1419, together with Tristão Vaz Teixeira. He returned the following year with Bartolomeu Perestrelo, Columbus's father-in-law. The settlement of the island began immediately, although Prince Henry authorised it and communicated the fact officially only six years later, in 1425.

In 1877, a Spaniard, Marcus de la Espada, produced a sixteenth-century document[25] which attributed the discovery of the Azores to a Sevillian mendicant friar; this was proved to be a forgery, but as in the case of the Canaries and Madeira, it was then generally accepted that ancient mariners, Phoenicians or Greeks, had landed on one Island of Azores. It is also quite likely that Portuguese sailors had navigated those waters at the time of the rediscovery of Madeira and the six years of this island's secret settlement, but the silence that enveloped the position of the archipelago is, to say the least, strange.

The planisphere of Meciá de Viladestes (1413), which showed an almost correctly positioned Madeira, indiscriminately scattered imaginary islands everywhere, some of which could have been the Azores, as one of them is called São Jorge, patron saint of Portuguese cavalry and the name of Lisbon's castle. Of the others – Lobo (sea-wolf), Colombos (sea pigeon), Coelhos (rabbits), Cabras (she-goats), Ventura and Brazil – only the last two deserve any attention. The legendary Island of Good Fortune (Ilha da Bem-aventurança) must be linked with the original name that was given by the Portuguese: 'Ventura'. Brazil was the name given to a tree from whose bark and wood a dye for the conservation of fishing lines and nets was extracted. The Portuguese maps of this time were so well concealed that no-one has ever been able to unearth them, but the planisphere of Gabriel Valsequa of 1439 is explicit: 'these islands were found by Diogo de Silves, pilot of the King of Portugal, in 1427'. The discovery of the island was kept such a close secret by the Portuguese that as late as 1960 it could be stated that the Azores had been found by the 'great navigator' Gonçalo Velho Cabral. The Portuguese Navy also baptised one of its ships with this name. Gonçalo Velho was a friar of the Order of Christ who in 1431 and 1432, had been sent to set up the first settlement on the islands and explore the high land.

The *Bianco Map* of 1436 places the 'imaginary' islands alongside the nine 'true' islands discovered by the Portuguese, which were officially called (by royal charter of 1460) Santa Maria, São Miguel, Graciosa, São Jorge, Jesus Cristo, São Dinis, São Luis, Santa Iria and São Tomé. The names of these last five islands were changed in 1471 to Terceira, Pico, Faial, Santa Cruz (today Corvo) and Flores discovered only in 1451 by Diogo Teive, who I shall mention later.

Until recently, it was also thought that the Archipelago of Cape Verde had been discovered by Luigi Cadamosto; in fact, this had been done by the Portuguese Vicente Dias (of Tavira according to Columbus), who had landed on an island that he baptised São Cristóvão on his second voyage in 1445. On the *Bianco Map* of 1448 this island is confirmed as 'authentic', and appears with the name which it kept till 1460. On his return to Portugal, Cadamosto lied, claiming that he had made the discovery at an earlier date (I shall refer to this fact in more detail in pp. 136–7). This island was officially discovered by Diogo Gomes only in 1460. Diogo Afonso discovered the island of São Tiago two years afterwards, and the other eight followed later.

SAILING WEST – BRAZIL

In 1471, the Portuguese Fernando Pó (who some writers wrongly think to be Spanish) found the island which now bears his name. In the following year, João de Santarém and Pedro Escobar discovered the Archipelago of São Thomé and Principe. The rest of the Atlantic islands, farther to the west (except the islets or rocks of St Peter and St Paul, which had already been found before 1438, and the Brazilian island of St Mathew, which was discovered in that year) would come to light only at the turn of the sixteenth century – Ascension in 1501 and St Helena in 1502, both found by João de Nova on the 'return from Mina' (see below and Figure 5.1); another one, on the same parallel south as Australia and the River Plate, took the name of its discoverer, Tristão da Cunha, who was sailing to India in 1506 with the Viceroy Afonso de Albuquerque on board.

The 'return from Mina' needs further explanation. After João de Santarém and Pedro Escobar had reached this port (today Elmina, to the west of Cape Coast and Accra in Ghana) in 1471, the fort of St George was built in order to guarantee the safety and provisioning of the ships that were sailing farther south. On their return to Portugal, the ships ran into headwinds and currents (as had already happened on the Guinea coast), so that they had to double Cape Palmas and sail west in a wide arc before tacking north: this was known as the 'return from Mina'. It was thus that the Portuguese discovered Brazil and the first Antilles. Sailing west from the Azores, they discovered the Bermudas and the Sargasso Sea, which was already located on the *Bianco Map* of 1436 near the coast of America, but which was explored only later.

After the death of Prince Henry in 1460 and the succession of his nephew, Prince Fernando, to the Mastership of the Order of Christ,[26] the Portuguese set up a base at Mina and for three years explored the Gulfs of Benin and Biafra. They had always tried to find a seaway to the west and it was on one such voyage that they discovered the Islands of St Mathew off the coast of Brazil in 1438 – over fifty years before Columbus's first voyage. It was later given the name of St John and nowadays it is known as Fernão de Noronha Island[27] as it was donated to Noronha by King Manuel I after the official discovery of Brazil in 1500.

In 1452, Diogo de Teive, discoverer of the westernmost island of the Azores, found another island, which the sailors thought was the legendary 'Seven Cities'[28] (see below) forty years before Columbus's first voyage. Due to the secrecy that was maintained in regard to the geographical coordinates, it is still unknown whether this island was Bermuda or one of the Barbados islands (the Lesser Antilles). It was more likely to be Bermuda, as the Sargasso Sea was already being mentioned by that time. In fact, whoever made for the Azores by that route would pass through the Sargasso Sea and the many expeditions that left the Azores to sail west would definitely cross it. The same Diogo de Teive, sailing with Pedro Vasques de la Frontera, may have landed on the 'Ilha dos Bacalhaus' (Cod Island, Newfoundland) in 1452.[29] In 1472, he discovered another island which is thought to be Haiti. If it really were – and not another one farther south – it would mean that he had been there twenty years before Columbus.

THE DISCOVERIES BEFORE COLUMBUS 59

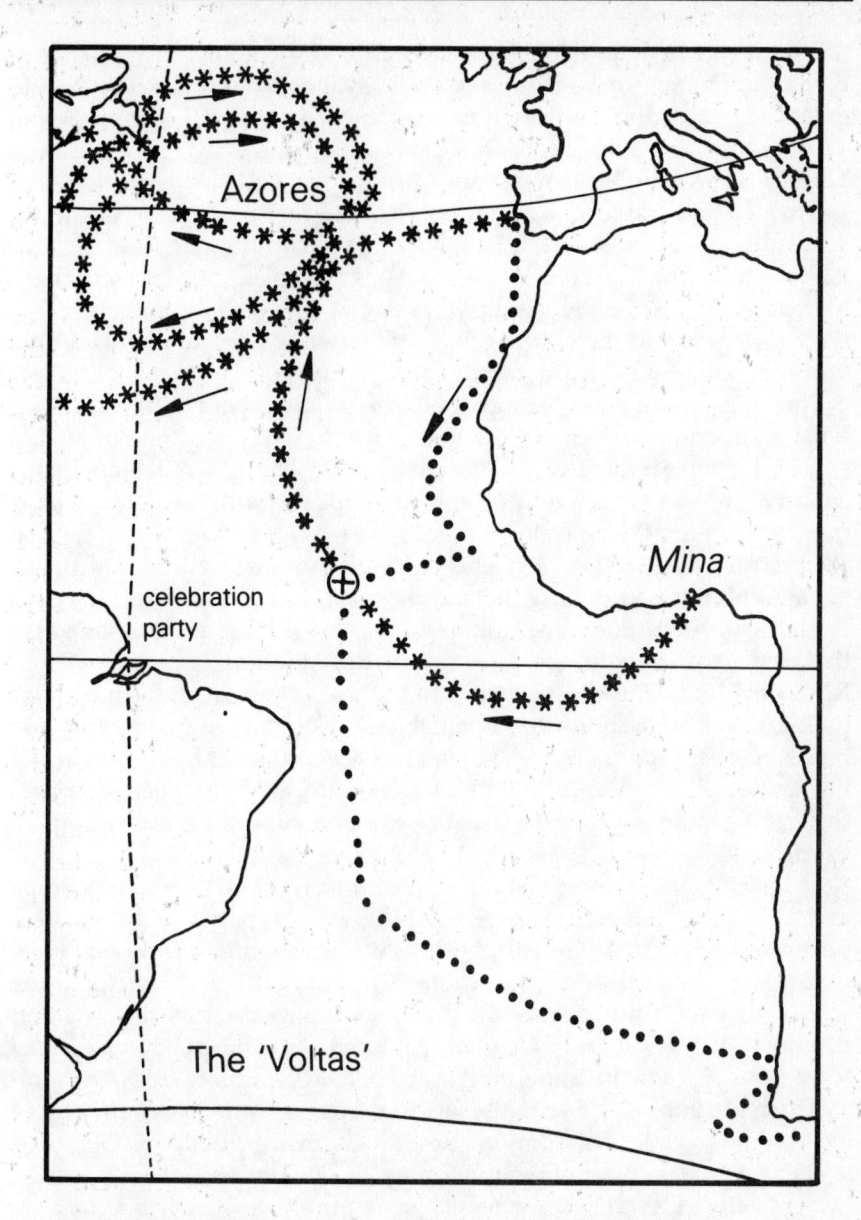

5.1 The 'return' from Mina.
During their voyages of discovery, the Portuguese usually returned by sailing in a wide arc or in circles, not only to take advantage of favourable winds and currents, but also to try and discover new islands or lands. This map shows the 'return' from Mina and from the west and Vasco da Gama's first outward voyage to India, during which the sailors celebrated their proximity to Portuguese territory – Brazil – after passing the islets of St Peter and St Paul.

• • • Vasco da Gama's first voyage to India.
∗ ∗ ∗ 'Arcs' of discovery.

The navigator João Afonso do Estreito and a colonist of the Azores, Fernando Dulmo (or de Ulm), received royal consent on 24 July 1486 to explore an island to the west that was known as 'Sete Cidades (Seven Cities)' and King John II granted Dulmo one of the two captaincies of Terceira Island.[30] Departing from Mina in the Gulf of Benin in 1487, Pedro Vaz da Cunha, 'the Bizagudo', headed for Brazil. João Fernandes Andrade sailed from Guinea the following year for the same destination. Although we know that these two voyages took place, we know nothing of their outcome due to the policy of secrecy. It is certain, however, that they did not end in shipwreck as the two navigators returned to Portugal and were rewarded.

The case of Pedro Vaz 'Bizagudo' deserves special attention in view of what happened to him on another voyage. The King of Benin had visited Lisbon in 1488 to ask for help in crushing a rebellion in his country. He was knighted by John II and baptised with the name of John in honour of the Portuguese monarch. On his return to Senegal, he was escorted by a squadron of twenty caravels commanded by Pedro Vaz 'Bizagudo'. Before reaching the African coast, however, he assassinated the African King. For some unknown reason, he was not punished for this act; according to the chroniclers Rui de Pina[31] and Garcia de Resende[32] 'the King did not want to involve others in this crime', and took into consideration 'the invaluable service that the navigator had rendered the Crown'. What 'invaluable service' would be a justifiable reason for Pedro Vaz da Cunha to be pardoned for such a crime? Everything points to the fact that it was the survey he carried out in Brazil in 1487.

In 1500, Pedro Álvares Cabral was 'sent to India' with thirteen ships and 1200 men. But strangely, everything was prepared to announce a new discovery. One of his supply ships was so old that it was easy to see that it would never be able to round the Cape of Good Hope; on the westward leg of the voyage, the fleet had fair weather all the way and dropped anchor off Vera Cruz on 23 April, but the policy of secrecy forced the Crown to announce that such a deviation from the pre-established route was due to the fact that the fleet was blown off course by terrible storms. This diplomatic lie was made public in 1500, five years after the ratification of the Treaty of Tordesillas, which we shall discuss later. Cabral was in reality sent purely to discover Brazil officially, since it was already well known to the Portuguese.[33] One of Cabral's crew was one 'Master' João, who sent the news back to King Manuel I by way of the old nau mentioned above. Was this 'Master' João Fernandes do Estreito who had already navigated in these waters twelve years earlier, or the mysterious João Coelho already mentioned? A part of the message he sent to the King read as follows:

Regarding the position of this land Your Highness, you will be able to ascertain it if you ask for the mapa-mundi in possession of Pedro Vaz; but that mapa-mundi is old and does not tell us whether this land is inhabited; Your Highness will also find Mina on that mapa-mundi.

The map in question, therefore, must have been made after 1471, as Mina was discovered in that year, but it would show only an outline of the coast, seeing that it does not indicate if Brazil was inhabited. This is obviously a map made in 1487, during the voyage of Pedro 'Bizagudo' to the west and we can clearly deduce from 'Master' João's letter that other voyages had been made, during which it was seen that Brazil was inhabited.

In another letter sent to the King on the same ship, Pedro Vaz de Caminha (who was not a farmer) often uses the expression 'planted . . . has yielded'. Agriculture was unknown to the Indians of Brazil, so Caminha was obviously talking about the good results obtained from planting of European or African species that was carried out before 1500 as a part of the Portuguese policy of experimental agriculture.[34]

The French historian Jean de Léry[35] wrote that the Antilles and Brazil was the fourth part of the world already known to the Portuguese; 'depuis environ octante ans que elle fut premièrement découverte' (i.e., 1483).

Finally, we have a text from Duarte Pacheco Pereira,[36] who relates:

in the third year of Your Reign, in the year of Our Lord fourteen hundred and ninety eight, from where Your Majesty sent us to discover western lands, sailing over the great ocean, from where such great terra firma with many and large islands adjacent to it has been found and navigated.

As this cannot be Newfoundland which was explored by the Portuguese between 1471 and 1473 and was known as the 'Ilha dos Bacalhaus' ('Island of Cod') – since this has hardly any adjacent islands and no large ones at all – the author is undoubtedly referring to the islands of Trinidad, Barbados and the Windwards. These explorations were thus undertaken before King Manuel's official order in 1498 (when Vasco da Gama reached India). It was without doubt the 'land of João Coelho', the Archipelago that Columbus would rediscover twenty years later, in 1492. The Portuguese continued to sail those waters, even after the signing of the Treaty of Tordesillas. This was the case regarding Florida, the discovery of which the Spaniards attributed to Ponce de Leon[37] but which was reconnoitred by Gaspar Corte Real between 1494 and 1500.

On the 'return from Mina', António Leme also sailed west and found three islands which are presumed to belong to the Archipelago of the Antilles; the same happened to João Coelho, whose feat is hinted at on the *Maiolo Map* and referred to clearly in a complaint made by Estêvão Frois in 1514. Frois was accused by the Spaniards of undertaking discoveries in the part of the world attributed to Spain by the Treaty of Tordesillas. In his defence, he alleged that he was merely navigating in areas already discovered by 'João Coelho, from Luz, in the suburbs of Lisbon, twenty-one years ago'. This means that in 1493 João Coelho continued with the voyages that he had started in 1487.

In fact, the Viscount of Maiolo's map of 1504 (which is now to be found in Fano in Italy) called the whole of the north coast of Brazil 'Land of Gonçalo Coelho', which proves that a navigator by the name of Coelho (maybe the brother of 'Master' João) had sailed those waters. Given the time it would take for the news to cross the Atlantic to Lisbon and then be leaked to Italy, plus the tardiness of the cartographers in making their maps, the voyages registered by Maiolo must have been undertaken much earlier.

GREENLAND AND NEWFOUNDLAND

But it was not only Central and South America that attracted the Portuguese sailors in their maritime explorations westward. King Alphonse V had come to an agreement with King Christian I of Denmark and Norway (1448–81) for a joint expedition to the west. Although nautical science was unknown to the Danes (Vikings), they had already reached Greenland (which would not have been very difficult for them owing to the short intermediate distances between the coast of Norway and the Shetland and Faroe Islands and then on to Iceland and Greenland). Rounding Cape Faver and sailing north to the Arctic Circle, they may have found Baffin Island and, to the south-west, Labrador. If they did arrive there, nothing indicates that it was before 1477, the year of the Luso–Danish expedition, which was able to plot its course scientifically with the improved Portuguese cosmography.

Between this date and the first voyage of Columbus, the discovery of America by Scandinavians is mere conjecture. Everything indicates, however, that the Vikings had explored the Island of Thule, which Ptolemy spoke about and which in the fifteenth century became known as Friesland and then Iceland. Dr Lofus Larsen[38] has tried to prove that the goal of the Luso–Danish expedition was Newfoundland. But there had been previous Portuguese voyages to the west. These include the one of Diogo de Teive in 1452 and another of João Vaz Corte Real undertaken in 1471 and followed by a second one in 1473 with Álvaro

Martins Homem to the same islands, both of these expeditions setting out from the Azores. A later expedition of Corte Real to Newfoundland did not return, so his son, Gaspar, set out in 1494 to try and find out what had happened. This led to him making other expeditions, the last of which was in 1500, after the King had bestowed benefits on him. Like his father, he disappeared, as did his brother, Miguel, who went in search of him. Miguel, however, left some signs of his voyage through an inscription on a rock at Dighton (see pp. 150–3), which will be dealt with later; the discovery of the 'Island of Cod' was officially attributed to Gaspar Corte Real in 1500. One of Gaspar's expeditions, which set out from the Azores and took place before 1474 (the year in which the King granted him one of the Captaincies of Terceira Island as a reward for the several discoveries made by him) was documented by the Portuguese but not made public.[39] Similar secrecy surrounded the expeditions of João Corte Real and Álvaro Martins Homem in 1473 and the Luso–Danish expedition in 1477, on which Columbus claimed to have taken part. The next expedition took place only in 1492, the year in which Columbus reached the Antilles. This voyage of maritime exploration, one of the most daring of the time, was commanded by Pedro de Barcelos and João Fernandes Lavrador. Setting sail from the Azores, they discovered the peninsula to which Lavrador gave his name (Labrador) and sailed along the whole of the north–northeast coast of this western region of Canada.

An American writer, Samuel Eliot Morison,[40] attempted to refute the fact that the Portuguese reached America before Columbus. But even if the voyages which were undertaken in the utmost secrecy to the American continent and islands were nothing more than a hoax by all the Portuguese navigators and Italian cartographers, how can one deny the expeditions carried out in 1473 and 1492 which reached Newfoundland and explored Labrador?

Notes and References

1. When Lisbon was conquered from the Moors, the English, Aquitainian and Breton Crusaders were camped to the west of the city walls and those of Flanders and Cologne to the east. They murdered and robbed and many of their victims were (Christian) Mozarabs, who for mercy implored with cries of *'Mariam Bonam'*. *In Vita Sancti Theotonii* 17, the biographer emphasises the compassion that the Saint showed for the Christian Mozarabs who had lived under Muslim rule and whom the Crusaders treated like infidels. According to the monk Osborne: 'against all human and divine right, they even killed the bishop of the city by cutting his throat . . . whose church, completely razed by the Moors, still had three stones standing as if to

remember its destruction' (see Oliveira, José Augusto do, 1936; Castilho, 1936).
2. Estimate in Girão (1937).
3. See Quintela (1830), vol. 1112.
4. 'and I forbid anyone to dare mistreat my sailors': *Des cobrimentos Portugueses*, vol. I, p. 6.
5. Mentioned by Straban, Eutropius and Aurelius Victor; see Shulten (1948).
6. In Caius Sallustius Crispus, *Historiae*, referring to the 'speech of the Consul C. Cotta to the people', in which he notes Servius. The Roman historian Sallus was born in 86 BC and died in 34 BC; after serving as Governor of Africa in 46 BC, he wrote *The Cataline Conspiracy* and *The Life of Jugartha*.
7. The circumnavigator of Hanon (c. 270–190 BC) is not mentioned because the Island of Cerne where he supposedly founded a colony would be to the west of the coast of Africa and not to the east if it were the Canaries. As this is neither a spelling mistake or an error of translation, this subject deserves future research.
8. The date of the Portuguese expedition was long a subject of debate; Boccaccio referred to it (National Library of Florence, R.B. no. 50), but either he made a mistake in writing the date or his copier falsified it as MCCCXXXXI (1341) is written instead of MCCCXXXVI (1336): 1341 written correctly should be MCCCXLI (see Peres, 1943, vol. 15).
9. Luís de la Cerda was the great-grandson of King Alphonse X of Castile (known as 'The Wise') and grandson of St Louis, King of France.
10. See Serrão (1975); Kreschmer (1909); *Arte de Navegar* (1772), with critical annotations by Armando Cortesão, Fernando de Aragão Aleixo and Luís de Albuquerque; Mota (1957).
11. Friar João de Montecorvino was born c. 1247 and died in China in 1328.
12. Marco Polo (1254–1324) began his travels in the East in 1271.
13. *Alguns Documentos da Torre do Tombo*, p. 127.
14. Pereira, Duarte Pacheco (1505), vol. 1, Chap. 32.
15. Barros, João de (1552; 1945–6), Book V, Chap. 100.
16. Risch (1930), vol. II, Chap. 11.
17. Gois (1556), Chap. VII.
18. Lisboa (?1540), *Introducão*, p. xxx.
19. Rego (1958), The supposition of Duarte Leite, in his article '*Talent de Bien Faire*' (Prince Henry's motto), that Henry's main motive was commercial greed and not religious zeal is unacceptable. Bensaúde (1943b) describes the Prince as being 'an ascetic governed by the sincere enthusiasm of a Crusader, pledged to the defence of Christianity'.
20. A Hebrew from Majorca making compasses in 1350; his science came from Arab Seamen, who had been using such an instrument since at least the middle of the thirteenth century: the invention of the compass has been wrongly attributed to Flavio Givia, an Italian from Amalti, in 1302.
21. Sousa, Friar Luís de (1623), 1st Pr, Book VI, Chap. XV.
22. Freire, Francisco de Castro (1872), p. 10.
23. Bensaúde (1943a): the 'gaps and lacunas, whether accidental or deliberate, that appeared were almost always filled up with arbitrary or plundered hypotheses. One of the most momentous and harmful examples of this for

our history was the ease with which Humboldt took advantage of the silence imposed on Portuguese nautical science in the 15th century to fill that lacuna with his political propaganda in favour of the astronomy of Regiomontanus and Behaim'.
24. The 'Island of Timber' later became known as Madeira and the 'Island of the East' as Porto Santo.
25. *'Libro del conoscimiento de todos los reynos y terras y Señorios'* mentioned by Coutinho, Gago (1951–2).
26. Prince Fernando, the son of King Duarte, was the brother of Princess Joana, who was Queen of Castile and Leon through her marriage to King Henry IV, of Princess Leonor (who was the Empress of Austria through her marriage to Frederick III) and of Dom Manuel, Bishop of Guarda. Fernando was the father of King Manuel I (who succeeded King John II because the heir to the throne had died) and King John II's father-in-law.
27. Barros, João de (1552; 1945–6), Vol. I.
28. Lima (1946).
29. *A Viagan de Diogo de Teive e Pedro Vaz de la Frontera ao Banco de Terra Nova em 1452*, Naval Historical Archive, Vol. I.
30. *Livro 4 de Dom João II*, p. 101.
31. Historian, b.c. 1440 and d. 1521, see Pina (1790).
32. Poet and historian, b. 1470, d. 1540, see Resende (1545; 1798).
33. See Leone (1968); Sampayo (1971).
34. The Portuguese always tried to transplant all the silvicultural and agricultural products that they thought might grow on other continents. In Africa, where the natives ate only certain edible roots like yams and sorghum, they introduced maize, cassava, rice and sweet potato (the staple diet of the African today) in the sixteenth century and also certain European vegetables, exactly as the spices and Indian fruit trees were taken to Brazil; see Valles (1964); Cardoso (1989).
35. Léry (1563; 1927).
36. Pereira, Duarte Pacheco (1505).
37. Leite (1950); Peres (1943); Coutinho, Gago (1951–2).
38. Larsen (1924).
39. *Tombo de Pedro Aunes do Canto* (Ponta Delgada, Azores, n.d.).
40. Morison (1940).

6 The Diplomatic Treaties of King John II

By the Treaty of Alcáçovas in 1474, signed by Prince John (son of King Alphonse V and the future King John II) the Canaries became once and for all Spanish. King Alphonse V had taken Princess Juana, daughter of King Henry IV of Castile and Queen of the Realm on her father's death in 1475, as his second wife. But as she was only thirteen years of age and so too young for the marriage to be consummated, it was annulled – obviously for political reasons.[1] If the marriage had been confirmed, the history of the world would have taken a very different course. Under the rule of one sovereign – Alphonse V (known as 'the African') – the combined forces of Portugal, Castile and León would have been brought to bear on the Muslims much earlier; they would have occupied Granada and consolidated their positions in North Africa. Only after this would the Discoveries have been continued and the Treaty of Tordesillas, which divided the world in two between the Portuguese and Spanish, would never have been signed. But as the marriage was dissolved, Granada was conquered only later, in 1492, and Portugal continued to devote herself exclusively to the dominion of the oceans on her way to Brazil and the Orient. Christopher Columbus would also not have been sent to Castile and to the west as 'Cristóbal Colón, a secret agent of King John II'.

How can the sudden Portuguese loss of interest in the Canaries be explained? On the political chessboard where Portugal and Castile faced each other, with the Church of Rome as a referee, the Canaries were a 'rook' that the Portuguese sacrificed in order to set up 'checkmate'. The first piece moved towards this end was the rather ambiguous Treaty of Alcáçovas, through which, as we saw in Figure 6.1, Portugal retained the right to 'the seigniory of Guinea, Capes Bojador and Non, to the Indies, including all the adjacent seas, islands, coastline discovered and to be discovered, with their commerce, fishing and revenues, as well as the Islands of Madeira and Azores and Flores and Cape Verde'.

The second piece played was the Treaty of Toledo, signed in 1480.

This merely confirmed what Castile had obtained in the previous treaty, but Portugal specifically reserved for itself: 'the Islands of Cape Verde and all the islands that it has until now discovered and any other lands that are found or conquered below the Canary Islands and the west of Guinea', as West Africa was then designated.

Prince Henry, as we saw in Chapter , knew that India could be reached by sailing round Africa, and he was already considering the chances of getting to 'the other side of the world', by the west. He could never guess, however, that two enormous continents would block the path. Through the writings of Plato, it was thought that Atlantis had existed to the west of Europe long before the Greek civilisation and that if Knowledge came from the East, Happiness (contrary to the reality of life that the world of that time represented) was to be found in the West. This hypothetical theory had been defended by the navigators of the Arab Caliph Al-Mamanum when they came to Hispania in 827.[2] With the Treaty of Toledo, Prince John called his first 'check': he was not only reserving the road to the east down the coast of Africa, but was also opening the door for the official discovery of Brazil. With these two treaties, the future King John II was assuring Portugal of trade with at least half the world.[3]

In 1228, Raymond Lully, an astrologer from the Island of Maiorca, had proposed to the Holy See that the wealth of Islam, sustained by the eastern trade, be utterly destroyed by conquering the whole of the African coast from Ceuta to Cairo. A Venetian Jew, Marino Zacuto, later proposed (between 1307 and 1321) an alliance between Pope Clement V and the Great Khan of Mongolia against the Muslims (the Turks that controlled Egypt) in order to sweep them from North Africa and so be able to organise a squadron south of Suez in the Red Sea and open the way to India, which was also under Turkish domination. But all these dreams came to nothing and as there was no canal between the Mediterranean and the Red Sea it would mean that when the Portuguese rounded Africa the Italian trade with the east would be killed off. The respective strength of the two naval powers may be summed up thus:

Venetian and Genoese trade + Jews and Moors of the Middle East = Mediterranean seaborne trade

'Order of Christ' in Portugal + Jews and Moors of the Iberian Peninsula = Navy of oceanic discovery

On his return from the Kingdom of Benin, João de Aveiro had heard about a powerful sovereign, Ogane. Would he be the famous Prester John? King John II decided to send two explorers overland – Pedro de Montarroio and Friar António de Lisboa. But as they did not know the

Arab language, they had to retrace their steps. The King then sent two others, Afonso de Paiva and Pedro da Covilhã, both of Jewish origin and Arabic-speaking. Disguised as merchants, they reached Aden in 1488. According to information received, the former made for Abyssinia and the latter for the Malabar coast by sea, where he landed at Cananor and visited the cities of Calicut and Goa. From there, he went on to Ormuz in Persia, then he sailed to Sofala in Mozambique. On his way back, Covilhã went in search of Paiva, but was informed by two Jewish messengers of King John II (Abraão de Beja and José de Lamego) that he had died in Abyssinia. Covilhã handed the two Jewish envoys a report for King John which contained invaluable information regarding navigation in the Indian Ocean, where pilots could be found, the sources and prices of spices, and other details. He then went to Abyssinia, where he was extremely well received by the monarch and where he married and spent the rest of his life.

When King Ferdinand (husband of Queen Isabella of Castile and León) inherited the Crown of Aragon, he launched an attack against Granada, conquered the Alhama and forced King Ali to abdicate in favour of his son, Abu-Abdalah. Keeping a close watch on events, King John II of Portugal foresaw that the last bastion of Islam in the peninsula would soon fall. After the conquest of Granada, would the Castilians respect the terms of the Treaties of Alcáçovas and Toledo? Or would they persist in their expansion beyond the Canaries and dispute the 'lands discovered or to be discovered' with the Portuguese? Was not India the fulfilment of every ambition? Who would be able to stop the Castilian ships from following in the wake of the Portuguese on the sea route to the Orient?

The cosmographic knowledge of the Castilians was way behind that of the Portuguese, who as we know had developed their science at Sagres not only as regards the art of navigation (the 'return of the high seas' as well as the 'return along the coast'), but also as regards the geographic conception of the world. The Portuguese King already knew of the existence of Newfoundland (1473), Brazil and the islands to the south of the Sargasso Sea, the last named already being located on charts. In his game of political chess, therefore, he would have to sacrifice his 'queen' – North America, not yet absolutely confirmed but already perceived, between Newfoundland and Brazil. This would be the bait for the Castilians to swallow. Who would be able to put it on the hook?

John II knew that he would need an exceptional man to convince the Catholic Monarchs to give up any idea of competing with the Portuguese in the race to India, someone who would present them with a logical hypothesis that the Orient could be reached by sailing west, eliminating the need to round Africa. It would have to be a person of culture and education, able to speak with Kings and convince Admirals, daring and experienced in nautical science – attributes that were possessed only by the men of the 'Sagres School' and the Order of Christ. He would have to be a good linguist and conversationalist and yet maintain, till death, the vow of secrecy. He should not be thought to be Portuguese, so as not to arouse the suspicions of the nation's rivals; he would have to be patriotic and unswervingly faithful to his King: a nobleman and, if possible, a relative that would be willing to give his life with the same willingness as Prince Ferdinand (who had died in captivity in Fez). And finally, someone with sufficient intelligence to defend himself against the rigorous questioning that the Castilians would subject him to. Who would be able to play such a thankless and perilous role? Who would possess the fibre and the nobility to sacrifice his name, his origins, his family (and, possibly, his life) on this mission as secret agent?

In 1484, civil war broke out in Granada. The time was ripe for the Catholic Monarchs to conquer the whole of the south of Spain. King John had to make up his mind. He chose a young man who was closely linked to the Order of Christ (and, presumably, to the royal family). He had already sailed to the west from Mina and the Azores, had taken part in the Luso–Danish expedition and had had some contact with Genoese people. Let him pretend to be Genoese, concealing the name of his birthplace and his parents. Let him sign documents with his own name using a cabalistic siglum, but transformed into 'Cristóbal Colón'.

As it happened, the Spanish Monarchs were not able to take advantage of the situation in Granada. Alzagal proclaimed himself King of Granada in Malaga and lost the city only in 1487; only then did the Castilian conquest get under way – Baena (1489), Almeria, Cadiz and the cliffs of Alzujares (1490). Granada was finally besieged in 1491 and it fell the following year.

In this same year, Columbus began to fulfil his mission. He had convinced Queen Isabella of the feasibility of reaching India by sailing west; the sovereign authorised the equipping and provisioning of three naus, which departed from Palos, on the Atlantic coast. On discovering the first islands – San Salvador (in the Bahamas), Cuba and Haiti (today the Dominican Republic) – he cunningly called them 'Indias'. He then wrote to the Catholic Monarchs, announcing that he was going in search of Cypangu (Japan) and pretended to believe that the chief of the Caribs was the Great Khan.

However, instead of going straight to Queen Isabella on his return, Columbus went to see John II. As the King was not in Lisbon, where Colón had landed, he made the long journey to Azambuja to inform John II that his mission had been completed. The King treated Columbus coldly in public, as though he was upset by the news of the discovery. But what happened in their private meeting? Surely the King would have thanked his agent in the name of Portugal? The King sent Columbus back to his Spanish pseudo-lord, so that the oceanic exploration to the west would continue. The ocean is vast and nothing was known of North and Central America except Labrador and some islands perhaps Bermuda, Barbados, Haiti. Now Columbus had brought news of more islands and not a continent, which was inconvenient for Portugal, whose aim was the trade monopoly in the east. Islands are surrounded by water and water is the highway of ships. It would only need a very narrow 'Panama Canal' to upset the plans of King John II, who was trying at least to delay the arrival of the Spaniards in the Orient. He did not want the control of Oriental trade to become a race.

It was therefore becoming necessary to be more specific regarding the operational areas of the Portuguese and Spanish discoveries. For this to happen, it was essential that the Spanish sally westward did not grind to a halt. King John II's next move shows that Columbus had served him well, and undoubtedly proves that the navigator had given him the exact location of the islands he had discovered – south of the parallel of the Canaries. This, in turn, proves that Columbus had not sailed due west along the parallel from Gibraltar, or even from the Canaries, but had edged west–south–west into the zone which was closed to Castile by the Treaty of Toledo. He managed to deceive the pilots during the long journey and he discovered islands that were in the geographic zone not reserved for Spain. It was this news he gave to his true master.

King John II then invoked the treaty of 1480, and simulated a great interest in staking his claim to the Bahamas. Ostensively to back up this claim, he fitted out a powerful fleet under the command of Dom Francisco de Almada in order to occupy the islands discovered by his secret agent. Feeling that he had been cheated, King Ferdinand of Spain proposed immediate negotiations with Portugal. The first Portuguese envoy, Rui Sanches, returned without having reached any agreement. Pedro Dias and Rui de Pina successively confirmed the decision of the Portuguese sovereign, but also returned home without having achieved any conclusive results. To assure possession of the islands in the Caribbean, King Ferdinand sent the Bishop of Cartagena, Dom Fernandino de Carvajal, to speak to Pope Alexander VI in 1493, with a protest, and the Church issued five documents, two of which were Bulls entitled *Inter Caetera*. The pontificial decision established a dividing line 100 leagues to the west of the Azores, to the west of which the sea explored or to be explored would by right belong to Spain. Thinking that it would be benefiting the Catholic Monarchs, the Holy See had in fact handed the whole of the Orient to Portugal.

Playing his part to perfection, Columbus peddled the theory, attributed to Toscanelli, that there was a navigable passage to India and

Japan. To the end of his life, even after it had been proved that the Isthmus of Panama was unpassable, he insisted, even threatening with death anyone who contradicted him. Even after Amerigo Vespucci had astutely given his name to the new continent, Columbus persisted in his obsolete theory that there was a passage to the Orient.[4] The ignorance of the Spaniards was such that, following the discovery of the seaway to India by Vasco da Gama, the King of Spain ordered his navigators to find a passage through Argentina.

Columbus lacked neither intelligence nor geographical knowledge, nor did he suffer from attacks of insanity, despite the number of times he defended himself by simulating such infirmity. He simply remained faithful to his vow of secrecy that was one of the principles of the Templars: *Sigillum Militum Christi*. In his anxiety to obtain the concession of the right of way to the Orient by the west from the Pope, King Ferdinand did not even notice that the highly publicised fleet anchored off Lisbon had never been fitted to leave the Tagus: it was just one big bluff on the part of King John.

Portuguese cosmographic science was by now truly advanced. The longitudes of Ptolemy had already been rejected and the Arab rectification regarding the circumference of the world had been miscalculated through an error in measuring the degrees, but the experimental voyages of the Portuguese had given them an almost exact notion of the size of our planet and the intermeridional distances. The exact measurement of degrees was to be achieved only in the middle of the sixteenth century by the Portuguese Francisco Faleiro, but such precision was not needed for King John II to calculate that the line drawn by the Bull *Inter coetera*, 100 leagues to the west of the Archipelago of Cape Verde, would pass far beyond the Indian Ocean, which had already been sailed by Pero da Covilhã, on the other side of the world.

The width of Africa had been approximately calculated through the overland journeys made by Arabs, as mentioned by the chronicler, João de Barros. Covilhã had also calculated the width of the Indian Ocean on his voyage from India to the east coast of Africa, and through the study of the earth's curve that had been furnished by the calculations of navigation, the approximate size of the earth was known and would be confirmed by the voyage of circumnavigation led by the Portuguese Ferdinand Magellan and concluded by Juan Sebastian Cano some years later.

To have a notion of a meridian and its diametrically opposed line, a cosmographer would need only to observe an armillary sphere or a globe with the tracings of meridians and parallels. The sphere played an important role in the decisions of King Manuel I: he adopted it as his emblem, as had John II before him, and it became the standard of the Discoveries.

Returning once more to our figurative game of chess, King John II knew that the papal referee, Alexander VI, would not go back on his word. It was not now in his interest to sacrifice his 'queen' – North America. Now was the time to call 'checkmate', the 'sting' of his overseas policy: save Brazil to the west of Africa, to the south of the Canaries. For this reason, he treated with disdain an embassy of the Catholic Monarchs, the ambassadors of which 'were Dom Pedro Dayala, who had a pronounced limp, and Garcia do Carvajal, who was very foolish, and the King, after hearing their message, said that his cousins' embassy was a waste of time, both in the person of the ambassadors and as far as he was concerned'. He then sent Dom João de Sousa, Rui de Sousa, Estêvão Vaz and Aires de Almada to his Spanish cousins with a counterproposal, which was accepted in 1494 and led to the signing of the Treaty of Tordesillas. This treaty moved the dividing line of the world to 370 leagues to the west of the Archipelago of Cape Verde.

It is surprising that historians have never questioned the reasons that led King John II to chose exactly that number: 370. Why did he not suggest a round number like 300 or 400, or even 350, to the west of Cape Verde? There can be only one explanation. The Portuguese King was certain that the meridian of 370 leagues corresponded to the easternmost island of the Antilles.

It is essential to point out that when Columbus went to inform the King that he had fulfilled his mission, he was not in a position to give him the exact longitude of the islands as he had been unable to explore that area and had only found other islands 1800 km further to the west. This is proof, therefore, that the King of Portugal already knew, with geographic certainty, what existed beyond those 370 leagues before Columbus's first voyage. Even more surprising is the fact that Columbus, despite the policy of secrecy that forbade the divulgation of maps that showed Portuguese Discoveries, had a chart in his possession so that he would not make a mistake on his way to the Antilles (see Figure 6.1). On 3 October 1492 the log-book records:

> Some fish that looked like carp and a lot of seaweed, some of it very old, some of it fresh, and it looked like fruit on a bush: no birds appeared; the Admiral thought that the islands depicted on his chart had already been left behind.

These islands that 'had been left behind' could only have been those situated to the west of the 370-league meridian, but to the *east* of those that Columbus would discover 1800 km. further on (see Figure 6.2). And if he did possess a chart, he could have obtained it only from the Portuguese, with whom he had already sailed. Or King John II had given it to him when they had met in Lisbon in 1488. Figures 6.3–6.5 illustrate this thesis cartographically.

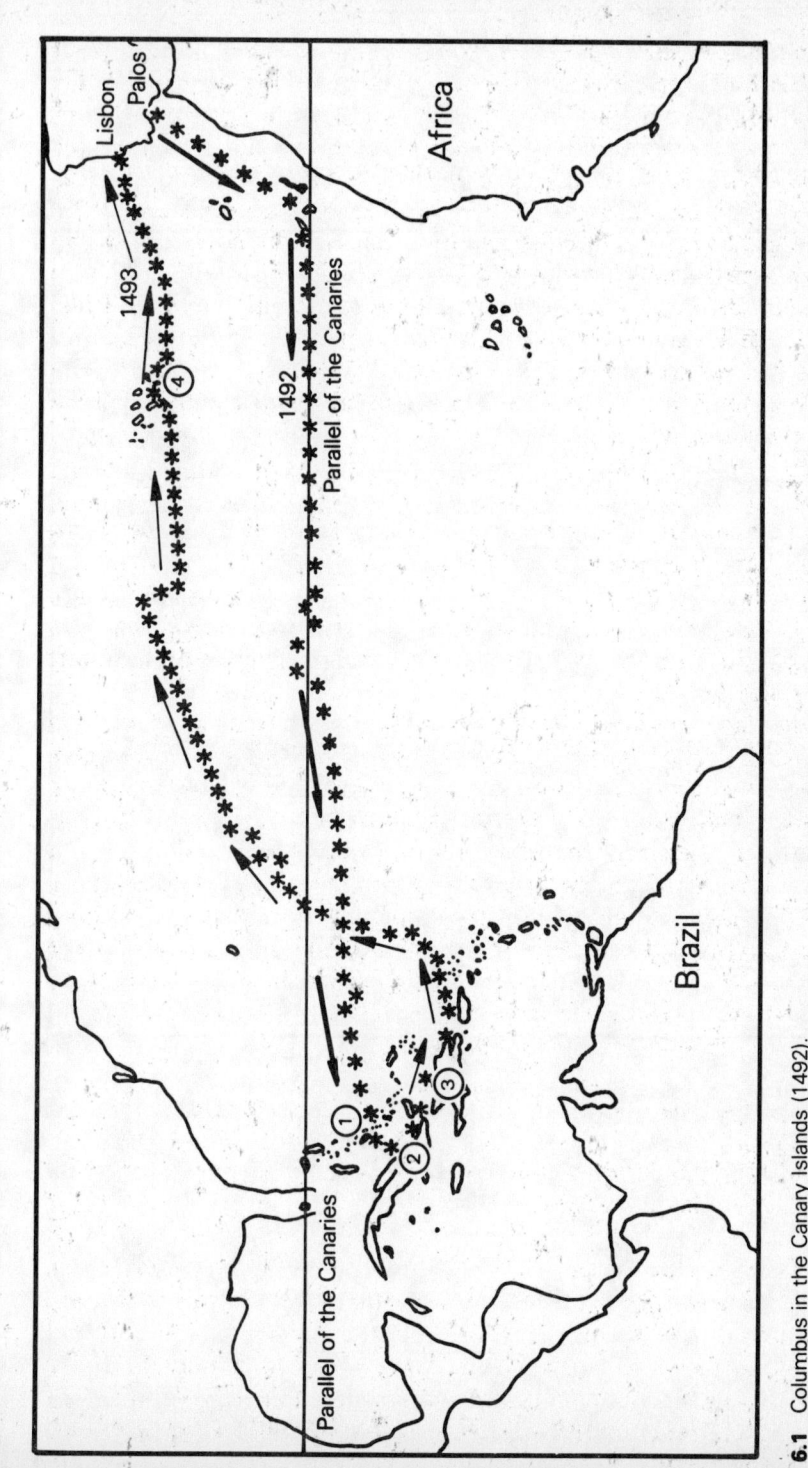

6.1 Columbus in the Canary Islands (1492).
On his first voyage in 1492, Columbus did not sail due west in an effort to reach Japan. He called first at the Canary Islands, the parallel of which defined the Portuguese and Castilian zones of maritime exploration. He sailed along the parallel of the Canaries at first, then tacked north-east, which disorientated the pilots, and veered to the south of the parallel of the Canaries. As a result, all the islands discovered were in the area reserved to Portugal.
∗∗∗ first voyage.

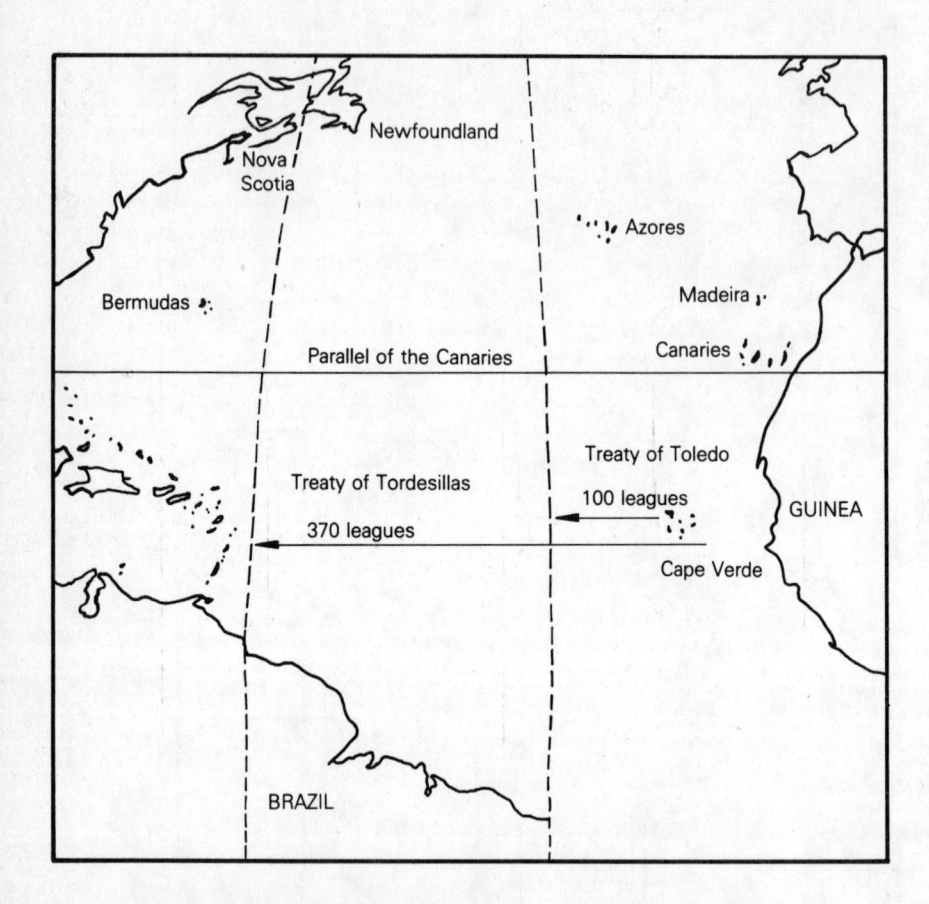

6.2 The Treaties of Toledo and Tordesillas.

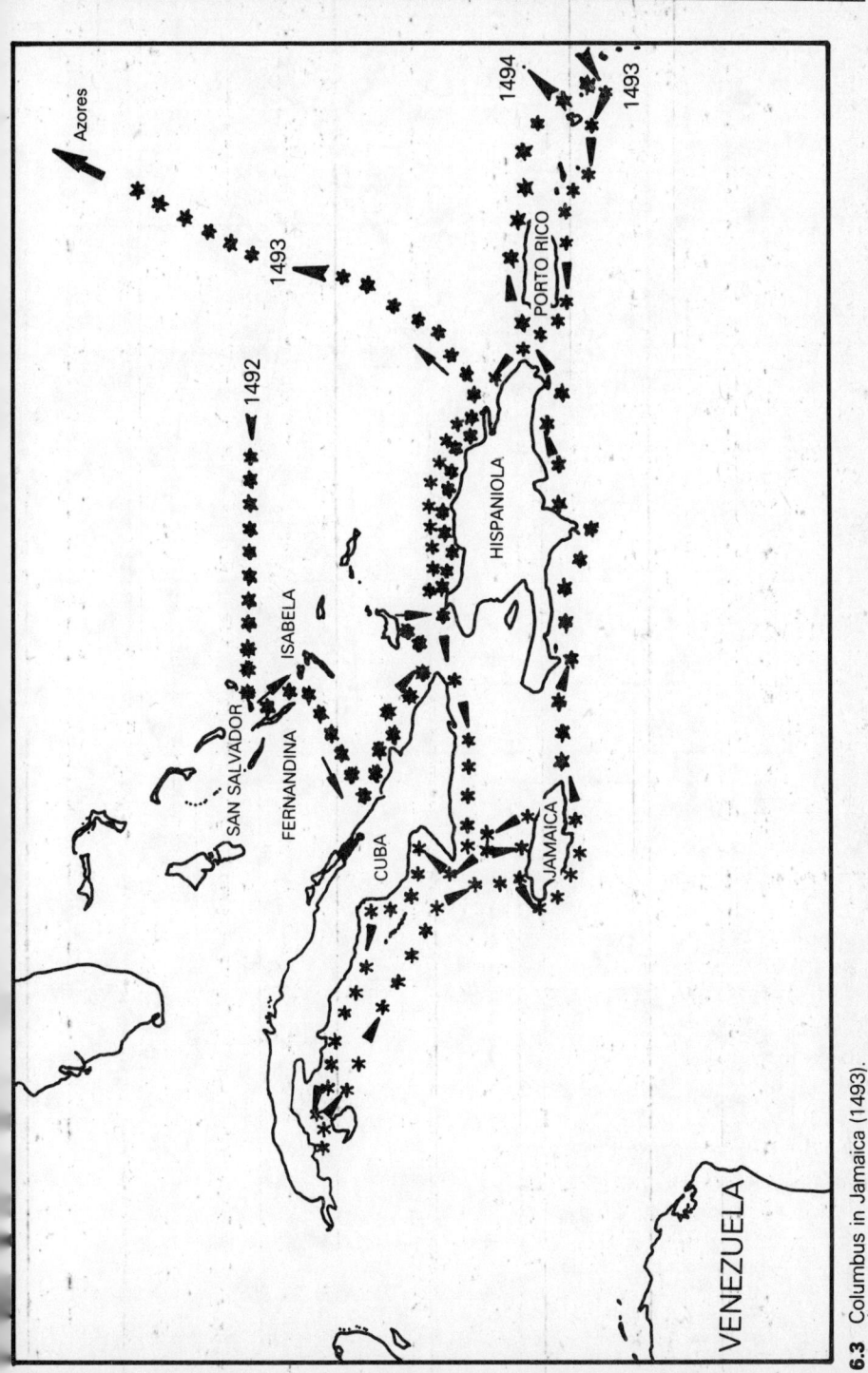

6.3 Columbus in Jamaica (1493).
On his first voyage, Columbus surprisingly returned by way of the Azores and Lisbon; on his second, he returned to Jamaica and other islands where he had already been, but did not attempt to sail further west.

✶ ✶ ✶ first voyage.   ✱ ✱ ✱ second voyage.

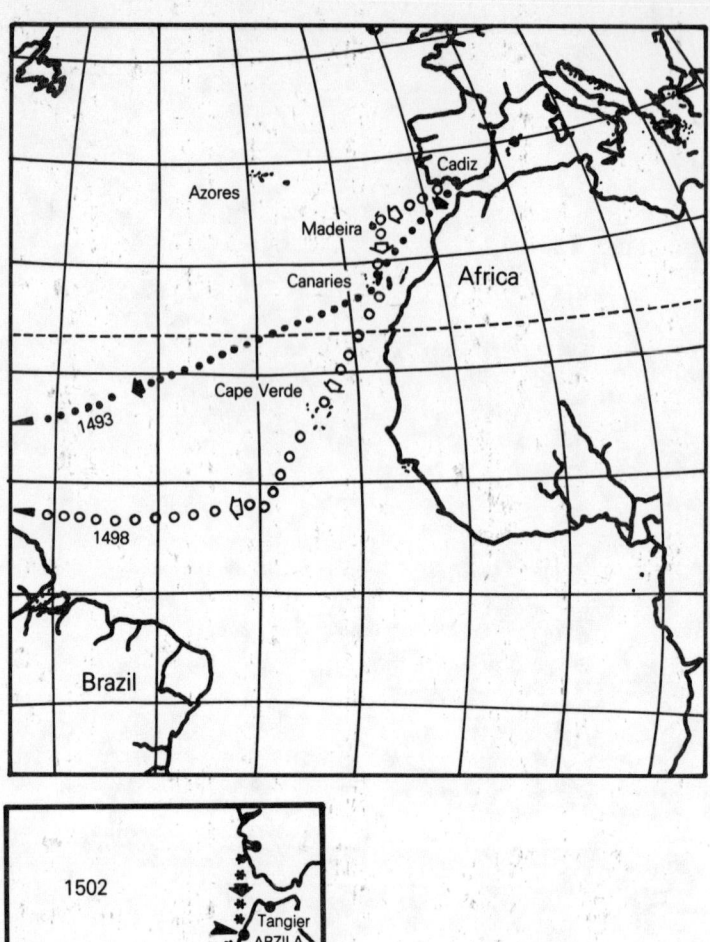

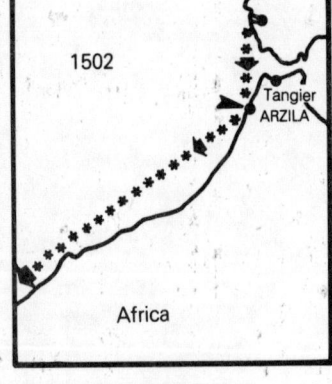

6.4 Columbus in Arzila (1502).

  a Columbus had given King John II a legitimate motive to rescind the treaty in force and sign the Treaty of Tordesillas. He thus did not hesitate to go south of the parallel of the Canaries.
  b On his fourth voyage Columbus surprisingly went to Arzila, where the Portuguese were said to be under siege by the Moors.

● ● ● second voyage.   * * * fourth voyage.   ○ ○ ○ third voyage.

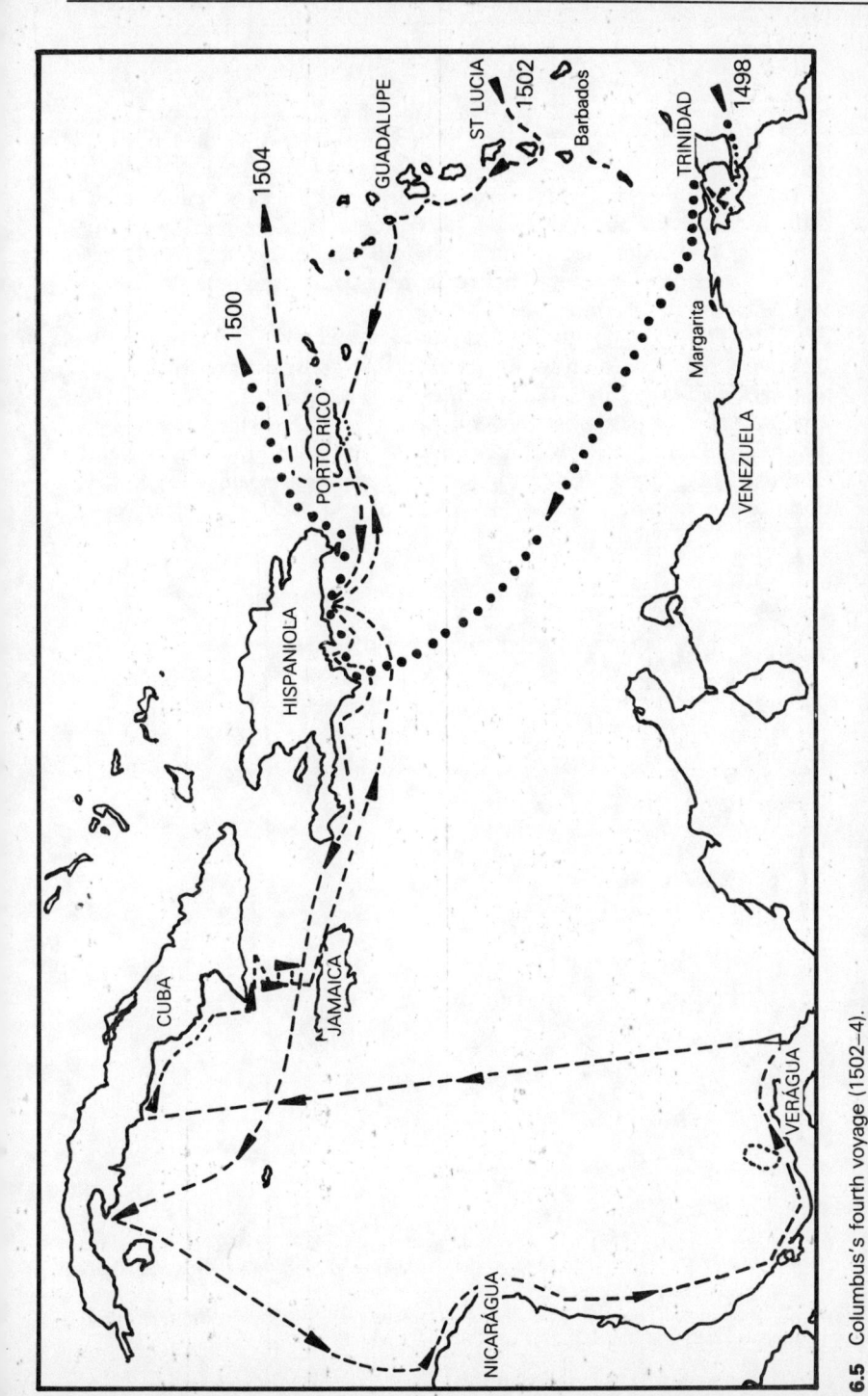

6.5 Columbus's fourth voyage (1502–4). On his third voyage, Columbus had had to contend with a mutiny; on his fourth voyage, his behaviour was surprising: in spite of knowing that the Portuguese had reached India and 'officially' discovered Brazil, he persisted in not going beyond Cuba in search of a passage to Japan and China. ••• third voyage. – – – fourth voyage.

Notes and References

1. Princess Isabel, sister of King Henry IV of Castile, married Prince Ferdinand, heir to the Crown of Aragon in 1469. King Henry IV died and Isabel was proclaimed Queen in 1474. The nobility revolted and, led by the Bishop of Toledo and the Marquis of Villena, proclaimed Princess Juana – born in 1462, married to King Alphonse V of Portugal in 1475 – as Queen. King Alphonse V was defeated in the resulting Battle of Toro in 1476. Princess Juana entered the Convent of Santa Clara in Coimbra, where she took the veil. She died in Lisbon in 1480.
2. See Bidez (1945); Hanotaux (1929) states that the Holy See already knew of this hypothesis, based on the assumption that the world was round.
3. See *'Legitimidade do Direito de Portugal às Terras Descobertas'*, in Peres (1940).
4. The only natural passages from the Atlantic to the Pacific were also discovered by Portuguese – round Cape Horn, in the south of Argentina, by Ferdinand Magellan in 1520, although in the service of Spain, and round the north of Canada and Alaska by João Martins and Maldonado in 1558.

Part II
The Hermeticism of Columbus

7   Documentation of Columbus's Origins

The documentary evidence of the Genoese origins of Christopher Columbus, published by the Italian Institute of Graphic Arts in 1931 at the expense of the Genoa Town Council, in fact proves only that the discoverer of America could not have been the 'Cristóforo Colombo' that they identified among his many namesakes who lived in Southern Europe in the fifteenth century.

Martin Fernandez Navarrete[1], who defended the Genoese thesis, says that Columbus was a very noble and knowledgeable man and Ricardo Beltrán y Rospide[2], who defended the Catalan thesis, was of the same opinion.

There is no doubt that Colón-Zarco (his own name, that Columbus 'left behind' when he assumed the identity of 'Cristóbal Colón', as we shall see) not only did everything possible to conceal his birth and youth, but also covered up his education and the scientific and classical knowledge that he possessed and which in those days were the privilege only of princes and prelates – and indisputably out of reach of a Genoese wool-carder's son. He distorted certain events and ordered his children only to sign the name Colón. The only signature he used was secret and cabalistic.

The life of Columbus appears to many of his biographers as a dense, impenetrable mist. Celso Garcia de la Riega[3] says:

> The homeland and origin of the first Admiral of the Indies, the day or, at least, the year of his birth, his childhood, education, youth and his life before he appeared in Castile; the class his parents belonged to; the date and place of his marriage in Portugal; his amorous affairs in Cordoba and other details of his existence remain enveloped in obscurity.

It is especially these 'other details' that the defenders of a Spanish 'Colón' try to avoid, and which is a testimony to the historical and sociological ignorance of those who insist on his Italian origin. Comfortably secure regarding the supposed historic authenticity of the Genoese documents, most historians did not understand that they could be fictitious; they took everything for granted and concluded that any contrary thesis would be groundless. Now that the Genoese thesis has been proved invalid because it is based on suspect documents, one question comes to mind: why were documents calling Columbus an Italian Colombo or a Spanish Colón tampered with? The reply is obvious. The authors suspected that Columbus was neither Italian nor Spanish, but could not accept the fact. This may also explain the systematic omission of the supposition that Columbus was Portuguese and the complete silence with which foreigners have treated Portuguese arguments.

The more honest writers on Columbus agree in proclaiming the absolute necessity of studying the 'contemporary sources' to find the truth; and that the Italian campaign to document the fact that Columbus was Genoese was carried out only at the end of the sixteenth century:

> If the life and the successes of the great Colón must be written with certainty and impartiality, it is necessary to examine the historians who were his contemporaries and who had personal contacts with him, such as Andrés Bernáldez 'The Bernel', Pedro Mártir de Angleria, Fernando Colón, his son, Friar Bartolomé de Las Casas and Gonzalo Fernandez de Oviedo.[4]

But what Bernáldez[5] wrote about Columbus refers only to the period after 1496 and, although he was the host of the navigator for some days, his work is replete with lacunae and tells us nothing about Columbus's birth or past. He merely mentions him as a 'foreigner' before the Catholic Monarchs, – which, in fact, he was.

In his *História General de las Indias*, Gonçalo Fernandez de Oviedo limits himself to describing the discovery of America. Friar Bartolomé de las Casas began writing his *História General de las Indias* on the island of Hispaniola in 1527. He rewrote it seventeen years later when Lady Maria de Toledo gave him access to the Columbus archive, which had been taken to that island. Las Casas wrote his text with the agreement of Fernando Colón and even copied some of his material. He rewrote it yet again between 1530 and 1563, after his return to Spain.

As Ramón Menéndez Pidal stated:[6] 'he was a despotic man who asserted himself unscrupulously when it suited his purposes . . . a domestic historian in the service of the Admiral's heirs, he furthered their cause'. Was this really the case, or did he tell a truth which was very inconvenient for the Spaniards? The apologists of a Genoese Columbus claim, of course, that Las Casas had falsified his writings.

The work of Pietro Mártir de Anghiera (Angleria) was printed in Alacalá de Henares (1530) and followed by the Amsterdam edition (1570) and by the Italian copies of the sixteenth century which are the only ones existent.[7] Anghiera wrote twenty letters to the Archbishop of Braga (Portugal) and in one of them mentions *'Colunus quidam'* (one Columbus). This correspondence was edited in a work entitled *Opus Epistolarum*, which consists of 812 touched-up anedoctes, many of which the author no longer fully remembered:

> It was later rewritten *by an unknown hand*, which gave most of the letters an absurd chronology, mixed up the dates of others and allowed serious errors to be inserted.
>
> *'Opus Epistolarum'* began to be suspect. One German critic became so sceptical that he thought the whole work a new case of fraud, similar to that of *'Centon Epistolario'*.[8]

Anghiera was born in Milan, lived in Saragossa in 1487 and was 'official chronicler of the Indias'. He wrote his *Décadas* in Seville and died in 1526. Carlos Salas[9] says about him:

> it is a pity that a man who was so witty and affectionate when writing was so careless and negligent in correcting his narrations and works, as Dom Juan Bautista Muñoz points out when advising the reader to be prudent regarding his conclusions and to take into account the errors and ambiguities that are a consequence of the futility and fickleness with which he wrote.

Las Casas himself, who was Anghiera's contemporary, states that 'his letters contain falsehoods'.[10]

It remains only for us to talk about Fernando Colón, the son and biographer of Christopher Columbus. António Ballesteros Beretta[11] says of him: 'Everything he wrote was to praise the deeds of his father, but he always concealed his nationality and mysterious past'.

Among the *Autografos de Cristobal Colón y Papeles de America*,[12] there are letters signed by Colón-Zarco to his son Dom Diogo Colón (his natural son and Portuguese) and to Friar Gaspar Gorricio (from Seville) that contain references to certain subjects that are completely different from those that came from the pen of Fernando Colón (his natural son and Sevillian), maybe because Fernando did, in fact, try to inflate the value of his father's work.

The 'Columbian Library' that Fernando assembled and which he left in his will to the 'Cabido Eclesiástico Hispalense' in 1539 contained 15 370 books. In 1544, at the time of the Inquisition, the Dominicans unsuccessfully brought a lawsuit in an effort to get possession of the library, and according to Bayerri Bertomeu (p. 100) only about 1000 volumes survive. The original biography written by Fernando Colón in Castilian later disappeared.

Alexander von Humboldt[13] was of the opinion that the *Historia del Almirante*, translated in Italy by Afonso Ulloa, was written 'from a very incorrect copy'. This incomplete and adulterated translation[14] was used to make a new translation into Castilian, which was published by Carbia: 'so negligently; these defects, therefore, may be the result of the carelessness or lack of intelligence on the part of the translators'.[15]

It seems that Fernando Colón, in his anxiety to present his father to the world as a genius who discovered America alone, without the scientific help of anyone, even his Portuguese masters, may have intentionally introduced falsehoods.

I have taken care to mention the sources of information of the authors that were contemporaries of Colón-Zarco and which have inexplicably been referred to as 'reliable sources'. But we must also consider Columbus himself, as he was his own first real historian. However, there is only one extract of the log of his first voyage written by Las Casas. The narrations of his second voyage and his report 'in the form of commentaries by Júlio César', which was his true diary, have disappeared.

Finally, Luis Colón, the grandson of Columbus took the *Historia* of his uncle Fernando to Italy, an act of crass irresponsibility; the notes that Colón-Zarco wrote in the margins of the navigation charts that he consulted and of a few other books are all that remain. They have been today compiled, analysed and microfilmed. Regarding the notes that Columbus wrote in the books that are a part of the 'Columbian Library', one of the most distinguished German palaeographers, Father Fritz Streicher[16] says: 'the passages attributed to Columbus, annotations supposedly from his pen, were written by scribes or unknown authors'.

Studying the documents of the highly-proclaimed *Raccolta*,[17] the basis of all the arguments that Columbus was Genoese, Streicher states: 'We are faced with numerous false documents . . . it is therefore time to submit the manuscript material that has survived to methodic doubt'.

The genuineness of the celebrated *Raccolta*, organised by the Italian government and scattered to the four winds by the Genoese, has also been refuted by several investigators, among whom is Rómulo Carbia,[18] who concluded that, 'The *Raccolta* does not include one single document that proves the Italian origin of Columbus or clears up the enigma of his birth or childhood'.

Referring to the works considered 'original sources', Father Ricardo Cappas[19] says that 'They printed in their books many things passed on through oral tradition and that were not very reliable, although in general they kept . . . a certain amount of truth'.

In his work *Cristófor Colón fou Catalá*, Luís Ulloa states that 'Most of the

Columbian texts have reached us disfigured by incorrect transcriptions, through unscrupulous printers or through translations of translations'.

Over the years, I have also studied more than eighty books and documents, and none of them proves anything regarding the birthplace or the youth of Christopher Columbus. A rational theory regarding the mystery of Columbus remains enveloped in the darkness of history. At this time, Europe swarmed with people with the surname of 'Colomo', 'Colom', 'Colón', 'Colombo', 'Coulomb' and 'Coullon'. As has already been mentioned, Cathars, Jews, New and Old Christians were identified with names that derived from 'Colomba' (the dove of the Holy Ghost). Columbus himself was named 'Colón' according to the country he was in, and the intention of a later translator to distort history. But, as we shall see, he never signed with any of those names: only and always Xpo FERENS./.

Notes and References

1. Navarrete (1825), Vol. I, p. 66.
2. Historian and palaeographer, see Róspide (1918) and (1919).
3. Riega (1914).
4. Navarrete (1825), Vol. I, p. 66. Gallo (1945) says the same thing.
5. Bernáldez (1570) describes the reign in 254 chapters; only 14 mention Colón.
6. Pidal (1943).
7. *'Bibliografia de Pedro Martir de Angleria'*, Boletin de la Academia Nacional de Historia, Vol. X (Quito).
8. See Bartolomeu (1961). The *Centon Epistolario* is a collection of letters that were later altered by Italian writers.
9. Salas (1917).
10. Las Casas (1530).
11. Beretta (1945).
12. Published by the Duchess of Alba in 1892.
13. Humboldt (1833).
14. See *'Le Historie della Vita e dei Fatti di Cristóforo Colombo per D. Fernando Colombo, su Figlio'* (Venice, 1571), an Italian translation of the lost original (partly copied by Las Casas (1575–6).
15. Navarrete (1825).
16. Streicher (1928). Vol. I.
17. *Reale Comissione Colombiana: Raccolta di Documenti e studi publicati della reale comissione Colombiana, pel Quarto Centenario della Scorpeta del America* (Rome, 1892; Genoa, 1931).
18. Professor of the University of Buenos Aires and la Plata; see Carbia (1923).
19. Cappas (1915).

8 The Language of Columbus

One of the biggest obstacles that the defenders of the Genoese thesis come up against is the fact that Columbus 'never wrote in Italian and although he did not write Castilian correctly, it was this language he used when he wrote to the Bank of St. George in Genoa'.[1] The author of this statement is the French American Henri Vignaud, one of the staunchest supporters of Columbus's Genoese birthright. Another supporter of the same thesis, the Italian Próspero Paragallo, confirms that conclusion.[2] The supposed 'Genoisms' and 'Italianisms' that according to the Peruvian Rómulo Cúneo-Vidal[3] appear in the *Letter from Columbus to Santángel* were totally refuted by Pedro Catalá y Roca,[4] who produced documentary proof that they were Spanish expressions at the end of the fifteenth century.

Let us now look at the fifteenth-century literary culture of Portugal. During the first dynasty (which was Burgundian, see Appendix 1) expressions from the south of France were common in court circles. This period (called 'Provençal') produced a large amount of poetry, but the preference of the nobility inclined towards 'books of lineage', 'chronicles of conquest' and 'books of chivalry'.

King Dennis included the poems of *Tristan and Isolde* and *Flowers and Whiteflower* in his *Book of Verse*. Prince Duarte had *Merlin* and *Tristan* in his library and Portuguese knights even adopted cognomens of the twelve knights of the Round Table, such as Arthur, Lancelot, Percival, etc. The Constable of the Realm, Nuno Alvares Pereira, chose Galahad as a model. The 'Legend of the Holy Grail' – the Grail being the mystical cup from which Jesus Christ had drunk at the Last Supper and in which Joseph of Arimathea received Christ's blood at the Crucifixion – became a symbol of the Portuguese knights, who were deeply imbued with the spirit of the Templars.

This literary trend of the Carolingian (*The Song of Roland*) and Breton (the Stories of King Arthur and the Knights of the Round Table) cycles was represented in Portugal by the cycle *of the Amadises*[5] (the work of

Vasco de Lobeira, a contemporary of King Alphonse IV), which was the first of its type in Europe. The *History of the Knights of the Round Table* and the *Search for the Holy Grail* was translated into Portuguese during the reign of King John I.

The 'Provençal School' was followed by the literary current that was known at the 'Spanish School', which lasted from the reign of John I to that of Manuel I. The works *Proemio* of Iñigo Lopez de Mendonza, Marquis of Santillana (1398-1458), written for the Portuguese Constable, Dom Pedro[6] and *Labyrinto* by Juan Mena (1411-56) were read in Portugal, whose court included many Spaniards who came with the retinues of princesses that married Portuguese kings and princes. All of King John I's children received a superb humanist education, particularly the future King Duarte and Prince Peter. The education of many noblemen, their heirs and ladies included the study of Latin and Greek, besides, in some cases, Hebrew and Arabic.

Latin texts were especially common among the nobility and the clergy. King Alphonse V had Master Matheus de Pisano, author of the *Book of the War of Seuta*,[7] as a professor and he also brought the Latinist and historian Friar Justo Baldino[8] from Italy to teach his children. The Italian poet and orator, Cataldo Aquila Siculo, also came to Portugal to teach Dom Jorge, the bastard son of King John II.

The Bishop of Évora, Dom Garcia de Menezes (see p. 166) was the object of admiration in Rome because of his Latin eloquence and erudition. Some Italian authors, whose works were only later translated into Portuguese, were read by the nobility, particularly those of the Florentine Dante Alighieri (1265-1321), such as *De Monarchia* and *The Divine Comedy*, which were replete with esoteric symbolism. The poems *Rime* and *Trionfi* of Petrarch (1304-74) and the mirror of the sexual life of the period, the *Decamerone* of Boccaccio (1313-75), were also known. While the erotism of the *Decamerone* was never adopted by Portuguese writers, the works of Dante and Petrarch inspired many literary compositions in fifteenth-century Portugal. The fact that the Portuguese learned Latin helped them to understand the other Hispanic Languages like Catalan, Aragonese, Castilian and above all Galician (which is nearly identical to Portuguese, see Appendix 2) and the Romance tongues.

Some of the bourgeoisie had learned Italian, especially those that traded with the Mediterranean countries, and there were so many Ligurian merchants in Lisbon that one of the city's streets was called 'the street of the Genoese'. Some Italian traders had also settled on the islands of Madeira and Porto Santo. The Portuguese crews that sailed the Mediterranean also learned to understand Italian, and it is natural that Columbus would have done so as well.

What is not natural is that he never wanted to use that language when he wished to write or speak to people of Genoa or other Italian cities.

It is a habit of many scholars to make notes in the margin of foreign books in the language in which the book is written. Columbus did so twice: a two-line annotation, without mistakes, and a longer one that has been transcribed as follows:

> *Del Ambra es cierto nascere in India soto tierra, he no ye ho fato cauare in monti in la insola de Feyti vel Ofir vel Cipango, a la qual habio posto nome de Spagnola; Y ne trouato pieca grando como el papo, ma no tota chiara, salvo de chiaro y parda, y otra negra; y ve ne asay.*

Many of these words are common to both Italian and Spanish, but the text is replete with errors of orthography and syntax and includes words that are exclusively Luso–Castilian because Columbus did not know their Italian equivalent, such as *'es cierto'*, *'tierra'*, *'pieca'*, *'como el'*, *'parda'* and *'otra negra'*. He also translated Italian sentences into Spanish in the same way as any other person learning a foreign language. This was the case when he was reading the *History of Pliny*, translated into Italian by Christóforo Landino in 1495. Columbus wrote the translation of the sentence *'La volpe no piglia e polli e quali hano beccato el fegato en figado de zorra'*, in the margin of the book. He wrote 'the fox doesn't catch the chicken that has pecked the fox's liver'; but for 'liver' he wrote the Portuguese word *'fígado'* and not the Spanish *'higado'*. He later translated the sentence *'non sente fame ne sete'* as 'I feel neither hunger or thirst', writing 'feel' with the fourteenth-century Portuguese *'sinte'* instead of the Spanish *'siente'*. If Columbus were really Italian, he certainly would not have needed to make such notes in the margin of the book.

While in Lisbon in 1474, Columbus wrote a letter to the Italian cosmographer Pablo del Pozzo Toscanelli in Florence. Toscanelli wrote a long reply, an extract of which has been transcribed:

> I am not surprised, for this and many other reasons, that you, who is so great in spirit, and the Portuguese Nation, which has always been so ennobled through so many heroic deeds carried out by so many illustrious men, are so interested that this voyage be undertaken.[9]

Professor Manuel Ballesteros y Gaibrós,[10] of the Universities of Valencia and later Madrid, had no doubts in stating that the mother tongue of Columbus was Portuguese, and concluded that 'he wrote in Portuguese as Toscanelli thought him to be Portuguese'; and adds: 'Colón could have done it in Italian, but maybe he did not wish to as he doubted his ability to use that language'. These conclusions are even more valid since Ballesteros y Gaibrós believed that Colón was Genoese.

In an attempt to refute this thesis, Angel de Altolaguirre y Duval[11] wrote: 'all the correspondence between Toscanelli and Colón was forged by Don Fernando'. This is a truly strange argument, since Fernando was Columbus's bastard son from an affair with a Cordovan lady, and was born and bred in Spain and the Antilles among his compatriots. Even if Fernando Colón did wish to insinuate that his father was Portuguese – and as I will prove he had strong reasons to deny this fact – this intention cannot be attributed to the letter in question, which is an independent and separate document. It was the cartographer Toscanelli himself (who had close contacts with the Portuguese court) who stated, in his own hand, that Columbus was Portuguese.

In Fernando Colón's *Historia del Almirante*, it is said that Columbus acquired his knowledge of cosmography and nautical science at the University of Pavia; this is clearly in error, as Pavia taught only Philosophy, Law and Medicine: there is also no record of any Colombo or Colón attending it in the fifteenth century.

António Ballesteros Beretta,[12] who thought Columbus had been born in Genoa, confessed that: 'a careful search through the University of Pavia archives drew a blank . . . Columbus never spoke or wrote Italian . . . Columbus's Latin suggests Spanish influences and it can be said that he only started learning Latius' language in the Peninsula . . . He only used Latin for short texts but used it badly'.

It is to be expected that Columbus, in the service of Spain, would wish to write his letters in the language of his adoptive country, but the philologist Professor Ramón Menéndez Pidal has conclusively stated: 'It is difficult to think that this consideration was important enough to lead him to write only in Spanish, even short exclusively intimate notes like the "Letters to Miguel Mulliart about the 29 Maravedis he owes me", written in 1494'. This means that Columbus did not even write a simple memorandum for his own private use in Italian.

The fact that Columbus did not write in Italian when dealing with his private business with the Bank of Genoa is inexplicable. Besides this, he asked a Genoese, Nicolo Oderigo, who had been in Spain as the bank's delegate and spoke Spanish, to act as his intermediary. He sent him a letter with the following recommendation: 'so that you will give it to Micer Juan Luís, with another communication informing him that you are the reader and the translator of my letter'. Columbus therefore needed a translator from Spanish into Italian. And he referred to the banker as 'Micer' and not 'Messer', as would be correct in Italian, and called him 'Juan Luís' instead of 'Giovanni Luigi'.

After analysing the texts written by Columbus, the Argentinian Rómulo de Carbia came to the conclusion that the navigator 'did not speak Castilian very well and the writings that are regarded as his were the work of secretaries and scribes. The only irrefutable handwriting of Columbus that we have are the marginal notes that he jotted down in some of his books and which reveal that he possessed a very limited vocabulary in Castilian and his grammar was also weak'. In speaking about this, Luís Astrana Marín[13] of the University of Buenos Aires and La Plata says: 'he is the fairest, wisest and most unbiased lover of things Spanish'. The American writer, Samuel Eliot Morison[14] defined Carbia's work as being 'the most extensive and precise study of the original sources'.

Luís Ulloa[15] emphasises the fact that the birthplace of the discoverer is in question because, among other things 'Columbus did not know how to write Genoese Italian, despite having lived in the city, according to the supporters of a Genoese Columbus, for at least 24 years and having spent the rest of his life in constant contact with Italians'. In the first French edition of his work *Colomb, Catalan*, under the heading '*Les Catalanismes indiscutables et le language de Colomb*', Ulloa indicates nineteen expressions of Catalan origin – although distorted – in the *Letter from Colombo to Santángel*. In the second Castilian edition, however, under the heading '*El Lenguage de Colón; sus idiotismos y sus catalanismos; su conocimiento desde niño del castellano*', this number is reduced to twelve. But after considering the study of Pestana Júnior,[16] even these twelve disappear as Pestana Júnior proved that the so-called 'indisputable Catalanisms' are all fifteenth-century Portuguese expressions.

Ulloa also considered Cesare de Lollis, co-author of the *Raccolta*,[17] the 'honest and wisest of all the defenders of the Genoese legend'. In an effort to explain why Columbus did not speak his native Genoese tongue, this Italian author alleges that 'he had forgotten it'. If, however, his thesis that Columbus was born in Genoa in 1451 and that he went to Lisbon in 1476 is right, it is difficult to believe that he would forget his mother tongue completely. Would he have spoken Portuguese in Lisbon and Madeira? Had he learned it in Genoa when helping his father to card wool? Nobody can forget his mother tongue after speaking it for the first twenty-four years of his life.

Lollis admits that Columbus certainly spoke Castilian by 1481. Not wishing to put the already weak 'Genoese' thesis at greater risk, Lollis decided that the navigator had had to speak 'Castilian' in continental Portugal and Madeira, as well as during his married life which, according to the 'Genoist' thesis, lasted nine years. Whoever had been born and had lived for so long in Genoa and in Portugal would have been obliged to speak Italian and Portuguese. To have been able to express himself in Castilian before this certainly shows that he attended a course in Italy, and a poor one at that, as he never learned the language well right up to the time of his death. If he had studied Castilian in Genoa when he was not carding wool, he could have improved his knowledge of the language in Spain or even in Portugal, as the language of the country up to the fourteenth century was Galaico–Português, which was very similar to the Castilian of the time. The similarity between Portuguese and Galician at that time is in fact so evident that some writers, basing themselves on popular legends and the linguistic peculiarities of Columbus's errors in Castilian, have even tried to put forward a fragile theory of a Galician Columbus.[18]

Finally, we shall quote again Ramón Menéndez Pidal. In his study *La Lengua de Cristóbal Colón*,[19] he came to the bold conclusion that in his writings elaborated in Spain, Columbus neither expressed himself in 'Italianised Castilian' nor in 'Hispano–Hebrew', but in a 'Portuguese–Spanish', or even better, in a *'Portuguised-Spanish'*:

> In the extensive Spanish writings of Columbus, the confusion arises from his introducing Portuguese forms and voices and not Italian . . . When he spoke, Columbus showed himself to be a foreigner . . . and the writings he left give us the same impression – that Spanish was not his mother tongue, but an acquired one . . . The Admiral had learned Spanish in Portugal, when the 'Castilianised' style started by Prince Pedro was in fashion.

This so-called 'Castilianised style' was in fashion in the third quarter of the fifteenth century, but only in the court and cultured circles, which means that if Columbus had moved in these circles he must have belonged to a noble or at least an illustrious family.

If spoken and written Portuguese were very similar to Galician up to the reign of King Dennis, it is no less true that it resembled Castilian in the literature of the fifteenth century. This is particularly noticeable in the *Book of Verse of Resende*, compiled by Garcia de Resende who, like other writers of the day, also wrote in Castilian. Menéndez y Pelayo said that the 'difference between the languages is purely incidental'. The father of Portuguese theatre – the New Christian Gil Vicente – wrote plays in Castilian. It was because of this that Professor Mendes dos Remédios[20] christened this period as the 'Spanish School' and considered it to be specifically courtly and the domain of cultured people.

The chronicler Friar Bartolomé de Las Casas also emphasises the high cultural level of the Colón brothers, Cristóbal and Bartolomeu. He mentions that they were well versed in 'geometry, geography, cosmography, astrology and seamanship', sciences that were in the range of very few people in those days. He also says they both knew Portuguese, Castilian and Latin, and possessed some knowledge of Greek and Hebrew. Why should Columbus's chronicler omit the fact that they spoke Italian?

Columbus wrote many letters in Latin, and also a book on theology, *Profecias*. Las Casas describes him as being 'well spoken and eloquent . . . with moderate seriousness and discreet conversation . . .

he was of respectable presence and appearance, learned and authoritative and worthy of complete respect . . . talking often with ecclesiastic and secular, Latin or Greek doctors, Jews or Moors'. His chroniclers state that he had read many Jewish works of cosmography and geography and in his book *Who was Columbus?* Cecil Roth, from Oxford University,[21] claims that Columbus not only had a knowledge of Hebrew religious and scientific literature, but also wrote specific notes (like those that are to be found in the margin of Pius II's *Historia rerum ubique gestarum*) referring to the date of Jesus's death that are literal translations of Jewish texts that had never previously been turned into Spanish.

Besides this, the *Tables of the Declination of the Sun* written by Zacuto,[22] based on the studies of the Jews Samuel and Jehuda Cresques and the Arab Aben Ragel (known as 'the Alchemist') and expressly ordered by King John II, were sent to Columbus from Portugal before he departed for his first voyage to the Antilles. (The Tables are at present in the Hebrew Theological Seminary Museum in New York.) These Tables not only helped Columbus to plot his exact course and positions, with calculations known only to the Portuguese, but also saved his life, when he was able to forecast an eclipse of the moon and so frighten some Indians that were about to attack him; but it is evident that these Tables would have been absolutely useless to anyone who was not conversant with the Hebrew tongue.

The Polish investigator, Professor Janina Klawe, of the University of Warsaw quotes the work of Professor António Rumeu de Armas, President of the Royal Academy of History of Madrid,[23] in which, after studying the writings of Columbus, he comes to the logical conclusion that 'Portuguese was the first language read and written by Cristóbal Colón' and confirms that, while in Spain, he had spoken a 'Portuguised' Castilian. He also adds that Columbus presented himself at the court of the Catholic Monarchs as Portuguese, and as such he was considered; but officially he was identified only as a 'foreigner'. Rumeu de Armas did not accept the Portuguese arguments as he thought they were not backed up with sufficient proof. He paradoxically concludes: 'It was convenient for Cristóbal Colón to be accepted as a Portuguese explorer, given the enormous prestige and great fame that the Portuguese pilots had enjoyed all over Castile since the time of Prince Henry the Navigator'.

The paradox is evident, as Columbus had always avoided being seen as Portuguese; I will prove in Chapter 30 that only the Catholic Monarchs, some influential noblemen of the court and certain Jews knew that he was Portuguese. Among the other Spaniards he was simply known as a 'foreigner'. Columbus may have consented to being – or even insinuated that he should be – thought of as Genoese.

Bartolomé de Las Casas comments: in this and other things that appear in his diaries, he seems to be the speaker of another language, as he does not get to the root of the meaning of Castilian vocabulary, neither of the way it is spoken'.

Our Columbus was not a Genoese wool-carder or weaver, nor did he work for a commercial company, as he could neither speak nor write Italian. His Spanish employers knew that he was not a fellow-countryman of theirs. Is it not then logical to admit the probability that Columbus was a Portuguese of high birth, entrusted with a secret mission and, in order to carry it out, changed his identity and learned Castilian?

To refute this thesis, the Catalan, Enrique Bayerri y Bertomeu argues that Columbus wrote nothing in Portuguese before making his way to Castile. This is, in fact, true, but he left nothing in Portuguese nor in any other language. There is no record whatsoever that he carried out his secret mission. If he had written in Portuguese, he would not have signed anything with the name of 'Cristóbal Colón'. And he would not have needed to conceal his identity by means of a cabalistic siglum.

Professor Janina Klawe also points out the fact that the Genoese have never carried out any research. After the conquest of Malaga 'La Real', in 1487, Columbus was in that town to make his proposal to the Catholic Monarchs. He was unemployed and living from the sale of 'colour prints'. Among the manuscripts in the 'Instituto de València de Don Juan' in Madrid, there is a notebook that includes a document entitled *Book of maravedis received by Pero de Toledo*, which mentions: 'To [name left or crossed out], Portuguese, on this day, thirty Castilian doubloons, on the orders of Her Highness, in the presence of the doctor of Talavera; they were given to him by Alonso de Quintanilla; this is the Portuguese that was at La Real [Malaga]; this happened at the departure for Liñares and Her Highness sent me personally'.

After the conquest of Granada, the Catholic Monarchs were at 'La Real', where Columbus met them for the first time. Janina Klawe concludes: 'There is no doubt that Queen Isabella de Castile knew who Columbus was, because when she authorised the Admiral's coat-of-arms in the "Provision" of 20th May 1493, the words "the arms that you used to have" can be found'.

Regarding this empty space and the nameless Portuguese, the Spaniard Rumeu de Armas says: 'What was the reason for the omission? Why did the payer, Pero de Toledo, ignore the name of the person received with such consideration in the camp at Malaga? But he identified him with such care that there is no room for doubt or hesitation'.

In answer to this, Janina Klawe states:

> Pero de Toledo, could not have ignored the name of Columbus and certainly wrote it in the document. The name must have been crossed out later by someone who did not want it widely known that the 'immortal sailor' as he was known in Spain, was Portu-

gue:.e. This has not happened to any other Castilian documents. Besides this, the letter that grants Columbus's brother, Diogo, naturalisation also omits the latter's country of origin. Why was he not recognised as Genoese?

When Columbus left Portugal for Spain, he made his way to the Monastery of La Rábida, near Palos (Huelva). Using the testimony of Garcia Fernandez (the Palos doctor who accompanied Columbus) Las Casas mentions the fact that Columbus and Father Juan Peres de Marchena held a long conversation in a language that was not Spanish. In regard to this, Ramón Menédez Pidal says:

> Ignoring the words of Las Casas [who censored Columbus's Castilian as bad, without, however, admitting that he was Galician, Leonese or Aragonese] and those of the Palos doctor [who said the same thing], one cannot really say that he spoke Galician, which is familiar to all Spaniards. We may see that the dialects of the west of the Peninsula – which could be noticed in the Admiral's speech – are not Galician but decidedly Portuguese. Yet one can also see that Portuguese was not the mother tongue of Columbus, either.

It must also be remembered that the Spaniards do not learn Portuguese with the same ease that the Portuguese pick up their neighbour's tongue. The former is resonant and open, while the latter is softer, replete with nasal dipthongs, and not only transforms but omits phonetic sounds. Spaniards never forget their mother tongue, and Columbus was always regarded as an *'extrangero'* in Spain, even by the Catholic Monarchs themselves. Pidal based his argument on the fact that while in Spain Columbus, on writing to Castilians or Italians, did not do so in their tongue, but in a 'Portuguised' Castilian. It is impossible to follow this line of thinking. If texts written by Columbus wholly in Portuguese cannot be found, neither can they in any other language, unless it is Castilian replete with Portuguese etymons. It seems evident that a secret agent who does not wish to disclose his nationality must express himself in another tongue. It is also evident that only out of necessity would Columbus have resorted to this, hoping that no-one would notice.

It is surprising that no wholly Portuguese texts are to be found among Columbus's writings. It once more seems evident that he did his utmost to destroy everything related to his origins and past; he certainly took the greatest care to conceal his nationality when he went to Spain. Pidal did, however, prove his integrity when he declared: 'the dialects . . . which could be noticed in the Admiral's speech – are not Galician but decidedly Portuguese'. He goes on:

It is difficult to say where Colón learned this faulty Spanish, as he wrote it before he settled in Spain. Very little is known of the Admiral's youth. Everything we know about him from his own hand or from those closest to him displays a constrained simulation and is for the most part false. [Columbus's] twenty-one years of residence among the Andaluzians and Castilians were not sufficient for him to lose his habit of using Portuguese words or expressions. And this habit was so ingrained that it crops up . . . [even] when one has the impression that Colón [has attempted to improve] his style. He didn't even repress [the habit] in his formal letters, which is indisputable evidence that Colón learned his Spanish in Portugal. . . . As far as Italian is concerned Colón never used it in either reports or documents. He always wrote in Spanish when he sent letters to Genoa or to his Italian friends, as, for example, to the Bank of St. George or to Nicolo Oderigo . . . In the same way, Colón wrote [in Spanish] to Friar Gaspar Gorricio, who lived in Seville and published pious works in Latin, . . . even though the latter was Italian.

Appendix 2 shows some of the many words used by Columbus, either because he did not know the Spanish equivalent or because they had no equivalent; in the latter case, the words are usually nautical, which is to be expected as Columbus sailed with Spanish seamen from 1492 to 1504.

Notes and References

1. Vignaud (1905); although in his Genoese thesis he commits some notorious errors Gallo (1942, p. 42), among others, thought him 'the soundest and most rigorous critic of Christopher Columbus'.
2. Paragallo (1893), Vol. 1, p. 10.
3. Defender of the Genoese thesis (his father was Italian), he founded the *Societá Ligure di Storia Patria*, see Cuneo-Vidal (1924).
4. Roca (1951), Vol. II, pp. 283–90.
5. Sousa, Dom António Faria de (1785), Pt 4, Vol. III, Chap. 7, pp. 360–72.
6. Son of Prince Pedro, Duke of Coimbra and grandson of King John I.
7. Pisano (1970).
8. Carvalho, Francisco Freire de (1846).
9. This letter is included in Aeneas Silvius, *Historie rerum ubique gestarum;* he was to be Pope Pious II. It was also transcribed by Bistrice in *Vite de Uomini Illustri* and copied by the chronicler Bartolomé de Las Casas, whose translation I have used here.
10. Gaibrós (1943), p. 59.

11. Duval (1903).
12. Bevelta (1945).
13. Marín (1929).
14. Morison (1945).
15. Writer and mathematician, director of the National Library of Lima and a member of the *Instituto Historico del Peru*; this quote is from Ulloa (1928).
16. A historian who was Minister of Finance; he defended the thesis that Columbus was Portuguese (Pestana Júnior, 1928).
17. Author of 3 volumes of the 1st part of the *Raccolta* (see Lollis, 1892–4).
18. See Arribas (1913); Riega (1914).
19. Pidal (1943).
20. *História da Literatura Portuguesa* (Coimbra, 1908).
21. *Mensrah Journal*, Vol. 18 (1940).
22. They were taken from the *Almanach Perpetuum* of Abraham Zacuto (see p. 121 Fig. 10.1 and p. 481) and, on the order of King John II, were translated into Italian, Portuguese and Castilian by the New Christian José Vizinho and printed in 1496.
23. Armas (1928).

9 The 'Foreign Columbuses'

No-one in history has been so discussed or vied for as Christopher Columbus, the Admiral of the Indies. Distinguished by an identical adulthood, but by a wide variety of childhood and youth, 'Columbuses' proliferated all over fifteenth-century southern Europe. This 'Columbian plurality' was created by the sentence: 'From Genoa I departed and in Genoa I was born'. For more than four centuries, the Genoese have striven to prove that the discovery of America was made by their fellow-citizen; their arguments were, however, so fragile that other claimants appeared in their wake, with various cities and nations putting up their own candidate as Columbus's progenitor.

If, as we have already suggested, Columbus was a secret agent of King John II, instructed in Portuguese nautical science and trained in seamanship and oceanic exploration, he could never have let his true ancestry or nationality be known. Never has espionage been more common than it was at the beginning of Renaissance period. It was practised at all levels of politics and by all social classes, in national, feudal, commercial and family orbits, at the instigation of the Crown, the nobility, the clergy, the bourgeoisie and even by the common people. Columbus presented himself in Spain as a foreigner, and he would have been easily recognised as such. His probable transit through or short stay in Genoa (never proved) and the epistolary contact he had (in Spanish) with the Bank of St George allowed the Genoese to insinuate that Columbus was a fellow-citizen. Then, several Italian investigators, taking advantage of the contradictions and lacunae in the Genoese thesis, tried to prove that other Italian towns were in fact Columbus's birthplace. Francisco de Uhajón, professed knight of the Order of Calatrava,[1] Giuseppe A. Rocca,[2] Luís Astrana Marín[3] and Giulio Salinério[4] have defended the thesis that Columbus was born at Saona (today Savona); they all based their theory on a sixteenth-century work *Memorie*, by one Verzelino, a jurist of Savona, and related by Arcipreste Andrea Astengo[5] and on notarial minutes created by Salinério. Like

Giustiniani in his *Psalterium*,[6] Salinério used his work *Anotaciones* to present some notarial minutes, the originals of which never appeared, as if they were genuine documents.

Abbot Francesco Cancellieri[7] defended a thesis of a Columbus from Cuccaro, Pietro Agnelli[8] said he was from Placenza, F. Brunet y Bellet[9] from Gogoreto, Rómulo Cúneo-Vidal[10] from Terrarossa and Fernán Perez de Oliva[11] from Terracossa di Moconesi. Henry Harrisse clearly proved that Columbus was not born in any of these places. He came to the conclusion[12] that Columbus had been born in Paris, of Russian parents and descended from the Admirals of King Charles VIII of France, and having for this reason commanded a warship of René of Anjou. Próspero Paragallo in his work *Cristóforo Colombo e la sua Famiglia*,[13] however, comprehensively disproved this.

The thesis that Columbus was born at Calvi, on the island of Corsica, was first put forward by Martin Casanova (known as 'the Abbot'),[14] and was immediately followed by A. Edouine Cesari.[15] Following their cue, Luís Franco y López, Baron of Mora[16] argued that if Columbus had been born on Corsica then he must be considered to be Aragonese, as the island then belonged to Aragon. Enrique Bayerri Bertomeu, referring to the work of Cesari, says that it was 'written with a unilateral spirit and limited horizons'; regarding Casanova's thesis he says: '*La vérité*, . . . of M. Casanova is everything people want, except the truth, concerning the origins of Cristóbal Colón'. Tomás Rodriguez Pinilla adds that: 'Casanova's proof and argument are deplorable, up to the point that we think of doing him a favour by not taking them into consideration'.[17]

Let us now take a look at some of the Spanish 'Columbuses'. Adrián Sánchez Serrano,[18] a parson in Argentina, defended the thesis that the Admiral was born at Olivade la Frontera in the Spanish Estremadura. Celso García de la Riega[19] in 1897 tried to prove that Columbus was from Pontevedra in Galicia. His thesis was supported by many other writers, among whom were Juan Salari,[20] Constantino de Horta Pardo,[21] Eva Canel,[22] José Perez Hervás,[23] Fernando Antón del Olmet,[24] Ramón Marcot[25] and Rafael Calzada.[26]

The thesis of Celso García de la Riega deserves special attention. In 1869, his uncle, Luís de la Riega discovered a charter, drawn up at the Monastery of Poyo in 1519, in favour of one Juán Colón. There were families of Jewish origin called Colón, although very few, in both Iberia and Italy;[27] a majestic tomb of a New Christian of that name is to be found in the parish church of Pontevedra.

Using the letter from Columbus to Nicolo Oderigo previously mentioned in which he refers to the despatch of some '*Cartulários*', Celso de la Riega announced[28] that he had bought a '*Cartulário*' that contained another charter dated 1496 concerning an estate which, as far as could be made out, belonged to one Cristóbal de Colón. He exhibited a document that had supposedly been signed by the Archbishop of Santiago, Dom Lope de Mendoza, on 15 March 1413, which contained a written order for a monetary payment in favour of *'Maese Nicolao Oderigo de Ianvua'*. Now Genoa (Ianvua) is mentioned in this document. It is highly improbable that his 'Nicolao Oderigo' was the same Nicolo to whom Columbus had written in 1502, as he would have been over 100 years old, but he could have been his father. However, the word *'Ianvua'* instead of *'Janua'* suggests that the Archbishop was a terrible Latinist. The document contains words written in a hand and ink that are different from the rest of the text; the word *'Maese'* is in a modern handwriting, it is out of context and the ink is smudged by a previous erasure; there are spaces between the letters of the word *'mandamos'*, being stretched out so as to be in line with the line below, above the word *'Nicolao'* which has also been clearly tampered with. It is, therefore, a fraudulent attempt to link that Galician Colón to the one that had sent the *'Cartulários'* to the Genoese Delegate.

The alteration was brought to light by Ulloa and by Manuel Serrano y Sanz,[29] and was confirmed by the palaeographer Father Fritz Streicher (SJ),[30] who concluded:

> the Galician documents are not only refutable palaeographically because they have been altered by means of interpolations, erasions and corrections at later dates, but also because they are totally lacking in proof as far as Columbus's origin is concerned.

The Academic Board which was formed for the purpose made a palaeographic study of the documents exhibited by Celso de la Riega and concluded:

1. 'The documents have been the object of systematic manipulation in order to modify them or include the names of people that figured in documents of Pontevedra between 1437 and 1525';
2. 'Through the type of the ink used, it can be seen that the alterations are of a recent date';
3. 'Through the uniformity of the alterations, it can be deduced that they were made by one and the same person';
4. 'Documents in this condition cannot be considered as crucial nor as support material for a serious historical investigation'.

The only conclusions reached are that Columbus may have been of Jewish origin and that when he was at a loss for a word in Castilian he resorted to Galician when the word was the same as Portuguese.

THE SUPPOSED 'SHIPWRECK'

Luis de Ulloa, born in Lima, Peru, in 1869, was the main writer responsible for the Catalan thesis put forward by late Spanish authors. Ulloa's work disproved the Geonoese theories, but were usually speculative.

The Catalan thesis is based on a text in Fernando Colón's *Historia del Almirante*, Chapter V of which apparently explains the appearance of Fernando's father in Portugal. Colón writes:

> while the Admiral was sailing with Colombo-the-Younger, as he had done for many years, they received news that four big Venetian galleys has sailed from Flanders. They went in search of them and came across them between Lisbon and Cape St. Vincent in Portugal. The galleys were boarded and a furious fight ensued . . . as the Admiral was a good swimmer and being two leagues or so from land, he grabbed an oar that was floating nearby and alternately hung on to it and swam. God . . . gave him the strength to reach the shore, although he was so tired and stiff that it took him several days to recover. As he was not far from Lisbon, where he knew he would find many of his fellow-Genoese, he made for the city with all haste and, as he was known to them, they made him so welcome that he settled in the city and married there.

This shipwreck was said to have happened in 1485, but in the time of Dom Fernando Colón, only two occurrences are known of attacks on ships in the area of Cape St Vincent and both were widely narrated.

The first took place on 13 August 1476, and was carried out by the corsair-admiral Guillaume de Casenove, known as 'Colombo-the-Elder', of the French family Coullon of Noiret, whose name in the thirteenth century was still spelt *'Coullong'* (long neck). The privateer was in the service of King Louis XI of France, who was allied to Portugal, and he attacked four Genoese merchantmen and a big Flemish vessel that had sailed from England (as narrated by the Venetian chronicler Marco Antonio Sabelico).

The second attack happened on 21 August 1485, and was made by another privateer, George Bissipat, known as George 'the Greek' or 'Colombo-the-Boy', who was also in the service of the King of France (now Charles VIII) and attacked four Venetian galleys coming from Flanders. But this time the Venetians surrendered without a fight and Bissipat brought his prizes into Lisbon (in the narration of the Castilian chronicler Alonso Palencia).

From Dom Fernando Colón's description of the naval battle in which his father took part, it would have been the first one mentioned above. In that case, however, it would have taken place in 1476 and not 1485, and off Cape St Vincent and not between this promontory and Lisbon.

In fact, Columbus exhanged letters with Pablo de Pozzo Toscanelli between Lisbon and Florence in 1474. He married Filipa Moniz Perestrelo in Portugal and his legitimate son, Diogo, was born in 1478 (see pp. 196–7). He lived in Madeira until 1481. When, according to the convenient official version, he presented his proposal to reach India by the west to King John II, in 1483, he was living in Lisbon. And he went to Spain in 1484.

Fernando (born 1488) must have known the age of his brother Diogo (born 1478). Could he have purposely omitted the years his father had lived in Portugal and thus reduced the minimum period of nine years to one, enough time to marry and sire his first son? If this were the case, Diogo would have been only one year older than Fernando, which we know is not correct.

We can now compare the Genoese and Catalonian theses (Ulloa's) concerning the 'wreck'. The authors of the initial Genoese thesis ignored the chronological disparity just described .

Genoese Thesis

1. As Columbus was already in Spain in 1486, the naval battle had been fought in 1485.
2. In 1485, Colombo-the-Younger did not capture Genoese galleys; he attacked Venetian vessels.
3. Columbus did not sail with Colombo-the-Younger but on a Genoese vessel that belonged to the shipowner Spinola and which transported merchandise for Negro.
4. The ship was not wrecked between Lisbon and Cape St Vincent, but two leagues or more off the Cape.

Catalan Thesis

Bayerri Bertomeu repeats the arguments put forward by Ulloa. The shipwreck happened ten years before, in 1476.

1. Columbus discovered America, on his own initiative, in 1477.
2. Columbus was born in Catalonia, of a noble family, and had fought for his country against King John II (known as 'the Faithless') of Aragon.
3. The need to conceal these facts was the cause of all mysteries and forgeries that are to be found in the documents concerning Columbus.

Bayerri Bertomeu then argues as follows.

4. Columbus was a privateer and the son of a privateer, the French Admiral, Guillaume Casenove-Coullon. So as not to be confused with his father, he called himself *Coullon-le-Jeune*, but his real name was Jean Baptiste Coullon, in Catalan Juan Bautista Coulon.

But if he were Jean Coullon, he could not be confused with Guillaume Casenove-Coullon because the latter was known as 'Admiral' in 1476 and Columbus received this title only in 1492. Besides this, Coullon-le-Jeune (known as 'the Greek') was definitely French.

5. Jean Coullon was a Catalan rebel who fought against King John of Aragon and who entered the service of René of Anjou during the Catalan revolution. When this was put down, Juan Coullon continued his privateering in the Mediterranean, first under the command of René of Anjou and then of King Louis XI of France. In 1473 or 1474, he joined the fleet of his father. In 1476, they attacked a Genoese fleet off Cape St Vincent and his ship was sunk.

The French privateers must have been pretty inept if their ships were sent to the bottom by merchantment. Note that Columbus was not picked up by any of the other ships. He swam to the coast. The only indication we have of this shipwreck is from Columbus's son Fernando, when explaining – in Spain – the appearance of his father in Portugal and insinuating that he was not Portuguese.

If Columbus was the son of Coullon and French, he was not Catalan, because to annex the province of a foreign country does not imply a change of nationality. If Dom Fernando wrote that his father had sailed with Casenove-Coullon for a long time, it would be the same as saying that Coullon had been a traitor to his King by serving an ambitious French monarch.

Neither the author of the *Historia* nor Las Casas saw the slightest danger for Columbus in declaring that he had sailed with Colombo-the-Younger. They never gave a thought to the idea that Columbus might have been Catalan and an enemy of King Ferdinand (known as 'the Catholic'). They did not even think of Casenove-Coullon as that.

It is clear that neither Fernando Colón nor Las Casas linked Columbus with Coullon or Colombo-the-Younger; neither did Anghiera, for that matter. The Spanish sovereigns, to whom Fernando Colón dedicated the biography of his father, never voiced the slightest suspicion that Columbus could have been a French privateer in the service of René of Anjou, their enemy.

6. On his deathbed, in 1506, Columbus repented having attacked Negro and Spinola's ship and decided to compensate them for the losses they suffered.

But why, if it was his own vessel that was sunk?

7. Possessing great moral qualities, Jean Coullon (or Juan Coulon) renounced his life of privateering and married a young lady of noble birth on the island of Madeira. But this was only after having been to Greenland in 1477. When the expedition returned, he either stayed behind or went back later and from there undertook a voyage to Labrador, Newfoundland and maybe south along the American coast to the Bermudas.

It is strange that Ulloa does not mention that the expedition to Greenland had been organised by the Kings of Portugal and Denmark in order to wed the Portuguese experience of 'high sea' navigation to the knowledge that the old Vikings possessed of those waters. Why would they have invited a French privateer to take part? How did he come by a ship to make his 'private' voyage?

It also seems that Ulloa ignored the fact that some of these lands had been visited by Portuguese mariners in 1473.[31] These voyages were recorded and the crews received their due rewards. And how does one explain that a shipwrecked privateer arrived on Madeira and immediately married the Donee's daughter – in the fifteenth-century?

8. Columbus did not ask for any support from the King of Portugal and as he did not speak Portuguese went to Spain, from where he intended to return to France. In Spain, however, he met the Duke of Medinaceli, the leader of the Catalan revolt before René of Anjou. It was this Duke who encouraged him to persuade the Catholic Monarchs to let him discover 'India'. Although he changed his name from Juan Coulon to 'Christo-Ferens Colón', King Ferdinand mistrusted him and the project came to nothing. He then left the Spanish court and made his way to Seville, from where he was to depart for France with his son Diogo. But he broke his journey at the Monastery of La Rábida to speak to Friar Juan Peres, in whom he managed to inspire confidence.

It seems that Ulloa also ignored the existence of the letter that King John II wrote to Columbus. Although the policy it refers to was deliberately falsified, the letter does exist. According to Ulloa, Juan Coulon lived in the Palace of Medinaceli in 1485–6. He moved to Córdoba, then

Cadiz, then Malaga and finally Salamanca, to the north-west of Toledo. After his request to be received by the Monarchs had been turned down in 1487, he made his way to France by way of Seville. It is difficult to believe that Colom was such a bad 'navigator'. Instead of going north either through Old Castile or Catalonia, where he could have met up with his compatriots if he was Catalan, he went in exactly the opposite direction. And it was at La Rábida that the Friars Juan Peres and Antonio de Marchena (who were not Catalan rebels but obviously Portuguese) gave him the help he needed.

We cannot censure King Ferdinand of Castile and Aragon for having mistrusted this Jean Coullon or Juan Coulon who, to conceal his real identity, had adopted the name of 'Christo-Ferens' (which, according to Ulloa, means Juan Bautista) Colón.

A DISTORTED MIRACLE

Fernando Colón placed his father's shipwreck between Lisbon and Cape St Vincent. This would put it near the Cape of Sines, the fortress of which was under the command of Dom Estêvão da Gama, father of Vasco da Gama – the real discoverer of the seaway to India. It is interesting to recall that Dom Estêvão was suspected of having taken part in the conspiracy against King John II and died in prison, this being the reason that we know nothing of the first years of Vasco da Gama's life. To the interior is the city of Beja and the town of Cuba, made up of one single parish called St Salvador, as well as the village of Vila Ruiva. The Duchy of Cuba stretched to the coast.

But if Columbus's ship was wrecked between Cape St Vincent and Lisboa, this would not necessarily mean that he was near the Portuguese capital, as the distance between the two places is 150 miles. Even half that distance in the fifteenth century could not have been considered 'not far from Lisbon'. Admitting that Fernando Colón made a mistake and that the shipwreck occurred in 1476 and off Cape St Vincent (according to Sabelico), just like the one in 1485 (according to Palencia), it must not be forgotten that the waters in that area are very choppy and cold even in August, being very different from those of the southern coast of the Algarve. The sentence 'he grabbed an oar that was floating nearby and alternately hung on to it and swam', is also hard to understand, as it seems highly improbable that while he was swimming the oar would follow him like a well-trained dog. And if he had not abandoned the oar it would have been extremely slow work swimming with only one arm. Besides this, two leagues measure at least six miles. A fully-clothed man, swimming with one arm, would take at least twelve hours to cover that distance. Columbus would certainly have frozen to death before this, because he had been in a fierce fight before this ordeal.

The perpetrators of the *Historia del Almirante* inserted the sentence: 'not far from Lisbon, where he knew he would find many of his fellow-Genoese'; it must not be forgotten that, in Chapter I of his book, Fernando had denied that his father's surname was Colombo and took care not to claim that he came from Genoa. He would be unlikely, therefore, to contradict himself four chapters later. Did Fernando Colón purposely exaggerate the drama in order to suggest a miracle? He said 'God gave him strength [he had saved Columbus for greater things] to reach the shore'.

In 1980, Senator Paolo Emilio Taviani produced a vast book,[32] which follows the thesis of the Genoese Giustiani in his *Psalterium* of 1516. Taviani claimed that the brothers Ugolino and Vadino Vivaldi rounded Cape Non in 1291. One of their ships sank, but the other one carried on to Mina, on the Ivory Coast, where the crew were imprisoned by the Christian subjects of Prester John (who in fact never existed there). A son of one of the Vivaldis sailed the East African coast in search of India (round the Cape of Good Hope). Taviani claims that Columbus took part in the attack on Tunis in 1477 and then went on a study and recreational voyage to Iceland in the same year. He obviously forgets to refer to the Luso–Danish expedition, but as Columbus mentioned the fact that Bristol merchants traded in Iceland, he concludes that the navigator must have sailed from that English city. Taviani transcribes part of a letter that Columbus wrote to King Ferdinand of Castile in 1505; he, however, omits[33] to note that the letter began thus: 'Our Lord has miraculously sent me here, as I landed in Portugal'. The words 'miraculously' (which he repeats twice) and 'landed' clearly signify that Columbus did not reach the Portuguese coast under normal circumstances.

The 'miracle' that Columbus refers to is a different one – that of being able to serve the King of Spain because (apparently) the King of Portugal had refused his services. Taviani deliberately omitted the rest of the text. First he says that King Fernando was addressed as 'Your Majesty' (a style not at that time used), when in fact he was 'Your Highness'. Secondly he terminates the transcription with the words 'I landed in Portugal'. Columbus's letter reads as follows:

> I say miraculously, because I presented myself before the King of Portugal, who is the most knowledgeable of all in the field of the discoveries, but God closed his eyes and ears and other senses so well that my proposals fell on deaf ears for fourteen long years.

The 'miracle' was the fact that King John II did not want to hear him. At least that was the justification that the King's secret agent

Colón-Zarco had to present to the Catholic Monarchs. Taviani has thus ignored Columbus's consideration of the King of Portugal as 'the most knowledgeable of all in the field of the discoveries', besides another serious revelation – God had closed the King's eyes and ears for fourteen long years. As Colón-Zarco left Portugal in 1485–6, it means that he had been in contact with King John II in Lisbon since at least 1472. Taviani was not interested in transcribing the complete letter, in his attempt to turn the wool-carder Cristóforo Colombo into the navigator Cristóbal Colón.

As far as the shipwreck is concerned, Taviani claims that 'the only valid argument' is that 'his' Genoese merchantman was shipwrecked between Cape St Vincent and Cape St Mary (in Faro, capital of the Algarve) near the town of Lagos, and that he was picked up by fishermen. If this is the case, the whole narrative of Fernando Colón is false. Taviani places Columbus about 220 miles from Lisbon, which in the fifteenth century (when there were no decent roads and no bridges over three large rivers), could have been closer to 330 miles, owing to the detours that were necessary.

We must here consider the contradiction of Columbus having taken 'several days to recover. As he was not far from Lisbon . . . he made for the city with all haste'. After journeying for several days, he arrived in Lisbon 'where he knew he would find many of his fellow-Genoese . . . and, as he was known to them, they made him so welcome that he settled in that city and married there'. How can it be explained that Fernando Colón, who stated categorically that his father was Colón and not Colombo and who had always concealed his father's birthright, fell into the trap of saying that he was Genoese? How can it be explained that Columbus, who had been sailing for many years with Colombo-the-Younger, was already known to the Genoese of Lisbon, who immediately helped him to set up house in the city?

In his *História*,[34] the Spanish writer Luís Arranz expresses the following opinion:

> Many historians consider this narrative [the shipwreck and arrival in Lisbon] to be a levity on the part of Don Fernando or a falsification committed by a third person . . . In merging these two episodes, it seems that Dom Fernando wanted to divert or distract one's attention to conceal something of greater importance.

Other historians, including Professor Janina Klawe of the University of Warsaw agrees that the whole narrative is an invention on the part of the Genoese. It is even possible that it suited Colón-Zarco to encourage this fable and that his son carried on in the same vein in order to please or help his father. It is probably an expedient that was created in order to conceal Columbus's true nationality and to camouflage his identity as a secret agent of the King of Portugal.

Following the death of his wife, Filipa Perestrelo, Columbus, already armed with his new *nom de guerre*, took his legitimate son Diogo to Cadiz in 1484. Being five or six years of age, Diogo started a new life; his father did not want to leave him where he was known, and from then on, his true name was to become a hindrance to him in his natural desire for normal relations with his father. Both of them, therefore, went through a metamorphosis of personality, as did Columbus's brothers, Diogo and Bartolomeu. None of them would be able to return to what they had been before (see Chapter 28).

What was the reason for this? Was it because Columbus was ashamed to admit that he was a wool-carder's son, which would agree with the Genoese thesis? It is possible. Was it because he feared that the Spaniards would find out than he had been a privateer, which would agree with the Catalan thesis? Improbable – as his other son, Fernando, did not hesitate to narrate this episode – but not impossible.

A close examination reveals an anxiety to coordinate facts as if they were pieces of one puzzle, but they are disparate pieces that are impossible to put together so as to make one picture. Taviani attributed to one Fisher, whose work he does not quote, the claim that the Portuguese chronicler João de Barros had recognised that 'the Italians were the first and true masters of the Portuguese in the art of navigation and that the knowledge of the latter [even in the fifteenth century] did not allow them to sail far from the coast'.[35] However, the Archipelagos of Madeira, the Azores and Cape Verde are in the middle of the Atlantic, well away from the coast. And João de Barros never said that.

We now need to consider the complementary 'Catalanist' thesis of Enrique Bayerri Bertomeu.[36] As we have seen, Bertomeu's work undermines both the Genoese and Ulloan theses, and many others that invent Columbuses from San Juan les Fonts, Tuy, Barcelona, Palma de Majorca, Biscay, Ampirdan, Tossa, Blanes, Lerida, Greece, Arbizóli, Burggiasco, Cosselia, Cuyeres, Ferroso, Oneglia and many other places. But Bayerri Bertomeu gets bogged down in the quicksands of Ulloa's speculations. Demanding proof from his opponents, he fails to display one single document. Ignoring (or pretending to ignore) the Portuguese theses, he superficially mentions some of them without even making a constructive commentary.

CATALAN NAVIGATION

Bertomeu attempts to justify Columbus's nautical experience through his intimate contacts with the great Catalan mariners, even though the world is ignorant of any great pre-fifteenth century Catalan feats of seamanship. For any such Catalan navigating genius, we must look to the island of Majorca and, even then, restrict ourselves to the wisdom of Raymond Lully, who was never a navigator but an alchemist and astrologer friar.

There were Catalan cosmographers, if we consider the Jews Abraham and Jefuda Cresques, who were Majorcans. But we have no knowledge of navigators who could have taught Columbus to overcome the winds and currents of the Atlantic in lateen caravels. This does not mean that there were none; it is just that no-one has ever heard of them.

In his work *The Art of Navigation*, Raymond Lully leads us to understand that the Majorcan sailors already knew the astrolabe, which is not surprising as the Portuguese had been using it for a long time. He also says: 'the influence of the Arab astrologers led to progress in Hispanic nautical astronomy [we must understand "peninsular" here and not "Spanish"]'. He follows up, however, by adding: 'many methods to determine longitude have been experimented, but without success'.[37] It would be impossible to navigate the high seas without the determination of longitude, which is not the same thing as crossing the Mediterranean, the 'inland' sea known by mariners since ancient times.

Catalan sailors and navigators appeared on the world scene only in the sixteenth century. Up to this time, their nautical knowledge, based on Lully, did not reach the level of the Portuguese 'Sagres School'. There were so many Jews on Majorca in the fifteenth century that the island was almost a ghetto. Most of them emigrated to Portugal at the time of the Castilian persecution. One of the great cosmographers of the time, Abraham-ben-Samuel Zacuto, author of *Sepher ha Yuhsin* (born in Salamanca in 1450), lived on the island. He came to Lisbon at the time of the Castilian persecution and rendered his services to King John II. It was in the orbit of what we may call the 'Majorcan School of Astronomy' that the 'Sagres School of Nautical Astronomy' really developed astronomical science.

Astronomical science had its origin in the 'School of Alexandria', with the contact between the Greeks and the Egyptians. An edict of Justinian closed the last pagan school in Athens in 529 and the Hellenic–Egyptian science passed on to the Arabs. They corrected and perfected it and then transmitted it to the Jews. Through translations, passed down from father to son, the science reached Provence with the exiles expelled from the Iberian Peninsula, but it was propagated only in the sixteenth century.[38]

THE ISLAND OF GENOA

Bayerri Bertomeu put forward the thesis that Columbus was born in Genoa. But not the Genoa we would expect: this one is the now disappeared Island of Genoa that had the name of the 'Island of the Ebro' up to the twelfth century, when a chapel with the name of St Lawrence of the Genoese was constructed there (see p. 5) (as the Genoese had helped Count Raymond IV of Beranger conquer Tortosa, see pp. 4–5). This little island was by the fifteenth century nothing but a sandbank with the ruins of the chapel of St Lawrence standing on it. According to Bertomeu, it was called the Island of Genoa for only 50 years, but he does not supply any proof. However, we know that there was a parish on the river bank that provided spiritual succour to the inhabitants of the island and the crews of the ships that dropped anchor in the river.

This parish was actually called St Lawrence of the Genoese in 1159, when the island was known by the name of *'Insula Sancti Laurentii Januensium'* – the Island of St Lawrence of the 'Genoese', not of 'Genoa'. The inhabitants of Tortosa called it the Island of Genoa for short, but by the fourteenth century it was already called the Island of the Ebro on Catalan maps. For Columbus to have said 'I left there [Genoa] and was born there' (see Part IV), he would have to be speaking about a name that disappeared two centuries before (that is, if it existed in the first place).

All these hypotheses come from the fact that Columbus was considered to be a 'foreigner' in Castile, and Catalonia was not a sovereign state. The Aragonese took Barcelona in 1472 and incorporated Catalonia into their kingdom. On the death of John II of Aragon, King Ferdinand ascended to the throne and his marriage to Isabella united the states of Aragon and Castile–León in 1479. An inhabitant of Tortosa, near the mouth of the Ebro, would never be regarded as a 'foreigner' by his own sovereign, the King of Aragon, who was also King of Castile. If Columbus were from Tortosa, therefore, he would certainly not have presented himself before his sovereign as some cunning adventurer of doubtful origin, but as a genuine citizen willing to offer India to his country.

According to his historian, Friar Bartolomé de Las Casas, Columbus stated in his letters to the Catholic Monarchs that he was not Spanish, but a foreigner. As far as Bertomeu is concerned, Columbus was a foreigner in Spain because he had been born on the islet in the River Ebro, which was to all effects Spanish territory. He also insists on the fact that the islet was called the Island of Genoa, even though the documents designate the whole area as the 'parish of St. Lawrence of the Genoese' and the island by the same name. Bertomeu argues that

Columbus always considered himself to be a foreigner and Genoese in relation to the Sovereigns because his mother had given birth to him in the middle of the river, among the ruins of the old chapel. This was the reason that Columbus insisted on the fact that he had been born in Genoa and had departed from Genoa – 150 years after the maps had designated the small, semi-deserted sandbank as the Island of the Ebro.

According to Bertomeu, the privateer Coullon was not a foreigner because he was French, but owing to the fact that he came into the world near Tortosa, a city inside Spain. In his opinion, Jean Coullon concealed his identity with the name 'Xpo Ferens Colón' because this was the symbolic sign of Juan Bautista. This interpretation is strange, as Christo Ferens means 'he who bears Christ' (like St Christopher, for instance, or a missionary, a saviour of souls, or even 'he who bears the Cross' – in this case Simon, who helped Jesus carry His cross to Calvary). To think of Juan Bautista (St John the Baptist) as a 'carrier', in the sense that he 'brought Christ into history' through the preaching of his Gospel, seems to be going a little too far.

When Bertomeu refers to Jean Coullon's voyage to Greenland with the Luso–Danish expedition, he takes care to omit all reference to the Portuguese participation and suggests that at this time Columbus did not call himself Jean Coullon but 'Joannes Scolvus'. This appears to be the Latin form of 'Jan Skolp', which in Danish stands for 'John the man of the coast' or the 'seasider'. As far as Bertomeu is concerned, 'Scolvus' is 'Scolmus' (i.e. 'Colón' misspelt). Bertomeu did not mention the fact that he had got his inspiration from a book by the Portuguese Pestana Júnior[39] which by 1928 had already been challenged by another Portuguese, Saúl Ferreira, in a work written with Ferreira Serpa and to which I shall refer in more detail later. Although mentioning this work[40] Bertomeu omitted to note that Pestana Júnior had suggested that Scolvus was a falsification of 'Scolmus' or 'Colmus', which, in his opinion, would be 'Colmo' – a Portuguese word that is almost synonymous with *'Palha'* (straw in English). Pestana Júnior advanced the hypothesis that *Xpo Ferens* represented *Siman* (Simon) and *Colmus* (*Palha*) – the Simão Palha who was the cousin of Felipa Moniz Perestrelo, Columbus's wife. Bertomeu also said that Columbus's son was not named Diogo but Jaime.[41]

To sum up, we may say that *Colón tal qual fué* is a book of indisputable merit as a bibliographical compilation, but we cannot accept the thesis that Columbus was simultaneously French and Catalan, born in the ruins of a chapel on the long-disappeared Island of the Ebro.

Ricardo Carreras y Valls[42] also keenly defended Ulloa's thesis. Regarding his work, Bertomeu says: 'A lot of fantasy on badly written and even more badly criticised documents. In his opinion, the birthplace of the Discoverer of America was Terra-Roja, on the right bank of the River

Ebro'. Referring to the views of Gonzalo de Reparaz in relation to another of Valls's works[44] Bertomeu says: 'It is a wise, sharp and above all, an implacable criticism of the mistakes, exaggerations and undocumented fantasies of Carreras y Valls'.

Wenceslao Aygnals de Izco[45] discovered that Columbus had been born in the State of Geneva, thus being Swiss. Stefen Zeromsky[46] and Réclus[47] claim he was Polish, as his real name, translated from Scolvus, would not be 'Skolp', which is Danish, but 'Kolmo' in the opinion of the former and 'Polen' according to the latter.

The Scandinavian historian, Thorwald Brynidsen[48] came to the conclusion that Columbus was American, while the Agency *Tass* informed the Soviet public in 1949[49] that the discoverer of America had been Russian.

Other theses make logical contributions, like the one of la Riega[50] which – although creating a Galician Columbus – defended the probability that the navigator was either Jewish or of Jewish extraction. Roberto Almagia[51] claims that he was a Jew or an Italian convert. Maurice David,[52] José Amador de los Rios,[53] Fritz Baer,[54] Salvador de Madariaga,[55] Simon Wiesenthal[56] and others are inclined to believe that Columbus was a Spaniard of Jewish origin.

Armando Alvarez Pedroso[57] was eminent among Spanish-speaking opponents of this movement. He put forward feasible arguments which were enthusiastically applauded by the supporters of the Catalan thesis, as they identified Columbus with the non-Jewish privateer Coullon. Pedroso bases his arguments on the fact that Columbus was described as a redhead (see pp. 471-2), but the colour of his hair and the Christian fervour that Columbus always manifested would not necessarily mean that he was not of Jewish descent: the hypotheses that Columbus was Portuguese have always admitted both his Hebrew culture and his possible Jewish blood, but at the same time his Christian beliefs have never been doubted.

Henri Vignaud was inclined to believe that Columbus was either Basque or Portuguese. But the Basques were still in the dawn of seamanship in the sixteenth century when compared to the hundred years of nautical studies carried out by the 'Sagres School'. The Catalan Ricardo Beltran y Róspide[58] was unable to accept the theses of Luis Ulloa and Bertomeu, and his critical and objective studies refute the Genoese, Catalan and Galician theories. In his monumental work, *Historia de España*[59] António Ballesteros y Berretta, who is also Catalan, comes to the same conclusion as Ricardo Beltrán Róspide: 'If it is possible to feel the conviction through our knowledge, there are sufficient reasons to place ourselves between doubt and certainty that the discoverer of America was not born in Genoa but came from somewhere in the west of the Peninsula, between Capes Ortega and St. Vincent'.

As Columbus could not have been Galician, since the whole theory of his having come from Galicia has been undermined, the only feasible hypothesis left is that Columbus was Portuguese. Several things point to this. His knowledge of cosmography and nautical science; his statement that he had made voyages to Mina and Greenland; his knowledge of previous voyages to the west and of the existence of western islands and the Sargasso Sea. There are also his noble origins, justified by his intimate relations with King John II, and his marriage; his humanistic culture and the fact that he knew Hebrew and Latin, certainly not common in the fifteenth century among wool-carders and bartenders. There are, his contacts with the Portuguese friars at the Monastery of La Rábida and his 'Portuguised' Castilian.

To all this may be added the impossibility of his renouncing his vows as a Portuguese of the Order of Christ, which obliged him to sign his name with a cabalistic siglum, and the secrecy that forced him to conceal his true identity, since he was a secret agent of his true lord – King John II of Portugal.

Notes and References

1. Uhajón (1892).
2. Rocca (1892).
3. Marín (1929).
4. Salinério (1864–6).
5. Astengo (1885).
6. Giustiniani (1536).
7. Cancellieri (1809).
8. Agnelli (1892).
9. Bellet (1892).
10. Cúneo-Vidal (1924).
11. Oliva (1635).
12. Harrisse (1887).
13. Paragallo (1889).
14. Casanova (1881).
15. Cesari (1932).
16. Mora (1886).
17. Pinilla (1884).
18. Serrano, Adrián Sánchez (1928).
19. Riega (1897), (1899) and 1914.
20. Salari (1912).
21. Pardo (1912).
22. Canel (1913).
23. Hervás (1913).
24. Olmet (1916).

25. Marcot (1919) and (1920).
26. Calzada, Rafael (1892).
27. There was someone in Italy called Colón – and not Colombo. A Jewish Rabbi, José Colón, still today considered one of the greatest ever scholars of the *Thalmud*, died in Mantua in 1840. Living in Italy and not being called Giuseppe, we may deduce that he came from the Iberian Peninsula. There was a notable family of New Christians by the name of Colón that lived in Galicia in the fifteenth and sixteenth centuries, one of whom is buried in a magnificient tomb in Pontevedra Cathedral.
28. Riega (1914).
29. Sanz, Manuel Serrano y (1919).
30. Streicher (1929).
31. By João Álvares Corte Real and Álvaro Martins Homem.
32. Taviani (1980).
33. Taviani (1980), pp. 63–4.
34. Arranz (1986).
35. Taviani (1980), p. 73.
36. Bartolomeu (1961).
37. Corrêa, Jácome (1929).
38. Duhem (1940); Ventura (1944).
39. Pestana Júnior (1928).
40. Bartolomeu (1961), pp. 226–7.
41. 'Jacob', 'Jacobo', 'Jacopo', 'Giacome', 'Jaime', 'Jacques', 'Iago', 'Tiago', 'Diogo', and 'Diego' are different ways of translating the same name; however, I shall show that Christopher Columbus and the notaries of his time made a distinction in Latin between *Jacopus* for Giacome and *Didacus* for Diogo.
42. Valls (1930).
43. *Bulletin del Centre Excursionista de Catalunya* (Barcelona, 1928).
44. Valls (1928).
45. Izco (1852).
46. Zeromsky (1922).
47. *Géographie Universelle*, Vol. 15 (Paris, 1936).
48. Brynidsen (1934).
49. Tass Agency (1949). The details in nn. 45–49 are all taken from Bertomeu (1961).
50. Riega (1899).
51. Almagiá (1918); see also '*Ció che é definitivamente acquisto alla sciencia e ció che puó essere indagato intorno alla vita, ai viaggi ed alla acoperte di Cristoforo Colombo*', at the International Congress of Historical Sciences (Warsaw, 1973).
52. David (1933).
53. Los Rios (1876).
54. Baer (1936).
55. Madriaga (1952).
56. Wiesenthal (1972).
57. Pedroso (1946), Vol. X.
58. Róspide (1918), (1925) and (1948).
59. Beretta (1948), Vol. III, 3rd Pt, pp. 120–1.

Part III
Empirical and Scientific Navigation

10 Portuguese Nautical Cryptography

In 1474, King Alphonse V published an order that 'increased the privileges of shipowners'. The experience acquired by Portuguese seamen had taught them that the bulky merchant galleys were unsuitable for navigating the high seas. A vessel with a narrower beam and rigged with lateen sails was needed, so that it could tack and sail on any point of the compass.[1] This type of lighter and faster ship had been constructed in Portugal since at least 1225, as is mentioned in the charter of Vila Nova de Gaia. When King Alphonse II conquered Faro in 1249, he had used heavy ships (galleys with oars and sails) and ships which had no oarsmen and were stepped with two large masts and a small one. The usual ship for faster sailing, however, was the Arab-type caravel.[2] The caravels were bigger in the reign of King John II and could sail in any sea. It was with one of these that Bartolomeu Dias rounded the Cape of Good Hope.

The Castilians continued to use heavy vessels. In 1492, Columbus described his ship, the 'Santa Maria', as an ex-galley. In the reign of King Manuel I, the Portuguese naus that would cross the Indian and Pacific Oceans were much bigger than the 'Santa Maria' and the caravels of the early Discoveries, rigged with their own type of sails and very different from the Mediterranean ships. Shipbuilders from all over Europe came to Portugal to learn how to design and build these new ships that the Portuguese were constructing.[3]

Columbus had his ship 'Pinta' rigged in Portuguese fashion so as to be able to tack. Transcribing Columbus's logbook, Las Casas explained that the ship carried square and lateen sails, which made it more manoeuvrable and, with its extra rigging, more like a caravel. Tradesmen of all types were called to prepare the Portuguese armada. The Tagus was a forest of sails, while shipyards lined the river banks, the ribbing of ships under construction looking like the skeletons of so many whales.

Sails were made of cloth that was as light and as impermeable to the wind as possible; the rigging was a mixture of linen fibre and tow. But it needed more than navigational knowledge to undertake long voyages. Many practical problems had to be overcome, such as the provision of water, which became undrinkable when in contact with the raw wood of the barrels or with the pitch that was used to caulk them. Thus tanks lined with slate were invented. Rewards were given to whoever came up with solutions for the many problems that arose while at sea. The Order of Christ and the Jewish shipowners financed this boom in naval construction.

Following the pattern of the meetings of experts at the so-called 'Sagres School' of Prince Henry, King John II created the 'Junta of Mathematicians',[4] which counted on not only the participation of Portuguese but also of foreign experts in an effort to speed up scientific progress. The Portuguese had long abandoned coastal sailing; they used the astrolabe,[5] inherited from Arab–Jewish science, the cross-staff, which they perfected, and the *toleta de marteloio* (which shows the distance travelled along a set course when tacking).

The Greek astronomer and geographer, Claudius Ptolemy, had already proved in his *Almagest* (second century) that the world was round. This fact was still ignored in many European courts in the fifteenth century, but it had been recognised in Portugal since at least the reign of King Dennis.

At the time of King John II, the armillary sphere was already a royal emblem and it was sufficient to compare it to the maps of the world – drawn according to the information supplied by overland travellers and already showing China and Japan, though grossly distorted – to have a clear notion that the Far East could in no way be on the other side of the Atlantic, i.e. in the Antilles.

Martin Behaim (from the Jewish name Benhaim), from Nuremberg, came to Portugal in 1481; he studied the *Tables of the declination of the Sun* made by the Jew Abraão Zacuto with two other Jews, Master José and Master Rodrigues (see Figure 10.1). Three years later he was authorised by the King to sail to the Congo with Diogo Cão, to verify the use of the astrolabe and the *toleta de marteloio*. In the Azores he married Joana de Macedo, who was daughter of the New Christian João Huerter, first Donatary Captain of the islands of Fayal and Pico. In 1491 he met Ieronimus Monetarius in Lisbon and later accepted a map by this cartographer.

According to the Marquis of Jácome Corrêa,[6] 'this map of Monetarius is another kind of a confidence trick, similar to the one of Toscanelli . . . brought to Portugal by Behaim who, for this and other reasons, fell into disgrace at the Portuguese court and died shortly after, completely forgotten'.

Behaim made a globe in 1493 on which half of the world was missing and he was discredited in Portugal due to his lack of scientific knowledge. The same year he went to Nuremberg to show his map. There he has a statue, depicting him wearing magnificent fifteenth-century Gothic armour, as if he were an illustrious knight.

Baron Alexander von Humboldt[7] ignores the complete field of Portuguese nautical science and, arguing that the Portuguese Discoveries were a matter of luck, attributes to Martin Behaim the glory of having taught cosmography to Columbus. Humboldt forgets that Behaim had been a student of Camil Johan Mueller (or Müller), cosmographer and alchemist of Monterregio, also known as 'Molyart', meaning 'the lead of the art' (*moli* + *artis*) – the base metal with which alchemists wished to carry out the '*chrysopaeia*' and the '*argyopoeia*', transforming lead into gold and silver (see p. 448) and that Müller of Monterregio had studied the *Tables of the Declination of the Sun* of Abraão Zacuto, made in 1476 (see pp. 481–2).

In 1493, King John II summoned Behaim to Lisbon and asked him to obtain the support of Emperor Maximilian I of the Holy Roman Empire for his request to legitimise his bastard son, a move that his queen, Leonor, opposed. Behaim departed in 1494 and returned one year later, at the time of the King's death. This was one of the 'other reasons' that led to his loss of prestige at the Court. He died in 1506, the same year as Columbus. This brief biographical note shows the problems of hypothesising that Columbus learned cosmography with Behaim. Even if Behaim was in Portugal, why should Columbus have studied with him and no-one else? Von Humboldt's work, therefore, does nothing to clear up the doubts established. How and when did Behaim have the chance to teach nautical cosmography to Columbus?

When Martin Behaim arrived in Portugal, he found the astrolabe, to determine latitude by the Polar Star, and the *toleta de marteloio*, to read the rhumbs and measurements of the equinoctial points, in common use. The *toleta de marteloio* made this calculation easier in the northern hemisphere, but after Álvaro Esteves had crossed the equator in 1471 it became necessary to use new methods. The Polar Star was not visible from the Tropic of Capricorn and the Southern Cross, although it had already been seen by Portuguese sailors, had not yet become the central point to calculate a ship's position. So the navigators began to calculate latitude by the meridional height of the sun. Knowing the declination of the sun and its meridian, latitude could be determined. Thus the *Tables of the declination of the Sun* (see Figure 10.1) were elaborated. The ones that Columbus took on his first voyage were – for a special purpose – commended by King John II, who had ordered them from the Jew Samuel Ben Zacuto, son of Abraão Zacuto. (They are written in Hebrew and are to be found today in the Hebraic Theology Seminary Museum in

New York.) They were based on studies of the Jews Samuel and Jehuda and the Arab Aben Ragel, (known as 'The Alchemist').

Two comments are in order here:

1. The Genoese wool-carder had also learned the *Tables of the declination of the Sun*, which were still unknown to the rest of the world, and mastered Hebrew to the point where he could unravel their complexity.
2. The King, who had ordered the Portuguese to keep their nautical science a closely guarded secret, ordered a copy of the *Tables* from a Jew and handed it over to a foreign adventurer who could have passed it on to a rival sovereign.

There is no plausible explanation for these two facts unless, of course, Colón-Zarco were a secret agent of John II and a Portuguese of Jewish descent. Again we have the duality: Colón-Zarco, secret agent, and Cristóbal Colón, rejected as a fool: one real person with two distinct personalities.

In 1428, Prince Henry's brother, Pedro, made the acquaintance of a cosmographer named Paolo del Pozzo Toscanelli at the court of the Medicis in Florence. Some years later, during the meetings that Cosimo de Medici held in his palace, Toscanelli struck up a close friendship with the Portuguese Fernando Roriz, also known as Fernando Martins, a canon who lived in Rome. In a letter to Roriz dated 25 June 1474, Toscanelli traced an interesting map on which he indicated an island approximately located on the coordinates of Porto Rico or Haiti, to which he gave the name of 'Antilia' and which the Portuguese had called the 'Island of the seven Cities' after the second voyage of Diogo de Teive. Toscanelli also claimed that Cypangu (Japan) lay 625 leagues to the west. This theory was based on Ptolemy, whose book had been published two years before and in which he claimed that the world was round and that India could be reached from the west. This theory, moreover, had already been accepted by Pope Pius II in 1458 in his Bull *Terra jam fere omnes consentiunt rotundam esse*.[8]

Toscanelli's map, plus the ones of Behain and Monetarius, were treated with total indifference by the Portuguese. Why? Was it because they depicted lands, some which the Portuguese had already located, that were pure imagination on the part of the cartographers? Or was it because the navigators of the Order of Christ already knew of a continent between Europe and Asia? It was thought that it would be much slower and more dangerous to sail round the south of Africa than to cross the Atlantic to reach India. Why did the Portuguese disdain the shorter route? Did they happen to possess a better knowledge of the American continent than the rare documents of today lead us to sup-

10.1 The *Tables of the declination of the Sun*.
Ordered from the Jew Samuel Ben Zacuto, son of the astrologer and physician Abraão Zacuto, by King John II for Columbus's first voyage. They were based on the studies of the Jews Samuel and Jehuda and the Arab Aben Ragel (known as 'the Alchemist').

(*Rabbi Joseph Toledano, ruler of the Lisbon Synagogue in 1972*)

pose? Or was it simply because the calculations of latitude (and sometimes of longitude) of those foreign cosmographers was totally wrong? How could these cartographers have elaborated their maps if they were not navigators and it was only the Portuguese that scientifically sailed the Atlantic? Obviously through information gathered from intermediaries coming from Portugal. Judging from the number of mistakes on these foreign maps, however, these spies of the 'nautical industry' were by no means experts.

Unfortunately, owing to the secrecy of the Order of Christ, it is only these maps that have survived till our days. The more exact ones that the Portuguese used were deliberately hidden and reappeared only in touched-up editions in the sixteenth century. There is no doubt that the dubious and fantastic works of Toscanelli and Behaim were rejected out of hand by the Portuguese from the beginning. The latter were soon to prove that they were right and demonstrate that their own methods were more exact, namely through the works of Francisco Faleiro and Pedro Nunes.[9] It is essential to emphasise the fact that the possibility of reaching India by the west, expounded by Ptolemy and accepted by Pope Pius II, was unknown to the Catholic Monarchs and their counsellors, to the point that Columbus had to wait for seven years before he was allowed to put forward his theory, which was the same as Toscanelli's, whose map he presented to back up his argument. Columbus had already sailed with the Portuguese and knew that Toscanelli's theory had been rejected by them – not through cosmographic ignorance, but because they doubted the existence of a western passage and that Japan was so near. In spite of this, Columbus succeeded in convincing the Castilians to accept a rejected thesis, using a map that he knew was wrong. Columbus himself corrected it in relation to the position of Iceland and the Antilles but, feigning semi-insanity, was to defend the theory until the end of his life, even after it had been proved to be wrong. How can such persistence on Columbus's part be justified, unless through the necessity of his maintaining the oath he had made before his true lord, the King of Portugal?

While the Portuguese disdained a map because they knew it was wrong, the Castillians hesitated in accepting it because they found it difficult to understand the principles of its implicit arguments. European cosmographers knew of Ptolemy's and Eratosthenes's[10] calculations regarding the circumference of the earth. For Ptolemy the *miliarium* was 1563m and the league was 6612m. For the Portuguese and Columbus, the *miliarium* was 1800m and the league 5400m.

The Marquis of Jácome Corrêa[11] wrote:

> But Colón knew the measurement of a longitude degree in 1492, as well as of the Portuguese peripheral league – 16 2/3 per degree; in Spain, in the reign of Philip II [1566–98], it seems that they ignored the fact that the league could not be 4 Roman miles . . . And the Italians in the service of Portugal did not know the distance of a league. If this historical fact is conjugated with the information of Pigaffeta, for example, who wrote in the log when he passed through the Straits of Tierra del Fuego with Magellan that they were 110 leagues or 400 Roman miles long, a conclusion he always

came to when comparing leagues and miles[12], we will have discovered Columbus's crime. He took false information abroad – the measurement of a degree and a league, with which the geographical coordinates of any newly-discovered land were calculated.

Did the author not mix the names, writing 'Columbus's crime' instead of 'Magellan's crime'? If both the Spaniards and the Italians were ignorant of the Portuguese coordinates in the reign of Philip II, and if Columbus concealed them, how can we accuse him of indiscretion? The fact that even the Italians in the service of Portugal were ignorant of them is really significant. If Columbus knew them, he must have been Portuguese.

The constant Portuguese voyages to the west and the patiently collected information made it clear to the Order of Christ and the King that there was neither a flourishing Oriental civilisation nor a spice trade in the western lands (we must not forget that even if the westward expeditions from Madeira and the Azores did not find land, the sailors knew lands were not far off and that they were inhabited): 'the west winds bring bamboos and the corpses of strange-coloured men, indications that there are lands in that direction'. But the Asiatic ships that Pero (as Pedro) da Covilhã had described (see pp. 28–167) and which were known to sail the Oriental seas were never sighted.

On the other hand, the calculation of the circumference of the world showed that, by voyaging west, India was further than those western lands – some already found, others announced by the above-mentioned bamboos and corpses. It was, therefore, impossible to believe that the spice-producing lands of Cypangu, Ophir, Cathay and Industan would be found in these empty seas. Was there an impassable barrier? The Portuguese could not be sure that this barrier really existed, as they had not explored the whole of the American coast, but they were convinced it did. If there were a passage, why had they not met Asians coming from the west? After all, they had a highly developed sea-borne trade and widely proclaimed oriental wisdom. And why did the Asians continue to export their merchandise to the Mediterranean via the Red Sea, Cairo and Suez by means of Arab intermediaries?

The theory expounded by Columbus to the Catholic Monarchs, therefore, was in complete contradiction to the scientific knowledge and the presentiments of the Portuguese. By convincing the Castilians, who were at first slow to understand him and then reluctant to support him, the future Admiral of the Indies led the Catholic Monarchs to hand the highly-sought prize of the real India on a plate to the King of Portugal. If Columbus had sailed to Greenland, as he claimed, he must have done so with the Portuguese between 1476 and 1477. He mentioned the year 1477, and I shall deal with this subject later. Possessing the cosmographic and nautical knowledge he did, he could never have accepted the theory of Toscanelli, nor Behaim's cartography. Yet he wrote to Toscanelli, pretended to believe in Behaim's obvious mistake and preached prophetically in regard to the deeds, nations and peoples described in Marco Polo's book. Why? Because they spoke about the lands that Castile coveted; because they led people to dream that they could be reached by sailing west.

Note that on his very first voyage, Columbus did not hesitate. He headed straight for the zone where the Portuguese claimed to have already found islands. Bartolomé de Las Casas, whose father sailed with Columbus on this first expedition, mentions in several passages of his *História General de las Indias* that the inhabitants of the Antilles and the American continent itself said that they had already seen: 'bearded white men like ourselves some years before'. Who else could they have been except the navigators of the Order of Christ?

What a shame it is for us today that the King banned the divulging of the routes the Portuguese mariners ploughed through the oceans. What an interesting story we could piece together from the logbooks of Diogo de Teive, the Corte-Reals, the Coelhos, Bezerra Fagundes, the Corrêas, João da Fonte, João Fernandes de Andrade, Afonso Sanches, Álvaro Fernandes, Pedro Fernandes de Barcelos and João Francisco Lavrador, every one of them navigators and discoverers of western lands before Columbus's first voyage. What a pity it is that we know only of their deeds but have no written reports that allow us to relive them.

Columbus was a great navigator and cosmographer. His routes were correctly charted, according to the Portuguese measurements of the time – developed secretly by the Portuguese mariners, as I shall later show. So when he sailed to the Antilles, he knew that he was in the maritime zone closed to Castile as he was south of the parallel of the Canaries. Instead of sailing due west from Palos or from one of the northern Canary Islands (as would have been natural), if he really were steering according to Toscanelli's map, which placed Antilles much farther to the north, or if he were searching for the southern continuation of the already-discovered Newfoundland, he first veered north-west and then west–south–west towards the north of South America – the 'Lands of Gonçalo Coelho'. His intentional route, planned with a predetermined target, took him at least 5 degrees south of the 27th parallel into the Portuguese zone of the Atlantic.

He had already sailed the waters of Guinea and the Ivory Coast, so he would certainly not have navigated the Atlantic 'chancing his luck', as has been proved by the calculations he made during the voyage. On his return, he made for the Azores, where he spent ten days, and then Lisbon. On his arrival in Portugal, he journeyed to Azambuja, twenty-five miles north of Lisbon, to inform the King of his 'discovery'. There can only be one explanation for this – on this first voyage, with a crew that hardly knew the Atlantic (only two Portuguese were aboard: João Árias, son of Lopo de Tavira, and Bernaldim, servant of Afonso, a crew member of the pilot João Rodrigues de Mafra) and with the Catholic Monarchs closely watching his every move, he could not fail. He would not get a second chance. He had to return to Spain with the announcement that he had discovered land. He therefore sailed to the exact place where he knew he would find land and at the same time would not anger his King, on whose orders he had sailed.

Ptolemy had propounded Aristotle's theory that it was impossible to round Africa in order to reach India. How could he have known that by putting forward these theories and making hypothetical maps he would later unwittingly defend the Genoese and Venetian trade in the Mediterranean and the Iberian Peninsula? The Majorcan cosmographer Raymond Lully, or Lulo (1233–1315) had divulged the map of the Venetian Marino Sanuto (1306), which contradicted the principle of Aristotle and Ptolemy, and admitted the fact that Africa could be circumnavigated. However, he mentioned a 'torrid zone' in which man would be unable to survive, thus continuing, unintentionally, to defend the Italian seaborne trade. The *Laurentino Portulano*, which supported Sanuto's theory, summarised the Portuguese Discoveries up to 1346 and depicted Africa as a huge island, appeared in 1351.

King John II strictly imposed the Templar-like policy of secrecy while in charge of the overseas policy during the reign of his father, Alphonse V, and he decreed a royal ban on the writing down of routes and the drawing of maps. The world learned only of lands 'found or discovered', and nobody but the Portuguese found or discovered lands until the appearance of Columbus, when information was leaked through indiscretion or bribery – and outside the circle of the Order of Christ. The Viscount of Santarém (see p. 158) disclosed that there were three Portuguese navigational charts of this period, all of which have since disappeared, two of them being before Columbus's time – 1424 and 1444. Andrea Bianco, a disciple of Fra Mauro, drew a map in 1436 on which the Sargasso Sea appeared off the coast of America to the west of the Azores, although it was not correctly located because it was still unknown. Bianco drew a second map in 1448. Gabriel Valsequa presented a map in 1439 which showed the Azores. Cardinal d'Ailly's book, *Imago Mundi* (of which Columbus owned a copy that he annotated in his own hand) is of the same date. Two years later, Enneas Piccolomini, who later became Pope Pius II, published, while secretary to the Bishop of Siena, *Rerum Ubique Gestarum*. These works were followed by the anonymous Catalan and Genoese maps of 1450 and 1457, Fra Mauro's planisphere of 1459, the pre-1460 map of Cristóforo Soligo, Gracioso Biancasa's maps of 1467 and 1469, an anonymous Portuguese map of 1471 and Toscanelli's map of 1474.[13]

In 1513, Arias Perez Pinzón (son of Martín Alonso Pinzón, the navigator and companion of Columbus) stated before the inspectors of Seville that: 'both I and my father, Martin, together with a curator, had been amused by the configuration of the land of the western Atlantic on the portulani in the Vatican Library. It is known that those Spanish allusions are almost worthless in the face of modern history and they only allude to the monuments reproduced here. They outlined places and coasts with figures of mere fantasy, at the whim of the illuminator, on the coast of Africa, or in the western hemisphere on the globe of Behaim'.[14]

The 'monument reproduced' in this text is the western hemisphere of Behaim's globe, drawn in 1492 – a geographic folly of pure imagination. In fact, the only maps that were almost exact were those drawn by the Portuguese sailors and with which they had crossed the high seas and discovered lands long before Columbus. Even in the time of Pinzón's son in the sixteenth century, the cosmographic ignorance of the Castilian mariners was notorious.

The Portuguese had long known about the phenomenon of the declination, which was a part of their Arab–Jewish inheritance. When Diogo de Teive explored the western seas, 150 leagues to the west of the Azores, in 1452, he penetrated into the aclinical zone of the compass and noticed a magnetic variation. This phenomenon, which went unnoticed in Mediterranean navigation, was a torment for sailors lost at sea as they could navigate only by the stars, which were not always visible. Columbus used the same Portuguese methods to calculate the discoordination of the compass needle, methods that were practically unknown to other European sailors.

In relation to Italian cartography, the Marquis of Jácome Corrêa says:

> Moreover, the geographic nomenclature of those originals [later than the ones invented by Behaim] is clearly Portuguese and that of the d'Este Library in Ferrara is accompanied by the story of its acquisition for 12 gold ducats, having taken 10 months of work [from December 1501 to October 1502]; and which Cantino himself debited at 20 ducats and 3 liras when he handed it over to Francisco Caetano, the agent of the Duke of Hercules, in Genoa on the 19th November, due to the difficulty he had had in obtaining the planisphere.

This Cantino was the agent of the Medicis and the d'Estes in Lisbon. He was in constant contact with Portuguese sailors and drew charts for their masters in Ferrara and Florence.

Columbus's Spanish contemporaries accused him of erring in his navigational calculations. As some of his bearings were correct, however, some defenders of the thesis that Columbus was Spanish have suggested that he used purposely distorted measurements so that the Portuguese would not be able to work out the Castilian routes if they happened to get hold of the documents. Not even present-day historians have come up with such an idea. As the discoveries were in the exclusive Spanish zone, the information would have been worthless to the Portuguese; moreover, there were so many islands in the waters of Central America that it would have been easy to find them even without Columbus's charts and bearings. What does seem strange is that even the Spaniards did not know exactly where they were located.

We must remember that a *stadium* measured 185 metres according to Herodotus; in turn, a Roman mile measured 8 *stadia* – i.e., 1481 metres. In the fifteenth century, each degree was thought to measure 50 *stadia* – i.e., 92 562.5 metres. At the same time, the Arab degree measured 122 826 metres, which was the equivalent of 200 000 *côvados* or 50 Roman miles, each mile measuring 245 652 metres.

The Portuguese mariners conjugated the Arab measurements with those of Ptolemy, producing a harmonious unitary conversion that was more correct in its peripheric reckoning. This was the result of an old meteorological study allied to the measurement of a degree and the comparison of a shadow at the equinox on twelve hour days and at 32 degrees of latitude, during the days of progressive decrease and reduction, according to the declination of the sun during the year – a unique result of the experimental observation of the Portuguese. For the navigators of the Order of Christ the mile measured 1653 metres, the maritime league 3 miles and the equinoctial degree was the equivalent of 18 secret Portuguese maritime leagues. This secret was, moreover, amplified in public documents – leagues and degrees always being multiplied by two.

On his first voyage, Columbus wrote that 'the equinoctial degree measures 56 and two thirds miles': he was obviously referring to the Portuguese mile of 1653 metres, and not the Roman mile of 1481 metres. It is clear that with these measurements the degree did not reach the 950 leagues (i.e., 6167.5 km that, according to the Spanish chronicler Herrera, was the minimum distance covered by Columbus between the Canaries and Cuanahani. Allowing for any deviation from the course set, owing to currents or the need to take advantage of favourable winds, Columbus's calculation of the league of the degree was the equivalent of that used by the Portuguese sailors of the time.

The league measured by Columbus therefore, was the Portuguese league of 5400 metres which almost nobody else used – if the Castilians did not use it, the Italians did so even less. Herrera, who accompanied Columbus on his first voyage, calculated a distance of 6167.5 km after sailing those 950 leagues. Columbus was therefore talking about the secret Portuguese maritime league and Herrera the league of the degree, or normal league, used by navigators of other nations.

The Italians Cadamosto and Nola, who sailed with the Portuguese, were not 'initiated' into the nautical–cosmographic techniques of the Order of Christ because they were foreigners. So how is it that Columbus knew these secrets? These secret measures, designed to protect the Portuguese nautical cryptography, known to and practised only by the Portuguese navigators, were the result of the royal decree of 1475 that forbade the elaboration of routes and cartographic documents. As these documents were essential, however, secrecy was also essential. Degrees always appeared multiplied by two in Portuguese public documents; it was for this reason that Columbus had placed the Antilles at 42 instead of 21 degrees North. Being ignorant of the fifteenth-century Portuguese methods, foreign investigators have accused Columbus of registering half-degrees instead of full degrees owing to misreading the astrolabe.

In fact, the astrolabes of that time were able to register quarter-degrees; what Columbus registered was full degrees. The accusation that he misread the astrolabe is completely unfounded, since he could not have systematically erred to such an extent and have managed to continue sailing. We can only come to the conclusion that Columbus had been trained to navigate using the secret Portuguese calculations, and that if he had made any voyages they had been on board Portuguese ships. And it was the Portuguese – and not the Castilians, who had not yet discovered anything – who tried to cover their tracks so that others could not follow them: Columbus had no reason to fear the persecution of his masters and fellow-countrymen.

Columbus learned the calculations of Portuguese navigation, and used them accordingly. On sailing straight to the Antilles, he knew that he would satisfy the ambitions of both Kings: he would fulfil the mission that his true master had charged him with and would give his pseudo-master new lands to explore, even though they were not India. It was because of this that when he fell into disgrace, he wrote an enigmatic reference to the Catholic Monarchs in his *Diario de a bordo*: 'What is happening to you now is your reward for the services you rendered to other lords'. He did not express himself in the singular, 'the other lord', who would have been the King of Portugal – his true lord.

Columbus claimed to have sailed to Mina after leaving Madeira (Porto Santo), which must have been in 1481 or 1482, as he was on Madeira again in that year on his way back to Lisbon, where we find him in 1483. But he also claimed to have discovered some islands to the west. The Catholic Monarchs knew this and refer to the fact in the *Capitulaciones*, signed on 17 April 1492: 'The things requested that Their Highnesses give and authorise Don Xpóval Colón are in lieu of what he has already discovered in the seas and Oceans and of the voyage that, with the help of God, he will make in the same seas'.

In fact, when Columbus handed the chart of his ship, the 'Santa Maria', to Martin Alonso Pinzón on the 'Pinta' on 25 September 1492, they held a long conversation, according to Las Casas, 'about some islands that should be somewhere ahead'. Who could tell him that? This statement of Las Casas has been mistrusted by many investigators. Yet Columbus was not mistaken, as they were already navigating on the 28th parallel – and the Great Bahama is on the 27th.

If Columbus were not Portuguese, and if he were not familiar with Portuguese nautical cryptography, King John II would never have called him in 1488 to assist Bartholomeu Dias elaborate the route and the charts for his voyage to the Cape of Good Hope. The latitude of the Cape and its distance from Lisbon appears on the itinerary and chart in Portuguese miles and in the cryptographic form (being multiplied by two). Why would the King of Portugal, with so many navigators, cosmographers and cartographers in the Order of Christ, have found it necessary to call on a former wool-carder to help one of his most notable navigators elaborate a secret itinerary? Certainly Columbus neither erred in his calculations nor feared the Portuguese. On his westward voyages, which took him more than 100 leagues west of the Cape Verde Islands and to the north of the parallel of the Canary Islands, he could never have been a target for persecution on the part of the Portuguese; he had Portuguese crew members on his first expedition and the Spanish were not afraid that they were spies.

Even on much later expeditions, many Portuguese captains and pilots sailed in the service of Spain. In the sixteenth century, the crews that circumnavigated the globe under the command of the Portuguese Ferdinand Magellan, who was killed and replaced by Sebastian del Cano, were mostly Portuguese. Duarte Barbosa was the captain and Estêvão Gomes the pilot of the 'Trinidad'; Luís Afonso de Góis (of the family of Columbus's wife) was the captain and Vasco Galego the pilot of the 'Victoria'; José Serrão captained and João Lopes de Carvalho piloted the 'Concepción'; and João Rodrigues de Mafra and João Serrão were the pilots of the 'Santo Antonio' and the 'Santiago' respectively.[15]

Toscanelli's map places Antille, which was already known to the Portuguese, at latitude 23. Columbus was heading for this island, bypassing the Bahamas, the fact that he was looking for it in this area proves that he knew very well where it was located, thus contradicting Toscanelli's map. If he had been on voyages of discovery to the west before his first in the service of Castile, had he done so before or after the Luso–Danish expedition to Iceland in 1477–8? If he had done so after, between 1478 (the date of his marriage to Philipa Moniz Perestrelo on

Madeira) and 1483, when he came to Lisbon, had he sailed with João Fernandes de Andrade (called 'of the Arco', the name of the zone in Madeira he had bought from João Gonçalves Zarco)? The latter had requested King John II 'to fit out a ship for an expedition for those places', and set sail on 9 August 1484. If this is the case, a mistake has been made regarding his coming to Lisbon, as his wife died in that same year. It is more likely that Columbus sailed with João Coelho, with whom Columbus claimed he had been to São Jorge da Mina in 1482–3, as Coelho went there at that time. He could have gone to the Antilles during those two years, as Estêvão Frois declared in a letter to King Manuel I: 'These islands belong to Your Highness, as João Coelho discovered them over twenty years ago'. These islands must have been some of the Antilles.

Even had Columbus lied, he could have known about them through Afonso Sanches, who returned from a voyage to the west at death's door and died in Columbus's house on Porto Santo, leaving the latter all his documents. In his *Memórias*, Columbus states: 'For the undertakings to the Indias, I did not use reasoning, nor mathematics nor *mapa-mundi*'. So sure was he that he would find land, he later said that it was: 'as if I already had it locked away in my room'.

For this expedition to the west, either made by him or a result of information gleaned from other navigators, to be real, did it take place before the shipwreck referred to in the work of his son, Fernando, when he tried to explain how his father came to Portugal in 1476 (see pp. 100)? Columbus claims to have begun his life at sea at the age of fourteen. If he had sailed westward, he could have done so only with the Portuguese who, using the Archipelagos of Madeira and Azores as supply bases, were the only ones to sail those waters, and they already knew 'the islands of the Guinea return', presumably the Antilles.

That leaves us with two hypotheses to consider. Columbus already knew the Antilles, as claimed, or had learned of their existence through the narratives of the Portuguese navigators with whom he had contacts. In either case, he would have known that the islands were inhabited by semi-savages, some, like the Caribs, cannibals. He would have at least heard of the corpses that were sometimes found floating in the sea by the crews of the caravels. He thus knew that there was land to the west, but he also knew that it was not Cypango nor Cathay nor India. Despite this (as already mentioned), he wrote the famous document that was signed on board by 'all those that know how to', in which they recognised that they had reached Asia and accepted the penalty of having their tongue cut off and a fine of 10 000 maravedis for whoever contradicted him in the future. From then on, Columbus would have to live with this enormous statement, against everyone who raised their voice in opposition. This was to be his martyrdom until his dying day, because of his obedience and loyalty to his King.

On his third voyage (1499–1500), after being informed that Vasco da Gama had reached the real India, he limited himself to sailing round a small part of the island of Trinidade and the Punt de Hierro in Venezuela. And here he surprisingly claimed to have landed on 'terra firma', even though he did not have a shred of evidence to prove it. Why did he make such a claim? And what was the reason that he did not sail from the Canaries straight to the Antilles as usual, but this time veered southeast in the direction of the Portuguese-owned Cape Verde Archipelago? Because that continent of South America and the Windward Islands had already been discovered by the Portuguese, and he had already been informed of the position of that mainland.

Referring to that third voyage, Las Casas relates that: 'The Admiral has repeated that he wishes to go south [from the Island of Haiti] because he thinks he will find islands and lands . . . and wants to see what the intentions of King John of Portugal are when he says that there is terra firma to the south . . . that King John said would certainly find famous lands and things within his limits'. An astonishing statement that no-one wanted to interpret.

How can it be explained that the King of Portugal went as far as to supply secret information to a Genoese in the service of Spain, his rival on the seas, before 1500? This is further evidence that the American continent and the Windward Islands had already been discovered by the Portuguese. For Columbus to be in the royal confidence can be explained only by the fact he was the King's secret agent.

It is necessary to verify that only on his fourth voyage (1502–4), after Vasco da Gama had opened the seaway to India in 1498 and Pedro Álvares Cabral had 'officially' discovered Brazil in 1500, did Columbus decide to sail south from Cuba. He sailed along the coast of the present-day Honduras, Nicaragua, Costa Rica and Panama and then returned to Spain, where he died in disgrace two years later. He had only gone 'to see terra firma' that King John II had indicated to him in order to find favour in the eyes of the Spanish monarchs. But right to the end he persistently refused to sail farther west than Cuba in search of a passage to Japan, China and India. Banished from the Court of Spain, for whom he had opened the doors of an empire that covered almost half the world, Columbus lost everything except the honour of remaining faithful to his fatherland.

Notes and References

1. Pestana Júnior (1928).
2. Mendoça (1898); *O Padre Fernando de Oliveira e a sua obra Náutica* (Lisbon, 1892), Appendix; *O Livro da Fábrica das Naus* (sixteenth century). See also Fonseca, Quirino da (1978), Vol. 2.
3. As was clearly demonstrated at the Centre at the Monastery of Jerónimos, under the supervision of Admiral Teixeira da Mota, during the 17th European Exhibition of Art, Science and Culture, held in Lisbon in 1983.
4. The first Portuguese 'Junta' met in the house of Pedro de Alcáçova, among its members being the Jew Moisés, Father Fernão Álvares, Rodrigo das Pedras Negras and Don Diogo d'Ortiz de Villegas.
5. The round astrolabe was invented by Abuiçag Arzaquiel in the reign of Alphonse X (known as 'the Wise', grandfather of King Dennis of Portugal). It was later described and analysed by Rabiçag.
6. Corrêa, Jácome (1929).
7. Humboldt (1833).
8. *Historia Rerum Ubique Gestarum*, Chap. I.
9. Pedro Nunes, pioneer of 'experimental science' with his work *Tratado de Sphera*, refers to the calculations of the Sevillian Moor Gebre, mentioned above, and the study of Montenegro, which were translated in Nuremberg in the sixteenth century: 'however, I wrote the Geometria dos Triângulos Espaciais a long time [before] they sent us the books of Gebre and Montenegro from Germany'.
10. Greek astronomer, mathematician and geographer, Eratosthenes was born in 276 B.C. and died in 196. He was called by Ptolemy II to administer the famous Library of Alexandria. Among numerous works, he wrote *Geographies*, some fragments of which transcribed by various authors, including Pliny, Straban and Macrobius) have come down to us. The details mentioned here came from Macrobius (Book XX); see Claudien (1885); Letrone (1851).
11. Corrêa, Jácome (1929), p. 255.
12. If the Straits were 110 leagues (18 leagues of 5400 metres per degree) this would have made it 594 kilometres in length at that time and 580 kilometres today. But 440 Roman miles would have been 729 kilometres according to the Castilian mile, and 651 according to the Italian mile.
13. It is possible that Fra Mauro's planisphere did not result from a leak of information but was an order, as it had been offered to Prince Henry. The anonymous Portuguese letter existent merely explains the negligence of its owner who lost it, despite the royal order of 1475 to destroy all such documents.
14. Corrêa, Jácome (1929).
15. Cordeiro, Luciano (1876).

11   The Enigma of Columbus's Voyages

The chroniclers João de Barros and Fernão Lopes de Castenheda[1] narrated the successes of the fifteenth-century Discoveries, even though they wrote many years later. They had to base their chronicles on oral traditions and the rare documents that came their way. This was the reason that they made a number of chronological mistakes and omitted some details regarding the true identity of various navigators, either because they did not know them or they did not think them worth mentioning. In other cases, they were closer to the truth than some of the coeval documents.

In relation to the voyage of Bartolomeu Dias, for instance, on which he rounded the Cape of Good Hope, those chroniclers state that it had taken place in 1486. However, the letter in which King Manuel I grants privileges to Vasco da Gama for his services rendered in India, refers to the fact that all the African coast to the present-day Great Fish River, which flows into the Indian Ocean, was discovered in 1482 and measured 1885 leagues. Was the '2' in the King's letter a spelling mistake? As it was an official document, this would be odd. In his work *Esmeraldo de Situ Orbis*, Duarte Pacheco Pereira relates that Dias returned to Lisbon in 1488. Why would he have waited for six years to bring such news back to the King? Why was Vasco da Gama's fleet ready only in 1498? Or is the '2' really a badly written '7'? If this is the case, the date is then 1487. These successive hypothetical questions are put to point out that mistakes can appear in documents that are considered to be undoubtedly authentic.

What can we, therefore, say about the Columbian documentation, copies of scribes, many of them translated into Italian and then into Castilian, the translators having destroyed the originals. The document signed by King Manuel I is authentic and was based on another equally authentic one of his predecessor King John II. Yet neither is trustworthy as far as the text is concerned. This is because Diogo Cão had sailed from Lisbon in 1482 and on reaching Cabo Lobo (Santa Maria, Angola) after exploring the River Congo, thought he was near the long-desired *'Promontorium Prassum'* (Cape of Good Hope). On his return to Portugal in 1484, King John II showered him with rewards and sent him to round the African continent. Cão arrived at Cape Cross in 1485 and as the land veered south–south–west he was convinced that he had reached the southernmost point of the continent.

The Portuguese ambassador to the Vatican, Vasco Fernandes de Lucena, informed the Pope that the Portuguese navigators were but a few days away from the famous cape. Such was the disillusion of the King when he learned that there was still a long way to go to reach the cape that he bestowed no rewards whatsoever on Bartholomeu Dias when he claimed he had rounded Africa; he feared that he would be rewarding another navigator who had made the same mistake as Diogo Cão. So it seems that when King Manuel I referred to 1482 and the rounding of the Cape of Good Hope, he was confusing the date of the highly rewarded voyage of Diogo Cão with that of Bartolomeu Dias. Either the '2' is there by mistake or it is illegible.

The Marquis of Jácome Corrêa[2] wrote that: 'A sailor like Columbus could not have taken part in nautical events in Portugal in the 15th century, especially as a foreigner, without leaving some trace of his participation. Following the discovery of the Antilles, his identity is recognised by the chroniclers more as the son-in-law of Bartolomeu Perestrelo and the brother-in-law of Pero Corrês da Cunha rather than as a navigator'. If Columbus were Portuguese, he could have sailed the seas of Guinea with Bartolomeu Dias, serving as a boatswain or assistant pilot. In this case, his name would not have appeared in the chronicles as it was not the habit to identify the crew in the fifteenth century. After Columbus had been named Colón and considered Genoese, later historians would have been unable to find out his previous identity.

If Columbus really were Portuguese and King John II's secret agent, he could very well have sailed, as he claimed, with Bartolomeu Dias and João Infante; he could have been in Guinea and Mina and, also as he claimed, could have taken part on the Luso–Danish expedition in 1477.

In 1475, a Flemish ship piloted by a Castilian sailed down to the Slave Coast, with the aim of trading with the local natives. But the pilot hugged the coast to such an extent that it ran aground on some rocks. It was in this year that King Alphonse V forbade the presence of foreign ships in the waters south of Cape Verde. According to Duarte Pacheco Pereira, Pedro Gonçalves Neto found the abandoned vessel some time later and expropriated the cargo. The thirty-five Flemings and the Castilian pilot had been devoured by local cannibals. Christopher Columbus, therefore, could not have taken part on this last foreign expedition into Portugal's African waters.

In view of the possible competition from foreigners sailing along the coast in the wake of the Portuguese discoverers, King Alphonse V demanded that his captains and crews take a vow of secrecy (already normal in the Order of Christ) and dismissed all the foreigners that were in his service. On 15 August of the same year (1475) the Catholic Monarchs granted the shipowners of Seville 'privateering licences' so that they could sail down to Madeira and attack Portuguese shipping.

In reply to this provocation, King Alphonse V had cannon placed in the prow of his ships, just as João Gonçalves Zarco (the pioneer of naval artillery) had done some years before. He also managed to persuade three Popes – Martin V, Eugene IV and Nicholas V – to excommunicate anyone, whether foreigner or Portuguese, who sailed in those areas without his consent. In 1480, however, thirty five Castilian ships sailed beyond Guinea and attacked the factory of Mina, stealing the gold that some Portuguese merchants had acquired from the local natives. From then on, the few adventurers caught in those waters were immediately executed, as happened to Diogo de Lepe.

On 6 March 1480, Alphonse V signed a treaty with the Monarchs of Castile, which stipulated that any foreigner that dared go to Guinea or São Jorge da Mina would immediately be thrown into the sea without any form of trial. On 6 April, the King wrote to the authorities of Mina (*Livro dos místicos de Dom Afonso V*), informing them of the decision taken and ordering them to carry out his orders.

Columbus may have gone to Guinea twice and São Jorge da Mina once in order to, according to his own words, confirm the previous cosmological observations carried out by José Visinho (a Jew). In accordance with the chronological details of the Genoese thesis, this could have happened only after 1476. In this case, Columbus was not a foreigner in Portugal and had royal approval. In fact, various foreign families, many of them of Jewish descent, had obtained royal authorisation to settle in Azores and on Madeira as colonists. Other foreigners had even managed to get permission to sail with the Portuguese.

Fernando Colón wrote that 'Gonzalo Fernández de Oviedo mentions in his "Historia de las Indias" that the Admiral had a chart in his possession in which the Indies were described by someone who had discovered them earlier, which had happened in the following way: a Portuguese by the name of Vicent Dias, resident in Tavira, having passed Madeira on his way from Guinea to the island of Terceira [Azores], saw – or thought he had seen – an island, which he was sure was really land'. It is obvious that Fernando Colón, thinking that this island was one of the Antilles and wanting to attribute their discovery to his father, used the expression 'or thought he had seen'. His intentions are pardonable. Fornari's work, however, didn't hesitate to add that Vicente Dias had communicated this fact to a Genoese merchant by the name of Lucas de Cazzano, who fitted out a ship so that his brother

Francisco could discover it, with the permission of the King of Portugal. But, it seems, he discovered nothing. It is clear that such permission was never granted and is mentioned by Fornari so that he could fit a Genoese into the picture (see p. 183).

Among the foreigners who, before the 'rule of silence', were granted that permission were Luigi Cadamosto and Antonio de Nola, both Italians. Sailing with João Alvares in 1445, Cadamosto explored a part of the River Casamansa in Guinea. He set sail again in 1456, this time with the pilot Vicente Dias, who told Cadamosto that he had discovered an island some time before (in the archipelago of Cape Verde) and had named it St Christopher. Alphonse V was to donate this island four years later to his brother Fernando, Administrator (or Master) of the Order of Christ.

On arrival in Lisbon, Cadamosto claimed that he had departed from Lagos at the beginning of March and, sailing in the direction of the Canaries, had sighted Cape Branco

> having this cape in sight, we put out to sea; there was a violent south-westerly storm on the second night; so as not to turn back, we tacked west–north–west, if I am not mistaken, in order to trim the sails and get round the storm, for two nights and three days. We sighted land on the third day . . . We called the first island on which we disembarked Boa Vista, as it was the first one we saw; another, which seemed to be the largest of the four, we baptised Santiago [St James] because it was on the day of St Philip and James that we dropped anchor off this island.[3]

Professor F.G. Wieder, supported by Fontoura da Costa, attributed the discovery to Vicente Dias in 1445. The latter is a much more likely hypothesis:

1. setting sail from Cape Branco, it is impossible to sight the Island of Boa Vista in three days, even in a very fast modern sailing vessel;
2. Cadamosto alleges that, in order to get round the storm, he sailed for two nights and three days 'west–north–west, if I am not mistaken', thus manifesting an incredible uncertainty for a navigator, even more so since he said that the storm was south-westerly and he should have been able to calculate the direction he was sailing when he sighted the island;
3. the name Boa Vista was given only later (in the Sixteenth century); the letter with which the King donated the island to his brother, Prince Ferdinand, designated it as St Christopher and it is dated 3 December 1460;
4. the main argument of Professor Wieder is based on an inscription on

the *Bianco Map* of 1448, which depicts an authentic island to the south-west of Cape Verde on the lower part of the map:

ixola otenticha
X elonge a ponente /500//mia
(authentic island X at the west/500 Roman miles)

It is natural to assume that Cadamosto, knowing that islands are the peaks of submerged mountains and that up to that time the ones that had been discovered were the continuation of a cape or in a group, decided to try his luck at finding some new islands in the area that his pilot Vicente Dias had claimed to have already discovered. He did so six years after Dias. By using the adjective 'authentic', Bianco wished to emphasise the existence of the island three years after its discovery by Dias – i.e., three years before Cadamosto's alleged discovery.

A collection of Travels, *Paesi Nuovamente Retrovati et Novo Mondo da Alberico Vesputio Florentino intitulato* was published in Vicenza in 1507. It was widely read and was printed in Latin and other languages. It must be recalled that Portuguese chroniclers of the sixteenth century, who wrote the history of the Discoveries of the second half of the previous century, owing to the policy of secrecy in force in the fifteenth century, had access only to the reports until then made public, very often due to Italians who wrote for the glory of their homeland. The Portuguese limited themselves to copying these publications, describing the 'discovery' of the island of Boa Vista by Cadamosto, who had until then not gone beyond exploring the Guinea coast and the entrance to some rivers. The most amazing text of all is the one by J.B. Ramusio, *Delle navigazione e viage, etc.*, a part of the above-mentioned Travels collection. Ramusio claimed that Cadamosto had met a Genoese, Antonieto Usodomar, making voyages of discovery in the Atlantic in the company of 'squires of Prince Henry'. (There is no record of such in the documents of the time, which registered the rare 'authorisations granted to foreigners'.) And, it was claimed, it was with this same 'gentleman' Usodomar that Cadamosto had discovered four islands of the Cape Verde Archipelago in 1456.

It is interesting to note that Columbus was the only writer who mentioned the fact that Vicente Dias had been born at Tavira (Algarve) and attributed the discovery of the island of St. Christopher to Dias. Columbus visited the archipelago in 1498, but never made the slightest reference to the alleged discoveries of Cadamosto, Usodomar and de Nola – which would have been very strange on the part of a Genoese. It must also be remembered that while those two Italians were authorised to sail with Portuguese pilots, they were completely unaware of the secret geographic measurements used by the Portuguese.

Diogo Gomes, 'who possessed a navigational quadrant', also wrote the following complaint in relation to the discovery of the Island of Santiago:

> From the Port of Zaia, Antonio de Nola and I were on the second day of our voyage to Portugal, when we saw some islands. As my caravel carried more sail, I was the first to arrive. On seeing some white sand and it seeming to be a good port, we both dropped anchor. I told Antonio that I wanted to be the first person to set foot on the island, and this I did. There was no sign of life. We called the island Santiago [St James] and it still bears that name today.
>
> From there we passed by the Canaries and went on to Madeira. As I preferred to sail to mainland Portugal by way of the Azores, even though it mean tacking against a head wind, while Antonio de Nola stayed in Madeira to wait for more favourable winds, he arrived in Portugal ahead of me, and asked the king for the Captaincy of the Island of Santiago, which I had discovered; and the King granted his request and he kept it till his death. And after a lot of hard work I arrived in Lisbon.[4]

This complaint of Diogo Gomes, presented in his *Relações*[5] was related to the cosmographer Martin Behaim, who wrote it down in Latin. It was later published by Valentim Fernandes (see p. 246). The historian Jaime Cortesão[6], rejected all the interpretations, disbelieving Diogo Gomes's complaint. However, the discovery of the island of Santiago by Diogo Gomes effectively undermines Cadamosto's claims.[7] Taviani goes so far as to claim that Cadamosto and Nola had come to Portugal in order to teach the Portuguese how to navigate in the Atlantic Ocean.

The other islands of the Cape Verde Archipelago (the so-called 'western' isles) were discovered by Diogo Afonso (de Aguiar?), squire of Prince Fernando, brother of King Alphonse V, Duke of Beja and Master of the Order of Christ. The presence of these 'authorised foreigners' on Portuguese ships was a cause of concern to the future King John II who, not wishing to make the same mistake, decided, like his predecessor, to ban them from the ocean of Discovery. There was one exception, however – Fernão Dulmo (or d'Ulm). In order to find an explanation for this, we must go back to the reign of Alphonse V and distinguish the foreigners that would colonise the Azores and those who simply signed up for service on Portuguese ships in order to get rich in the African trade.

The Fleming Jácome of Bruges (married in Orense, Galicia, to Antonia Dias Darce) came to Portugal in 1431 and lived in Oporto until King Alphonse V authorised him to colonise the (until then) uninhabited Island of Jesus Cristo (Terceira) in the Azores in 1450. On his own merit, he was appointed as Donatary-Captain of the island, being aided in its administration by the navigator Diogo Teive. Jácome of Bruges went to Lisbon in 1463 and mysteriously disappeared between then and 1466. He left a daughter, Helena Gonçalves, and two sons, both sailors and known merely as Gabriel and Pedro Gonçalves. None of them used either their paternal name, Bruges, or their maternal name, Darce, choosing another surname of origin for reasons unknown.

Before this, and on the initiative of Jácome of Bruges, Joz van Hurtere (son of Bailio de Wynendale) and a large number of Flemings had colonised the Azorean islands of Faial and Pico and Willem van der Haagen settled on the Island of São Jorge. In 1453, Fernão Dulmo, whose name was originally van Olmen and whose father, born in Ulm, Germany, had settled some years before on the Island of Madeira, took a lot of colonists to the Azores.[8] Although the Portuguese called them foreigners and nicknamed their children 'Dutch' or 'French', etc. – they considered these children to be Portuguese as they had spent their lives in Portugal and spoke the language fluently.

It is convenient to refer to another historical confusion here: the chronicler João de Barros mentions that a Frenchman called João Baptiste spent some time in Cape Verde and had been granted the Captaincy of the Island of Maio, when that of Santiago was wrongfully given to Nola. But everything points to a mistake on the part of the writer due to his being misinformed orally, as there is no document that confirms such a donation and the first donataries (who occupied the island while it was still deserted) and the chroniclers of the time had no knowledge of it either. It was this mistake on the part of João de Barros is his *Décadas da Ásia*, written about a hundred years after the event, which Luis de Ulloa used as base to invent Jean Coullon, French privateer, and also Catalan named Juan Bautista, later called Cristóbal Colón (see p. 102).

Many Portuguese navigators received captaincies, donations and rewards for their discoveries in the archipelagos discovered, among whom were Afonso Gonçalves Baldaia, Álvaro Martins Homem, João Álvares Corte-Real, João Afonso do Estreito and others. It was natural, therefore, that the colonists of the islands of Madeira, Porto Santo and the Azores, even though of foreign descent, wished to sail with the Portuguese. Despite the fact that they were already considered to be Portuguese, they sailed to the west only after the royal decree of 1475 and then always under the command of an 'old Portuguese'. The route south was closed to them.

If the future Admiral of the Indies had been a foreigner in the eyes of King John II, he would have been able to sail in the service of Portugal only if the King authorised it and then only under the orders of a Portuguese, as had happened to Fernão Dulmo. If Columbus were a foreigner in the service of Castile, he would be able to sail to those waters 100 leagues to the west of Cape Verde and to the north of the parallel of the Canaries, without the Portuguese King being able to raise any objections. He would be forbidden only to sail south. If he were Genoese or Catalan, why did he not seek support in Italy or Catalonia, especially in the former where Ptolemy's theory was more widely known? Being Portuguese, his duty was to fulfill the mission his King had entrusted to him.

The financial investment for the undertaking he proposed was nothing out of the ordinary. The war with Granada and other political problems, plus the backwardness of the Castilians in the field of cosmography and mathematical navigation were, in fact, the main hindrance for an attempt to reach India by the west. If Columbus were Genoese, why did he not sail in the service of Genoa and its flourishing commerce? If he were Catalan, champion of the rebels in the struggle against Castile according to the defenders of the Catalan thesis, why did he not seek backing from the prosperous Catalonian merchants who could count on the support of their own navigators and the French of Anjou?

But no. Columbus stubbornly insisted on sailing in the service of the Catholic Monarchs. It was not very difficult to get the money to fit out three ships; what was difficult was to persuade the Castilians to accept the idea he proposed. In order to persuade them, he resorted to the influence of a Portuguese friar and a strategy that always produces results with ambitious people – he led the Catholic Monarchs to understand that he was disillusioned with waiting for seven years and was consequently going to try and sell his ideas to Spain's eternal rivals, England and France. It was almost blackmail. On referring to the choice the Castilian Monarchs had made, in a letter written to King Ferdinand in May 1505, Columbus confirmed his lie of 1489: 'I received letters from the Kings of England and France, which Queen Isabella read and communicated to her council'. This letter was written shortly after the Queen's death and at a time when all of her former counsellors had also died and could not, therefore, deny Columbus's claim.

If these letters, which were in reply to the ones his brother Bartolomeu had handed to the two Kings in question, were refusals to his proposal, Columbus would not have shown them to King Ferdinand. If they were acceptances, his wish to sail in the service of Castile would be an undisputed fact. Yet he did not show them to the King. Had he lost them? Did he set so little a value on royal missives at that time? It must be noted that Columbus did not bother to go to England and France personally. Only once did he leave Spain, in 1488, and that was to pay one visit to Lisbon to speak to King John II – who referred to him as 'our special friend'. To write to those two Kings without duly identifying himself, without testimonials regarding his nautical knowledge or proving that he had already made some discoveries, without illustrious signatures to back up his proposals would have been a sheer waste of time. And Columbus never referred to his studies and never signed his name.

After the Treaty of Toledo (see pp. 66–7), the King of Portugal was not interested in sending more expeditions to the west. His only interest was to send the Castilians on a wild goose chase in a west–south–west direction in search of the Far East that he knew was much further than anybody thought – if there were even a passage. His aim was to force the Catholic Monarchs to sign another treaty (of Tordesillas) that would assure him of a free run to the Orient. He would get possession of Brazil, thus cutting off a hypothetical route to the Orient, and Newfoundland, thus blocking a northern route. Columbus's mission was to obstruct the central route. King John II was aware that the Castilians would turn to the sea as soon as they had conquered Granada. And he wanted them to sail westward. The widely-propagated proposal of Columbus to the Portuguese King was nothing more than a bait to whet the Castilian appetite. The reason for the Portuguese refusal was based on the excessive ambition of the proposer and not on the impracticability of the undertaking, which could have raised doubts in the minds of the Castilians.

It becomes obvious, therefore, that the well-rehearsed proposal of Columbus was a ruse. When Columbus reached the Antilles, as already mentioned, he shuttled about among the islands, boasting that he had arrived in Japan. But he never made the slightest attempt to sail farther west, even after the Portuguese had discovered the sea route to India. And he knew that there was 'terra firma' to the south.

In 1483, after easily crushing the Aragonese revolt against the Inquisitor Torquemada, the Catholic Monarchs were free to go ahead with the conquest of Granada, where they had already taken Alhama and caused the deposition of King Ali. This Muslim kingdom was also split by the dispute between Boabdil (Abdallah) and Alzagal, who became king of an independent Malaga the following year. The seven years that the Portuguese king's secret agent took to persuade the Castilian Monarchs were certainly times of uncertainty for both King John and Columbus. But at the end of those seven years, the Portuguese had reached the present-day Cape Cross in south-west Africa and Columbus had been heard by the Castilian Junta, on the eve of obtaining Queen Isabella's agreement to take an expedition west in search of India. If Columbus were indeed Colón-Zarco, a Portuguese, a secret agent in the service of his King, the puzzle begins to take shape.

Notes and References

1. King John II died in 1495 and João de Barros was born four years later. He started writing his *Décadas da Ásia* (1932; 1945–6) in 1531. He wrote another book during a plague and gave it a Greek title, *Rópica Pneuma* (Spiritual Market); it was condemned by the Inquisition in 1581. Fernão Lopes de Castanheda died in 1559 and his *Chronicles* were published between 1551 and 1601.
2. Corrêa, Jácome (1929).
3. Cortesão (1931).
4. Vasconcelos, Frazão, de (1956), pp. 65–73.
5. Pereira, Gabriel (1900); Godinho (n.d.), Vol. 1, pp. 69–115.
6. Cortesão, Jaime (1931), Vol. I, pp. 360–84.
7. Lima (1946).
8. Menezes, Manuel de (1949).

12 The Problem of the Legendary Thule

In his work *Historia*, Fernando Colón copied the following narration from his father's documents:

> In February 1477, I sailed 100 leagues beyond the Island of Thule; the southern part of this island lies at 73 degrees North and not at 63 degrees as some people think; it is also farther west than the western line of Ptolemy; the island is as big as England and the English, mainly from Bristol, trade there. When I was there, the sea was not covered with ice. But the tides were so great, that there were places where the sea rose and fell as much as twenty-five fathoms a day. It is indeed the Thule that Ptolemy spoke about, it is where he indicated, it cannot be any other except the one they call Frisland today.

This text is a source of perplexity among investigators, as for some (like the Marquis of Jácome Corrêa) Thule is Greenland, while for others (like Pestana Júnior) it is Iceland. Geographers are unanimous in condemning Columbus for his mistake in placing 'Thule-Iceland' at 73 and not at 63 degrees North at its southern limit. They also emphasise Columbus's error when he says he navigated 100 leagues to the west of the meridian of Iceland (20 degrees West), as this would have taken him into the interior of Greenland. They also dispute the fact that he sailed north, as the island measures 2 degrees from south to north and it would be impossible to sail 100 leagues (which would have taken him beyond 75 degrees North) in February, because of the Arctic ice.

Those that think that 'Frisland' was Greenland launch Columbus westward along the 60th parallel to the coast of Labrador, which was 'officially' discovered in 1495 by the Portuguese Pedro de Barcelos and João Fernandes Lavrador, the latter, as we know from p. 63, giving it its name, thus making him the discoverer of the continent of North America five years before Columbus found the Venezuelan South American coast. They contest, however, the discrepant calculation that places the southern limit of this 'other Thule', as he called it, 13 degrees further north (73 as against the true 60 degrees) when it should be only 3 degrees (Iceland – 63 degrees N). Why would Columbus have miscorrected Ptolemy?

Whoever reads Fernando Colón's text carefully will see that his father did not correct Ptolemy in respect to latitude, but corrected the mistake of 'some people'. Immediately after, he gives us to understand that Ptolemy is not included in those 'others'. So everything points to the fact that Columbus agreed with Ptolemy regarding latitude, and corrects only his longitude, moving it 'farther west'. Yet if he agreed with the cosmographer regarding latitude, why did he change the 63 to 73 degrees? This is a question that will be dealt with later in the book.

Historians pore over the Portuguese–Catalan *mappa mundi* on which Columbus refers to the fact that he made Greenland in twenty-five days – obviously departing from Denmark – and shake their heads in rebuke as a degree in those days measured 8⅓ modern leagues (30 miles today) and that would correspond to 2480 km as the crow flies,[1] when the navigator had calculated approximately 2655 km, without taking unfavourable winds and currents into consideration.

Let us analyse Columbus's text without any preconceived ideas. Let us begin by the fact that he was guided not by the Portuguese–Catalan *mappa mundi* but by Ptolemy's map (he had already corrected the longitude) which included the legendary Island of Thule. Leaving the latitude degrees for another study, let us try and calculate the distance he sailed. When Columbus said '100 leagues beyond the Island of Thule', coming from the Danish coast, he does not mention a change of course and does not specify whether he is travelling north, west or south. It is therefore logical to conclude that he continued in the same direction – i.e., north-west. When he referred to the 'southern part' of the island, the latitude of which he apparently did not correct, it is also clear that he was talking about Thule and not the land that he was to find '100 leagues beyond'.

As he also states that it was 'as big as England', it means that he either sailed round the island or a good geographer had told him. And as good geographers were non-existent at the time, as can be seen by the quality of the cartography, the island must have been Iceland. Firstly because this island is circumnavigable and Greenland is not, unless one penetrates into the Arctic, which was then impossible. Secondly because English merchants traded there, which means there were things to trade and people to trade with. At that time Greenland had no mercantile motivation. Despite this, the dispute about whether Columbus was referring to Iceland or Greenland goes on. Some critics say that Iceland is much smaller than England. Columbus clearly meant England itself and not 'Great Britain', which would make his mistake less serious.

Ireland had already been known as 'Antille' and the 'Island of Seven Cities', and Bartolomeu Pareto's map of 1455 that showed Antille to the west of the Azores caused great confusion. Ireland was distinctly separated from England by the Irish Sea, even though it was governed by the English crown. As Portugal and England had been close allies since the end of the fourteenth century, Columbus must have known that Scotland was then an independent kingdom. Although the political boundary between England and Scotland extended along a south-west–north-east line from Solway Firth to Berwick, the generally accepted geographical frontier was that which separated the ancient Caledonia from England, made up of the enormous estuaries of the Rivers Clyde and the Firth of Forth and a narrow neck of land between the two. The geographers of the fifteenth – and even of the second half of the sixteenth – century showed England and Scotland to be completely separated by a channel (in fact non-existent) running east–west. Was there, in remote times, a channel that has since silted up? Continental cartographers, copying their British counterparts, always drew the channel stretching from sea to sea.

Not having circumnavigated England, therefore, Columbus limited himself to expressing his ideas in accordance with what he had learned from contemporary cartographers: Iceland was about the same size as England. If Columbus had wished to refer to Greenland, he would not have said 'as big as England', but 'bigger than the three British Isles together'. Even if he could not have sailed round it, sailing along a part of the coast would have been enough. On occupying present-day England, the Romans called the whole of Great Britain 'Britannia', but the name disappeared during the Germanic invasions and the expression 'Great Britain came into use only in the second half of the seventeenth century. Columbus, therefore, did not err in distinguishing England from Scotland and Ireland.

Following the merger of the English and Scottish crowns, the channel disappeared from the maps. Regarding the fact that Columbus referred to 'floating ice', which has surprised historians, the phenomenon has been confirmed by Finn Magnussen, whose studies have proved that the winter of 1477 was extremely mild.[2] As far as the latitude of the Island of Thule on Ptolemy's map is concerned, Columbus rectified the calculations of the cosmographer, who had placed Iceland 'within the limits of his west', while it was actually further to the west.

It is therefore, the longitude that Columbus speaks about when he says: '[If] the Thule that Ptolemy spoke about is where he indicated, it cannot be any other except the one they they call Frisland today'.

Geographers who have studied Columbus, alarmed by his 73 degrees North for the southernmost point of Thule were not careful enough to notice that he was referring to longitude. Although Columbus corrected Ptolemy's error of longitude, the error in latitude remained uncorrected. This could well be a mistake in Fernando Colón's copy, recopied by Las Casas, translated into Italian and then into Spanish. Or, as happened to King Manuel I's letter in which he mentions the rounding of the Cape of Good Hope and wrote '1482' instead of '1487',[3] someone could have inadvertently changed Columbus's '73' for '63'.

But is Fernando Colón not accused of exceeding himself when he praises his father to the skies as a genius? Why should the blame be laid on the shoulders of Las Casas and the translators? If modern geographers point out the mistake, could not Fernando Colón himself have equally misinterpreted his father's letter? Looking at Ptolemy's map and seeing that Columbus had seemingly corrected it, Fernando could have changed the number with the best of intentions. After all, he was an experienced seaman. It was supposed that the legendary Thule, 8 degrees West/73 degrees North on Ptolemy's map, was more or less situated where the small island of Jan Mayen (19 degrees West/73 degrees North) is to be found. Columbus corrected the longitude to approximately 19 degrees West. Would he, in fact, have committed the mistake of placing it at 63 degrees North?[4] If we wish to accept the possibility of an accidental change of numbers or admit that Fernando Colón wished to correct a slip of his father's, Columbus will have been proved to be extremely precise in his Atlantic navigation: everything points to the fact that an error was made in one of the various phases of transcribing and translating the original and subsequent texts.

Everything then falls into place. Columbus stated, and correctly, that the Island of Thule must have been Frisland, the present Iceland – already inhabited, visited by English merchants and as big as England. Columbus would never have made a mistake to the extent of 10 degrees in latitude. One must not forget that the Iberian Peninsula measures 8 degrees in latitude and France 10 degrees. This area of the North Atlantic was frequently crossed by English and Scandinavian ships in the fifteenth century and an experienced sailor like Columbus would never have made such a mistake. But bad scribes and translators could have done so.

What about those 100 leagues? For his critics it was a grave miscalculation. Columbus referred to the southernmost latitude of Iceland, but did not say that he had sailed 100 leagues from that particular point. Seeing that he mentioned the size of the island, he must have at least partially sailed round it. One thing is certain – his original course was north-west. He was not sailing in quest of India by the west. He had no intention of sailing towards the setting sun, but to the already-known Frisland. If he

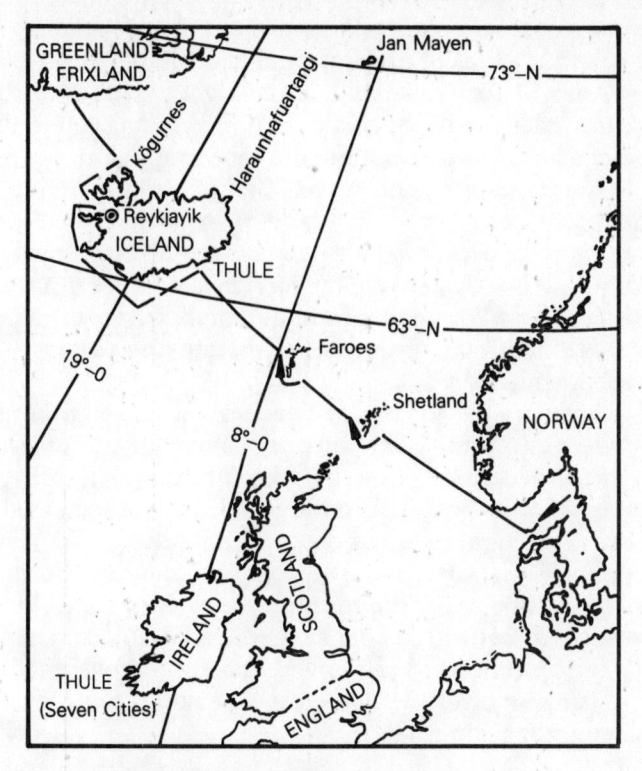

12.1 Columbus in Iceland and Greenland (1477).
Columbus's voyage to 'Frisland/Thule/Iceland' and 'Frixland/Vinland/Greenland'
with the Luso–Danish expedition.

set sail from Cape Reykjannes to the north-west, he headed for
Kangerdlussuak in Greenland and sailed exactly 100 leagues (i.e., 540
km). If he set sail from Cape Horn (north) to the north-west, he made
the coast of Greenland after covering 270 km (i.e., exactly 50 Portuguese
leagues of 5400 metres, the 'secret league' of the 'School of Sagres').[5] In
this case, he would have multiplied the longitudinal distance by 2, as he
did when he made his own calculations of his first voyage to the Antilles
– to the surprise of his Castilian collaborators and modern historians.

On either of the two north-western courses, whether sailing from
Reykjavik or from Cape Horn (north), he would have covered the 100
leagues registered. But as he made his calculations in secret Portuguese
leagues, I am inclined to believe that he sailed the 50 leagues and
multiplied them by 2. In this case, Columbus would have set sail from
Denmark and taken a north-west course for Shetlands and Faroes (see
Figure 12.1). Off Orsefa Jokull (Ingólfshövdi) in Iceland, he sailed naturally to the west–north–west, rounding the enormous promontory of
Reykjannes and on to the city of Reykjavik.

Here was where Columbus mentioned the English merchants, and where he would have taken on fresh food and water. But he did not sail north-west straight for Greenland, thus covering the 100 leagues in foreign measurements. He obviously continued sailing round the island by crossing the Faxa Fiord, rounding the vast promontories of Snaefellsjokull and Skor/Breidhavik and on to Kogurnes and Hofn/N. This is a much more probable course, as it would have given him an idea of the size of the island and exactly ties up with the 50 Portuguese leagues multiplied by 2. It is only because the foreign geographers have ignored the existence of these Portuguese leagues that they have been perplexed by Columbus's apparent errors in calculating the distance between Iceland and Greenland.

If one wishes not only to criticise but also find the truth, one cannot allude to twentieth-century measurements, nor to the ones used by non-Portuguese geographers of the fifteenth and sixteenth centuries and which are known today. The critic of Columbus must penetrate into the spirit of the Portuguese Juntas of Mathematicians and the navigators of the Order of Christ; he will have to get to know the Portuguese of those days better, and study the scientific methods of the Golden Age of the Discoveries. Above all, he will have to read the documents closely, and not make a preconceived judgement that the 'discoverer of America' could have made an error of 10 degrees (the north–south extension of France) when sailing the Atlantic.

Notes and References

1. Kioge Bay – Greenland: 65°N/40°W; Fano Bay – Denmark: 55°N/8°80 E; Skagerrak-point at 57°30 N/8°00 E.
2. Finn Magnussen carried out these investigations from 1832–41 for the Royal Society of Northern Antiquarians.
3. See pp. 133–4.
4. All geographical coordinates refer to the southern coast of the islands.
5. See p. 128.

13 More Theses on the Discovery of America

In 1473, João Vaz Corte-Real and Álvaro Martins Homem received the donatories of Angra and Vila da Praia on the Island of Terceira in the Azores for having discovered the Island of Cod (Newfoundland) (i.e., south of Greenland, on the American Continent, Canada)[1] (see Figure 13.1). If Columbus lied and did not travel to Frisland–Iceland and Frisland–Greenland in 1477, he could have based his story on information collected in Lisbon about the voyage of those two Azoreans. If he were a foreigner, however, who would have risked giving him that information? Having taken the oath of secrecy of the Order of Christ, the navigators certainly would not have been forthcoming with geographical references, and the ordinary seamen would not have possessed such cosmo–geographic knowledge. Even if he were Portuguese, Columbus would have had access only to those secrets if he were in the confidence of the Master of the Order of Christ. He would have had to have been both Portuguese and highly respected.

Lofus Larsen[2] has narrated the epic of the Viking mariners and their 'discovery' of America. He considers the date of Columbus's voyage in 1477 to be wrong, and is of the opinion that he took part on the expedition of Corte-Real and Homem in 1473. Unfortunately, no itineraries or descriptions of this voyage have come down to us owing to the policy of secrecy. We know only that Corte-Real and Martins Homem set sail from the Azores in 1473. If we examine the expeditions of Diogo de Teive, Afonso Sanches and João Vaz Corte-Real, as well as the later ones of Álvaro Fernandes, Gaspar Corte-Real, Barcelos and Lavrador, we shall see that the first-named set sail from Madeira and all the others from the Azores. The westward-bound navigators headed south-west or north-west when setting sail from these islands so as to avoid the Sargasso Sea, whose enormous, legendary algae supposedly trapped ships. It is logical to conclude, therefore, that the 1473 expedition of Martins Homem and Corte-Real had taken a direct north-west course and had not sighted Iceland.

13.1 The Corte-Real inscriptions.
Inscriptions that Edmund Delabarre identified as being made by the Portuguese Miguel Corte-Real, who went to America in 1511 in search of his brother Gaspar, who had disappeared in 1500 after having received rewards from King John II for a previous discovery that the 'policy of secrecy' erased from history. They were both sons of the Azorean navigator João Vaz Corte-Real, cousin of Isabel Moniz Perestrelo, mother-in-law of Colón-Zarco. The set-square can be seen below the cross on the right.

(*Dr Luciano da Silva*)

Lofus Larsen confused this voyage with the Luso–Danish expedition of 1477, which took place in the reign of Alphonse V in agreement with King Christian I of Denmark and Norway. The Luso–Danish fleet was fitted out in 1476 and must have reached Iceland and Greenland the following year. The following inscription can be read on the Mercatur and Frisius Globe of 1537: '*Quid populi ad quos Joannes Scolvus Danus pervenit circa annum 1476*'.

As already mentioned, the crews were not made up exclusively of Portuguese and Danes. Francisco López de Gómara[3] says: 'There were also Norwegians, such as the pilot Jan Skolp, and Englishmen like Sebastian Cabot in the crews'. As we know, this information became the basis of an improbable Catalonian thesis: that Joannes Scolvus was Cristóbal Colón (see p. 110). I think we should opt for Skolp and not become embroiled in a series of names that prove nothing. Would Columbus have changed his name to Scolvus? If he were Portuguese or Danish he would not have been named, as only those two pilots, Skolp and Cabot were mentioned, precisely to emphasise the presence of foreigners on the expedition.

Lofus Larsen presumed that there was another Luso–Danish expedition during Alfonse V's reign. It seems unlikely, as there is no knowledge of any other.

With regard to the voyage of the Azoreans João Corte-Real and Martins Homem in 1473, on which they discovered the eastern coast of

North America, it seems highly improbable that they first made for Denmark before going on to Iceland, Greenland and finally Newfoundland. Like all other expeditions of discovery to the western Atlantic, they would have headed westward from the Azores. In order to formulate his hypothesis that the Danes and Norwegians had also discovered Newfoundland on that Luso–Danish expedition, Larsen was obliged to amend not only the traditional route of the Portuguese discoverers but the date of the voyage as well. And the Portuguese had not accepted the invitation of the Danish King in order to learn how to sail; they went so that they could carry out some practical corrections of possible errors of cosmography.

Besides this, Columbus could not have been more than twenty-two years old in 1473, and if he were Genoese he would still have been on the island of Madeira working as a travelling salesman in the pay of his fellow citizens. If he were Portuguese and truthfully claimed that he had been to Greenland in 1477, it could only have been on the Luso–Danish expedition. He did not have his own ship on which he could sail around the seas correcting errors of Ptolemy, and if he were the privateer Jean Coullon he would not have anything or anybody to attack or rob in Greenland. Columbus's name is not mentioned, along with the rest of the Portuguese participants, and I cannot therefore go along with Lofus Larsen's thesis.

In fact, there is no evidence of any other similar expedition, nor of any non-Portuguese voyages. If there had been, they would have been proudly announced immediately. Only the Order of Christ imposed absolute secrecy on its navigators: two seamen who had sold information to a third party were killed by agents of King John II.

At the Congress of Nancy (1875), Madier de Manjou, analysing an announcement of Gravier de Rohan, came to the conclusion that 'the inscriptions on the Dighton rock [see Figure 13.2 and p. 152], which de Rohan had attributed to Norman pirates, were Portuguese words with autochtonous engravings superimposed on them'. In fact, what can be read is:

MIGUEL
CORTEREAL V DEI
HIC DUX IND
AD 1511

('*Miguel Corte Real voluntate Dei hic
dux indiarum anno Domini 1511*)

There is a full-size replica in front of the Lisbon Navy Museum, a gift of the Luso-Americans of Massachusetts. According to the study of the

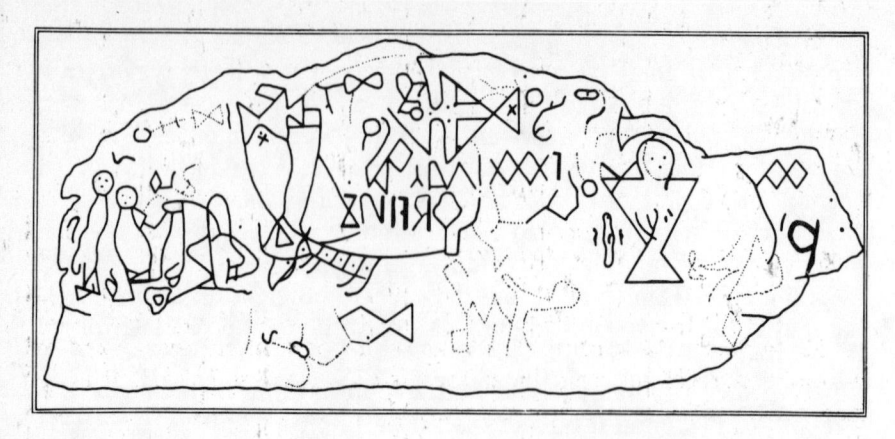

13.2 The Dighton Rock inscriptions.
Inscriptions on the Dighton Rock, which Rafne ascribed to the Vikings, on the assumption that they had discovered Labrador in the ninth century. They have nothing in common with Viking runic inscriptions.

(Drawn by the author from a drawing by Dr Luciano da Silva)

Luso-American, Dr Manuel Luciano da Silva,[4] the first graphological description of the Dighton rock was made by John Danford in 1680, followed by the study of James Winthrop in 1788. It was only in 1913 that Professor Edmund Burke Delabarre began to take an interest in the rock and five years later managed to identify the name of Corte-Real, the crosses of the Order of Christ, the Portuguese coat-of-arms, a triangle and the date. In 1951, another Luso-American, Joseph Dâmaso Fragoso, made another study and defended the thesis that the inscriptions were not Viking. The triangle, like the square and the compass, was one of the several symbols of the Templars that were adopted by the Order of Christ. They signified the cosmographer, the mathematician and the architect and can be seen in the emblems of Queen Leonor (1458–1525) and the printer Valentim Fernandes, as well as in the decoration of the Monastery of Christ in Tomar.

The work of Luciano da Silva is above all noteworthy for the interesting study he made on the Nautical Chart of Zuana Pizzigano (parchment nr 25924 of the collection of rare manuscripts of Sir Thomas Philips of London, studied by the Portuguese historian Professor Armando Cortesão from 1945 to 1950), which is today in the James Ford Bell collection of the University of Minnesota, USA. In *The True Antilles: Newfoundland and Nova Scotia*,[5] Luciano da Silva reproduced the four islands that Pizzigano placed in the Atlantic: Antilia, Satanazes, Saya and Ymana. He emphasised the fact that the cartographer spelt *'Antilia'* in Portuguese and not in Italian, *'Antiglia'*, and clearly demonstrated that these islands were not in the area of the Central American Antilles, between

10 and 23 degrees North, but further north and that, due to the imprecision of the fourteenth-century geographical coordinates, he placed them in a zone between 40 and 50 degrees North. When pairing up Satanazes with Saya and Antilia with Ymana, Luciano da Silva identifies them with Newfoundland/Avalon and Nova Scotia/Prince Edward Island and points out the striking similarity of the coastline and the angles of inclination (see Figure 13.3).

As the cartographer Pizzigano was never a navigator and no Italians sailed around that part of the world in the fifteenth century, we are bound to conclude that his *Nautical Map* (the date of which must be '22 August 1474' and not '1424' as seems be written) was based on information received from loose-tongued Portuguese sailors. It is, in fact, a fascinating comparative analysis, which proves that Pizzigano placed the Antilles off the eastern seaboard of Canada and not in the Caribbean. It also proves that the Portuguese had discovered America before 1474, i.e., eighteen years before Columbus discovered it 'officially': everything points to the fact the it was not the first time the Portuguese had sailed those waters.

The thesis of Luciano da Silva is also outstanding for other extremely valid discoveries, above all in identifying the eight-arched tower in Newport as a construction of the Templars (see Plate 2), being a reproduction of the octogonal altar in the Monastery of Christ in Tomar, the dome of which is replete with the 'initiation' symbols of the knights of the Temple.

In the original Portuguese edition of this book, I dismantled the hypotheses that the Vikings (a word that means 'men of the bays') discovered North America, basing my argument on the fact that they did not have lateen-sailed ships and so could not have sailed against unfavourable winds. They were excellent oarsman, but in ships that had no keel and with oarlocks only two or three feet above the waterline they would certainly have sunk in the heavy seas they would have encountered between Greenland and Labrador (see Figure 13.4). The Viking theses have been based on the sagas and never on documentary proof, except for a few maps that are obviously incorrect. A Vatican document shows that Vinland – which the Scandinavians claimed was the present-day New York – was a fishermen's village in Greenland (see Figures 13.5 and 13.6).

We do know that João Vaz Corte-Real received rewards from King Alphonse V for having discovered land in the Atlantic and also that the Portuguese sailed in a wide arc on their return voyages from Guinea and on round trips from the Azores in the hope of making new discoveries. This fact leads us to Andrea Bianco's map of 1436, which shows the 'Baga Sea' (*'questo xe mar de Baga'*), later known as the Sargasso Sea, between the Azores and the American continent. As Bianco had never

154 EMPIRICAL AND SCIENTIFIC NAVIGATION

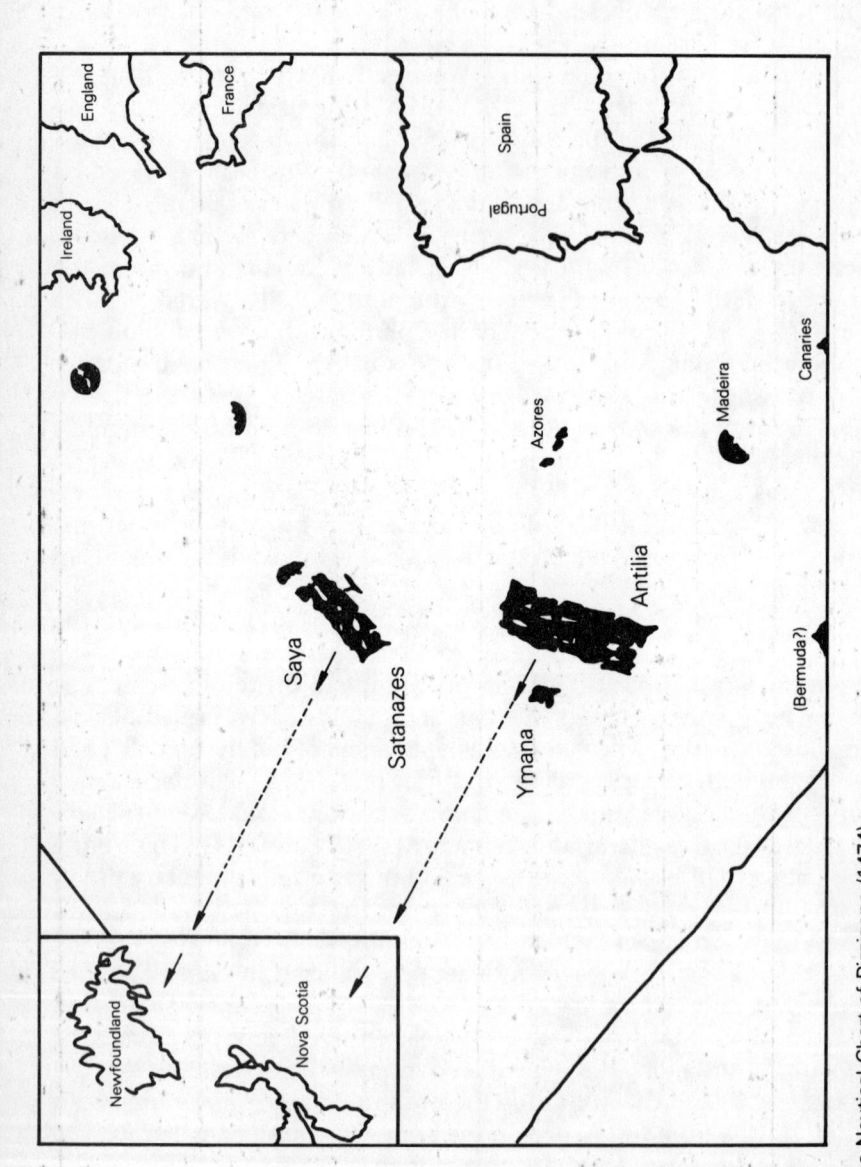

13.3 The Nautical Chart of Pizzigano (1474). Showing the comparison of Satanazes/Saya with Newfoundland/Avalon and Antilia/Ymana with Nova Scotia/Prince Edward Island.

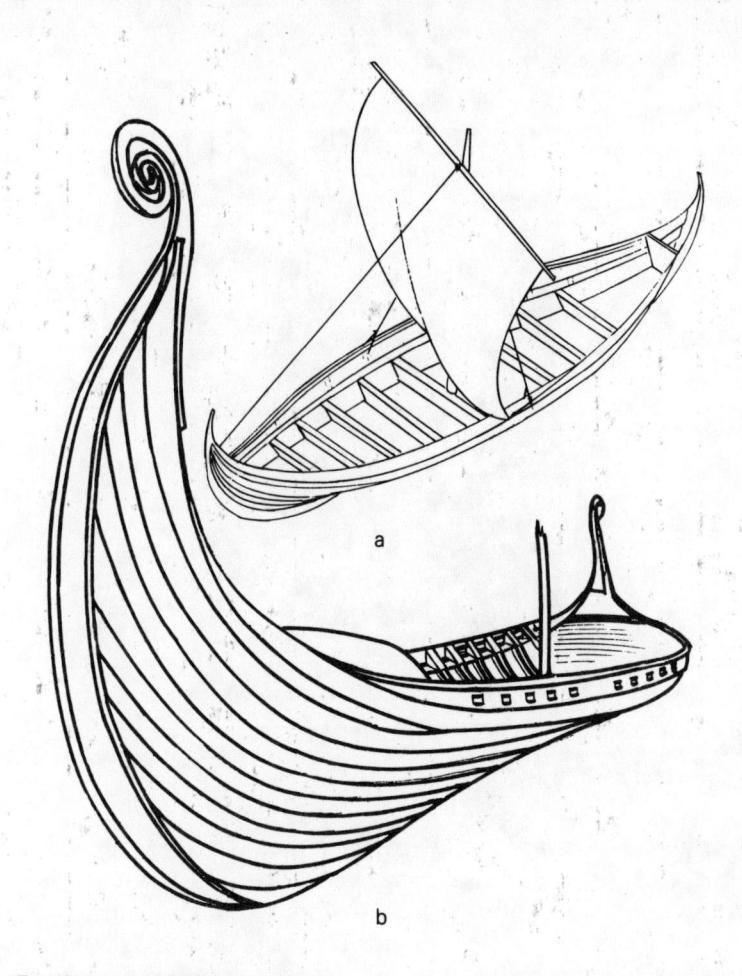

13.4 Twelfth-century Viking warships.
   a The twelfth-century Viking boat at Oseberg, restored in 1926. The oarlocks are only 80cm above the water level.
   b Viking warship (*skudelev*), rebuilt at Roskilde. The gunwale is less than 1m above the water level, so the ship would have been submerged in heavy seas.

crossed the Atlantic, he obviously based his map on indirect information, deducing the constantly distorted names and places picked up from Portuguese seamen. The Baga Sea is a pure Portuguese expression. Baron Alexander von Humboldt[6] says the name '*baga*' comes from '*Vagos*' (which is pronounced '*Bagos*' in some parts of Portugal), a flourishing town to the south of the River Vouga. Formalioni[7] claims that the Baga Sea was 'the sea sailed by the mariners of Vagos'. On the other hand, the Morais Silva dictionary[8] registers '*baga*' as being 'a small

13.5 The runic Ielling Rock.
First-century Viking inscriptions on the Ielling Rock of Karlevi (Island of Oeland). This type of runic alphabet had disappeared by the ninth century.

fruit, similar to a grape'. The seaweed that is commonly known in Portuguese as *'sargaço'*, nowadays *'golfo'* (from where we get the 'Gulf Stream') has a culm hundreds of metres in length and small grape-like berries that act as buoys. It was originally thought that this sargasso, or gulf-weed, would 'engulf' a ship and drag it to the bottom.

In the reign of King Alphonse V, Vagos was considered to be the most important town south of Aveiro (Talabriga at the time of the Romans and later Aviarium), where there was a 'Port of Vagos' within the old city walls. The name came from the Latin word *'vacus'*, which meant both 'empty' or 'ample' and must have given the name to the Vouga, the river on which Aveiro stands. The town was a centre of boat-builders and fisherman who became caravel owners and navigators. The bar of Aveiro took the name of Vaga, and it moved either north or south due to silting up. The 1367 map attributed to Pizzigano designated Vagos as *'Baga'* and the *Bianco Atlas* of 1436 calls the place *'Ibera Baga'* or *'Libera Baga'*. According to the encyclopaedia of Maximiamo Lemos,[9] the Romans had consecrated this area to Bacchus, the God of Wine and the Latin *'baga'* also means grape. Coincidentally, the Portuguese used to

13.6 The map of Siurdi.
The map of Siurdi (or Sigurdur) of Stephani, which was unknown in the sixteenth century, presented by de Mahieu as a proof of the Viking discovery of America. The Vinland promontory can be seen in the bottom left-hand corner. But Vinland, as shown in the Portuguese edition of this book, was a fishing village in Greenland (simplified by the author).

call sargasso, which is abundant along the coasts of Iberia, North Africa and the Canaries, the 'sea grape'.

It seems that João Vaz Corte-Real sailed westward from the Azores and discovered some lands on his return journey, as King Manuel I bestowed certain favours on him on the 12 May 1500. His sons Gaspar, Miguel and Vasqueanes were also navigators. Gaspar may have reached Newfoundland but did not return. Miguel requested permission from the King to go in search of his brother, which was granted, but he too disappeared. Vasqueanes, in turn, wished to go to the aid of his brothers, but the King refused his petition as he had already sent two ships on this mission and they had returned empty-handed.

The maps of the beginning of the sixteenth century show Greenland with the name of *Terra Laboratoris* and *Cavo* (Cape) *Laboratore* (*sic*), the latter being a toponymic distortion for which the Azorean historian Ernesto do Canto gives a complete explanation, based on a file of 1506 that is to be found on Terceira Island. This shows that Pedro de Barcelos had received, 'some time previously', a royal command to discover lands to the west in the company of João Fernandes Labrador, the two of them spending at least three years on their mission. Do Canto comes to the conclusion that the discovery of Labrador had occurred between 1492–5. But Duarte Leite, a defender of the Genoese thesis, stated that it had been between 1495–8, basing his claim on the *Cantino Map* of 1502, which affirmed that Labrador had been discovered during the reign of King Manuel I and undoubtedly by a Portuguese.

It has been proved that the inscriptions on the Dighton Rock were made by Miguel Corte-Real and his companions in 1511. Did the survivors of one of the Corte-Real expeditions erect the 'Templar Tower' at Newport (see Plate 2)? It certainly is not a Viking construction. We shall leave these conjectures for now.

No discovery in history has been coveted as much as that of the Americas; the Italian Amerigo Vespucci, in his attempt to eclipse the voyage of Columbus and forestall Pedro Alvares Cabral in his official 'rediscovery' of Brazil, made some very extreme claims. Vespucci, a celebrated Florentine who was born in 1451, was a mate on a Spanish ship of a fleet commanded by Afonso de Hojeda. That was enough to claim the discovery of the land that was always known as 'Hojeda'. During the long judicial process that Diogo Colón (the Admiral's legitimate son) brought against the Spanish crown, the *Pleyto contra la Corona* (see Chapter 17), Vespucci, as a witness at the court, declared that to realise that voyage he used a map drawn by Columbus. He then wrote the *Letters of Americo Vespucci to Pedro Soderini, perpetual standard bearer of the Republic of Florence, about two voyages made on the command of his Serene Majesty, the King of Portugal*. These letters were translated from the Italian and published in the *Colecção de Noticias para a História e Geografia das nações ultramarinas*,[10] but the Viscount of Santarém[11] stated that were no references to these voyages in either Portuguese documents or among the manuscripts in the Paris Royal Library.

Vespucci claimed that he had come to Lisbon at the invitation of King Manuel I, but there is no copy of the letter written by the King and none of the contemporary chroniclers refer to it, which is strange as he says that he was 'an Italian in the service of the king of Portugal'; all King Manuel I's letters exist except this one. The truth is that Vespucci offered his services only to the Catholic Monarchs. He was never in the service of the King of Portugal and lied when he wrote that the King had promised him rich rewards if he would undertake the exploration of the 'Terras de Santa Cruz' (Brazil) and that he set sail from the Tagus with a fleet of three ships in May 1501 and soon after met the fleet of Pedro

Álvares Cabral. While Vespucci was in Lisbon, he heard about the voyage of Commander of the Fleet, Fernão de Noronha, that departed from the Tagus in 1503. Someone supplied him with a detailed course of a voyage from Portugal to Brazil, as he described it in his fourth letter. But the indiscreet pilot told him nothing else, so the Italian, learning that some vessels had been lost on the voyage, invented the shipwreck of four naus and the death of the Commander of the Fleet, whose name Vespucci did not know but whom he took the liberty of describing as 'superb and mad'. In fact, only two naus sank and Fernando de Noronha discovered, on this voyage, the island of St John (later given his name), which lies 50 leagues from Brazil, on 24 July. Through royal letters patent of 16 January 1504, King Manuel I made Noronha Donatary-Captain of the island.[12]

In a letter dated 4 September 1504, Vespucci told Solderini that he had discovered All Saints' Bay on 11 February 1502 and that he had waited there for the commander (anonymous). As the latter had not turned up, Vespucci had decided to go to Cape Frio. In fact, All Saints' Bay had already been discovered by Cristóvan Jacob; Vespucci had never sailed with the Portuguese and even less had he been to Brazil.

A friend of Vespucci, Friar Giovanni Giocondo, translated some of his letters into Latin and showed them to King Louis XII of France. He then handed them over to an Alsatian, Mathias Ringman, a teacher of Latin who worked as a proof reader for a printing-press in Saint Dié, who agreed to publish a book entitled *Cosmographiae introductio*. The book contained only the letters of Vespucci that claimed that Columbus had discovered merely islands while the Florentine had been the first to set foot on the American continent. And Ringman added: '*et quarta pars quam quia Americus invenit Amerigen quasi Americ terram, sive Americam nuncupare licet*'. The book was published in 1507, a year after Columbus's death.

In a letter that he sent from Seville to his son Diogo in Hispaniola on 5 February 1505, Columbus wrote: 'Diego Mendez left here on Monday the 3rd. Following his departure, I spoke to Amerigo Vespucci, the bearer of this letter, who is going there to deal with things concerning navigation. He has always done his utmost to please me; he is a good man; . . . he is going with the hope that he will be able to do something to help my cause, if it is within his possibilities.

From here, I don't know what advice I can give him as far as I am concerned, because I don't know what the people there want. See for what service [Vespucci] can be used out there and he will fulfill it, but make sure everything is kept secret so as not to arouse suspicions . . . Show this letter to the Military Governor so that he can see how best to use his services'. Vespucci used the letter to such good effect that following Columbus's death King Ferdinand entrusted him with the post of Admiral of the Fleet. By 1505, Columbus had discovered 'terra firma, – Venezuela (1498), Nicaragua and Panama (1503) – but Vespucci did not hesitate to write that Columbus had discovered only 'islands'. (He also never mentioned the fact that Columbus was a fellow Italian.)

The book containing Vespucci's letters was followed by another work, *Globus Mundi*, in which the name 'America' came to stay: 'a quarter of the world, having been discovered by Amerigo, can well be called Amerigen, which means that land of Americo or America'.[13]

In 1501, the chronicler Pedro Mártir de Anghiera wrote a true narrative of the voyage of Vicente Pinzón, who had been to a place which he called Cape Santa Maria de la Consolacion or Consolation Point. Some years later, however, after Europe had been informed of the 'official rediscovery' of Brazil by Pedro Alvares Cabral in 1500, Pinzón unashamedly claimed to have discovered an enormous river in January of that same year, and that it was the Amazon. The now aged Anghiera wrote a new narrative in 1516 that was longer and exaggerated the duration of the real voyage. This was enough for some Spanish writers of the nineteenth and twentieth centuries to claim that Pinzón had discovered Brazil:

1. The place that Pinzón had called Consolation Point and which the Catholic Monarchs had donated to him was Cape St Augustine in Brazil, which was to the east of the line that was drawn with the Treaty of Tordesillas.
2. Having sailed from the Portuguese island of St Vincent in Cape Verde, Pinzón saw, after sailing about 300 leagues to the south-west, that the Pole Star had disappeared behind the northern horizon.
3. Sailing 50 leagues north, Pinzón came across a 'majestic river', the current of which was so strong that the sea in front of the estuary was of fresh water – exactly as the Portuguese had already described it.

The historian Duarte Leite[14] countered these arguments with the following conclusions:

1. If the cape discovered by Pinzón were that of St Augustine in Brazil, the Catholic Monarchs could never have donated it to him as it was not in their zone to start with.
2. If Pinzón had sailed those 50 leagues in January, he would have noticed that the Pole Star describes an arc above the horizon and is clearly visible, as it would have been in the Amazons for two hours before nightfall and two hours after daybreak. This proves that Pinzón imagined that the Amazons were much further to the south.
3. The 'fresh water' phenomenon also happens with the River Orinoco in Venezuela. However, the latitude that Anghiera refers to in his

second narrative, and was not mentioned in the first, does not even correspond to this river.

Pinzón also exceeded himself when he declared that he had seen local natives of a gigantic stature, as there are no Indians of this type in Brazil; only the Caribs of Guiana are tall. Moreover, none of the details in Anghiera's second version corresponds to any zone of Brazil.

Only the Portuguese sailed the 'Great Ocean' in the fifteenth century, scientifically planning their routes that were part of Prince Henry's 'Plan of the Indias', and the Portuguese Colón-Zarco was the first to take Spaniards with him in the last decade of this century.

Notes and References

1. A discovery made officially only in 1498 by the former's son, Gaspar Corte-Real.
2. Larsen (1924).
3. Gómara (1859).
4. Silva, Manuel Luciano da (1971).
5. Silva, Manuel Luciano da (1972).
6. Humboldt (1833) Vol. III, p. 88.
7. *Nautica dei Veneziani*, p. 48.
8. Silva, António de Morais (1891).
9. Maximiamo Lemos
10. Vol. II, letter 1, p. 14.
11. Santarém (1976).
12. D. João II, Book 37, f. 152.
13. The following can be read on a birth certificate in the Florence Archives, reproduced in *'Illustracione e note'* to the Baudin ed. (1892): 'Amerigo son of Nastasio Vespucci, born 9th March 1451'; in another document, published with the title of *Le Toscanelli*, the signature *'Amerigo, of Nastasio'* can be found. It must be pointed out that 'Amerigo' is a variant of *'Albericus'* and *Almericus'*; see Duhem (1940).
14. See Dias (1932).

14 The Mystery of the Seven Cities

On his *Toledo Map* of 1230, Abul-Hasan placed an island on the latitude of the south of Greenland and on the longitude of Iceland and gave it the name of 'Seven Cities'. Included in a manuscript of one Angelo and annotated in Arabic, the map is based on an Arab–Gothic legend which tells the story of a Bishop of Merida who fled to the Island of the Seven Cities or Antilia. When the city of Merida was under siege by the victorious Arabs after the Battle of Guadalete, Maximo, the Bishop of the city, escaped to Setúbal, to the south of Lisbon, in the company of other ecclesiastics and they set sail for that island.[1] Some historians identified it as Thule or Tyle (Iceland). On the big map attributed to the Columbuses that is to be found in the Paris Library and which was executed between 1488 and 1492, there is an island placed in the area of Iceland which has the following caption: 'This is the Island of the Seven Cities, now peopled by Portuguese and on which, according to Iberian sailors, silver can be found in the sand'.[2]

How is it possible that there was still talk of a Portuguese colony on Iceland in the fifteenth century? The Portuguese never forgot this legend, and it was because of it that they gave the name of 'Seven Cities' to a lake that was formed on the Island of São Miguel in the Azores after a violent volcanic eruption in the fourteenth century. How is it possible that a Genoese or a Catalan or Galician Columbus and his assumed brother Bartolomeu knew of the ethnic origin of the Icelandic population, and of the existence of silver in the sand? As Columbus had lived on Madeira and in Lisbon and knew the archipelagos that the Portuguese had colonised, how can we attribute such an unlikely piece of information to him? Would he have known of the eighth-century legend? Yet as he declared that he had been in Iceland in 1477, how could he have written such a legend on a map of his dated 1488–92? Did he meet the descendants of Portuguese on the island? It is unlikely that either he or his brother made the chart that is in the Paris Library (which I have not seen). The identity of the authors of the chart is doubtful, but at the same time it is easy to see if it is the work of the Columbus brothers through the coordinates and place names.

Were there already seven cities on Iceland at the time of the Arab invasion of the Iberian Peninsula? It seems not. In this case, which land is referred to in the legend if not Thule? Would it be Antilia, which means 'in front of Thule'? And why should it not be Ireland? On his second voyage, Diogo de Teive said he went in search of Antilia. Could he have thought he had found it when he sighted Cape Clear in Ireland? Ireland had only seven 'cities' at that time: Corcaigh (Cork), Luimneach (Limerick), Port Lairg (Waterford), Port Slaney (Westford), Baile Athja Ciath (Dublin), Dun Dealgan (Dundalk) and Belfast. The Bishop of Merida would, of course, have asked his captain, Sacoto, to take him to a Christianised nation and not to an uninhabited island.

On his first voyage of 'discovery' to Cypangu and India in 1492, Columbus went in search of Antilia – 'not the northern one but the western one'. That was the name he gave it and today the archipelago is known as 'The Antilles'. How could he have drawn a map that was so wrong, and in the year that he discovered the archipelago? It is remarkable that experts could attribute the anonymous legend to him.

Precious stones and metals – gold and emeralds, esoteric symbols of the Orient – Sun and the Chalice – Grail – were the bait that Columbus dangled in front of the Catholic Monarchs to justify his attempt to find the East by sailing west. Through the stories of the Greek and Roman voyages to Britain and Ireland, Columbus may have been aware that those islands were rich in tin and silver, but if he had been to Greenland and Iceland he would have seen volcanic rocks, snow and glaciers. He would never have placed 'Seven Cities'–Antilia on the parallel of Ireland, let alone the parallel of Thule–Iceland. Another fact must be emphasised. The Portuguese were searching for the way to India in quest of the highly prized and expensive spices, and Columbus enticed the Castilians with promises of precious stones and metals. Through information already received from Diogo de Teive, Columbus knew that he was heading for an island where 'white men with beards had already landed'. His chronicler Las Casas stated that he had got this information from his father, who had sailed with Columbus on his first voyage. He named the archipelago *Barbados* (bearded men) even though the local natives did not have beards. But they did have gold and pearls. Later the colonialists called *'barbados'* a variety of thorn bush. Columbus knew that Diogo de Teive claimed to have discovered the 'Island of the Seven Cities', when all the archipelago of the Azores was already known, Teive himself having found the last island (Flores). The 'Seven Cities' that Teive said he was looking for on his second voyage could not have been Thule, which lies at 73 degrees North, as he sailed west from the Azores and headed for the Sargasso Sea. It was, of course, a secret expedition, as the place named as its destination was wrong. The new island had already been given the temporary name 'Seven Cities', so that it would be thought that they were heading for Ireland. A Galician by the name of Pero Vasquez de La Frontera sailed with Teive. Could it have been through him that a cartographer received information to draw the map in the Paris Library? It is doubtful, as Pero Vasquez knew very well where he had been. But the information must have been the result of espionage and this must have been the reason that the author called the island 'Antilia', said it was inhabited by Portuguese and that silver was to be found in the sand. If it were neither Iceland nor Ireland, where had the Portuguese already settled?

Meanwhile, Columbus continued his farce of the search for Cypangu, Ophir and Cathay. The map in the Paris Library could never have been made by Columbus, simply because – as has been repeatedly stressed – he had often sailed with the Portuguese.

Following a theory of Raymond Lully and based on the fact that the world was round, Toscanelli predicted that the Orient could be reached by sailing west (see Figure 14.1). This theory was well known to the Portuguese, but they did not pay much attention to it. It is said that Toscanelli made a proposal to the King of Portugal to carry out such an undertaking and the King rejected it, just as the Junta of Mathematicians had rejected Behaim's and Toscanelli maps. How could this proposed discovery harm Portuguese expansion to the Orient? To answer this question we must look at the internal situation of the kingdom during the reign of King John II.

THE CONSPIRACY

According to Garcia de Resende,[3] King Alphonse V returned from a hunt in July 1454 and entered the apartments where Queen Isabel was waiting for him. There 'he took her with such passion' that he broke the emerald of her ring when he threw her on to the bed: 'Take it as a good omen, my Lady, and may it please our Lord that you will conceive a son that you will esteem more than all the emeralds in the world'. His prediction came true and the future John II was born in March 1455.

On ascending the throne in 1481, John II convened the Cortes at Évora, at which he determined that the Alcaides of various cities should swear an oath of fealty to him. This meant that they lost many of their privileges and became mere delegates of the central power.[4] Their previous warrants had given them generous rewards and a private income, but they were draining the resources of the Crown that were needed for public works and to finance the Discoveries. Deprived of their privileges, the Alcaides and the nobility reacted by not following

THE MYSTERY OF THE SEVEN CITIES 165

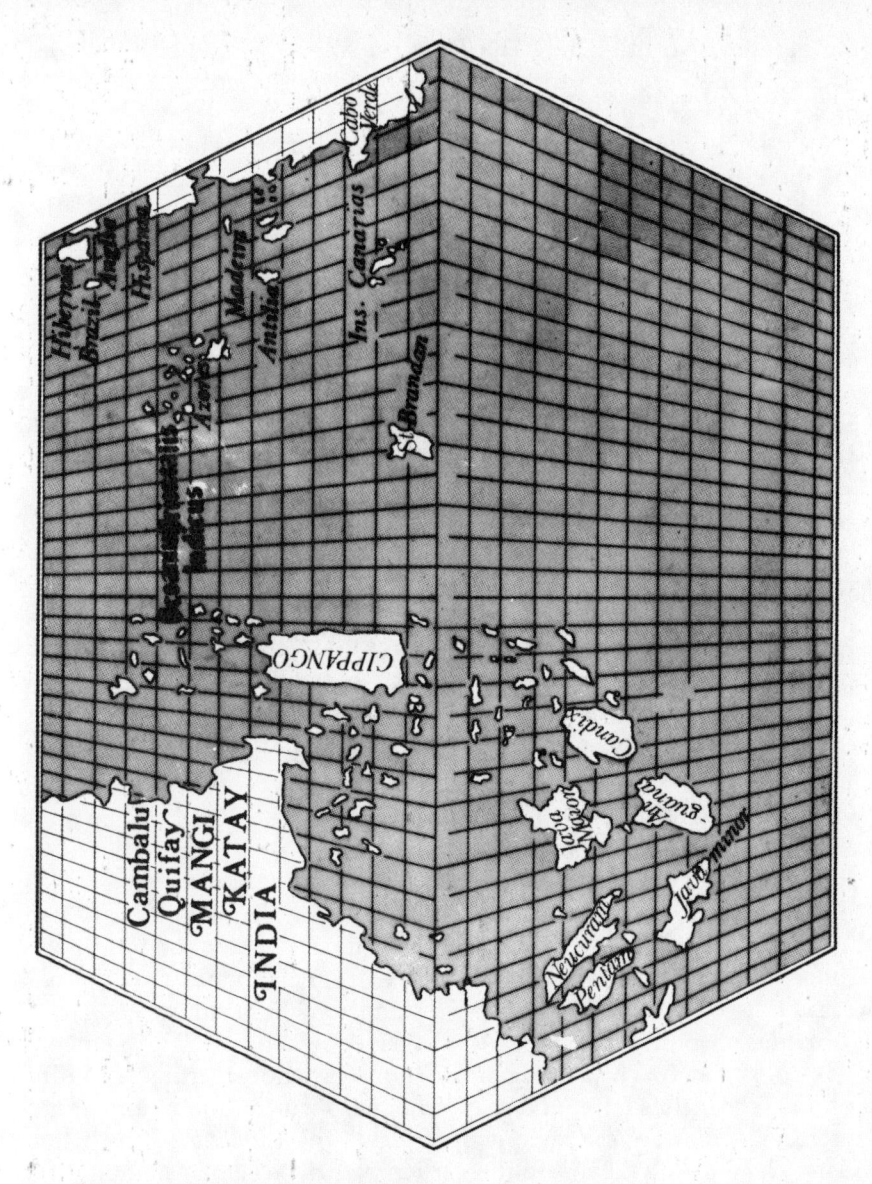

14.1 The map of Toscanelli (1474).
Despite the fact that Toscanelli already knew of the existence of an island on the parallel of Cape Verde that the Portuguese called 'of the Seven Cities', he identified it as Cypangu, on the way to India. He scattered imaginary islands all over the central Atlantic, which he designated as the Western Indian Ocean, among which was the invented island of St Brendan. This map, with the wrong calculations of Martin Behaim, allowed Colón-Zarco to convince the Catholic Monarchs and their Junta of Mathematicians that India could be reached from the West (simplified by the author).

the new laws to the letter. The People's Attorneys immediately complained to the Sovereign and he decreed the punishment of all those that broke the law.

A conspiracy was planned against the King led by his brothers-in-law, the Dukes of Viseu and Braganza.[5] In order to assure success in the case of civil war, Duke Fernando of Braganza sought aid from Castile. As this aid had to be compensated, he ventured to defend the interests of the Castilians and requested a revision of the Treaty of Toledo. He wrote a letter to Queen Isabella, using as a pretext 'that the exclusion of the Castilians from the Guinea trade was an infringement of people's rights; if John II denied this right, as he certainly would, Castile would have every reason to declare war, in which case I and my friends would refuse to support the King by alleging the injustice of the war. Not only would we refuse to support the King, but we would allow foreign troops through our lands'.[6] This was high treason and when the King learned of the contents of the letter from a secret agent in 1483, the Duke was tried and executed in Évora. The conspiracy widened, however, now led by Diogo, Duke of Viseu. But it was denounced by one of the conspirators. The King summoned the Duke to his presence and stabbed him to death with his own hands. Many more noblemen were immediately arrested and sentenced to death. Among them were Dom Fernando de Menezes, Pedro de Albuquerque, the Bishop of Évora, Dom Garcia de Menezes, Dom Guterres Coutinho and Dom Fernando da Sylveira, who fled to France but was tracked down in Avignon by one of the King's agents and stabbed to death. As new attempts against the King's life were denounced in Setúbal, many other conspirators met similar fates. This conspiracy made it more necessary for King John to divert Castile's attention away from the African route to India. Someone had to persuade the Catholic Monarchs to accept the theory of Raymond Lully and Toscanelli; someone who could assure his neighbours that he had already discovered lands to the west; someone whose culture and courtly manners would allow him to be received and heard. This someone was Colón-Zarco. The Portuguese Canon Fernando Martins sent King John II Toscanelli's letter and map. The map, which did not apparently interest the Portuguese sovereign, became the *leitmotiv* of the whole plan to lead the Castilians up the garden path. The subject matter of the letter enclosed with Toscanelli's map is, however, truly interesting, as it proves that the Portuguese knew of Antilia before 1474: *'Sed ab insula Antilia vobis nota ad insulam nobilissima Cippangu'* – 'Antilia, *which you already know*'.[7] What was to be found in the west if it were neither Ireland nor Iceland?

CYPANGU

Toscanelli took care to include Antilia (discovered by the Portuguese) on his map, but he wrongly placed Japan nearby. Japan had been known since the time of Marco Polo – the very noble island of Cypangu. On his return trip from India, Pedro da Covilhã gave the Order of Christ a vague idea of the size of China and the position of the island of Cypangu. From what they knew about the size of the world the Portuguese were aware that Japan could not be anywhere near the Antilles. But the rest of the world was ignorant of this fact. During the whole of the sixteenth century, when the naus of the Portuguese were crossing the seas of the Orient in all directions and when the Portuguese cartographers had charted the coasts, their foreign counterparts were still turning out maps with astonishing errors in the conception and shape of the different continents. It is also interesting to note that Toscanelli thought that Antilia was situated at 28 degrees North. At the end of the fifteenth century, however, it was placed at 23 degrees North.

In the case of Toscanelli's location of Antilia, based on 'the island which you already know' and calculated by navigators at the dawn of cosmographic nautical cosmography, a mistake of 5 degrees would be acceptable. On the other hand, a mistake of 10 degrees, as attributed to Columbus in relation to the latitude of Iceland, would be inconceivable – at least for a Portuguese. It is thus almost certain that the future Admiral of the Indies had not made a mistake in the location of Thule–Iceland, just as he had not been mistaken in calculating the position of the Seven Cities–Antilia (which was now on the same parallel as the Guinea coast), except for a small margin that was usual at those times. (A small margin of error that, following up the work of Teive and others, Columbus corrected.)

ESPIONAGE

Let us consider medieval cartography and analyse the main reasons for its notorious imperfections and the lacunae in cosmographic knowledge. When the first Portuguese navigators, who had certainly learned a great deal from the Jewish astrologers and Moorish mariners, began to sail the seas, always describing circles in the hope of discovering new lands, they did not do so at their own free will – as fishermen, they had no need to leave sight of land as fish abounded in the coastal waters. Oceanic exploration, long before the time of Prince Henry, was a royal initiative, based on the support of the Order of the Temple, and became more intensive from the reign of King Dennis. It was during this reign that the policy of secrecy was imposed on the 'brothers' of the Order of Christ.

Germano Correia recounts the following story:

> Between 1583 and 1594, everyone in the Kingdom of Portugal would swear that there was no more fervent Catholic in Goa than a certain Dutchman in the retinue of the Archbishop and a servant of his confidence. The man, who was and always had been a Lutheran heart and soul, not only took in the prelate, but also managed to enter the secret archives of the Tower of Tombo of Goa, from where he took copies of all the itineraries and ships' logs of the best Portuguese pilots and handed them over to Flemish pirates.

Germano Correia did not mention the name of the spy, but Gerrit Van Lent of Brussels has identified him as being Jan Huygen Van Linschoten, whose book (which is to be found in the Plantin and Moretus Museum in Antwerp) contains the following note: 'This work launched the Dutch on the path of colonialism'.[8]

When King Dennis had pinewoods planted in order to have timber to build ships, he ordered that anyone found cutting down a tree should be hanged on the spot. He imported pinasters from Landes and Gascony to supply masts and planks and used the local umbrella pine, together with the cork oak, for curved pieces (prow and broadside ribs), as we saw in pp. 47–8. It was at this time that stories of enormous dangers that were to be found on the high seas were propagated. Legends of sea and land monsters, evil spirits, islands that were hell on earth, submerged volcanoes and stormy seas that devoured ships and their crews were rife. Stories of chivalry were being replaced by accounts of travels to exotic and paradisial lands and by the emotion of tragic shipwrecks. The legend of St Brendan became especially popular among Portuguese seamen.

THE MYTH OF ST BRENDAN

The historian de Mahieu boldly stated that there was a tenth- or eleventh-century manuscript, *Navigatio Sancti Brandani* that narrated with great precision 'the Canaries and the Island of Hell, with the volcano Teide, as well as the Sargasso Sea [that] makes it possible that the holy abbot and his monks reached Florida'. In Catholic hagiology, there is simply a legend that identifies St Brendan as an abbot of the Monastery of Clesaifert that had ventured out into the Atlantic in search of the Island of Happiness, but accompanied by friars and not by experienced sailors. It is obvious that this story was inspired by *God's City*, which

St Augustine wrote following the conquest of Rome by the Barbarians in 410. Sir Thomas More later wrote *Utopia* in 1518, followed in 1623 by Friar Tomazzo Campanella with *The City of the Sun* and Francis Bacon with *New Atlantis*. The supposed voyage of St Brendan places the 'Island of Happiness' at the latitude of Ireland, and does not describe any of the Canary Islands.

De Mahieu cites the year 565 as the date of St Brendan's voyage. It happens that it was precisely in 1565 that Viscount Andrew of Bordeille, Lord of Brantome, sailed from the Barbary Coast for Malta. The only place he could have visited on the Barbary Coast was Mazagan, which was still in the possession of the Portuguese. It was situated north-east of the Canaries on the same parallel as Madeira and a little to the south of the present-day Casablanca.

Brantome narrates his voyages in a *Biography* and specifically states that he went: 'to see the world; Italy, Scotland, Spain and Portugal, where I wore the habit of the Order of Christ with which the King of Portugal honoured me!'[9] He described his journeys to North Africa, Malta and Greece and became known as the 'French Plutarch'.[10]

The legend of St Brendan was also inspired by the above-mentioned flight of the Spanish clergy from the victorious Muslims which, according to Martin Behaim, took them to the 'Island of the Seven Cities', led by the Bishop of Oporto.

From the reign of King Dennis onwards, the Portuguese sovereigns deliberately spread stories of the terrible dangers which faced oceanic explorers in order to scare off foreigners. On the other hand, any Portuguese that discovered new lands were promised rewards and promotion to the nobility. The legend of St Brendan was still alive in the minds of seafarers (and, above all, in the spirit of Italian cartographers, who were avid for information about the Portuguese Atlantic voyages and which they could obtain only clandestinely). As they were not eyewitnesses to the Discoveries, they placed islands indiscriminately on their charts, according to the rumours that abounded. A *Nautical Chart* of Pizzigano thus imagines some 'Fortunate Isles of St Brendan' – the Archipelago of Madeira with its islands of Porto Santo and Deserta, both out of position and out of proportion. He also depicts the Archipelago of the Canaries with the Island of Hell (see Figure 14.2).

Pizzigano wrote a caption on the left of his map that referred to malevolent divinities that stopped navigators penetrating into the western seas. But he shows only three islands to the north of Madeira, to which he gives the same name – Braçir (Brazil). He also shows one vague island of the Azores. De Manhieu claims that this map was made in 1467. But it is not dated and the only details that exist are due to a doubtful copy of one Buache de La Neuville, which is undated.

After the Islamic conquest of North Africa, the Moors used to go to the

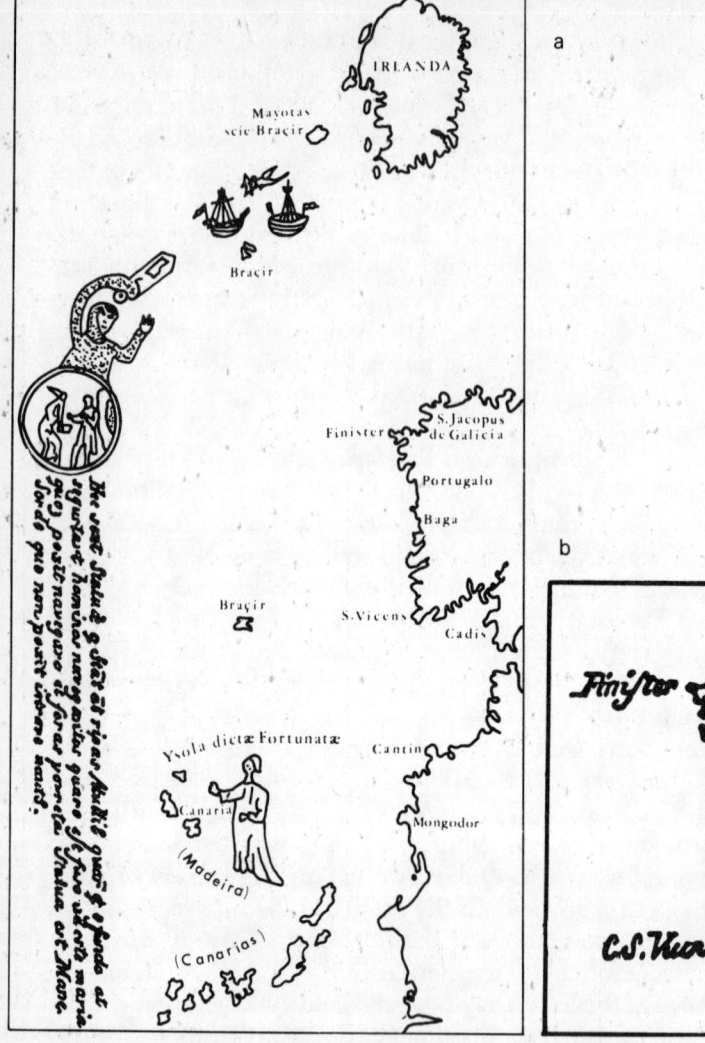

14.2 The Nautical Chart of Pizzigano (1467).

    a Partial and imaginary map of the North Atlantic which Buache de Neuville probably copied from an original attributed to Pizzigano and made in 1467. In the original, the cartographer called the archipelago of Madeira *Ysole dictae fortunatae* and the Canaries *Isole Ponzele*. The archipelago of the Azores is depicted as just one island, 'Braçir' (Brazil), with two other islands to the north with the same name. In badly written Latin, the author inserted a caption on the left that refers to the existence of malevolent beings that impeded navigation to the west. To the south of the city of 'Portugalo' (Oporto) is another designated as 'Baga' (or 'Boga') situated at the mouth of the River Vouga. The city of Lisbon was omitted, despite its importance, as was the River Tagus. The mention of 'Baga', therefore, suggests that the 'Baga Sea', discovered by the Portuguese after the 'refinding' of the Azores, was already known. It is not possible otherwise to explain the emphasis given to 'Baga', the mouth of the River Vouga (simplified by the author).

    b Enlargement of the area of Pizzigano's Chart showing Portugalo and Baga.

Canaries to get slaves. The islands were rediscovered by Iberian navigators at the beginning of the fourteenth century. On 15 November 1344, Pope Clement VI granted Don Luís de La Cerda, a Castilian nobleman – great-grandson of Alphonse X (known as 'the Wise') of Castile (see Tree VII, p. 409) and grandson of St Louis, King of France, the country for which he fought as an Admiral in the war against the English – the fief of the Archipelago for an annual payment of 400 gold florins to the Church. King Alphonse XI of Castile protested at once, claiming to be the heir to the islands (see pp. 48–9). In February 1435, King Alphonse IV of Portugal claimed that the islands had been explored by the Portuguese, who had returned with 'trophies', and that he was therefore intending to send a fleet there in order to assure possession of them. The map that Pizzigano presumably made in 1467 thus does not contain anything new, as it was made at least twenty-three years after the Portuguese voyages.

Other maps appeared, among them being the maps of Bianco in 1436, Pareto in 1455, Benincasa in 1467 and Martin Behaim in 1492, all of them out of date in relation to Portuguese navigation. Some experts have attributed the toponyms of the Archipelagos of the Canaries, Madeira and the Azores to the fertile imagination of cartographers, but this is not so. It is possible that there were goats (*cabras* in Spanish and Portuguese, *caprae* in Latin) on the Canary Island of Capriria, but some of the islands were inhabited by the Canarii tribesmen. None of the islands of the Archipelagos of Madeira and the Azores, however, were inhabited by mammals, apart from some seals. The Portuguese usually gave the names of saints or phyto-geographic characteristics to newly discovered or rediscovered lands and islands, or even the surnames of the discoverers. The *Bianco map* – also based on the legend of St Brendan – baptised some of the islands of the Azores Corbomarino, Conici (or Conigi), San Zorzi, Bentilez (or Benturas), Colombi, Brazil, Caprala (or Caprara) and Lobos, and the islands of Madeira with the names given by the Portuguese discoverers. The great navigators of Portugal were not in their 'trade' by chance. All of them were descendents of known or unknown mariners. Some were direct descendents of landowners or high officials, but their appearance on the high seas was certainly due to their immediate ancestors. 'Conigi', therefore, may come from 'Coelho' and 'Caprala' from 'Cabral'. And the Marinhos (whose coat-of-arms has the same crest as that of the Dornelas who were connected to the Zarcos) may have adopted the name of the island that was found by one of them: *Corvi Marini*.

I do not claim that all these facts are correct; I merely intend to point out that all the names on the maps had a Portuguese flavour, despite the attacks of historians like D'Avezac[11] and de Mahieu, who did their utmost to show that French navigators had been the forerunners of the Discoveries (including the legendary St Brendan who, if he lived, was

a

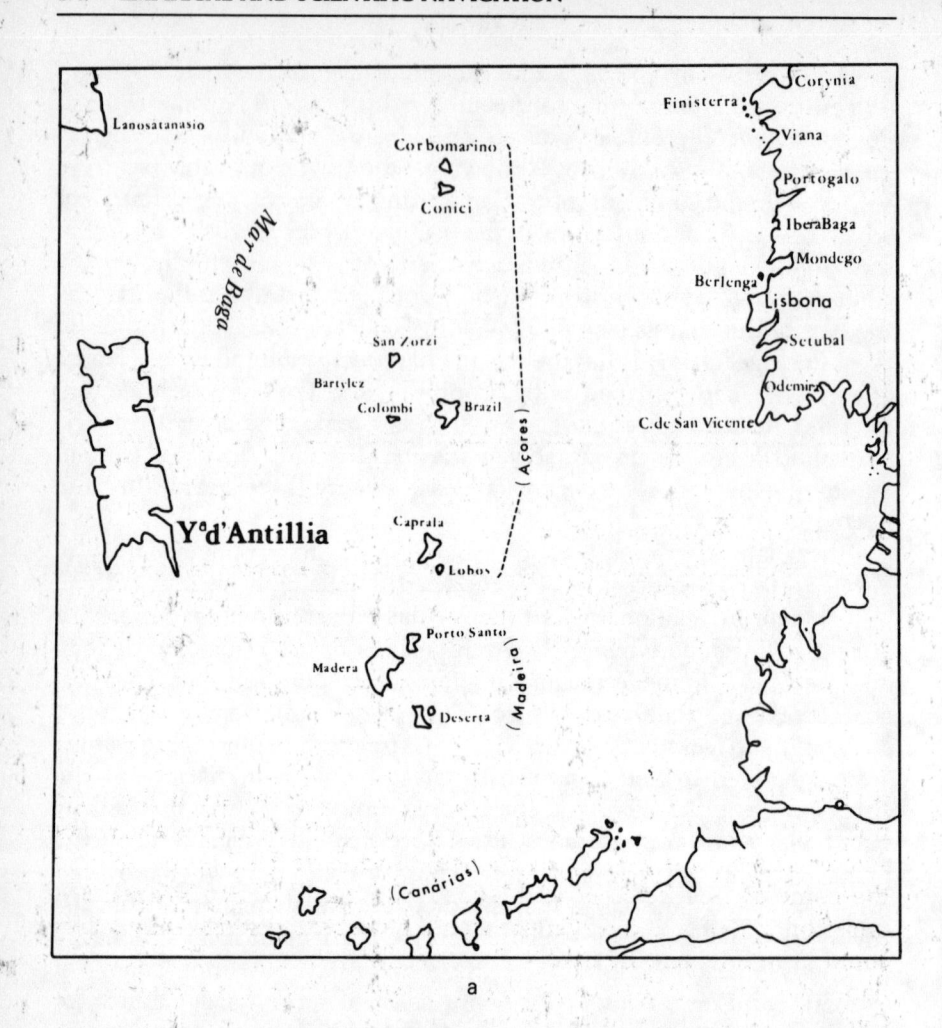

Irish). (For the toponymic evolution of some of these place names see Appendix 3.)

Pizzigano showed an 'infernal' island in the western sea on the parallel of Finisterre; but Bianco places his farther north and called it Lanosatanasio (or Manosatanasio), while his Antilia is farther south than Pareto's (see Figures 14.3 and 14.4). Behaim's is to the south of the Canaries, as is Toscanelli's who said it was 'already known to you' in his letter to Canon Fernando Martins in 1474, 'you' meaning the Portuguese. If Manosatanasio corresponds to Newfoundland, then the Portuguese discovered America before Bianco made his map in 1436. And it was surely because of this that the 'retranslations' of Andrés Barcia[12] and Navarrete[13] omit the phrase 'that you call the Seven Cities, which you already know' and 'has pearls, precious stones and solid gold' (see

THE MYSTERY OF THE SEVEN CITIES   173

[Map showing the Azores: Corvo, Flores, Graciosa, Terceira, S. Jorge, Faial, Pico, S. Miguel, Santa Maria, with coordinates 40°/21° and 25°/40° at top, 37°/21° and 37°/25° at bottom]

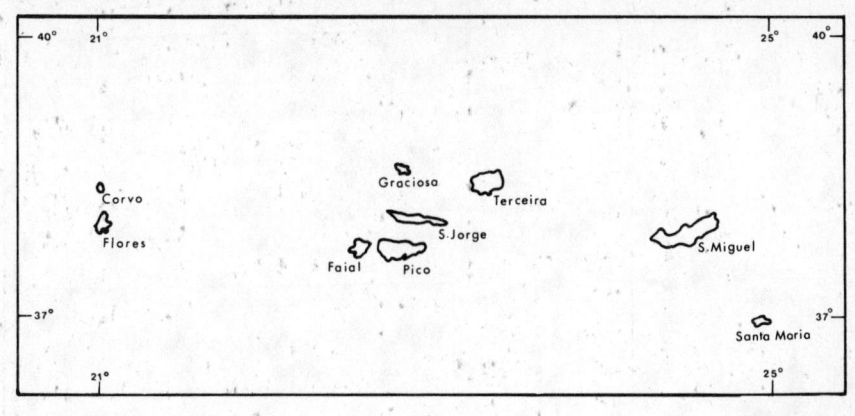

b

14.3 The *Bianco Atlas* map (1436).

    a Map no. 5 of the Andrea Bianco *Atlas*, on which the author refers to the 'Baga Sea' and a Portuguese port by the name of 'Ibera Baga' (at the mouth of the Vouga). If 'Manosatanasio' (Satan's Hand) corresponds to Newfoundland, it could mean that the Portuguese discovered America before 1436 (simplified by the author).

    b The real 'layout' of the Azores, horizontal and not vertical, as conceived by Bianco (simplified by the author).

p. 166). Yet this omission would have been unnecessary if they had noticed that Antilia on Toscanelli's map corresponded to an island of the Azores and that (as already mentioned) much later the Portuguese gave the name of 'Seven Cities' to a lake on the Island of São Miguel. The omission of Barcia and Navarrete would have been justified only if they could have foreseen the truth – Toscanelli and the Portuguese gave the name of 'Seven Cities' to the 'other Antilia', which the former called Cypangu, wrongly thinking that it was Japan.

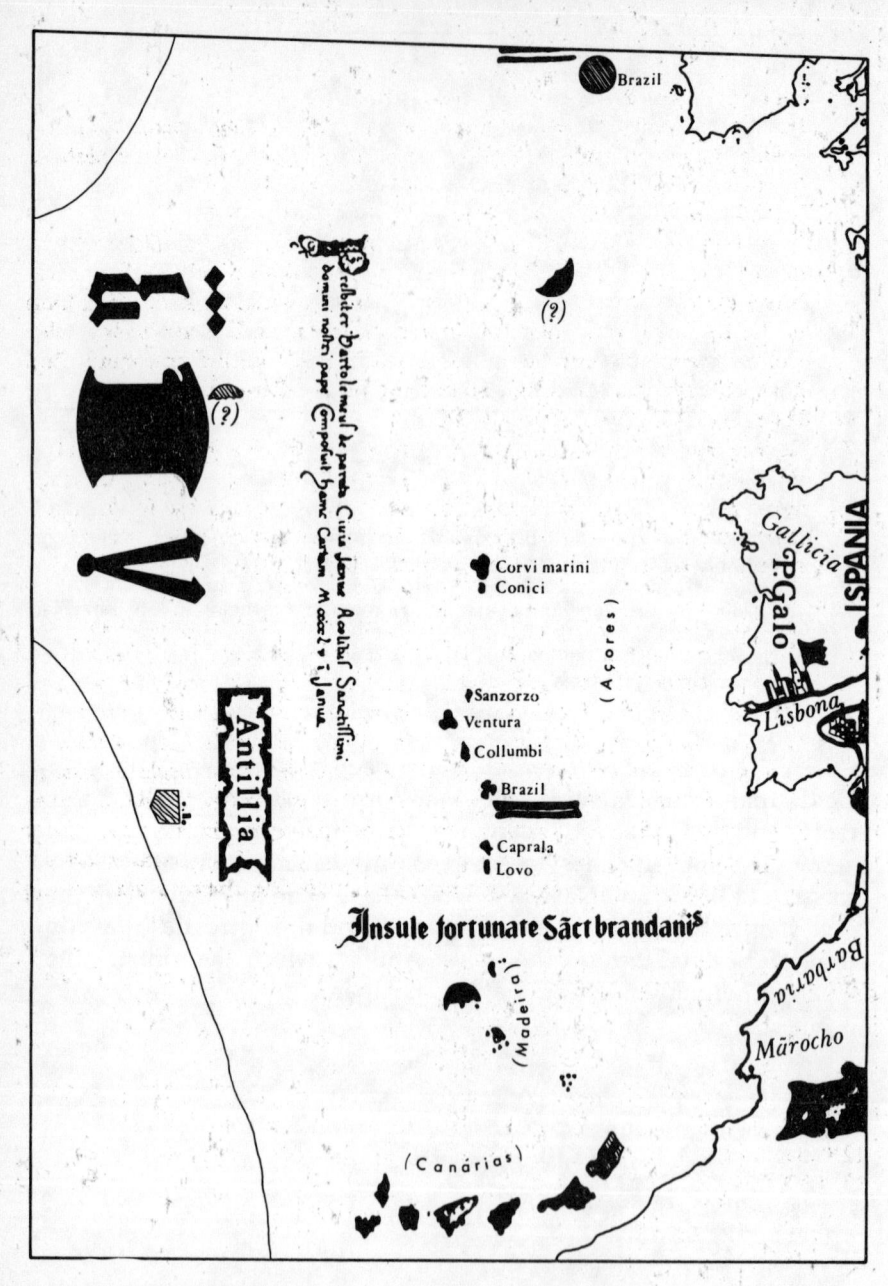

14.4 The Pareto map (1455).
Part of Bartholomeo Pareto's map of the Atlantic Ocean, with the Azores and Madeira vertically aligned and with 'Antillia' placed approximately on the parallels of Lisbon and Cape St Vincent.

Notes and References

1. See Purificaçao (1642), Book III, heading 4.
2. 'Hec septem civitatum insula vocatur, nunc portugalensium colonia afecta est gromite (?) citantur hispanorum in (?) repriri inter arenas argentum perhibetur'.
3. Resende (1545; 1798), Chap I.
4. Resende (1545; 1798); Pina (1790).
5. Dom Diogo, 4th Duke of Viseu and 3rd of Beja, and his sisters Isabel and Leonor were children of Prince Fernando (who had succeeded his uncle Prince Henry), son of King Alphonse V, who was the father of King John II. The King was thus their cousin and on marrying Princess Leonor also became their brother-in-law; and as Isabel had married Fernando, 2nd Duke of Braganza, the King also became his brother-in-law.
6. Resende (1545; 1798), Chap XXXIX.
7. Pestana Júnior (1928). Dias (1932) proves the authenticity of Toscanelli's letter and map, plus the *Note* that accompanies the two documents, thus showing that the argument of Vignaud (1901) is groundless. To quell any other doubts, here is an integral transcription of this *Note*, which was lamentably distorted in the French and Italian translations:

   > *A civitate ulixiponis per occidentem in directo sunt viginti sex spacia in charta signata quorum quodlibet habet miliaria 250 usque ad nobilissimam et maximam civitatem Quinsay* [illegible] *hoc spacium est fere tercih pars loci spherae que civitas est in provincia Mangi sive vicina provincie Katay in qua residencia terre regis est. Sed ab insula Antilia vobis nota ad insulam nobilissimam Cippangu sunt decem spacia. Est enim illa insula fertilissima auro, margaritis èt gemmis et auro solido cooperiunt templa et domos regias, itaque per ignota itinera non magna maris spacia transeundum. Multa fortasse essent apertius declaranda, sed deligens considerator per hoc poterit ex se ipso reliqua prospicere. Vale dilectissime. Data florencia 25 junio 1474.*

8. In Dias (1932, Vol. V) see *Itinerario – Voyage ofte Schipvaert van Jan Huygen Van Linschoten naer oos ofte Portugaels Indien – 1579/1592* (p. XXXIX), on the recommendation of Gerrit Van Lent.
9. John III; the order no longer had its former vigour.
10. *Dictionnaire de la Conversation et de la Lecture* (Paris, 1838) Vol. VIII, p. 309.
11. Avezac (1845), pp. 27–30.
12. Barcia (1794).
13. Navarrete (1825).

Part IV
Biographical Problems

15 Enigmatic Lineage

The Italian Pietro Mártir de Anghiera (also known as Pedro Mártir de Angleria), Chaplain and Almoner of the Catholic Queen Isabella and contemporary biographer of Columbus, is considered to be one of the 'original sources' of Columbus's life. In a letter written to Count Giovanni Borromeu on 14 May 1493, Anghiera refers to Columbus's arrival in Castile on 15 March as follows: *'Redit ab Antipodibus occidius Christophoms quidam Colonus, vir ligur'* (A Ligurian, one Christophoms Colonus, has come back from the Western Antipodes). This latter is no. CXXX (p. 72) of the *Opus Epistolarum* that was published in Alcalá in 1530.

The word *'ligur'* in Latin does not mean 'Genoese', but a person born in the huge province of Liguria. It was only in this letter, in one other written to Cardinal Ascanio in 1493 and in Book I of the 1st *Década* that Anghiera considered Columbus as being Ligurian. This is far from indicating Columbus's place of birth, and at the same time is a firm indication that Columbus wished to avoid giving any clue as to his true origins.

As he could not reveal his Portuguese birthright, he was quite willing to be seen as Ligurian and Anghiera was merely giving voice to the *'vox populi'*. What Anghiera clearly states is that this Ligurian had neither a Genoese nor a Spanish name. As Anghiera was Italian by birth, he must have spoken that language and as the chaplain of the Catholic Monarchs he must have also spoken Spanish; he also knew Latin, as all his writings are in that language.

The name of Columbus in Castilian was 'Cristóbal Colón; if he had been Genoese, it would have been 'Christóforo Colombo'. The Latin translation of *'Culan'* or *'Colom'* or *'Colón'* later appeared as 'Columbus', which would be correct if his name had been Colombo. However, the Latinist Anghiera made sure that he translated the name to *'Colonus'*, thus not giving it the Latin meaning of 'pigeon' (*columbus*). Besides this, the translation of 'Cristóbal' in Latin would be *'Christoforus'* as Anghiera translated the Portuguese word *'Christófam'*, may be spelt *'Cristóvão'*.

The diphthong 'am' or 'ão' and its respective sound are exclusively Portuguese, this would result in 'Christofom' or 'Christovom'.

Anghiera insists on giving Columbus the name of 'Colonus' and never 'Columbus'. This is correct, of course, as it is the Latinised form used by the humanists. The Spanish called him Cristóbal and adopted the abbreviation 'Xpoval' when writing; the Italians wrote 'Xpoferus' for their Christóforo and Columbus himself opted for 'Xpoferens'. Why would he use the Latin *ferens* after the Greek *Xpo*, that means Christ?

I shall prove that, while other people wrote his name, Columbus wrote a message.

When Anghiera, who had met and spoken to Columbus, wrote the latter's name, he certainly did so intentionally, as he knew that Columbus was not Spanish. He was also sure that Columbus was not Ligurian, as he never wrote the word 'Columbus', which sounded Paterine.

Pope Alexander VI, who was Spanish and to whom Columbus wrote in Spanish in February 1502, sent a letter (no. CXXVI of the *Opus Epistolarum*) to the Archbishop of Braga in which he refers to Columbus as 'Colom'. On 3 May 1493, the Pope had issued the Bull *'Inter Caetera'* in which he referred to *'Christophorum Colom, dilectissimum filium Hispaniae'*. It is interesting to note that the Pope did not consider Columbus to have been born in Liguria, neither did he mention Castellae, but 'Hispaniae'. In the fifteenth century the Catholic Monarchs were sovereigns of Castile (and Aragon), but not the Sovereigns of Spain, a title that came into being only with Emperor Charles V in the sixteenth century. The word 'Hispania' referred to the whole of the Iberian Peninsula.

The Pope, therefore, knew that Columbus was not Spanish and seems to admit that he could be Portuguese.

Anghiera's *Opus Epistolarum* was printed in Alcalá de Henares in 1530 and a second edition appeared in Amsterdam forty years later.

Several copies of these editions were later made in Italy, like the Vatican copy of 1657, and the name of Christophoms Colom was always interpreted as 'Christophorus Columbus'. The original manuscript has disappeared.

The Duke of Medinaceli wrote a letter to Cardinal Mendoza on 19 March 1493, in which he referred to Columbus's arrival in Castile in 1484. He said: 'I do not know if Your Excellency is aware that I had one Cristóbal Colomo, who came from Portugal, in my house for a long time (two years)'. He did not call his guest 'Colombo'.

Pope Alexander VI was, in fact, the first person to translate 'Cristovam' into 'Christophorus', which is quite correct in Latin. The corruption of 'Colonus' into 'Columbus' occurred in 1504.

Some of Anghiera's *Epistolas* entitled *Libretto di tutte le Navigazioni di Re de Spagna* were edited by the Venetian Angelo Trevisano, and two main alterations were made – to the name and nationality of Columbus. Where he found 'Christophoms Colonus', Trevisano copied 'Christoforus Columbus' and 'Ligurian' he changed to 'Genoese'. It is known that Columbus did not object to being considered Italian, since he would not be identified as Portuguese, but with Trevisano's distortion his ill-defined identity began to take shape.

Amerigo Vespucci's voyage and his much discussed 'baptism' of America naturally overjoyed the Italians, who immediately dubbed themselves as the greatest navigators in the world. In his work *Paesi nuovamente retrovati et Nuovo Mundo de Alberico Vesputio Fiorentino intitulato*,[1] Francesco Montalboddo boasted that the Italians had been the only ones to discover new lands since the Romans. The mariners of the rest of the world had done nothing but find what the seamen of Rome already knew. Despite the fragility of its arguments, Montalboddo's work was widely read in the sixteenth and seventeenth centuries and as it made constant reference to 'Cristofóro Colombo, Genoese,' it cemented the intentional mistake that painted everything in the local colours.

Many historians, among them Simon Wiesenthal, have concluded that the birthright attributed to Columbus was a forgery. Having gone to Rome in order to check the authenticity of the documents, the Italian authorities informed him that they were kept in the Vatican and could not be consulted.

Anghiera refers to 'Christophom Colonus' as a plebeian because he did not know any noble family by the name of Colón or Colombo in his native Italy, but only modest craftsmen that were possible descendants of Patarines. As far as Anghiera was concerned, the Admiral that the Catholic Monarchs had raised to the peerage came from nowhere, like a rabbit out of a conjurer's hat. This supposedly plebeian origin also appears definitively in the writings of the Genoese Agostino Giustiniani (1476–1536), Bishop of Nebbio (Corsica), who gave Anghiera's 'Christophorus Colonus' a completely new identity in his historical commentaries entitled *Psalterium hebraeum, graecum, arabicum et chaldium, etc.*:[2]

> in our times, through the notable daring of Christoforus Columbus, a Genoese, an almost different world has been discovered and a consortium of Christians united. This Genoese, born of humble parents, has explored more lands and seas in our time than nearly all the mortals of the last few centuries.

His short phrase *'humilibus ortus parentibus'* cemented the legend of a plebeian Columbus, ten years after his death.

Fernando Colón stated in Chapter II of his *Historia de Almirante*:

> We can only say that, due to the many errors and false statements that were to be found in Giustiniani's history (*'Psalterium'*), the Seigniory of Genoa – having taken the false statements into consideration – decided to punish those that possessed or had read the book and did everything in their power to have the copies they had distributed returned to Genoa, so that it could be banned by public decree.

The Genoese authorities, however, duped Fernando Colón, as the work was republished in Paris in that very same year (1516) (there is a copy in the Beriana Civic Library in Genoa) and included the following details (p. 51, lines 6–8): 'But when on his deathbed, Columbus did not forget his beloved homeland and left the Bank of St George, which the Genoese consider to be their most illustrious institution and the pillar of the Republic, a tenth of his income while alive and forever'.

The work was published in Genoa again in 1807 with the title *Castigatissimi Annali con la lora copiosa Tavola della Excelsa et Illustrissima Republica di Genoa*. This edition, with the title of 'Columbus's Gratitude to his Homeland' (p. ccxlix, lines 3–4 and 6–12) clearly explains the source of the false statements:

> Doctor Francesco Marchesi and Giovanni Antonio Grimaldi, ambassadors at the Court of the Sovereigns of Spain, returned (to Genoa) at this time ... And these ambassadors confirmed the report of the voyages of Columbus, who, called Christóforo, came from plebeian stock, his father being a wool-weaver, and he was a silk-weaver and, despite this, rose with such glory and such dignity to a position never achieved by any other Genoese, because he was the inventor of the voyages that are made between Spain and the Indies, i.e. the new world.

And regarding the income left *sine die* to the Bank of St. George, it added:

> although the above mentioned Bank, for some unknown reason, did not take up the offer nor did anything to obtain the said income.

FORNARI'S WORK

In the first chapter of the *Historia*, Fernando denies that his father was called Colombo. And because his name had been translated as Columbus in Latin, he added:

> We could cite many names, for example, which (not without reason) were given to indicate the fact that the marvels that the Admiral performed would turn out as foreseen: because if we notice the surname of his ancestors, we shall say that he was really a *Colombo* or *Palomo* when taking the grace of the Holy Ghost to the New World that he discovered, showing that, according to the baptism of St. John the Baptist, the Holy Ghost in the form of a dove demonstrated that he was the beloved son of God, who was unknown there, and as he carried the olive branch and the oil of the Baptism over the waters like the dove of Noah so that he would be able to unite the people of those places with the Church, as they were submerged in darkness and chaos, the name of Colón came to him, because it means *member* in Greek; since his first name was Cristóbal, he thought it sounded authentic, i.e. Christ, in whose hands those people would be placed; if we wish to give his name its Latin pronunciation, it is *Christophorus Colonus*.

'Columbus', therefore was a mistranslation of 'Colonus'. And even though it was not 'Colombo', the name would suit him, as he had played a role that was similar to the one played by the evangelical dove of the Holy Ghost and the biblical bird of Noah. Fernando emphasises the fact that colon means 'member'. He also says that 'Cristóbal' corresponded to 'Christ', – i.e., according to the thesis of Saúl dos Santos Ferreira and Ferreira de Serpa, which I follow in this book, '*Salvador*' (Saviour).

Don Luiz de Colón, grandson of Cristóbal Colón and third Admiral of the Indies, was the nephew of Fernando Colón, from whom he inherited the manuscript of the *Historia del Almirante*. Don Luiz gave his uncle's original manuscript to a Genoese, Baliano de Fornari, so that it could be published in Italian, Latin and Castilian.

Taking the manuscript to Venice, Fornari encharged one Giácomo Bautista Marino to publish the book. Marino, however, delegated the task to a third party, Giuseppe Moleto, who finished the work on 25 April 1571. The original manuscript disappeared, but the blame for the first distortions inflicted on the *Historia del Almirante* must be laid on the shoulders of these publishing 'middlemen' and González Barcia, the author of the first translation in Castilian in 1749, a work replete with errors.

Why did the Genoese Fornari not try to print the book in Genoa? Was he afraid that nobody would be able to identify Cristtoforo Colombo as the Admiral of the Indies? Since Fernando began his book by saying that his father was not called Colombo, he would never have mentioned a cousin of that name in the middle of it.

Given the secrecy that the Colóns had always maintained regarding their family and birthright, Fernando Colón would never have written: 'Juan Antonio Colombo, a cousin of the Admiral', which would have been an invitation to interested parties to bombard Juan Antonio with questions about Colón's origins. And Fernando would never have claimed to have been the grandson of a 'Domingo', as the Genoese writer had insinuated when he inserted the last two lines of Chapter LXIII of the *Historia del Almirante*,[3] in order to contend that Columbus had given the name of Santo Domingo to a bay neither in honour of a saint nor of the two towns in the Duchy of Beja, but 'in memory of his father, who was called Domingo'.

THE *CAPITULACIONES* DIPLOMA

In the fifteenth century, social strata were much less permeable. The *Capitulaciones* was the diploma drawn up in Granada on 30 April 1492 and signed by the Spanish Sovereigns and by which the navigator, who had left Portugal with the name of Cristóvam Colom, is referred to by the name of Cristóbal Colón. This diploma was granted before the Admiral left for the Antilles and confirmed by the *Provisión* of 20 May 1493, which is in the Columbian Museum of Seville and reproduced in Navarrete's *Collección*:[7] 'to some recompense for what you have *already* discovered in the ocean seas and for the voyage that you now, with the help of God, will make across them'. The work includes a description of the coat-of-arms that the Catholic Monarchs bestowed on Columbus, in which the fourth quarter is described as 'the arms that you *already* have' through your family; these family arms were made up of five anchors arranged in form of a sautor (diagonal cross), and the diploma refers to him as 'noble Don Cristóbal de Colón (see p. 484 and Figure 34.6).

So the Catholic Monarchs knew that Columbus was a nobleman and did not hesitate to bestow on him, besides the posts of 'Chief Admiral of the Ocean Sea, Viceroy and Governor of the Indies and Terra Firma', the titles of 'Knight of the Golden Spur', 'Grandee of Spain', and the privilege of being allowed to wear his hat in the presence of the sovereigns, which was granted to only the highest ranking lords of the realm. If the Monarchs confirmed the fact that he already had a coat-of-arms, it was because they had proof of his nobility. A nobleman was easily distinguishable in those days by his family links and relations and a plebeian would never have been able to deceive anyone. The Sovereigns knew what Columbus was from his bearing and culture, and from the proofs of his identity that he presented to them. It would have been fairly easy for King John II to furnish the Admiral with credentials that he could display whenever he wished, but if Columbus had been the son of a Genoese wool-carder and tavernkeeper, he would not have possessed a coat-of-arms for the Catholic Monarchs to *confirm*.

Several facts prove that Columbus was of noble birth. King John II addressed him in a very amiable letter as 'our special friend' (see p. 228) and let him be seated in his presence at a royal reception; he married Filipa Moniz Perestrelo, daughter of the Donatory-Captain of the Island of Porto Santo, whose ancestors already had a coat-of-arms, as did his wife Isabel Moniz Barreto, related to notable noblemen and kinsman of the daughter of Dom Nuno Alvares Pereira, Constable of the Realm and Count of Barcelos, the most important personage in the Court (the title of Marquis did not yet exist and that of Count was the second, after Duke, in the scale of nobility).

Taviano (see pp. 105–6) wrote that the Portuguese chronicler João de Barros had claimed that Columbus was Genoese. But, in his first *Década*[8] (1552), Barros merely mentioned that 'According to what everyone says, this man [Columbus] is Genoese'. The expression 'everyone' means the authors that, by around 1515, had written of the Admiral (Trevisano's *Epistolas* and Montalboddo's work). The chroniclers who had written of Columbus included:

1. Fernando Colón's friend, Las Casas who, trying to help Fernando in his process 'against the Crown' without making any clear statement, limited himself to saying: 'I heard that he was from Liguria' (a region that included parts of Monferrato, Piedmont and Milan down to the Genoese coast). He had heard this from Anghiera.
2. Oviedo[4] declared: 'according to what I learned from his fellow-countrymen, he came from the province of Liguria . . . some say from Savonna, others say from Nervi, others from Cugureo', but he had got this information when he was in Naples between 1497 and 1511.
3. André Bernáldez, Curator of the Palaces of the Archbishop of Seville (ex-Bishop of Palencia, Diogo de Deza, see p. 414 and Tree IX, p. 411), referred to Columbus in the following sentence: 'There was a man from Genoa, a vendor of books and prints', but later contradicted himself by writing 'Don Cristóbal de Colón, of great and honoured memory, was born in Milan'.[5]

The Portuguese historians mentioned so far all lived later than Rui de Pina and based their writings on his work *Crónica do Rei D. João II*, in which he wrote a chapter under the title 'Descobrimento das Ilhas de Castella *per* Collombo' (Discovery of the Islands of Castile by Columbus). And while all the other Portuguese chroniclers wrote 'it is said that Columbus was Italian' (or Genoese, Milanese or from Savonna), Rui de Pina states, without a shred of evidence: 'Christopher Columbus, Italian'. However, in a letter that King John II wrote to Columbus in 1488 he addresses the navigator as 'Cristovam Colom'.

It is normal for historians, when recording past events, to do no more than copy or summarise other works. But Rui de Pina is a special case. While, for example, King John II rewarded Diogo Cão because he thought the latter had rounded the Cape of Good Hope, he refused to reward Bartolomeu Dias in the same way when this navigator actually performed the feat. In order not to displease his monarch, Pina referred to the fact that the Cape had been rounded but did not mention the name of the navigator in question.

Domingos Gomes dos Santos of the Portuguese Academy of History wrote:[6]

> The knowledge of each person in regard to European history . . . is manifestly limited and not free of errors. From a critical point of view, Rui de Pina is a typical example of a palace historian, always ready to praise, but at the same time reticent . . ., an example being the laconic reference made to the deaths of Prince Afonso (the son of King John II who was killed in an accident) and Prince Pedro in the Battle of Alfarrobeira.

The same type of criticism is levelled at Pina by other Portuguese historians such as Dias Dinis (*As Crónicas Medievais Portuguesas – Adulterações de Rui de Pina*), and Manuel Braamcamp Freire (*Crítica e História*).

THE ORIGINS OF COLUMBUS

Fernando Colón is also alleged by Fornari to have said: 'so that some, who up to a point wish to lower his reputation, say he was from Nervi; other say he was from Cugureo and other Buyasco, which are all small places of the city of Genoa and its coastline; and others, who wish to further exalt his name, say he was from Savonna and others say he was Genoese'. This is another dubious text. Although the writer tried to show Fernando Colón as possessing a deep knowledge of the topography of the Genoa coastline and its towns, he overlooks the fact that it is not a person's place of birth that raises his prestige, but his social position, or the post that he holds. It seems that Fornari considered a Genoese tavern-owner to be more illustrious than, for instance, the Lord of Nervi.

Fernando Colón added: 'those that know most regarding his origins maintain he was born at Placenza, where some honoured members of this family are buried in tombs bearing coats-of-arms and epitaphs of the Colombos [a clear distortion of Colonna] because this, in fact, was the surname of his ancestors; he himself shortened his name so that it resembled the old . . . and so called himself Colón, depending on the country in which he started a new life'.

But if, in one chapter, Fernando repudiates all the writers that claimed that Columbus was called 'Colombo' and not 'Colón' ('member', 'arch' in Greek), how could he then claim that his family name was Colombo? Fernando Colón took pains to point out: 'those that know most regarding his origins maintain he was born in Placenza'. But who would know more of the Admiral's origins than his own children? Those, therefore, that suggest that he came from Genoa were either ignorant of his origins or wished to attribute the glory of being the birthplace of the 'discoverer of America' to that city.

Fernando Colón was unusually cultured and erudite. He assembled one of the biggest libraries of the time, in order (according to the executor of his will, Marcus Felipe) to 'gather every book of every language and subject that can be found within and without Christendom'.[7] Being a scholar of classical Latin texts, he frequently quoted them. In the last chapter of his *Historia del Almirante* he supposedly wrote:

> some people wanted me to claim that the Admiral came from an illustrious line, even though his parents had had the misfortune to reach the depths of great poverty; and they wanted me to show that they were descended from the *Colon* who Cornelius Tacitus, at the beginning of the twelfth book of his work, said had taken King Mithridates to Rome as a prisoner, which led the people to honour *Colon* with the rank of Consul and the Eagles.[8]

Fornari must have inserted this text into the *Historia*, as Fernando Colón would never have made such serious mistakes with historical facts. In his *Annals*,[9] Tacitus relates the episode of Mithridates's capture in the year 50, but makes no mention whatsoever of any 'Colon'. He clearly refers to the Governor Junius Cilo and to Julius Aquila, the former having been rewarded with the insignias of Consul and the latter with those of the Prefecture.[10] What was the reason for such a distortion? Surely because Fornari had found an allusion to the 'Colonnas' of Lombardy in Fernando Colón's text and thought it prudent to erase it.

He missed, however, one highly compromising statement: 'his ancestors being of the royal blood of Jerusalem, [the Admiral] thought it best that his parents should not be so well known'. Here is a confession of his Jewish descent and a veiled reference to the dangers of the Inquisition.

The surname 'Colonna' comes from the Latin word *columna* (*colomna* or *colonna* according to etymological modifications).[11] Fernando clearly stated that his father 'shortened his name', so that he was called 'Colom' in Portugal and 'Colón' in Spain, but the name was shortened from *colomna/colonna*, and not from the Italian *colombo* (pigeon).

This is a clear indication that the Admiral descended from the Colonnas of the branch in Lombardy, the Italian state in which Placenza is located. There has never been a tomb in this city bearing the arms of the Colombos, neither has a family of this name belonged to the nobility in any part of the world. It is remarkable that nobody has so far thought to make a logical analysis of Chapter I of the *Historia del Almirante*, written by Columbus's son and most intimate chronicler.

Notes and References

1. Venice, 1507
2. There is a copy, dated 1516, in the Beriana Civic Library, Genoa.
3. p. 251.
4. Oviedo (1547; 1749).
5. Bernáldez (1570).
6. Santos (1932).
7. Hernandez y Muro, *El Testamento*, p. 227.
8. Aquila was the officer's name and not an insignia (p. 474).
9. XII, 15 and 21.
10. The people gave Cilo the nickname of *Cilon*, but not '*Colon*', a word that does not appear anywhere in the text: 'the insignias of Consul were granted to Cilo and those of the Prefecture to J. Aquila'. Another part of the text mentions that the troops of the Kingdom of Cotys '*were commanded by a Roman cavalry officer, Julius Aquila*'.
11. The first Colonna adopted his name from the Column of Trajanus, which was near his house in Rome.

16 The Descent of a Weaver

There are today eighteen Italian towns and cities that claim to have been the birthplace of 'the discoverer of America', these claims stem from the fact that the surnames 'Colombo' and 'Columbo', were to be found everywhere, despite the fact that the Inquisition had extinguished the Cathar sect of the Patarenes, most of them congregated in weavers guilds 200 years before (see pp. 3–4). In their search for a Cristóforo among these ubiquitous Colombos, the defenders of the Genoese thesis scoured the archives of the region of Liguria. In a search limited to twenty-seven different places, Henry Harrisse[1] found 124 Colombos with the first name of Bartolomeu, Doménico, Giácomo, Giovanni and Cristóforo, although the last one mentioned was not so frequent.

It was not very difficult, therefore, for the 'Genoists' to settle on a Cristóforo that had a brother called Bartolomeu. Later – as will be shown – they had only to transform a Giácomo into Diogo and have the three navigators of the Columbus family. But they found it necessary to find a paternal grandfather with a double personality, in the form of Giovanni Canajole de Colombo of Quinto, in the suburbs of Genoa, and Giovanni de Colombo who lived at Moconexi and was the brother of one Luce (who sired a Giovanni) and Benedita. They amalgamated the notarial minutes concerning these two progenitors and chose Doménico, son of one Giovani, whose offspring satisfied the necessary conditions.

Doménico, who lived in either Genoa or Savonna, alternated between the trades of weaver, cheese-maker, wool-carder, tavern-owner, and, finally, weaver once again. This suggests an unnatural fusion of disparate personalities that at times share the same name. He also had a brother called António who owned a tavern.

Doménico married one Susana (daughter of Giácomo de Fontanarossa and sister of one Coagino), who gave him four sons and a daughter: Cristóforo, Giovanni Pelegrino, Bartolomeu, Giácomo and Bianchinetta (who married Giácomo Bovarello and gave birth to one son, named Pantaleone). António, in turn, also had five children: Benedicto,

189

190 BIOGRAPHICAL PROBLEMS

∞ = in wedlock
ϕ = out of wedlock

a

Doménico Colombo, wool carder and weaver
 ∞
Susana Fontanarrosa

- Bartolomeo Colombo, wool carder
- Giovanni Pelegrino, who died young
- *Cristoforo Colombo*, wool carder and tavern keeper
- Giácomo Colombo, wool carder
- Bianchinetta
  ∞
  Giácomo Bovarelo, cheese maker
  — Domenico Pantaleone

António Colombo, tavern keeper, brother of the above
 ∞
 (?)

- Giovanni Colombo, tailor
- Mateo Colombo
- Amigheto Colombo
- Tomaso Colombo
- Benedicto Colombo

b

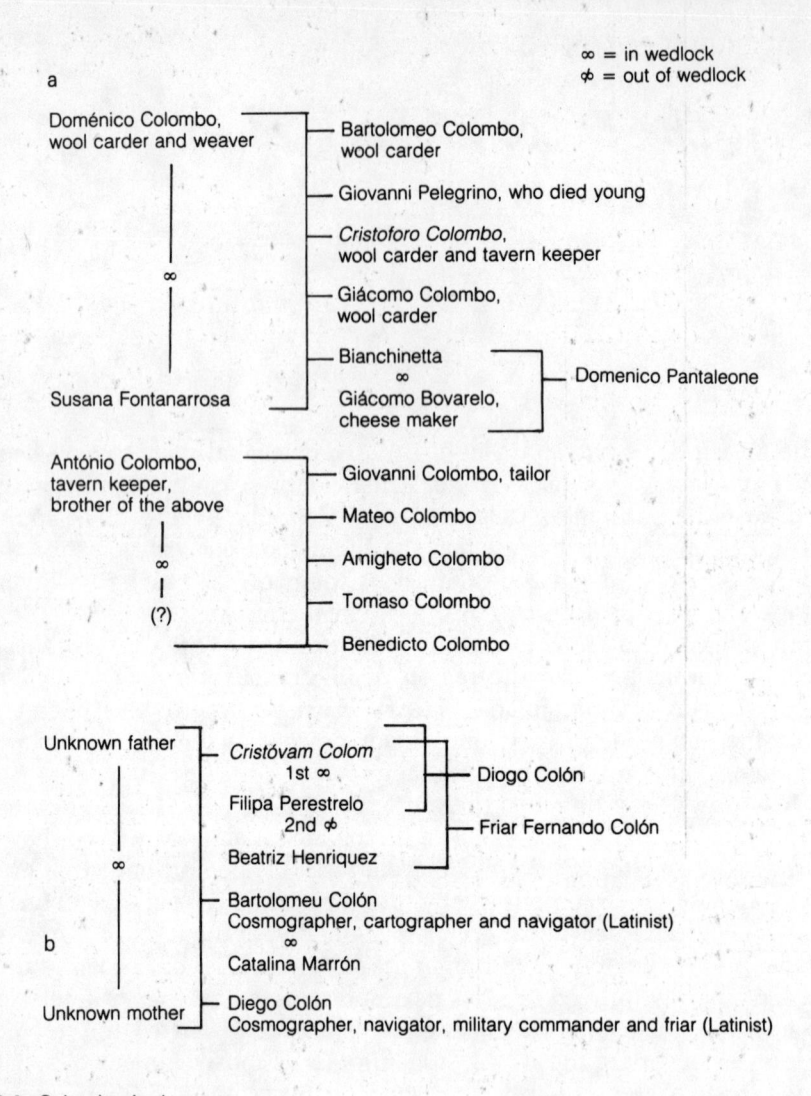

Unknown father
 ∞
Unknown mother

- *Cristóvam Colom*
  1st ∞
  Filipa Perestrelo
  2nd ϕ
  Beatriz Henriquez
  — Diogo Colón
  — Friar Fernando Colón
- Bartolomeu Colón
  Cosmographer, cartographer and navigator (Latinist)
  ∞
  Catalina Marrón
- Diego Colón
  Cosmographer, navigator, military commander and friar (Latinist)

16.1 Columbus's descent.
  a Genoese account.
  b Coeval account.

Tomaso, Mateo, Amigheto and Giovanni (see Figure 16.1). Doménico's many contract and judicial disputes were registered by thirteen different notaries of Genoa and by four from Savonna; he appears as a resident in various buildings in both cities and their respective suburbs, which he let, sub-let, sold or ceded, constantly changing both his place of residence and his trade. Could it really have been one and the same person, always on the run from his creditors?

NOTARIAL DOCUMENTS OF GENOA AND SAVONNA

Let us now have a look at some of the thirty-nine notarial documents, copies of which the Italians allow researchers to consult.

1. On 18 January 1455, the Monastery of St Stephen rented a house in the *'borgo* [suburb] *di Sancto Stefano'*, on the outskirts of Genoa, to Domenicus de Columbus.
2. On 22 September 1470, in Genoa, Domenicus de Colombo authorised his son Christoforus to appoint an adjudicator to decide on a debt owed to one Geronimo de Porte.
3. On 28 September 1470, also in Genoa, Domenicus, acting as guarantor, authorised his son to accept responsibility for the payment of the debt of 35 lira to the said Geronimo.
4. On 31 October 1470, in Savonna, Dominicus authorised his son, aged 19, to accept responsibility for the payment of wine he had sold without presenting accounts to the producer, Pedro Balascio de Porte. His son, therefore, would have been born in 1451.
5. On 25 August 1472, again in Savonna, Dominicus de Colombo and his son Christoforus recognised a debt of 40 lira to Giovani de Signorio.
6. On 7 August 1472, in Savonna, Susana Fontanarubea, in the presence of her sons Christoforus and Johannis Pellegrino, authorised their father, Dominicus de Columbo, to sell a house.
7. On 23 January 1477, Dominicus de Columbos, wool-carder, citizen and inhabitant of Savonna, with the authorisation of his wife Susana, sold a house in the suburb of St Stephen, near St Andrew's Gate, in Genoa, to André de Cairo.
8. On 25 August 1479, according to the draft of a document presented by Hugo Asseretto,[2] Christoforus Colombo made a sworn declaration before the notary Vintimiglia in which, on being asked his age, declared: *'etatis annorum viginti septum vel circa'* (27 or thereabouts), and also announced that he had to depart *'die crastino pro Ulisbona'* (for Lisbon the following morning). The draft was discovered in 1904.
9. In 1856, the Count Roselly de Lorgues[3] discovered a document dated 1484 which states that Jacopus de Colombo, son of Dominicus, entered the house of Luchino Cadamartori in Savonna as an apprentice weaver.
10. In a document of 25 August 1487, Jacopus de Colombo, already over 19 years of age, appears as a 'cloth weaver in the house of Dominicus de Colombo' in Genoa.
11. On 22 July 1489, in Savonna, Dominicus de Columbus, as the father of Christoforus, Bartholomeus et Jacobus and the administrator of

the assets of their mother, ceded to his son-in-law Bovarello the house in the suburb of St Stephen, near St Andrew's Gate, which had been promised as the dowry of his daughter Bianchinetta.

12. On 17 November 1491, Genoa, Jacopus de Columbus sanctioned the payment of a debt of his father, Dominicus, to Nicolo Rusca.
13. On 11 October 1496, Mateu, Amigesto and Giovanni (three of Antonio's five children) gathered in the presence of the Genoese notary Piloso and decided that Giovanni would have to go to Spain to look for *'Christoforus Columbum almirantum regis Ispane'* (Admiral of the Kingdom of Spain) in order to collect a debt of his father Dominicus, a weaver; they arranged to share both the expenses of Giovanni's trip and the amount he managed to collect.
14. On 26 June 1501, in Genoa, the son of one Corrado attempted to bring a court case (for a debt of Dominicus de Columbo, a weaver who had lived in Savonna) against Christoforus, Bartholomeus et Jacopus, heirs of the said Doménico, 'absent beyond Pisa and Nice, in Provence, and are at present somewhere in Spain' (*'in partibus Hispaniae commorantes'*).

With these fourteen notarial documents drawn up in Genoa and Savonna, the defenders of the thesis of a Ligurian Cristóforo, wool-carder and tavern-owner, were able to fit him into the shoes of the navigator Cristóbal Colón. Let's see if their arguments stand up to logical criticism.

A We may begin by noticing that the Latin name *Jacobus* (Giácomo in Italian) was translated into *Jayme* in fifteenth- and sixteenth-century Castilian. While 'Jacob' corresponds to 'Iago, Tiago, Jaques, James' and even 'Diogo', it has been proved that the Iberian 'Diego' or 'Diogo' was translated into Latin as *Didacus*, and not *Jacopus*, both in Iberia and Italy. This can be seen in a document drawn up in Savonna on 30 March 1515, in which Leone Pacaldo, 'procurator of the great Didacus Colón', delegates his notarial powers to Antonio Romano. He was referring to Diogo Colón, the Admiral's son, who supposedly tried to settle a bank debt of his father's. It shows that at that time the name Giácomo did not correspond to Diogo, as it certainly did later.

B The first six documents are the only ones that mention Christófoiro, living in Genoa or Savonna, either as a wool-carder or tavern-owner, up to August 1473. But in this case, this cannot be the same man who, just one year later, was already a navigator, cosmographer and cartographer and was exchanging letters between Lisbon and Florence, in Latin, with the scholar Toscanelli.

C Document 2 proves that Cristófoiro was born in 1451, which is a crucial element of the Genoese thesis as it annuls the consistency of rival theses.

D Let us note the wording of document 8.[4] It is strange that a normal 27-year-old man, on making a sworn statement, should not have a precise notion of his age. That the notary should have accepted the vague expression *'vel circa'* is also surprising, as the swearer was a legitimate son in the presence of his father. It is evident that the writer (Asseretto) was able to base his evidence only on the document of 31 October 1470, which mentions that Cristóforo had 'reached the age of 19', the age when one was able to accept judicial responsibility. But this does not mean that Cristóforo was 19 at that time; he could have been older. But the object was to confirm that Columbus had been born in 1451 and so found it necessary to use the expression 'or thereabouts'.

E In document 9, it is difficult to understand how, in 1484, the weaver Doménico sent his son Giácomo to the house of one of his competitors in Savonna to learn his trade, instead of teaching him himself, especially as we see, in document 10, that Giácomo was working as a weaver in his father's house in 1487 and remained there, according to document 12, until at least November 1491. So how can it be that in March 1494 (a mere two years and four months later) Giácomo was already a navigator and had been appointed 'Governor and Commander of the Fleet' on the Island of Isabella while Columbus went overland to the mines of Cibao?[5] When he returned to Spain in 1495, Diogo Colón, already a Latinist and an expert on Biblical texts, also manifested the intention of becoming a priest.

F Document 11 points out a curious fact: in 1486, Doménico ceded to his son-in-law Bovarello the same house that he had sold (according to document 1) to the said Cairo nine years before (1477). Had he rebought the house in order to give Bianchinetta her promised dowry, in spite of the fact that he lived in various parts of Genoa and Savonna and was always pursued by his creditors?

400 years later the Italian Marcelo Staglieni claimed to have identified this house in the *'borgo de Sancto Stephano'*, on the Via Mulconti, near St Andrew's Gate, as the one where the 'discoverer of America' had been born. The city authorities duly put a commemorative plaque on the wall. But if this same house (according to document 1) was let to him only in 1455, how did Cristóforo come to be born there four years earlier?

G Let us examine document 13. In 1496, the three sons of Antonio (Doménico's brother) knew that their first cousin Cristóforo was 'Admiral of the Kingdom of Spain'. They could, therefore, conclude that he had become important and rich. If so, why did they fear that they would not be able to collect their uncle's debt in full? It was not a large sum, so why would they risk spending enormous sums of money to travel to Spain when they were so unsure of how much Giovanni would collect from the Admiral? As they would have to trust their brother's word, why would they have needed a notary?

H From document 14 of 1501, it can be seen that the son of the said Corrado (who had known and traded with Doménico's family) was aware that the weaver's three sons were in Spain, but did not know that Cristóforo was an Admiral and Viceroy. As he was unable to locate the three men, why would he have paid for the service of a notary? It must be remembered that both the cousins of the navigator and the notary Piloso had supposedly been aware of Cristóforo's triumph five years earlier and such news would have spread through Genoa like wildfire at the time.

How can one explain the silence of the Genoese in regard to such an extraordinary feat performed by one of their fellow-citizens? No one bothered to celebrate it: no relative or neighbour of the Colombos of Genoa or Savonna linked the wool-carder and tavern-keeper Cristóforo Colombo with the celebrated Cristóbal Colón. The three navigators in the service of the Catholic Monarchs never gave any indication that they had any Italian relatives that deserved the least attention.

The Genoese Bank of St George had been corresponding with Cristóbal Colón since 1490, from the time that the latter had wanted to raise a loan in order to discover India from the West and within the close financial relationship that existed between Spain and the Republic. In 1493, the ambassadors Francesco Marchesi and Giovanni Antonio Grimaldi had returned to Genoa bearing the news of Columbus's success. Moved by rumours from Spain that he might have been born in that city, attempts were made to find his ancestors. But no one knew who he might have been:[6] not even the notaries.

This vacuum is even more inexplicable when it is seen that the weaver Doménico, in a minute drawn up on 30 September 1494, was a witness to the will of one Catalina Bernaza. He was, therefore, alive and totally unaware of the ill-conceived legend that his three sons had been suddenly projected from the world of wool-cards, looms and taverns into one of navigation, cosmography, cartography and classic studies. In fact, this theory rebounded in the faces of its expounders, as Fernando Colón never managed to find any relatives or anyone who had ever heard of them.[7]

I About the minute of the notarial deed of 1479 (document 8), presented by Hugo Asseretto in 1904, Luis de Ulloa expresses the following view:

It is sufficient to run one's eyes over the facsimile of the 'Asseretto document' to notice three distinct handwritings. Even if the notary that drafted the document had frequently changed the scribe, this cannot guarantee us the authenticity of the document . . . If we look closely, we shall see the name *Xforos* written in different ways, including once a 'C' instead of 'X'. The link of the 'o' or the 'r' and this to the 'u' was done three or four times, so it can give the appearance of being *Xristoferens* . . . The inventor of the Genoese minute of 1479, as much as he tried ad hoc, was not very skilful. The ruse of the 'Genoists', therefore, collapsed due to their own audacity.

According to Luis Arranz Marquez (p. 237 of the *Introduction* to his work on Columbus's *Diario de a bordo*)[8] the 'Asseretto document' was not accepted as a proof, as there was a loose page that had been stitched into the notarial book at a later date.

Based on documents that never came to the notice of the public, Julio Salinério wrote a work in 1606, in which he transcribed some notarial minutes (among which are mentioned in documents 13 and 14), that were supposedly drawn up in Savonna, and another in which certain anonymous creditors claimed possession of an uninhabited building that was said to belong to the 'Columbos' who, according to the neighbours, were in Spain.[9] The originals of these minutes have never been seen by anyone.

THE REAL DATES OF COLUMBUS'S BIRTH AND ARRIVAL IN SPAIN

Alexander von Humboldt[10] noted that the theories regarding the date of Columbus's birth covered a period of twenty-five years and mentioned the calculations of the following researchers: Ramusio (1430); Bernáldez, Napione and Navarrete (1446); Spotorno and Robertson (1447); Willard (1449). According to the 'Genoists' Columbus came into the world in 1451. Nobody has suggested 1448.

The date of birth of the Genoese wool-carder and tavern-owner Cristóforo Colombo is not of the slightest interest; what is of interest is the date of birth of the navigator who took the name of Cristóbal Colón in Spain. This can be easily deduced from Columbus's entries in his logbook and in the letters that he wrote, as well as from the details that appear in his son's *Historia del Almirante*.

It can be proved that Columbus went from Portugal to Spain in 1484. In a letter written in 1500[11] he said: 'it is seventeen years since I came to serve these Princes with the Enterprise of the Indies; the other eight were spent embroiled in disputes and, at last, I advised them of the fraud'. This leads us to 1484 and adding eight years to this brings us to 1492, the year he received authorisation to sail for the Indies. Columbus also adds: 'I made this conquest through God's grace seven years ago'. Writing to Juana de la Torre, Prince Juan's Governess, in 1500[12] Columbus said: 'seven years were spent in talk and nine doing many things of outstanding merit'. These sixteen years bring us once more to 1484.

Let us now calculate the date of the Admiral's birth. Fernando Colón (Chapter IV of the *Historia*) quotes a letter from his father in which he states: 'I first went to sea at the age of 14'. In a letter to Cardinal de Mendoza dated 19 May 1493, the Duke of Medinaceli said that Columbus had stayed at his house for two years before entering the service of the Catholic Monarchs. Columbus wrote in his log on 14 January 1493: 'during these six or seven years of great distress, arguing to the best of my ability, as to how much could be done'. He therefore spent two years as the Duke's guest from 1484 to 1486 and six or seven arguing his case. Columbus wrote in his log on 21 December 1492: 'I served before the mast for 23 years without a break'. Adding those twenty-three to the seven on land and his fourteen years of age we get forty-four. Subtracting those forty-four from 1492 gives us the date of 1448. In a note dated 1501[13] Columbus states: 'I went to sea at a very tender age and have been at sea ever since . . . I have been in this trade for forty years'. If we add those forty to the thirteen before he went to sea then subtract the resulting fifty-three from 1501 we again come to 1448.

Columbus distinguished the overall period in which he devoted himself to the Discoveries, to which he gave the generic name of the 'Enterprise of the Indies', and the time in which he was in the service of the Catholic Monarchs. It is obvious that he had been in the service of Portugal before this. It is only the anti-Portuguese theses that maintain that Columbus had explored the Atlantic at his own expense before going to Spain, as we know that this exploration was forbidden unless in the company of the Portuguese and that Columbus was never rich enough to be able to do such a thing.

Fernando Colón transcribed Toscanelli's second letter to Columbus in 1474, in which he stated: 'I have received your two letters, together with the many things you sent me, and with them I received a large reward'. As Columbus was in Lisbon at this time, it could only have been the Prince John (future King John II) who rewarded Toscanelli for the services he had rendered Colom. This fact destroys the credibility of the shipwreck in 1485 or in 1476 (see p. 100).

In the letter known as *Littera rarissima*, dated 7 January 1503 Columbus, having fallen into disgrace, lamented: 'it is my bad luck that the twenty years of service that I rendered with so much hard work and danger have done me little good, as I do not have a roof over my head in Castile'. If we go back twenty years from December 1502, we come to 1482. With two years of inactivity, we have 1484. In the same letter he adds: 'I came here to offer my services at the age of 28'; and later: 'God gave you the keys . . . to the padlocks [he wrote the word *atamientos*] of the Ocean seas,[14] which were closed with such heavy padlocks . . . [God] has now given you his reward for this work and danger, serving others [lords]'. He is insinuating that his first Lord, King John II, had never rewarded him for the years in which he was in his service. In any case, Columbus could not have been only twenty-eight in 1486, and even less so in 1492. He was not, therefore, referring to the date he entered the service of the Catholic Monarchs, but the one when he started the 'Enterprise of the Indies' in 1476. And if we subtract twenty-eight from that date we again get 1448.

The Christoforus that was born in Genoa in 1451 and who was a wool-carder and tavern-owner in 1473 could thus not have been the 'discoverer of America', cultured and experienced in the ways of the sea, born in 1448, who had first gone to sea at the age of fourteen and who had then spent forty years before the mast without a break.

MURATORI'S DOCUMENT

In 1733, one Muratori, in his book *Rerum Italicarum Scriptores*, included a crudely - written opusculum that he attributed to a Genoese (Antonio Gallo, Chancellor of the Bank of St George of Genoa) and presumably drafted in 1499. This opusculum (two pages only)[15] is entitled *De Navigatione Columbi per inacessum antea Occéanum Comentariolus*. This banker and notary claims that 'it was Bartolomeu Colombo who conceived the idea of sailing West, when he was in Lisbon observing the discoveries of the Portuguese beyond S. Jorge da Mina on the maps that he drew to earn his living'. He refers to Columbus with disdain, saying that he possessed only primary education and that it was his brother Bartolomeu who had taught him cosmography and navigation; he also said that he had written the *Comentariolus* with a map drawn by Columbus himself in front of him and using information that had been supplied by his father, the wool-carder Doménico Colombo.

Gallo may have been influenced by the same words of Giustiniani's *Psalterium Hebreaecum, etc.* (1516) and *Castigatissima Annali, etc.* (1537), both published in Genoa. Gallo's original disappeared, and only four much later 'copies' have been found, all of them suspect. How likely is it that a son of a weaver – and a foreigner – would have been able to make prohibited maps for the Portuguese who were charting the coasts of Africa? The only connection that existed between Gallo and Columbus was the fact that the former, as he claimed, may have bought a copy of the Latin edition of a map from Alonso Sánchez (from Huelva), who accompanied Columbus on his first voyage, and which was supposedly made by Columbus and Sánchez. However, neither the misty relationship between Columbus and Sánchez nor between Sánchez and Gallo can be taken at face value, since Sánchez sold maps he himself had made as the work of Columbus.[16]

SUSPECT DOCUMENTS

Bernáldez (see p. 82) gave lodging to Columbus for several days and came to the conclusion that he was Genoese, as he spoke very poor Castilian and denied that he was Portuguese. He referred to Columbus in his chronicle in the following terms:

> There was a book-seller from Genoa called Cristóbal Colón, who was working in Andalusia, a clever man without being very versed in letters, very able in the art of Cosmography and dividing up the world . . . He left me some of his writings in the presence of Sr. D. Juan de Fonseca, which I accepted and joined to some others that were written by the highly honourable doctor Chanca and other noble gentlemen who went with him on the already mentioned voyages . . . from which I wrote this about the Indias.

It is natural that someone who had been at sea since the age of fourteen was not versed in letters, Latin and Castilian. He would not have had the time to learn these two languages correctly.

Just as Bernáldez said that Columbus was 'not very versed in letters', Columbus could have retorted that Bernáldez was 'not very versed in sciences'. Columbus appeared to Bernáldez as a salesman of illustrated books, allowing himself to be seen as a Ligurian and simple merchant. The noblemen that the chronicler refers to were 'Caballeros' of the same class as Dr Chanca.

The fact that Columbus went to the 'Cura de los Palácios' to expound his views regarding the 'division of the world' and the Indias is curious. Did he hope that the royal chronicler would promote the possibility of discovering the Orient by sailing west in the court? With those writings in poor Castilian and with the information supplied by others after they had been (with or without Columbus) to the Antilles, he wrote what he could about them. He finished his work in 1513 (i.e., twenty years after Columbus had made his first voyage for Castile, had been appointed Admiral, had received other titles and had been raised to the peerage). The chronicler was ill-informed about the diploma that made Columbus a Lord; he either did not know that Columbus was already a nobleman or he did not like him, as he impersonally introduces the Admiral to history with the expression 'there was a book-seller'.

Bishop Giustiniani must have been inspired by Bernáldez's mistake when he created a Genoese and plebeian Columbus, a hero of his *Castigatissimi Annali, etc.*, written three years after Bernáldez's *Chrónica*. Details about Columbus were bandied about everywhere.

There is documentary proof that a Doménico Colombo, who had two sons called Cristóforo and Bartolomeu, lived in Genoa in the fifteenth century; there is nothing strange in this, but the story of a weaver's apprentice that became a travelling salesman and then an experienced and cultured navigator who discovered America should convince nobody.

THE MISSING DEED 'MAYORADGO'

The firmest base on which the Genoese thesis is built is the bestowing on Columbus of an 'Entailment', known in Castilian as *Mayoradgo*. The Spanish Sovereigns granted Columbus a Royal Deed allowing him to found an entailment on 27 April 1497. For this reason, Columbus had allegedly drafted a minute, which was confirmed by public signing of the deeds on 22 February 1498. Prince Juan of Spain was appointed to sign the Royal Deed and assure that the articles of its foundation were fulfilled. This Prince, however, died on 4 October 1497, so he obviously could not have signed the deed five months later. Martin Fernandez de Navarrete[17] states that despite an intense search that he and his investigators made in the Spanish Academy of History, no trace was found of the original of the institution of the Entailment or of any other coeval certificate which was said to have been registered by a Sevillian notary, Martin Rodriguez.

In order to analyse the contents of the *Mayoradgo*, Navarrete had to copy the only existent, suspect, document which was a part of a lawsuit of succession that was brought by one Baldasario Colombo, a Genoese who claimed he was Columbus's heir seventy-two years after the Admiral's death. This document is entitled *Mayorazgo*, instead of *Mayoradgo*, which is surprising, because the new word, with a 'z', was adopted only half a century after the deed's date.

Navarrete, who reproduced the document of 1498 in his work '*Colección de los Viages and Documentos de Colón*'[18] and also published an opusculum in which he defended the authenticity of the *Minute*, ended up by later confessing his doubts in relation to the genuineness of the document to the archivist of Simancas, Don Tomás González.[19]

The fact that the writer had forgotten or ignored the death of Prince

Juan in October 1497, thus ruling out the possibility of the signature of the 'royal confirmation' of April 1498 being his, led Navarrete to wonder whether it was not an orthographic error and that the signature of Prince Juan should read Princess Juana, the Prince's sister. Even if the Princess did write her brother's name instead of her own or signed in an illegible hand, however, it must not be forgotten that in those days a married princess could sign a document only together with her husband. As Princess Juana had married in 1496, protocol demanded that her husband's signature appear alongside hers.

THE FAKE DEED 'MAYORAZGO'

Let us now look at some extracts from the *Minuta de Mayorazgo*. The fourth sheet of the copy itself disappeared and was replaced by one in which both the paper and the handwriting were different. Columbus began with a long defence, stating that the Catholic Monarchs had conferred on him the posts of Admiral, Viceroy and Governor of the 'islands and lands' that he had given to Spain, as well as the right to a tenth of the riches found and possession of one-eighth of the territories, besides the salary corresponding to the post held.

1. In the first clause, Columbus names his legitimate Portuguese son, Diogo, as his heir. The end of this clause reads:

   And if Our Lord so wishes, if there are no legitimate male heirs after this Entailment has been in the possession of the said heirs for some time, that the nearest relative to the person that had inherited it and whose line has lapsed that comes into possession and inherits the said Entailment is a legitimate male heir who is called and has always been called, on the part of his father or ancestors, *Colón*. No woman shall ever inherit the said entailment, except if here or in any other part of the world there cannot be found any male heir of my direct line that is called, and his ancestors were always called, *Colón*.

   At this point, there is no reason to doubt the authenticity of this copied text, although the care that Columbus took in order to exclude from his inheritance all those that were not of his direct line and were not called 'Colón' is surprising. He clearly proscribed anyone called 'Colombo', and at the same time did not mention the city of Genoa, saying instead 'here or in any other part of the world'.
2. Columbus later referred to the distribution of his goods:

in such a way that a tenth of all this income be given to the most needy of my direct line whether they be here or in any other part of the world, from where they shall be brought and taken care of.

The defenders of the Genoese thesis should have been nonplussed by this constant omission of Genoa as his birthplace.

Columbus supposedly drafted this Minute in 1497. Among the hundreds of Italian families with the name Colombo, his would certainly have been one of them. It is possible that one or both of his parents were still alive; but he ignored them and his brothers completely. More to the point, he excluded them because they were not 'Colón' but 'Colombo'. He also determined that any other 'Colón' that was not of his line was also to be excluded from his inheritance.

3. Columbus mentioned Genoa later, but with different intentions:

> I also ask my son Diogo or the person that inherits the said Entailment that he has and always maintains a person of our line in the city of Genoa and that he has a house and wife there and that an income be placed at his disposal so that he may live honestly as befits a person of our line, and that the said person creates roots in that city as if he had been born there, as he may receive from that city help and favours that he may need. . .

4. Then Columbus added a sentence that is completely out of place and makes no sense:

> as I left there and was born there . . . that you know how to look after yourself and your assets; there, in the Bank of St. George, any money is safe as it yields six percent, and Genoa is a noble city and powerful at sea.

As can be seen, Columbus did not even hint that he had a family in Genoa, but wanted someone of his family, called 'Colón', to settle there as it was a powerful republic linked to the crown of Castile by close commercial relations. He maybe supposed that by having a descendant of his there he would be able to defend the interests of the rest of the Colón family, even if Columbus himself fell into disgrace.

This forecast of the future seems very strange, as if it were written by someone who already knew Columbus's life. Even at his zenith, did he fear and foresee the disgrace in which he would spend the last years of his life? It may be said that he expected the Catholic Monarchs to realise at any moment that he was in the service of the King of Portugal. Was he looking for protection in Genoa knowing that he could not expect any help, for his descendants, from Portugal?

At the time of the 'confirmation' by the signing of the deed that instituted this Entailment in February 1498, did he fear that the Sovereigns would place him in disgrace because the Portuguese would find the way to the Orient and thus prove the failure of his plans? If this document were genuine, it would be good publicity for the bank.

5. It seems that Columbus was obsessed by the future signatures and seals of his heir's contracts, stipulating that both his descendants and his brothers 'or any other party that inherits the Mayorazgo' were to use the seal of his arms, 'which I shall leave on my death' and which the said party

> will use to seal . . . and after having inherited and taken possession of it, will sign with the signature that I customarily use, which is an .X. surmounted by an .S. and an .M. surmounted by a Roman .A. and another .S. surmounting that, followed by a Greek .Y. surmounted by an .S. with its strokes and commas, as I hereby demonstrate . . . And the said party shall only write Admiral, even though he shall have other titles that will be granted by the King or won, and this is to be understood as the signature and not the text, where he may write all the titles he wishes.

It is here that the writer shows complete ignorance of the rules regarding the granting of a coat-of-arms. Columbus had no powers whatsoever to determine whether any of his heirs would use his arms, as these were granted by the Spanish Monarchs for his exclusive use, and only they could approve of their transfer to a third party. The writer was obviously ignorant of the fact that the cabalistic siglum spelt out Christopher Columbus's personal identification and clearly thought that anybody else could use it as their signature. Neither the Admiral's son nor any of his descendants would have committed such indiscretion. He then goes on to present *all the letters* of the siglum *between points*, when in fact only the 'SSS' were so flanked, this being a symbol with a specific function: the 'SSS' represent 'Sephiroth' (plural of 'sephirah', a sphere) in the temuric Cabala and the points indicate the 'mirror method' for the duplication of the letters, as I shall demonstrate in Chapter 22. Columbus, who conceived the siglum, would never have made such an error. Besides this, he says: 'will sign with the signature that I customarily use'. The expression 'I customarily' is not the same as the example he gives later 'as I do now'; it indicates a newly-adopted habit. Only an attack of amnesia would have led him to make such a statement, as he had already been using the seal for at least six years, as can be proved by the letter he wrote to Rodrigo de Escabedo from Hispaniola on 4 January 1493.

The disposition that passes on his title of Admiral to his heirs is no less suspect, as these titles never can be passed on. What is more, the sentence 'even though he shall have other titles that will be granted by the King or won' shows amazing foresight shown by the author, foreseeing that the Queen would die before him and her husband so that his heir would only have the King, and not the Monarchs, to grant him titles.

6. There is one other doubtful detail that should be noted: there appears an unwarranted reference to the Bull *Inter coetera* of 28 June 1493:

> of a radius, or line, that Your Highness had drawn a hundred leagues beyond the islands of Azores and another hundred beyond those of Cape Verde.

That extract from the Papal Bull is totally irrelevant in regard to the institution of the Entailment, for the following reasons:

(a) it does nothing to justify the first voyage of Columbus, as it is of a later date;
(b) the Bull in question had been annulled by the signing of the Treaty of Tordesillas on 7 June, 1494, which altered the 100 leagues to 370. Columbus could not ignore that, because he had been the responsible agent.

The document also reads: 'that Your Highness had drawn'. Which Highness? King John II of Portugal? If it was the Holy Pontiff that 'had the line drawn', then the text should read 'Your Holiness' and not 'Your Highness'.

On the other hand, the King of Aragon and Castile had nothing at all drawn: the Catholic Monarchs simply accepted the right to explore the seas outside the enlarged area reserved to the Portuguese, in accordance with the decision of the Pope.

The introduction of the extract from the Bull into the institution of Columbus's Entailment seems intended to demonstrate an intimate knowledge of the politics of that time – and an incorrect one. It is clear that the claimant to the inheritance was referring to King Ferdinand of Spain. However, all official documents of Castile and

Aragon of that time were always signed by both Sovereigns and always 'Your Highnesses' in the plural. Was this document an exception, or a mistake due to the ignorance of the writer? Although he was a foreigner, Columbus always addressed the Catholic Monarchs as 'Your Highnesses'.

7. Worried by the possibility of having the dispositions set out in the document altered, Columbus allegedly wrote:

> And I thus implore the King and Queen, Our Lords, and Prince in return for the services I have rendered them and in the name of justice . . . that they do not consent nor will they consent that my entailment of the *Mayorazgo* and my will be altered . . . and that it remains as I have set it out forever, because it is in the name of Almighty God and the roots and stem of my line and the memory of the services that I have rendered Their Highnesses that, being born in Genoa, I came here to serve them in Castile and discovered for them the Indies and the above mentioned islands to the West.

How could Columbus have implored the Monarchs and Prince Juan to make sure that the provisions of the *Minute* were honoured if the Prince (of whom his son Diogo had been page) had died two years before the document was allegedly drafted? Only his ghost could have watched over the Colóns' interests.

8. Columbus added that he had discovered the Indies so that the Monarchs could spend the wealth obtained from them on the conquest of Jerusalem; if this did not happen, it would be up to his son Diogo to go there 'alone, with the power invested in him . . . and invest his wealth in '*logos*' [bank shares] of St. George of Genoa and therefore increase it'.

Both Fernando Colón, in his *Historia*, and his father, in his *Profecias*, stressed the exultation that Columbus manifested for the conquest of the Holy Land, transporting his 'Templar vocation' for the construction of a 'New Jerusalem' in the lands to the West. But the writer completely misunderstood this, and thought that Columbus's aim was a new Crusade.

9. Columbus once more insisted that his heir 'worked for the honour and progress of the city of Genoa, neither being against the Church nor against the King and Queen, Our Lords'. He recommended that he build a church on the island of Hispaniola and the best and most highly organised hospital, like the ones in Spain and Italy; and that masses be celebrated for his ancestors. The *Minute* ended with the date '*Jueves, en 22 de Febrero de 1498*', and Columbus signed it with his cabalistic siglum under which he wrote 'El Almirante'.

The Admiral, who had been in Portugal six years before, could not have ignored the existence of the biggest and best organised hospital in the world at the time – the All Saints' Hospital that had been built in Lisbon on the orders of King John II (see p. 31).[20] The writer however, knew only Castile and Italy.

The textual date of the institution is *'Jueves, en 22 de Febrero de 1498'*.

This date intrigued historians, as no other document of the time, whether it be signed or authenticated by notary or royal scribe, bears the day of the week on which the act was carried out. *'Jueves'* (Thursday) has raised suspicions because, while not following the norm, it lends the text a pseudo-authenticity, as if to prove that it was not written at a later date but on the day indicated on the document. It is a simple matter to check on which day of the week a certain day fell; the expedience itself seems to be an attempt to distort the truth.

10. Here we have a remarkably verbose Minute of a will that contains about 5000 words, in which Columbus not only unnecessarily repeats certain phrases but also long, almost identical sentences. It seems that he was already mentally unsound in 1498; however, he showed extraordinary mental capacity in 1502 and 1504, when he carried out his longest exploration in the Caribbean, solved serious disciplinary problems in the colonisation of the islands and faced up to bitter intrigues in the Spanish Court.

THE *ROYAL DEED* AND THE *MEMORIAL* TO DIOGO

The *Royal Deed* of the Spanish Monarchs, dated 28 September 1501 confirms the right of Don Cristóbal Colón to institute an Entailment that had been granted in 1497. This document, although it proves the existence of a Minute elaborated four years before, contains nothing that confirms that it is a follow-up to the text presented in 1497.

Shortly before leaving on his fourth voyage in 1502, Columbus wrote a *Memorial* to his son Diogo. This is one of the documents whose authenticity has never been contested. In it, Columbus states that he had just drafted his instructions in relation to the Entailment and does not make the slightest allusion to the *Minute* of 1497–8 or to the *Royal Deed* of 1501. As this *Memorial* differs from the *Minute*, it seems natural that Columbus would have referred to it, either cancelling it or confirming it; but he ignored it completely.

1. In his testamentary *Memorial*, Columbus merely manifests his wish that one-tenth of the income from his assets be used to help any

descendant of the name *Colón*, 'here or in any other part of the world' (including the West Indies), but makes no mention of investing another tenth in the Bank of Genoa, nor does he make any mention whatsoever of that city.

The authenticity of this legal deed is indisputable.

2. This document, which contains about 400 words, is around an eighth of the length of the *Minute* and does not refer, as is normal, to the revocation of previous provisions and does not mention Genoa.

   (a) In a brief retrospective, Columbus says: 'when I found the Indies for them [the Monarchs] and Their Highnesses agreed that I should receive my share of the said Indies, islands and terra firma, that are to the West of a line they had drawn through the islands of the Azores and Cape Verde, one hundred leagues'. In fact, when Columbus went to serve the Spanish Monarchs, the Treaty of Toledo was still in force. The expression 'they had drawn' can be construed as 'someone had it drawn' without a definite subject, but 'they' must be interpreted as the Sovereigns. It seems to be a historical mistake on the part of Columbus or the result of his reluctance to mention the Treaty of Tordesillas, which he never once mentioned to the sovereigns. But, as we saw above, he wrote that it was Highnesses and not Highness who had indicated the distance of 100 leagues; and on locating the Indies to the West of the 100-league line he was not committing a geographical blunder, as even the 370-league line was to the East of the Indies. Columbus was referring to 1492, but the writer of the *Minute* did not understand this, and so expressed the action in the present tense.

   (b) In this *Memorial* document, Columbus never mentions a 'Genoese homeland' or the '*logos*' of the Bank of St George. He stipulates the nature of his bequests to his legitimate son and heir Diogo, to his bastard son Fernando and Fernando's mother Beatriz, to his brothers Bartolomeu and Diogo. But he makes no mention of any Genoese relative, let alone a visionary conquest of Jerusalem or the building of a church and hospital on Hispaniola. But he did recommend his son that he maintain in a chapel (without mentioning which) three chaplains to say three masses per day in honour of the Holy Trinity, Our Lady of the Conception and for 'the souls of all the deceased faithful and for the souls of myself, and my father, mother and wife', the late Filipa Perestrelo, the Portuguese mother of Diogo.

   (c) Columbus also stipulated: 'I ask Diogo, or whoever inherits my estate, to settle all the debts that I hereby leave here in a memorial, as they are discriminated, and any other debts that seem justly owed'. However, the 'memorial' (which will be known as the *Memorando* from now on) that Columbus referred to was not attached to the will. A sheet of paper that was neither drawn up by the same notary nor signed, so it may have been exchanged for another document, appeared later.

(d) The notary Pedro Ennoxedo appointed the witnesses to the will and ended with the date 'XXV August fifteen hundred and five' and the words *'Cristo Ferens'* and not *'XPO FERENS'* as the Admiral used to sign, which seems to confirm that it is a copy of the original. But it was drawn up only on 19 May 1506 (the date at the top) – i.e., nine months after having been drafted by Columbus and on the eve of his death:

Andrés Bernáldez relates: 'Colón died at Valladolid in 1506, in the month of May, at a good old age of about seventy'. (In this case, he would have been born around 1436, and not in 1451 like the Genoese wool-carder). Madariaga came to the conclusion that it was a printing mistake, *'setenta'* (seventy) being written instead of *'sesenta'* (sixty), which would give us the date of about 1446, which more or less coincides with 1448, the year which I believe is correct.

Columbus's last moments are related by Friar Bartolomé de Las Casas:

> the Admiral's gout got worse by the hour, aggravated by the harshness of the winter and by his anguish at seeing himself forsaken, dispossessed, his services forgotten and his rights in danger . . . his last words were *'In manus tuas, Domine; commendo spiritum meum'*. He died in Valladolid on the day of the Ascension, which in that year fell on the 20th May in 1506.

A short time before Columbus passed away, his brother Bartolomeu went to see him, but having to go and greet the King and the new Queen, left him in the company of a confessor friar. When Columbus died, they did not ring the bells of the city, but only those of his parish, as though he were an ordinary man in the street. The chronicler of Valladolid did not even record his death.

THE *MEMORANDO* OF DEBTS

Let us now look at the apocryphal *Memorando* concerning Columbus's debts, which was not attached to the will.

> Roll of certain people to whom I wish to be given, from my estate, the amounts mentioned in this memorial, to the full amount

mentioned. The amounts must be paid so that the receivers are not aware of the identity of the payer. First to the heirs of Geronimo de Puerto, father of Venito de Puerto, Chancellor of Genoa, twenty ducats or their equivalent. To Antonio Vazo, Genoese merchant who usually lives in Lisbon, two thousand Portuguese reals, which are little more than seven ducats at the rate of three hundred and seventy reals to the ducat. To a Jew that lived near the gate of the Jewish quarter in Lisbon, or to whoever *sends* the priest, the amount of half a silver mark. To the heirs of Luis Centurión Escoto, Genoese merchant, thirty thousand Portuguese reals, which are about seventy five ducats. To these same heirs and also to the heirs of Paulo Negro, Genoese, one hundred ducats or their equivalent, one half to be given to the former heirs and the other half to the latter. To Baptista Espínola or to his heirs if he has died, twenty ducats. He was the son of Micer Nicolao Espínola of Locoli de Ronco and, it seems, was resident in Lisbon in the year fourteen hundred and eighty-two. To this said memorial . . . I as notary swear that it was written in the same hand as the will of the said Don Cristóbal, which I undersign in good faith. Pedro de Azcoitia.

Let us analyse the debts, which he listed as follows (1 ducat = 370 reals):

	'Ducats'	'Reals'
Puerto	20	7,400
Vazo	7	2,590
Jew (1/2 mark)		114
Escoto (75 + 50 ducats)	125	46,250
Negro	50	18,500
Espínola	20	7,400
		82,254 = 411,270 escudos
		US$ 2,848.7 today

The exchange rate noted by Columbus himself is correct (if the document were genuine) and agrees with that fixed by the monetary reform of King John II.[21] However, there was a change in the monetary values during the reign of King Manuel I, so the exchange rate quoted by Columbus can be placed between 1482 and 1495.

Columbus's total debt at the exchange rate of 1483 would have been 82 344 reals, while in 1506 it would have been much less due to the devaluation (2.8 per cent) of the currency:

'1482'	'1506'
82,254 (US$2,848.7)	84,557 (US$2,928)

Columbus's creditors would have been cheated of US$79.3 at today's rate of exchange.

As payment would have been made in Portuguese currency, the creditors must have been in Lisbon at the same time as Columbus. It was therefore easy for the writer of the *Memorando*, having access to Columbian documents, to copy the record of the debt (if genuine) before destroying the original. In order to make the *Memorial* look genuine, they could have inserted the document into it.

Columbus received about 250 000 dollars from the Catholic Monarchs. His annual income, at a time when the cost of living was much lower than it is today, was over 41 000 dollars, five years before he discovered America. He made a fortune after 1492. He received a percentage of the riches he deposited in the royal coffers. He could have easily paid off his old debts, but according to the Genoese it was only after eleven years – and in *articulo mortis* – that he settled them. But it seems unlikely that a man of his ability would have been forced to flee from Portugal because of debts totalling 2,848.7 dollars. It seems equally unlikely that, after becoming wealthy, he would have remembered these debts when on his deathbed many years later and obliged his heirs to pay them off.

Even if those IOUs did exist, it is probable that the writers of the *Memorando* added the word 'heirs' to the names of the creditors, as some of the latter could have died in the meantime and it would have made the document up-to-date.

(a) It must be noted that Columbus took special care to stress that all his creditors were Genoese, except the Jew, and that certain debts had been contracted in Lisbon, at least until 1482. Would he really have waited twenty-four years before thinking of settling his debts, and then have chosen a time when he had hardly any money to do so?

(b) From the above-mentioned notarial document of Genoa dated 22 September 1470, we know that Dominicus de Columbus authorised his son to take the responsibility for a debt of 35 lira contracted from one Ieronimus de Porte; and another dated 28 September 1470 of the same month confirms the court decision and that the payment of the debt was carried out. In the *Memorando*, Columbus quotes the name of the son, Venito, but orders the debt to be paid to his father's heirs. Why should Columbus, feeling death approaching, worry about paying off a debt that he had in fact paid thirty-five years before?

(c) In 1478, the Genoese merchant Ludovico Centurione had remitted, through his son-in-law Baptista Spínola in Lisbon, 129 ducats to the agent Pablo di Negro, also established in Lisbon, for the acquisition of 2400 *arrobas* (an *arroba* was about 15 kg) of sugar in Madeira for despatch to Genoa. Negro contracted Christóforo Colombo, who went to buy the consignment from Eurógio Catalá, but the deal was cancelled because Negro gave Colombo only 103 ducats. For this reason Centurione sued Negro, who remained in Lisbon. It is obvious, therefore, that it was Negro who had swindled Centurione and that Colombo owed the latter nothing. It was on this same date that Christophorus declared that he 'had to leave for Lisbon the following day'.

(d) The only Antonio Vazo (*Hazo* in Italian) that is to be found in Columbian texts appears once in a notarial document dated 20 April 1448 (fifty-eight years before the *Memorando*), not as a merchant living in Lisbon, but as a notary from Genoa who bore witness to the fact that the brothers Dominicus and Antonius de Columbus had accepted responsibility for the payment of the remaining part of their sister Baptistinna's dowry. The writer of the *Memorando* seems to have been desperately searching for names.

(e) On examining the document more closely, we can pick up a clue as to the possible date of the original document: 1484. The sentence 'To a Jew that lived at the gate of the Jewish Quarter of Lisbon or to whomsoever sends the priest, half a silver mark' (114 reals or 16 dollars today) may be interpreted in two ways:

1. Being unable to locate the creditor, the word 'priest' is the *subject*, and would be a confessor, with the right to appoint someone to receive the payment;
2. The word 'priest' is a *direct object*, the confessor having been sent for the Jewish messenger as his services were needed.

It is thus possible to conclude:

– In the first case, we have an order for payment to an unknown person, carried out by an unknown person, as Columbus did not wish it to be made known who was paying the debt.
– In the second case, Columbus being in Valladolid at the time, the payment to the messenger who sent for the priest would not be an alternative for the reward for the Lisbon Jew.

Both possibilities are faintly admissible in a testamentary addendum. I assume, however, that the writers intentionally wrote 'sends' instead of 'sent'.

Let us look at the historical facts. On 20 May 1506, Columbus was on his deathbed, accompanied by his sons and most faithful servants, Diogo Méndez and Bartolomé Fiesco. A priest administered the Last Sacraments. This priest was not sent for by a Jew from Lisbon. The expression 'to whomsoever sends' in the *Memorandum* of 4 May of the same year indicates the future, which makes no sense at all. Columbus had moved to Lisbon with his wife Filipa and their son Diogo in 1483. They may have lived in a house in the Lisbon Jewish Quarter. This is possible, as many noblemen had occupied Jewish palaces in the Quarter since the reign of King John I, and the fact of Columbus's Jewish ancestry has never been refuted by sound arguments.

But to ask an unknown person, namely a Jew who was at the gate of the Jewish Quarter, to send for a priest indicates that Columbus did not live in the quarter itself, if it is not 'the gate of this quarter' that is referred to by someone in the quarter, but the gate where there was usually a Jewish messenger from the point of view of an outsider. To call any priest to a house would have meant an emergency of such gravity that the messenger would not have been rewarded so poorly.

Filipa Moniz Perestrelo died in 1484. On writing his testamentary addendum, Columbus wished to reward the person that had helped him in his time of need. The real addendum, however, (if it existed) would have read 'sent' and not 'sends'. The writers of the *Memorando* used the present tense to bring the 1506 addendum up to date. The past tense would have announced the fact that the priest was not going to administer the last rights to the dying Admiral, but had done so to his dying wife 22 years before.

Columbus went to Castile in 1484. Did he depart without paying off the debts he had so carefully made a note of? The addendum suggests that he intended to settle his debts, up to the point that he kept it. It is possible that he settled them when fortune smiled on him and then forgot to destroy the note. What cannot be admitted is that he wrote an order of payment of 16 dollars to an unknown person through an unknown person twenty-two years later.

It must be noted that Columbus wished to keep the name of the payer of the debts to this creditors' heirs a secret. This detail, which stands out in the IOU, leads us to two conjectures:

1. At the time of his death, the debts really existed and Columbus wished to settle them, but anonymously, so that his creditors would continue to think that he had not paid off what he owed and was a swindler.
2. The debts had already been paid, in which case the creditors would have been amazed to receive double payment.

Besides this, Columbus had a priest at his bedside during his last moments and did not need 'an anonymous Jew' or anyone else to fetch one.

It was not very difficult for the writer to discover those characters in order to invent the *Memorando*. The Madeira District Archive possesses documents that refer to the establishment of Antonio and Baptista

Spínola on the island before 1491, a fact that can be verified by the will of Maria de Bettencourt of that year.[22] They were descendants or close relatives of the Genoese merchant Nicolao Spínola, who was established in Lisbon and traded with the Republic through the Bank of St George. And around the year 1500, one Lucano Spinola moved from Spain to Madeira, where, in 1513 he petitioned King Manuel I for authorisation to use his Genoese coat-of-arms[23] and took the name of Espíndola. The Archive also has records that show that the merchant Sebastião Centurião, now with a Portuguese name, resident in Funchal, baptised his daughter Hipólita in the cathedral in 1554 and that Manuel Rodrigues de Negro, also a merchant, died in the city in 1561. Although the family was already Portuguese, it was known by the nickname of 'Negroni'; they were probably descendants of the Jewish family called 'de Negro' (see p. 26).

The person that drew up the *Memorando* could have consulted the archives of the Bank of Genoa and found the names of the merchants in Lisbon that traded with the Republic. Only the 'anonymous Jew' could have been mentioned in an old note of Columbus, and that in relation to the death of his wife. It was then enough to link the names of merchants with those in the notarial documents, from where he also got the name of the notary Vazo. The forged *Memorando* has absolutely nothing to do with Columbus's will.

The notary claimed to have seen the will and confirmed Columbus's handwriting and signature. But how could he have done this if Columbus had dictated it to another notary and had not even signed the document that replaced the original?

CONCLUSION

The only documents that refer to a Christopher Columbus as an employee of Genoese merchants, the *Memorandum* and the draft of a notarial document of 1479 (produced by Assereto), are of doubtful origin and clearly false. It can therefore be concluded that Centurione, Spínola and Negro were not a part of the Admiral's story. There is no doubt that he lived in Madeira, but was never mixed up in a doubtful trade of sugar. The forgery was based on the *Memorial* addressed to António Torres (30 January 1494) in which the Admiral mentions that he had commended 'fifty pipes of sugar from Madeira Island, because it is the best and the healthiest product of the world'.[24]

Notes and References

1. Harrisse (1864–6).
2. Assereto (1904).
3. Lorgues (1856).
4. A clause in the document of 1479 forged by Hugo Assereto in 1904 (see below).
5. Marques (1984) p. 176.
6. Document commented on by Zaz (1923).
7. Document commented on by Sáncho (1941).
8. Madrid, 1985.
9. *Raccolta, 'Annorazioni sopra Cornelio Tacito'*, Pt II, Vol. I, pp. 164–8.
10. Humboldt (1833).
11. The first words of a loose sheet, probably written when he was brought back in chains with his brothers on his 2nd voyage, see Varela (1984) p. 271.
12. Varela (1984) p. 264.
13. *El Libro de las Profecias*, in the Columbian Library.
14. A Portuguese word *'atamento'*, which Columbus misused; the Spanish word means 'shyness'.
15. Manuscript copy in the Genoa Library, vol. 22, col. 301/302; the part of Muratori's book that refers to Antonio Gallo is entitled: *Antonii Galli Genuensis Opuscula Historica De Rebusqestis populi genuensis et De Navigatione Columbi; Nunc primum in lucem efferantur et Mst* (sic) *Codice Genuensi.*
16. See Paragallo (1896).
17. Born in Avalos (Logroño, Spain) in 1765, died in Madrid in 1844; Director of the Academy of History, see (1915), which deals meticulously with the problem of the *Testamento de Colón* (pp. 334–48).
18. Navarrete (1825), Vol. II, p. 221.
19. A letter transcribed by Prof. Serrano y Sanz in the 'Revista de Archeologia, Bibliotecas y Museus', March, 1914.
20. The King gathered the income of 43 hospitals, each one under the protection of its patron saint, and centralised the health services of all the religious orders of the city and its suburbs in the capital. King Manuel I later had an even bigger hospital built in Goa, India (see p. 30).
21. See Resende (1545; 1798), cap. 57; *Descrição Geral e Histórica das Moedas Cunhadas em Nome dos Reis, Regentes e Governadores de Portugal; and Épocas de Portugal Económico*, Apêndice (Lisbon, 1929).
22. See *'Ementa dos Livros da Vereação da Câmara do Funchal'*, in *Archivo Histórico do Funchal*, Books III and IV, the research by Dr João Cabral do Nascimento.
23. Faria, António Machado de (1961) p. 199.
24. Varela (1984) p. 160.

17 Stratagems and Deceits

It is essential to explain the attitude of Columbus's sons Diogo and Fernando during the lawsuit that Diogo brought against the Spanish crown – '*Pleyto contra la Corona*'. The case revolved around the concept of the expressions 'native' (*natural*) and 'foreign' (*estrangero*). The adjective '*natural*' in the Iberian languages has two distinct meanings; it can either refer to 'nature' or to 'birthright'. In his will of 1506, (see p. 000), Columbus decreed that if his legitimate son and heir Diogo died without leaving a successor, his inheritance would pass on to his 'natural son Don Fernando', a bastard son being considered as a product of nature. When Columbus said 'to serve Your Highnesses by the will of God', he was considering the Sovereigns to be his 'natural' lords. This fact can also be read in the *Memorial* that he sent to the Monarchs from the island of Isabella on 30 January 1494 and which was delivered by António Torres: 'I greet Your Highnesses as King and Queen, my natural lords, in whose service I shall certainly die'. Diogo, who was Portuguese, became a 'natural subject of Their Highnesses, the King and Queen', according to documents, when he was granted his 'naturalisation papers'. In none of these cases, however, does the meaning have any relation with 'birthright' in the sense of nation, place of birth or family.

Columbus frequently complained of his status of a foreigner:

> I have been considered in the town, in the way people treat me and in many other things, to be a poor envied foreigner.[1]

> I am an envied foreigner far from home; Your Highness has always helped me and not wished me harm.[2]

> they have been artful in pouring their intrigues into the ears of Your Highness in order to injure me, through envy, because I am a poor foreigner.[3]

Who would have believed that a foreigner could have reached such a position?[4]

Following the death of the Queen in November 1504, Columbus realised that the King was not willing to honour the provisions of the *Capitulaciones* (see p. 484) and that he was in danger of losing his rights and the right to transmit them to his son Diogo. He looked for support from the powerful lords of Genoa, who exercised a great financial influence at the Court. He also wrote to the Archbishop of Seville: 'it seems that his highness [written without capital letters] does not wish to fulfill what he promised in writing and by signing with the Queen, who is now in Heaven, and I believe I am struggling against the odds ... as if I were fighting the wind; as I have done everything in my power, I now leave it to God'.

King Ferdinand took Germaine de Foix as his second wife on 10 March 1506. With the death of the Queen, his protector, Columbus found himself dependent on a King who had never forgiven him for not having opened up the way to the Orient as he had promised, and even announced. King Ferdinand seemed to give little importance to the feat of the navigator who had given Spain an empire that covered half the world. Columbus died two months later, without honour or epitaph. It was up to his sons to fight for their hereditary rights with every means at their disposal.

Diogo Colón had also lost his two closest protectors: Prince Juan, of whom he had been page, had died nine years before and Princess Juana had become insane. Although promising to respect the provisions of the *Capitulaciones*, the King merely handed over some money that he owed to Columbus but which he had until then unfairly withheld. And as Diogo was in the Antilles, he was not in a position to defend his cause.

THE LAWSUIT AGAINST THE CROWN

Diogo laid claim to the posts of Viceroy and Governor of the Indies, to which he had a right, for two years, but seeing that the King paid no heed, he requested permission to resort to the courts. Through a decree signed at Arrevalos on 9 August 1508, the Crown stripped him of his hereditary rights. Diogo then appointed Juan de La Peña, who served the Duke of Alba, to defend his case, but he achieved nothing.

Bartolomeu Colón and his nephew Fernando, Diogo's brother, returned from the Antilles at the end of 1509. Fernando presented himself with powers of attorney to represent his brother in the lawsuit, but the Crown again rejected Diogo's plea on the grounds that the 'Law of

Toledo of 1480' forbade a 'natural citizen' of the kingdom to hold a post for life and any privilege to be renewed. This means that when the Monarchs signed the *Capitulaciones* in 1492 and then later confirmed them, they had neither observed this legal provision nor had they revoked it with a later decree.

On 3 March 1511, Fernando Colón claimed, on behalf of his brother, that Columbus was not a 'natural subject' of the Crown when the *Capitulaciones* were signed, with the result that they had the 'power of contract' between the 'Kingdoms of Castile and Aragon' and a 'foreigner'. He insisted on the fact that his father had fulfilled his promise to the Catholic Monarchs, so the 'contract' should therefore be honoured. It was an astute move, as the King had never openly admitted that he had fallen into a well-placed trap when he had signed the Treaty of Tordesillas and so lost the race to, and the wealth of, the Orient.

The Crown Revenue Officer, Diego Porrás who, together with his brother had rebelled against Columbus in the Indies, argued that when the *Mayoradgo* (the document that was confirmed in 1501 but then disappeared, see p. 199) was signed, Cristóbal Colón was already a natural subject of the Catholic Monarchs and that the law forbade any foreigner from holding State posts. Given that a foreigner was automatically granted Spanish citizenship on marrying a Spaniard, Columbus was a 'foreigner' because he had married a Portuguese lady. His son Diogo, born in Portugal, was also a 'foreigner' up to the time he received his 'naturalisation papers'. This means that they were both foreigners when the *Capitulaciones* were signed and 'natives' when the Crown confirmed the *Mayoradgo*.

This argument finally convinced the King and he accepted Diogo's petition on 5 March. Fernando, on behalf of his brother, requested a declaration of the sentence, but this denied him the hereditary rights to the posts of Viceroy and Governor of the Indies. Diogo could have insisted, claiming that he was a foreigner, but it was not convenient that the 'dual nationality' of both himself and his father should become public knowledge.

On 29 December 1512, and after having consulted his brother, who was in San Domingo, Fernando appealed against the sentence, renewing the claim that his father, not being a 'natural subject' on his return from his first voyage, had fulfilled his contract with the Crown. He requested that the witnesses be heard, but as there was no stipulated time limit for the length of a lawsuit at that time, the case seems to have come to a standstill. His uncle Bartolomeu had meanwhile died and could not testify in 1514 and the King passed away two years later without having changed anything. Emperor Charles V (I of Spain) came to the throne and Fernando carried on the struggle.

It is essential to point out that the only documents presented by

Fernando were the *Capitulaciones* and the will of 1506. The pages of the contract drawn up by the Sevillian notary Martin Rodriguez had been torn out; the *Mayoradgo* and Columbus's will of 1502, which he had entrusted to Friar Gaspar Gorrício, disappeared from the Simancas Archive. Did these documents contain some clue or suggestion regarding the true identity of Columbus? How could Fernando have explained their disappearance?

Then the unexpected happened. Emperor Charles V, who had always shown signs of friendship to Fernando by financing the purchase of books for his library and appointing him as an adviser, suddenly forbade him, without the slightest verbal or written justification, to continue consulting the archives and documents of the realm, an act that has led researchers to conjecture that Columbus's son could have been responsible for these strange disappearances and even for the manipulation of documental evidence. It is thought that when Fernando entered the church his desire was to become a bishop, as his uncle Diogo had done, but they both became friars. On returning from the Antilles in 1495, Diogo planned to go to France and Italy, but gave up the idea due to the war with Charles VIII of France, a decision that met with the approval of the Monarchs. Fernando had better luck and visited Italy twice, in 1515 and 1530. He went to Genoa, Rome and Viterbo, not only (it is presumed) to plead with the Genoese authorities to use their influence at the Spanish Court, but also (it is known) to buy books for his library, which was the most important of the time. There is no record, however, that the Genoese exerted any influence on his behalf in the lawsuit.

Fernando Colón, who had always maintained that he knew nothing of his father's origins and wrote in his *Historia del Almirante* that he thought he was descended from the Colonnas, strangely informed the court that his father was a foreigner because 'he came' from Genoa. In this way, he could attenuate the disquieting accusations that Columbus and his elder son were Portuguese. Fernando was unable to prove this claim, just as he was unable to find any relative of his father's or anyone that had known him in Genoa. Diogo's identical efforts through notaries brought the same results.

It must be stressed that the chronicler Las Casas never claimed that Columbus was Genoese; he merely mentioned the fact that he thought he was because he 'had heard so'. Diego Menéndez, Pedro de Arana and Rodrigo de Barreda, companions of Columbus in the Antilles, all said the same thing, but they had heard of both Genoa and Savonna. And all of them were convinced that he came from noble lineage. It is also interesting to note that Columbus, who had sailed in the Mediterranean at the time of the expedition to Chios, spoke of Sicily and some places in Greece (although rarely), but frequently mentioned Portugal,

King John II, Lisbon, Madeira and the Azores. He never claimed to have been born or even lived in Genoa.

After Diogo Colón's death in 1526, his son Luís carried on with the lawsuit. It is here that Juan Martín Pinzón, son of Martín Alonso Pinzón, who accompanied Columbus on his first voyage westward and later turned against him, appeared on the scene. He claimed that it had been his father who had discovered the Indies, with the result that Columbus had proposed to cede him 50 per cent of all the income that might accrue from them. He then went on to claim certain rights that even his own father had never claimed forty years before, even when he was at loggerheads with Columbus.

Of the twenty-four witnesses that testified in court, none had been on the first voyage. The same can be said of the eight that testified in the Antilles in 1535, and four of these would have been six years old at the time.

Don Luís de Colón finally decided to cede the privileges of Viceroy and Governor of the Indies to the Crown; in compensation, he was allowed to keep the post of Admiral and received 10 thousand ducats 'interest and inheritance' income, the island of Jamaica with the title of Marquis and 25 000 square leagues of Verágua (Panama) with the title of Duke on 8 September 1536.

THE SUSPECT FINAL PARAGRAPH OF FERNANDO'S WILL

In Diogo's first will, dated 16 March 1509, he directs: 'bring the body of Filipa Moniz, his [Columbus's] legitimate wife, my mother, which is in the Chapel of Pity in the Monastery of Carmo in Lisbon and which belongs to the Moniz family'. But his second will, dictated in San Domingo in the Antilles on 8 September 1523, contains another version: 'that a thousand masses be said for the souls of purgatory . . . for my soul and for those of the Admiral my father, of my mother Filipa *Martinez* and of my uncles Don Bartolomeu Colón and Don Diogo Colón'. Would he have forgotten the name of his mother in those fourteen years? The scribe that wrote it down may have been hard of hearing. But whatever else Diogo Colón's wills contain, they certainly do not contain any mention of Genoa.

It is said that Fernando Colón, who in his *Historia* had denied that his father's surname was Colombo (but where Fornari had inserted 'relatives' of Colombo of Genoa) and had been born in Córdoba of a lady of that city, or so it was thought, had scribbled in the margin of a page of his *Registrum Librorum* (the catalogue of his library) the words 'Cristóbal Colón, Genoese'. But could it not be an annotation of his nephew Luís, who inherited his Library? Luís also came into possession of his uncle's will, signed in Seville on 3 July 1539, which contains the following passage, completely out of context, quoted here from a copy of Ballesteros:

> And because a buyer of books must carry them from one place to another, it is difficult to avoid using the services of Genoese merchants, these merchants are everywhere; seeing them, say that you are from the Fernandine library, established by Don Fernando Colón, son of Cristóbal Colón, Genoese, the first admiral to discover the Indies . . . and as the merchant is from the homeland of the founder [Fernando] ask him to help you in a way he can in those lands, because Fernando so determined and recommended the buyers to do so in his name, because he knows that they will always find people from his beloved homeland.[5]

This text, which is a recommendation to an anonymous book buyer and completely out of place in a will, is almost certainly a forgery; if it were genuine, however, Luís could free himself from the anathema of being a descendant of an Admiral who had betrayed the Spanish Crown and would thus not be ostracised and reduced to misery. He wanted to be a duke and a marquis and live in opulence. Not for him the stigma of being the grandson of that other Colón who was Portuguese.

THE LAWSUIT OF SUCCESSION

Another lawsuit, the *Pleyto Sucessorio* followed the 'Lawsuit Against the Crown'; it lasted much longer, and was fiercely disputed by the claimants to Columbus's inheritance.

Columbus was succeeded, as we already know, by his legitimate son, Diogo. Before marrying, the latter had two affairs. The first with one Constanza Rosa, by whom he had one son who died in infancy; the second was with a widow by the name of Isabel Samba, to whom he left a fifth of his assets in his will. He later met Maria de Toledo, a niece of the famous Don Fernando, Duke of Alba and great-grandson of the Catholic Monarchs, whom he married.

Two sons and three daughters were born of this marriage; Don Luís Colón, Don Cristóbal de Toledo, Maria (de Cardona), Juana (de la Cueva) and Isabel (de Portugal). The women's parenthesised names are their husbands. The first-born, Don Luís, married Maria de Mosquera, who bore him two daughters, Maria[6] and Filipa. The latter married her first cousin, Don Diego, son of Don Cristóbal de Toledo, but they had no children. Because of this, the Entailment passed into the hands of a Portuguese, Dom Jorge Alberto de Portugal, son of Jacob (Jaime) de

Portugal (Count of Gelves, descendant of the House of Braganza) and grandson of Isabel, youngest daughter of Diogo Colón.

Don Jorge married Bernardina de Vicentelo, who bore him a daughter, Leonor. The Entailment, therefore, which included the titles of Viceroy of the West Indies, Duke of Verágua, and the Marquis of Jamaica, passed on to Dom Nuno de Portugal, brother of Dom Jorge Alberto. Dom Nuno de Portugal had married Aldonza de Portocarrero, who gave him four children, the eldest of whom, Dom Álvaro de Portugal, inherited all the assets and titles.

One of the daughters of Don Diogo Colón, Juana, married Don Carlos de la Cueva. Her grandson, Don Carlos de Córdoba y Bocanegra, Marquis of Villamejor, had a daughter who gave origin to three generations of the Colón family.

A descendant of the first of these three branches married a distant cousin, Don Pedro Manuel de Portugal, who in 1647 married Teresa Maria de Ayala y Toledo of the second branch of the issue of Don Carlos de Córdoba's daughter. Being born of this marriage, Dom Pedro Nuno de Portugal inherited the titles of the Duke of Verágua and of La Vega, the Marquis of Jamaica, La Mota and San Leonardo and the Count of Gelves, Ayala and Villanoso.

It must be pointed out that a succession lawsuit was filed in 1578, based on the following arguments: as Dom Jorge Alberto de Portugal and his brother Dom Nuno had inherited the Entailment as grandsons of Isabel, Francisca, daughter of Don Cristóbal, claimed the same rights of inheritance. After thirty years of appeals and counter-appeals, the court rejected Francisca's petition in 1608. In spite of this, other petitioners filed lawsuits for the same reason. They were:

1. Francisca, her daughter Guiomar and her grand-daughter Ana Francisca.
2. Don Juan de La Cueva, son of Don Carlos and grandson of Juana, penultimate daughter of Don Diogo Colón.
3. Don Luís de Ávila, son of Maria de Toledo and grandson of Isabel, youngest daughter of Diogo Colón.
4. Dom Pedro de Portugal, son of Dom Nuno de Portugal and also grandson of Isabel.

It was only in 1664 that Dom Pedro de Portugal, descendant of the House of Braganza which was then reigning in Portugal, won the struggle that had lasted for eighty-six years and inherited all Columbus's Entailment.

In 1578, that Baldasário Colombo, born at Castel de Cuccaro, appeared on the scene with the fake *Minute* that had supposedly been legally recognised eighty years before. Brandishing this, Colombo contested the claim of Colón's descendants through the feminine line, not only arguing that Francisca was a woman, but that Dom Nuno was descended from another woman, Isabel.

In the period between 1539 and 1552, Don Fernando Colón's *Historia* had already been distorted by the Italian translation of the Castilian manuscript that disappeared (or had been deliberately destroyed). Taking advantage of the fact that the original of the rectified will of 1506 had disappeared, the *Minute* was immediately drafted to replace it. With the descendants from the feminine lines of the Colón family excluded from the inheritance, the Entailment would revert to the Bank of Genoa or to an imaginary relative of the Admiral who was not called Colón as was stated in the will, but Colombo, as suited the writer of the *Minute*. In the report of the court's findings, it was admitted that the *Minute* had been fabricated by one Luís Carvajal Buzón, 'a person of doubtful character who boasted of his ability to mutilate and disfigure documents'.[7]

According to the 'Genoists', Doménico Colombo had five children and a grandson; and his brother António had four sons. Doménico was still alive in 1494. Apart from Bartolomeu and Giácomo (Diego), were they all dead? Columbus always showed a certain affection for his relatives, including those of his deceased wife Filipa and his sister-in-law Violante (for whom his son Diogo showed the same affection) and in his will Columbus left Violante 'whatever is necessary for her mourning plus 10 ducats and the quittance from a debt of 7000 or 8000 maravedis'. Yet neither Columbus nor his sons showed the slightest interest in the Colombos of Genoa, who were brought into the limelight through the research of nineteenth and twentieth century Genoese historians.

THE 'MILITARY CODICIL'

Another document now needs to be considered, the *'Military Codicil'* (*'Codicillus More Militari Christofori Columbi'*), signed in Valladolid on 4 May 1506. It is drafted in Latin and in a style that philologists find impossible to attribute to Columbus, and the siglum of his signature is wrong; two weeks after signing the document the Admiral also had his will recognised. His will, which is indisputably genuine, makes no reference to the *'Military Codicil'* allegedly written a fortnight before. It decrees that the administration of his assets be executed by the legitimate male heirs of the descendants of his direct line or, in the absence of these, by the closest male relative of the last administrator (see p. 205). Only in the absence of any male issue of the Colón lineage would this administration pass into the hands of a woman. If the 'Military Codicil' were genuine, we would have to conclude that Columbus ignored the fact that the provisions of one's last will annul all previous provisions. Or did he suddenly have a change of mind – instead of excluding his descendants from the Entailment, leaving everything to the Republic of Genoa in the absence of direct male heirs, deciding to consider them as his successors?

After concealing his place of birth all his life, Columbus declared in the *'Military Codicil'*: 'my beloved homeland the Republic of Genoa'. He bequeathed it a new hospital in detriment to the female line of the family, obviously forgetting that women could bear sons and continue the Colón lineage. It was also forgotten that the posts of Admiral and Viceroy which were included in the *'Military Codicil'* could never be passed on to the Republic of St George as they were bestowed on Columbus by the Sovereigns in the *Capitulaciones* of 17 April 1492 and, therefore, were at the disposal only of the Sovereigns. If Columbus had drafted the *'Military Codicil'*, it would mean that he was unbelievably ignorant of the Spanish laws of his time, and at the same time manifesting a great fickleness of mind. It was the *Memorial* that served as a basis for Columbus's will made in 1505, and not the *'Military Codicil'* drafted two weeks before.

In the *'Military Codicil'*, Columbus mentions that 'His Holiness Pope VI' had honoured him with the gift of a small book of prayers 'that will console you in captivity, in battle and in adversity'. Pope Alexander VI may have given Columbus a small book of prayers, but nobody can believe that Columbus was so ignorant as to write 'Pope VI' and leave out the name 'Alexander'. We know that Columbus was only imprisoned once, on the return of his second voyage to the Indies, from the beginning of October to 25 November 1500, when he was freed on Royal orders. Sixteen months later, in February 1502, Columbus wrote his only letter to the Pope, requesting priests and medicines for the Antilles. But, strangely enough, he never had the courtesy to thank him for the book.

The exclusion of women in the absence of direct male heirs and the admission of a parallel male heir would have been to the benefit of the opportunist 'cousin' Baldasário. The *Minute* that proclaimed 'from there [Genoa] I left and there was born', the suspect *Memorandum* and this last *'Military Codicil'* serve only go to strengthen the case against Columbus being Genoese.

Baldasário introduced himself as being the son of Francisco Colombo and great-nephew of Doménico Colombo, who was Lord of Cuccaro and Conzano and who had died in 1456. According to the notarial documents of Genoa that are related to the *Minute*, Doménico Colombo was still alive in 1479 and was not Lord of Cuccaro. Baldasário was unable to prove that he had had uncles or cousins called Bartolomeu, Giácomo and Giovani in Genoa, but he claimed that his great-uncle had been the father of one Cristóforo Colombo, who would thus be his cousin.

THE LETTERS OF COLUMBUS

1. Before departing for his fourth voyage, Columbus wrote a letter to Nicolo Oderigo, Genoese representative for Castile, from Seville on 21 March 1502, from which the following extract is taken.

   > Sir, I can find no words to express how much I miss you. I have given the book of my documents to micer Francisco de Riverol, so that he can send it with another copy of the letters . . . please write to Don Diogo. Another book will be finished and sent by the same means and through the same micer Francisco. Through them you can make up another letter. Their Highnesses have promised to give me everything that belongs to me and allow me to pass it on to Don Diogo, as you will see. I am writing to *señor micer* Juan Luis and *señora madona* Catalina, who will deliver this letter. I am about to depart . . . If Geronimo de Santiesteban appears, he will have to wait until I return.

   It is apparently an autograph of Columbus, which he signed with his siglum.

   (a) The letter only mentions Genoese that knew Columbus in Spain. Reading between the lines, we can see that he was looking for support from the Genoese in the Court of the Catholic Monarchs. He makes no mention of dealings in foodstuffs. Accused by his rivals of being reluctant to look for a passage to the Orient (where Vasco da Gama had arrived in 1498) beyond Cuba and also devoting himself too much to the controversial colonisation of the Antilles in order to enrich himself, Columbus feared that the Spanish Sovereigns would not fulfill the provisions of the *Capitulaciones* nor transmit his posts to his son Diogo. He thus sent Oderigo letters and documents gathered together in a book, known as *Cartulários*, so that Oderigo could help him in the defence of his interests. It seems, therefore, to be more political than commercial.

   (b) It must be noted that, though writing to a Genoese, Columbus wrote in Spanish and translated the Italian names – Francesco de Riverol, Giovani Luigi and Ieronimo de San Stephano – into Spanish, besides preceding *'micer'* (*'messer'*) in Italian and *'madona'* with the Spanish *'señor'* and *'señora'*. This is the equivalent of writing *Señor Mr* John Louis and *Señora Mrs* Catalina in English, which Columbus would never have done if he were writing to a fellow-countryman.

2. In 1502 (probably in the second half of March), Columbus allegedly

wrote to his son Diogo, recommending that, in the case of his death, he should grant a pension to his brother Fernando's mother and to his aunt, Violante Nuñez. He informed his son that the Genoese bankers established in Spain, Francisco de Riverol, Francisco Dória, Francisco Cataño and Gaspar d'Espíndola, had loaned him money to finance the purchase of the eighth part of the merchandise that he was going to take to the Antilles on his fourth voyage, was well as 118 000 maravedis that he had spent in Seville, 15 000 in Xeres and 25 000 in Granada. This document, entitled *Memorial*, is neither dated nor signed and it is doubtful whether the handwriting is Columbus's.

3. On 2 April 1502, Columbus supposedly wrote to the Bank of Genoa announcing that he was about to sail for the Antilles and that, as he might not return, had instructed his son Diogo to pay the Bank a tenth of his annual income from the Indies 'forever' in order to repay the loan contracted for the purchase of 'wheat, wine and other victuals'. He informed the Bank that he had sent Oderigo copies of his 'privileges' and said that he would like the Bank to see those documents. He signed the letter 'Lord High Admiral of the ocean sea and viceroy and governor general of the islands and terra firma of Asia and India of my Lords the King and Queen, their Captain General of the sea and of their Council', followed by his cabalistic siglum. Was the income from the Indies to be infinite and were the victuals purchased at an astronomical price?

4. Another hypothetical letter was written to Oderigo from Seville on 27 December 1504, two months after returning from his fourth voyage. In it he mentions that two years earlier he had entrusted to Riverol a book containing letters and another *Cartulário* containing his 'Privileges', besides two other letters, one for the Bank of Genoa and one for Oderigo himself. He manifests surprise at the fact that the bank had never replied, despite Riverol having assured him that 'everything had arrived safely'. He adds that he had delivered another *Cartulário* to Francisco Cataño, together with a 'very good' letter from the King and Queen. The signature is similar to that of the letter to the bank dated 2 April 1502, except that Viceroy is spelt *'Viso-Rey'* instead of *'bisorey'* and the words 'of the sea' are omitted after the title 'Captain General'.

5. On 29 December 1504, Columbus wrote to his son Diogo, saying that he had sent a letter to the merchants Pantaleón and Agostín Italián with instructions to deliver money to him in the Antilles, having supplied them with letters of warranty from Riverol and Dória.

6. Ulloa categorically rejected the letters to Oderigo, the first because it was not written in Columbus's hand and the second (found in Paris)

because viceroy was spelt *'Viso-Rey'* and not *'Virrey'* as was the custom at the time, besides the fact that Columbus had entitled himself 'Lord High Admiral' when in other letters he simply used 'Admiral'.

(a) It can be seen that the letter dated 2 April 1502, sent to the Bank of Genoa, is signed *'bisorey'*, which appears just this once in Columbus's correspondence.

(b) In the letter of 1504, a mistake has been made in the use of *'el mar'* (the sea) in the masculine gender when the noun 'sea' is feminine in Spanish. In the roll of titles in the letter of April 1502, both the masculine *'el mar'* and the feminine *'la mar'* are written. Being a word that Columbus had heard almost every day for eighteen years, he would never have made this mistake.

(c) These are the only two letters, to the Bank and Oderigo, that are signed with such a profusion of titles.

(d) In the *Memorial*, which is neither dated nor signed, Columbus's sister-in-law, sister of Filipa Moniz Perestrelo, appears as Violante Nuñes instead of Moniz, which is the strangest of errors as he had lived with her and her husband Moliarte in Portugal and Huelva and he surely would not have forgotten the name of his wife, the mother of Diogo.

7. It is crucial to point out that only the *Minute of the Mayorazgo* makes any reference to these *Cartulários*, which suggests that, like the correspondence that accompanies or announces them, they were created in order to give more substance to the *Minute*. According to the Italian *Raccolta* and the reproductions that were made of the *Cartulários*, some pages have gone missing and some are blank. Besides this, the coat-of-arms granted to Columbus by the Spanish Monarchs is different from the original and the holder of arms could (and would) never have altered them without the authorisation of the sovereigns.

8. Ulloa, who made a detailed study of the material, was of the opinion that all these documents presented by the Italians were to demonstrate that Columbus had direct dealings with the Bank of Genoa and turned to the Genoese only when he needed help. The letter to Oderigo dated 21 March 1502, however, could be genuine as it makes no mention of debts contracted for the acquisition of provisions; Columbus was purely seeking political backing. But the authenticity of the others is questionable, some because of their doubtful origin and other because of the accumulation of errors that Columbus would never have committed. Plus, of course, the roll of titles that only someone making out a later case could have invented.

9. When writing to the sovereigns, Columbus addressed them as 'Very Christian and Very Illustrious Princes' or 'Very Illustrious and Powerful Sovereigns and Lords'. He addressed the Bishop of Badajoz as 'Very Reverend and Illustrious Lord', Juana de La Torre, Prince Juan's governess as 'Very Virtuous Madame' and Friar Gaspar Gorrício, his patron, as 'Reverend and Devoted Father'. Luís Santángel, the Sovereign's Secretary for Finances and Nicolo Oderigo were addressed as 'Sir'. When writing to the highest authorities of the Bank of Genoa, who were at that time almost as powerful as the 'Seigniory of the Republic', he addressed them as 'Your Very Noble Lordships' and wrote the letters in direct speech; these were the usual forms of written address in those days.
10. Let us now look at the way the governors of the Bank of Genoa addressed Columbus and his son Diogo in the hypothetical letters dated 8 December 1502:

> Most Illustrious and Noble Sir, *Cristóforo*, Lord High Admiral of the ocean sea, Viceroy and Governor General of the islands and terra firma of Asia and the Indias of the Serene King and Queen, and Captain General of their Council. Illustrious sir and enlightened and beloved citizen and very memorable Lord: through the offices of the respected Jurisconsult Nicolo Oderigo, returned from his delegacy in your Community at the court of their very excellent and Glorious Sovereigns . . . we were delivered a letter from Your Excellency, to which we gave our singular attention, verifying from it that Your Excellency, as is your way, remains very attached to your homeland, to which you show singular love and charity . . . for such sublimity on your part, of such very singular glory . . . that you show to have for the fatherland of your birth.

Very singularly, the Bank attributed his father's titles to Diogo:

> Very noble Don Diogo, Lord High Admiral of the ocean seas, highly honourable son, Illustrious and very honourable sir. His Excellency the Lord High Admiral, your father, through his letter delivered by hand of the respected Messer Oderigo.

(a) The whole content of these two letters reveals a bombastic but subservient gratitude, merely because Columbus allegedly instructed his son to repay the loan he had contracted from the bank, despite the fact that the reimbursement would depend on the tenth of the income that he hoped to obtain in the future. It does not seem, however, that this future was promising enough

to warrant such abject flattery in the face of the 'affection' manifested by the 'beloved citizen for his place of birth'.
(b) The writer, probably in the seventeenth century, did not understand that he had committed a gross anachronistic error by using a linguistic style that was in fashion about a century after Columbus's time.

11. Columbus had written to the Bank of Genoa on 2 April 1502, but on 27 December 1504 had not yet received a reply. In other words, he had asked for a loan in order to victual his vessels before sailing for the Antilles and when he returned it had not yet been granted, unless we are dealing not with a loan from the Bank of Genoa but simply a request for political backing, as could be deduced from the letter dated 21 October 1502. On 8 December 1502, however, the Bank supposedly wrote two letters, to father and son, which came to light 100 years later.

THE HANDWRITTEN NOTE IN THE *IMAGO MUNDI*

In Chapter IV of the *Historia*, Fernando Colón wrote that his father had sailed to São Jorge da Mina. There is no reason to doubt this, as Columbus probably took part in the expedition of 1481, which comprised eight ships[8] and which transported Diogo de Azambuja, who remained there in order to build the fortress.

Referring to a *Memorial* written by Bartolomeu Colón and handed to his nephew Fernando, the latter states that his uncle had sailed to the Cape of Good Hope and returned to Lisbon in 1488. He must therefore have voyaged with Bartolomeu Dias. Christopher Columbus had a book, *Imago Mundi*, by Cardinal Piere d'Ailly, in the margin of which is written, in Latin:[9]

> Note that in the year of Our Lord 88, in the month of December, Bartolomeu Dias, commander of three caravels that the King of Portugal had sent to Guinea [Coast of Africa] to find new lands, arrived in Lisbon; and he announced to the King that he had navigated 600 leagues beyond the point until then known, i.e. 350 to the south and 250 in the return north, to a place that we think is Abyssinia[10] and near the Cape of Good Hope.
> And the captain calculated by the astrolabe that this point was at a latitude of 45 degrees and that the last place he had reached was 3100 leagues from Lisbon; he charted this voyage and mapped out a route, describing league by league, in order to present it to his serene Highness, and I played a part in all this.

What amazes one in this text is the accuracy of the distances and the route taken. If we divide the 3100 secret Portuguese maritime leagues by two we get the 16 740 Km that separate Lisbon and the Great Fish River, which flows into the Indian Ocean and was reached by Dias. Due to the policy of secrecy, as already mentioned, only the Portuguese used these measurements. The route taken by this expedition was described by João de Barros.[11] But unlike other sixteenth-century writers, Barros maintained that Dias docked in Lisbon in December 1487 and not 1488. The note in the margin of d'Ailly's book shows him to be correct. Analysing the opening phrase *'Nota quod hoc anno domini 88'*, we see that the pronoun *'quod'* is accusative, corresponding to 'what' or 'this', thus giving the expression the sense of 'Note that in the year 88 [in which the note was written]; in the month of December [the previous month], Bartolomeu Dias, etc.'

Las Casas did not make a correct translation, and opined that the text was in Bartolomeu Colón's handwriting. He fell into this trap because Bartolomeu had declared that he had made that voyage; Columbus had never claimed to have been to the Cape of Good Hope. Las Casas thus came to the conclusion that the handwriting was Bartolomeu's. This has led some authors to put forward the hypothesis that Bartolomeu may have been his brother's scribe, which would not have been so unusual. But at that time Columbus was not yet suffering from the arthritis that was the result of his many years at sea and his writing was not so erratic as it became later in life. It is natural that, in the margin of a rare book that he esteemed, he would have wanted to jot down notes in his best and clearest hand.

When Columbus wrote *'que in Agrissibam acestimamus'* ('which we think is Abyssinia'), he speaks in the plural, but in the end he talks about himself: 'I played a part in all this'. He is not talking about an active part, but a consultative role, his last words meaning that he helped in the mapping out of the route and the report presented to the King.

THE KING'S LETTER

At the beginning of 1488, King John II still could not be sure that they were getting near the desired passage to the Indian Ocean. Meanwhile, the conquest of Granada seemed imminent, following the conquest of Malaga La Real by the Spaniards. King John was sure that once this had happened his neighbours would launch their navigators into the Atlantic in an effort to reach India and would not hesitate to rescind the Treaty of Toledo. The King had a job for Columbus. Apparently in reply to a request from his secret agent, he wrote the famous letter to him:

Cristóbal Colom
We, King John by the grace of God, King of Portugal and The Algarves, here and beyond the seas in Africa, Lord of Guinea, warmly greet you. We have seen the letter you wrote to us: and we thank you very much for the good will you have shown in our service. Regarding your coming here, and other matters concerning your work and intelligence, we need you and will be pleased to receive you and will be content with your reception.

And if you happen to fear our Justice for any reason, this letter will guarantee your safe conduct and that you shall not be arrested, detained, accused, cited or tried for breach of trust or any other crime.

We shall give orders to this effect. We beg you to come as soon as possible and hope that you will encounter no hindrance. We thank you and owe you much. Avis, 20th March 1488.
To *Cristóvam Colom* our special friend in Seville.[12]

For the Dukes of Verágua, Columbus's descendants, to have kept this letter (which is fully transcribed) in their library, is a proof that they were sure of its authenticity.

The future Admiral's name is written in two different ways: *Cristóbal Colom* and *Cristovam Colom*, terminating in 'm' in both cases. Did the King absent-mindedly forget that in Spain Columbus would be known as Cristóbal Colón and not 'Cristovam', the Portuguese name? Or was it a mistake of the scribe who knew who he was writing to?

How can such a 'special friendship' between a powerful King and a foreign wool-carder be explained in the fifteenth century? How many noblemen were there and what would they not have given in return for such intimate confidence and friendship on the part of their King at that time? The Monarch is so interested in Columbus's return that he granted him safe conduct and an amnesty for all his crimes. Foreign critics see these crimes as debts. The Portuguese think that they were linked with Columbus's participation in the conspiracy against John II. I can see a certain reason for their opinion.

Did Columbus take part in the conspiracy and then manage to keep a prudent distance between his head and the royal arm? And in order to be able to flee, would he not have appropriated some money though a breach of trust? In his anxiety to hasten Columbus's return to Portugal, the King took care to feign ignorance of the crime in question, forgetting and forgiving everything. Had Columbus's involvement in the conspiracy been revealed to the King only after he had entrusted him with the secret mission in the Spanish court? Was Columbus closely related to one of the princely conspirators?

Like everyone else, Columbus would have been aware that John II had shown no mercy in carrying out his 'justice'; he had had his cousin and brother-in-law the Duke of Braganza, head of the second most powerful House of the Realm, executed; and he stabbed the other cousin and brother-in-law, Dom Diogo, Duke of Viseu and Beja and Master of the Order of Christ (Columbus's half-brother, as we shall see).

It can be concluded that the importance of the mission with which the King had entrusted Columbus outweighed the criminal process. It was essential that Columbus continued in the service of Portugal; and Columbus was equally aware that his King needed him. He was fully confident, in spite of realising that the King would go to any length to eliminate the conspirators, some of whom had been tracked down and executed in France. Seville was by no means a safe refuge from the King's ire.

I believe that this was a kind of blackmail conceived by the machiavellian genius of King John II. The letter was meant to be shown in Castile in order to awaken the fear that Columbus was thinking of entering the service of the Portuguese Crown. The reference to the dangers that threatened the navigator in Portugal justified his reasons for going to Spain and offering his services to the Catholic Monarchs. This same form of blackmail would later be used in an effort to entice the Kings of England and France.[13] But the immediate task that motivated the sending of the letter was to collaborate in the report of the rounding of the Cape of Good Hope. Obviously the Portuguese navigators who had carried out the mission with Bartolomeu Dias did not need Columbus's help, as they were all steeped in scientific theory and tested and experienced seamen. But King John II had to set the scene before sending his agent back to Castile, so that the Sovereigns would waste no time in sending him westwards.

CONCLUSION

In fact, the history of Portugal, with its truly great epic of the Discoveries, does not actually need a Christopher Columbus to glorify it. The sea was already Portuguese before he began sailing it. If the Holy Land[14] of ancient peoples was where they buried their dead, thus making it the homeland of their ancestors, the Atlantic Ocean had become the Holy Sea of the Portuguese, so many of them having found their last resting place in its depths. The vast oceans that the caravels crossed in their quest to find India were already divided up. The Treaties of Alcáçovas, Toledo and Tordesillas drew limits (the *Terminus*)[15] on these sacred waters of the Portuguese, until then forbidden to foreigners. How many of these knew how to navigate scientifically before Columbus's historical feat? They needed a knowledgeable Admiral to lead them out of the darkness in which they still lived in the fifteenth century. As St Augustine once said: 'they have all passed like a nau that ploughs its way through the waves and, when passing, leaves no wake'.[16]

But if Columbus were Portuguese, why did the Portuguese not try to find his vanished track? Why did they leave him despised and abandoned, just as he had in Valladolid? Why did they not give him the place he deserved, alongside the throne of King John II and recognise him together with the Portuguese saints, heroes and scholars? The time has come to reveal the true Christopher Columbus.

Notes and References

1. *Historia de las Indias*, Vol. II, p. 342.
2. *Historia de las Indias*, Vol. II, p. 200.
3. *Historia de las Indias*, letter of 1499.
4. Navarrete (1825).
5. Beretta (1945), pp. 157–8.
6. Before his marriage he begot a son he named Cristóbal.
7. *Memorial del Pleyto* (1578), p. 74, no. 446/447.
8. Cardoso (1990).
9. *Nota quod anno Domini 88; in mense decembri appulit in Ulixbona Bartolomeus Didacus capitanius trium carabellarum, quem miserat Dominus Rex Portugaliae in Guineam ad tentandum terram et renunciavit ipso Domino Regio prout navigaverat ultra jam navigatum Cabo Boa Esperanza, quem in Agrisimbam aestemimos. Quod quidem in eo loco invenit distare per astrolabium gradus 45, quem ultimum locum distat ab Ulixbona leuchas 3100; quod viagium pictavit, scripsit de leucha in leucham in una cartha navigationis ut oculi visui ostenderet Domino ipso serenissimo Regi, in quibus omnibus interfui.*
10. At that time they used to give the name Abyssinia to the South of Africa, although they already knew the real country in the North of Africa.
11. The Portuguese league was 5.4km; 3,100 leagues were 16,740km. The present-day degree is 113.6km and there are 73.7 degrees between Lisbon and the Cape, i.e. 8,372.32km, which multiplied by 2 are 16,744.64km (Corrêa, p. 255).
12. The original was copied by Navarrete (1825), in the Archives of the Duke of Verágua. It was considered as authentic by Róspide (1919) and by the palaeographer Prof. Freire (1930).
13. See pp. 328–30.
14. Cicero, *De Legibus*, II; Tibullus, *Lares Agris Custodes*, I, 1.
15. Siculos Flaccus, *De Conditione Agrorum*, 5.
16. Sapient, V, 10, '*tanquam navis, quae pertransit fluctuantem aquam; cujos, cum praeteriet, non est vestigium invenire.*'

Part V
Columbus Revealed

18 Dispersing the Mist

Impossible est [?id] quod probabilitar dicitur, secundo totum esse falsum (It is impossible that something considered probable be completely false, author's free translation)

(St Thomas Aquinas, *Contra Gentes*, III, p. 9)

I have already mentioned the enlightened opinions of Martin Fernandez Navarrete, who received a lot of support regarding the need to study the life of Columbus, using contemporary chroniclers as a base: Andrés Bernáldez, Pedro Mártir d'Anghiera, Fernando de Colón, Bartolomé de Las Casas and Gonzalo Fernández de Oviedo.

I also presented various critical studies on the different theses in which unfathomable contradictions and disparities were revealed in the evidence produced by Columbus's chroniclers and the theories of subsequent analysts, especially the Genoese and Catalan.

The supporters of those theses had to opt for a rather unusual form of contestation to defend their opinions – by claiming that the contemporary writings concerning Columbus, when they did not fit in with the theories they defend today, must have been corrupted.

They first accept a hypothesis as genuine, then when logic confirms otherwise, come to the conclusion that the premises are wrong. This seems to be a case of *making a hand to fit the glove*.

Columbus's most important work is his *Libro de las Profecias* (*Book of Prophecies*), but as Friar Gaspar Gorricio was his secretary in the making of the book, some critics wish to attribute a number of biblical quotations to him, convinced that Columbus, as a wool-carder, could not have possessed such a vast knowledge of theology. The book is a remarkable work and worthy of the most careful interpretation and cryptographic study, in order to reveal its cabalistic messages. However, this book had thirteen of its pages torn out – 'the best prophecies of this book' according to the person who jotted down some notes in the margin and denounced the outrage. Assuming, mistakenly, that Columbus was one and the same as Colombo, the wool-carder, people were unwilling to accept the fact he could either have written such a work or possessed such a deep culture.

Las Casas relates the difficulties that Columbus faced in making himself understood by the Catholic Monarchs and the Spanish Board of Mathematicians and they 'summoned the experts most versed in cosmography to study Columbus's project', which was based on the fact that the world was round and that the Orient could be reached by the west. He added that 'the Board contained more men of law and theology than of cosmography' and also that 'the decision went against Columbus and the Monarchs dismissed him... in the light of the absence of mathematical sciences and information on ancient stories'. Bernáldez wrote: 'Christopher Columbus arrived at the Court of King Ferdinand and Queen Isabella and gave free rein to his imagination, to which they gave little credit,... and he explained the *mappa mundi* and aroused in them the desire to know more about those lands'. It is ironic that the map he used to convince everyone was the fantastic and erroneous map of Toscanelli (see p. 167). But Columbus, who had so many times claimed to have sailed with the Portuguese, took care not to attribute his knowledge to them. He limited himself to saying that it was due to 'Our Lord': '[He] gave me sufficient knowledge of astrology, as well as of geometry and arithmetic'. Despite their manifest scientific ignorance, very few of the Board's 'experts', gripped by the knowledge he proved to possess, supported him. When Princess Juana had to go to Flanders, Columbus gave her some useful advice regarding the voyage, as he possessed an authoritative knowledge of those seas. It is essential to read the letter he sent to the Castilian Monarchs in February 1502, in which he expounds concepts of seamanship that only an experienced mariner and geographer could have acquired. On all his voyages, Columbus showed that he had a profound knowledge of astronomy and proved that he had read the works of the ancient Greek, Roman, Arab and Jewish scholars by quoting them. Besides possessing a deep knowledge of geography, he was familiar with the influence of the longitude on the declination of the magnetic needle, which at the time was known only by Portuguese seamen. He spoke with authority about the Sargasso Sea, which until then had been navigated only by the Portuguese on their expeditions to the west of the Azores, and described the phenomenon of the east–west and the north–south currents of the tropical zone, the different degrees of freshness of marine flora and referred to the stability of the laws that governed the geographic distribution of seaweed. He discoursed on the configuration of islands and the geological reasons for the origins of the Carribean Sea and on meteorological phenomena. He described a seaquake and a volcanic eruption and explained the physical reasons for their taking place. And he was a cartographer. Testifying in the 'Lawsuit against the Crown', (see pp. 215–16) the navigator Hojeda said that on his first voyage, on which he was accompanied by Amerigo Vespucci, he had used a map made by Columbus, who had entrusted it to Bishop Fonseca, who in turn had indiscreetly handed it over to Hojeda and so ensured the success of the expedition.

Columbus was also versed in ornithology and corrected his shipmates in the identification of birds and their migrations. He was also able to distinguish the different species of fish like a trained ichthyologist, which was rare in those days.

On 4 May 1493, after his return from the Antilles, the Catholic Monarchs asked Columbus for his opinion on the alteration of the 'demarcation line' of the world so that they could communicate their views to the Pope, saying that Columbus 'knew more than any man ever born'. Following the Treaty of Tordesillas, on 15 August 1494, the Monarchs sent a letter to their locum tenens for Catalonia, Don Juán de Lanuza, recommending that 'to draw the said line [370 leagues to the west of Cape Verde] we need to have the opinion of experts'. Lanuza sought the advice of the cosmographer Ferrer, who in turn wrote to Columbus on 5 August 1495, ending one of his paragraphs with the words: 'even though you know more about these things when you are asleep than I do when I'm awake'. Columbus had no equal in Spain when it came to scientific knowledge.

António Ballesteros y Beretta [1] criticises Fernando Colón, saying:

> in order to exalt the scientific preparation of his father, he treated the experts that studied the Project [the Castilian Junta of Experts] as ignoramuses. A highly competent cosmographer, Fernando Colón endowed his father with a wisdom that he did not possess and, with his usual anti-Spanish opinions, denigrated Spanish science. Las Casas accepted Colón's arguments without question.

But we know that Columbus possessed a deep knowledge of the subjects he dealt with, just as we know of the lack of nautical knowledge from which the Spaniards suffered, as did other nations, up to the sixteenth century. Yet Beretta does supply us with new information: Fernando Colón, born in Córdoba and always living in Spain, was 'anti-Spanish'.

THE ORIGIN OF COLÓN-ZARCO'S NAME

The armillary sphere – the 'skeleton' of the geographic coordinates of the globe, with its imaginary lines of latitude, longitude and ecliptic – was tridimensionally made up of metal rings. It appeared on the frontispiece of geographical works and illuminated nautical documents, as well as other books, during the reign of King John II. His successor, Manuel I adopted it as his personal emblem; he had it engraved on each one of the numerous scutcheons that adorned his armour and it became the symbol of the scientific discovery of the world. In fact, the armillary sphere gave one a notion of the size of the earth; it suggested the possibility that the Orient could be reached by the west. But for those that already knew the calculations made by the overland and seafaring explorers of Africa and Asia, it indicated that Japan could not be located on the latitude of the Antilles. Surmounted by the Cross of Christ, it was the standard of a people of discoverers and missionaries.

Each curved segment of the armillary sphere, whether tridimensional or plane, was called a 'ring' or 'colon', a Greek word (*kolon*) that means 'member' or 'arch'. They are curved 'arms' that take the same name in anatomy – rising colon, transversal colon and declining colon. And 'colon' is the segment of the ring that joins the Iberian Peninsula to America. Did this induce someone by the name of Arco to change his name to Colon?

We must now analyse the work of the pioneers that have thrown light on to the mystery of Columbus's siglum (See Figure 18.1).

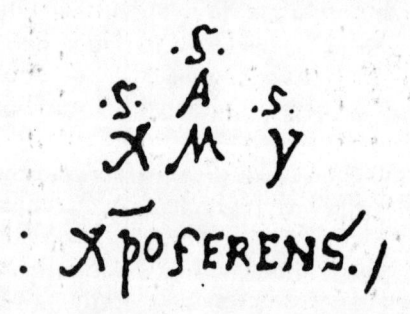

18.1 Colón-Zarco's siglum.

On examining Columbus's siglum, Afonso Dornelas (a descendant of the navigator João Gonçalves Zarco's brother)[2] verified that the oblique terminal line, preceded by a point, was called a '*colon*' both in Portuguese and in Spanish; it is a graphic sign of punctuation that separates two clauses of the same sentence. In former times it was called the 'imperfect' separator, today the semi-colon. Today's colon was known as the 'perfect' separator. As *Xpo* means 'Christo' in Greek and as '*Ferens*' means 'he who transports' in Latin, Dornelas came to the conclusion that the cryptographic siglum probably indicated a name. Translating '*Christo-ferens*' as 'Cristóvão' and then adding the Greek *colon* (./), he obtained '*Cristóvão Colon*'.

Following up this analysis, Saúl Ferreira saw that the graphic character (./) corresponded to (-) in Hebrew, meaning *'zarga* or *'zarco'*. 'Zarco' is a Portuguese surname, made famous by the navigator João Gonçalves Zarco who discovered the Archipelago of Madeira in the service of Prince Henry the Navigator and was raised to the peerage with the name of *Câmara*.[3] The origin of the name 'Zarco' was enveloped in a dense mist in the genealogies that appeared at the end of the sixteenth century, when everything possible was done to cover up the Jewish ancestry of any distinguished family, especially in the cases of the nobility and those that had close ties with the royal family. Totally incoherent origins were invented for both surnames and titles.

History cannot be restricted to one multifaceted version of events, based exclusively on written documents, many of which are falsified in the writers' private interests. It is essential that documentary research be complemented with a philosophical reflection in its widest sense. If a historical text is full of lacunae and doubts, one must resort to other sources of information, such as heraldic symbology, tumular inscriptions and ideograms to be found in works of art, which also transmit messages and are not mere fantasies of their authors. In this field, the investigator must be on his guard against the intentional distortions that were perpetrated during the Portuguese Inquisition, when the Jewish origin of many noble families was concealed by means of erasures and anthroponomical changes. And it must not be forgotten that nearly half of the population of the country had Jewish blood; from the sixteenth to the eighteenth centuries, some genealogists changed surnames in return for payment by those who feared that they would be denounced as descendants of New Christians. Although trying to be honest, Gomes Eanes Zurara, Diogo Gomes and the genealogist Alão de Morais, for instance, presented several contradictory versions regarding João Gonçalves Zarco, all of them based on what they heard. They suggested that the name 'Zarco' derived from a nickname (as he was strabismic), or *'zargo'* (from the Arab *'zarca'*), which means 'light blue eyes'. They invented the fact that he had only one eye, having lost the other one in battle. Or, according to Gaspar Frutuoso, he may have killed a Moor called Zarco. But he was already Zarcos, in Matosinhos, before fighting any battle. There were also other Jewish Zarcos in several cities such as Tomar, Santarém and Torres Novas and no descendants of João Gonçalves, discoverer of Madeira.[4]

No one seems to have considered the fact that it was common practice to join the last letter of the penultimate name to the first syllable of the last surname, as in the case of Hernandez Uñiga, which became Hernandez Zuñiga, and of Rodriguez Ola, which became Rodriguez Zola. Gonçalves Zarco, therefore, may have come (as I found out later to be the case) from Gonçalo *Estevez Arco*.

Why Zargo or Zarco? Transliterating the root *'arkh'* of the demotic Egyptian writing to Greek gave rise to *'archê'*, which in turn formed *'arch'* and *'arg'*, which are found in the Semitic group of languages such as Hebrew, Arab and the Hamitic tongues, to which modern Egyptian and Ethiopian belong. They also became a part of some Indo-European languages such as Persian, Greek and the Italic tongues, one of which was Latin. Portuguese comes from this last group.[5] We should also note when analysing the surnames 'Zarco' and 'Zargo' that the same character *gamma* (γ) in the ancient Greek alphabet has the value of a G or a C.[6]

Let us move from philology to history to demonstrate this phenomenon. The Hellenic city of Argos, a toponymic that comes from Egyptian, gave its name to the region of Argolis, on the Aegean Sea. According to Greek mythology, Argos had been founded by a colony of Egyptians under the leadership of Inachus in 1800 BC. On landing, Inachus found a cave inhabited by evil, 'amphibious spirits', against which he fought. Three hundred years later, Danais took another party of Egyptian colonists, among whom was Hercules, to Argos and they settled in the interior. Aesculapius (a scholar highly esteemed by the Hermeticists, alchemists and cabalists) later raised his temple near Epidaurus, on the Aegean coast.

The root *'arkh'* or *'archê'* means 'the beginning' in Greek (as does the character *alpha* in an esoteric and cabalistic sense), as well as 'priority', 'power' and 'command'. From it, come the words 'archangel', 'archaism', 'archbishop', 'archduke' and even names like 'Archimedes'. In its form *'arch'*, the same root gave the Latin word *'arcus'*, not only the offensive weapon *'arco'* (bow), but also triumphal 'arch', from which derives the word 'architecture'. It also produced the word *'archium'*, meaning a vaulted chamber (*camara*) or a box with a convex lid (*arca*), where writings of wisdom were kept and from which comes the word 'archive'.[7]

The hero Jason and his companions sailed in a ship called 'Argo' in the quest for the 'Golden Fleece', and were thus called 'Argonauts'. The name 'argonaut' is also given to the univalve mollusc of an arch-shaped shell which has two membranes that can be used as sails and six other membranes that can be used as oars. Some writers of Classical Antiquity, among them Aristotle, referred to this mollusc. Pliny, who described the coastline and the people of Lusitania, wrote a fable about an argonaut that landed on a beach and left its shell to explore it before returning to the sea.

The chest with a convex lid (*arca*) later became linked to the Jewish religion in the form of 'the Ark of the Covenant' and 'Noah's Ark'. This form also appears in the quarterdeck of the galleys, the convex covering of which was called *arca* or *camara*. This latter word was used by the Portuguese in their caravels and naus. From the same root came the adjective *'arcane'*, which designated the hermetic and secret operations in alchemy, such as the making of gold (*chrisopoeia*)[8] and the creation of the 'philosopher's stone', in the 'philosophical egg', which is also called *'câmara'* in Portuguese. I shall demonstrate later that it was not mere chance that led João Gonçalves Zarco to choose the name 'Câmara' when he was raised to the nobility, leaving his descendants to use the surname 'Zarco'. The Arcos were a distinguished Jewish family of Tomar, headquarters of the Order of the Temple and later the Order of Christ, where a large colony of Jews had settled and had built the Synagogue of the *Arco* (Arch), still standing in Rua Nova.

The Portuguese encyclopaedia says that the toponymy of the town of Tomar derives from the Saxon etymon *'tomian'*, which means 'to take'. In fact, Isaak Bitton has shown that TOMAR is the cabalistic *temurah* cryptogram (see pp. 43, 273) chosen by the Jews to designate the hill where the Templars' castle and monastery were built late in the twelfth century. The name derives from the reverse of the Hebraic word *RAMOT*, which means 'Blessed be the Lord who ascends to the highest'.

Referring to his father, Fernando Colón said in his *Historia* that 'in 1484, he met a neighbour of his from Madeira who had asked the King for a caravel in order to discover certain land that he swore he had seen every year and always the same way... just as they used to speak about the Azores'.[9]

What land would this have been? Brazil? One of the Lesser Antilles?

What is more important here is the name of Columbus's neighbour. It was one Fernão Domingos do Arco, father of another navigator, João Afonso do Arco, who went to Newfoundland. Originally surnamed 'Andrade', this family took the name of Arco because King John II authorised them to buy some land on the island that was called either Calheta do Arco or Arco da Calheta from João Gonçalves Zarco.

There was no 'arch' on the island, so it seems that the Donatary Captain himself gave the place its name. Was he known to his comrades as 'João do Arco' (John of the Ark)? The Andrades certainly acquired the surname.

João Gonçalves Zarco started the noble family Câmara, who took great care in the sixteenth century not to get caught in the tentacles of the Inquisition that led so many New Christians to the stake, although not all of them escaped.

PREVIOUS INTERPRETATIONS OF COLUMBUS'S SIGLUM

The upper part of Columbus's siglum has already been interpreted by many experts, who came to the most diverse conclusions – Figure 18.2 shows six interpretations.

.S.
.S. A .S.
X M Y

Supplex	Servus
Servus Altissime Salvatoris	Sum Altissime Salvatoris
Christi, Maria, Yosephus	Xriste Maria Jesus
Christophoro[10]	Xriste Ferens[11]
Salvabo	Sanctus
Sanctum Altissimi Sepulchrum	Sanctus A Sanctus
Xriste Maria Yesus[12]	Christus Maria Yosephus[13]
Shaday	.S.
Shaday Adonai Shaday	.S. Armiratus .S.
Yehova Molch Chesed	(Xristi) Maior Yndiarum
Xristo ferens[14]	Xristoferens[15]

18.2 Previous interpretations of Colón-Zarco's siglum.

Patrocínio Ribeiro[10] used the cabalistic method of the 'mirror', which I shall refer to in detail later. Seeking the Truth on the other side of the mirror, in accordance with the temuric alchemic cabala, he obtained the image in Figure 18.3.

·S·
·S· A ·S·
x m y
─────────── Xpo Ferens
x w γ
·ƨ· ∀ ·ƨ·
·ƨ·

18.3 Ribeiro's image 'on the other side of the mirror'.

Inverted, the M corresponds to a small *omega* O and Y to a *lambda* L. Ribeiro then found the letters

X Chi
M Omega
L lambda
S sigma (inverted)
A alpha

We know that Columbus had a command of Greek. *Xpo Ferens* is a Graeco–Latin word. On this basis, Patrocínio Ribeiro was able to use Latin and Greek characters according to their phonetic value, which was usual practice in the so-called western alchemic cabala, as it dispensed with the Hebraic alphabet, which was too complex for non-Jewish Europeans.

The five initial letters (S, A, X, M and Y) have O and L added to them, thus making seven. And phonetically the X has the value of a C.

In Latin inscriptions, the points signify suppressed letters.[11] Therefore, the six points would indicate that six words should be formed by those seven letters. But by this system the author did not come to an unequivocal conclusion.

Saúl dos Santos Ferreira[12] was of the opinion that '*Cristoferens*' had nothing to do with St Christopher (who, according to legend, carried the Child Jesus across the river on his shoulders). He concluded that in the language of the primitive Jewish–Christian church of Spain the word '*Cristóbal*' meant 'Christ–Lord', corresponding to 'the Saviour (*Salvador*) Himself'. As far as he was concerned, the siglum should read:

Save	Save	Save
X (Christ)	M (Mary)	Joseph

Placing the words in three columns and reading them from bottom upwards, he obtained

Save	Save	Save
Christ	Mary	Joseph

'Christ saved, Mary saved, Joseph saved' is the equivalent of '*Cristo, Maria et Joseph consalves*'. He thus noticed the word '*Consalves*' (i.e., Gonçalves).

He then thought of transliterating the characters of the siglum into Hebrew, so that the Hebrew *Kaph* would correspond to the Greek X, *Mêm* to M, *Yôdh* to Y, *'Aleph* to S (*Sámekh* or *Sîn* according to the circumstances), and this gave him a new reading (see Figure 18.4).

```
       S            ש
  S A S  =  ס א ש  =  שש אס ימה
  X M Y             י מ ח
```

18.4 Saúl Ferreira's transliteration of Colón-Zarco's siglum.

He considered *Xpo Ferens* to be '*Salvador*' ('Saviour') and the three Ss 'Gonçalves'. He added the ./ and the Hebrew word he had obtained in the siglum, which gave him the following reading:

'*Salvador Gonçalves Zarco, he that robbed disappear*'.

Obviously the author of this cabalistic-*notarikon* message (see p. 279) based it on the letter sent to Columbus by King John II (see p. 228).

TWO SEVENTEENTH-CENTURY MESSAGES

Saúl dos Santos Ferreira's conclusion brought the erudite bibliophile António Ferreira de Serpa into the field of Columbian research. Serpa raised the hypothesis of a correlation between Columbus and a pseudonym *Dom Tivisco de Nasao Zarco y Colona*, adopted by one Manuel de Carvalho e Ataíde, author of some genealogical memoirs.[13] Manuel de Carvalho e Ataíde was born in Pombal, an old Templar stronghold which, like Tomar, had a large Jewish colony in the fifteenth century. He was a cavalry officer, a commander of the Order of Christ and a Bailiff[14] of the Holy Office of the Inquisition. He was the son of Sebastião José de Carvalho and Luisa de Mello (bastard daughter of an Ataíde).[15]

Carvalho used that pseudonym, preceded by the title of 'Prior', when he wrote *Theatro Genealógico que contém as árvores de costados das principais Famílias do Reyno de Portugal e suas Conquistas*. The work was banned by King Peter II because it contained some inexact genealogical information.

In another book, *Periscope Genealogica* published some years after the *Theatro Genealógico*, Saúl dos Santos Ferreira and Ferreira de Serpa found once more, by the same process of gematric cabala (see p. 273),[16] other messages indicating that Columbus was a 'product' of Prince Henry 'the Navigator' and also a descendant of João Gonçalves Zarco and his wife Constança Roiz (or Rodrigues) de Sá, daughter of Cecília Colonna.[17] A 'product' is a non-specific word and, as Columbus could not have been the son of Prince Henry, Saúl dos Santos Ferreira admitted that he could have been the son of the *Prince* Fernando, Henry's nephew and heir of both his title of Duke of Viseu and his position of Master of the Order of Christ, and later Constable of the Kingdom, Duke of Beja and Lord of the Achipelagos of Madeira and Cape Verde.

The gematric transliterations of *Theatro Genealógico* and *Periscope Genealógica* opened new horizons. However, the books were published almost two centuries after Columbus's death. Could the cryptographic texts be valid? The opponents of the thesis that Columbus was Portuguese rejected this possibility, saying that it was merely a non-documentary seventeenth-century message.

Notes and References

1. Beretta (1948), Chap. II, (1945), pp. 117–205.
2. Álvaro Gonçalvez Zarco, bastard son of João Gonçalves Zarco. His son, Pedro Álvares da Câmara married Catarina Dornelas (daughter of Álvaro Dornelas of Madeira) and he begot Isabel who married Antão Monis Homem, Captain of Praia (Azores).
3. Morais (1688).
4. Tavares (1982).
5. Havelace (1928).
6. Martins and Figueiredo (1952).
7. *Dictionnaire de la Conversation et de la Lecture* (Paris, 1836): *Argo*.
8. See pp. 119–257 and 448.
9. P. 72.
10. Ribeiro (1927).
11. In *Donatus*, vol. I, chap. VI. 'Tres sunt positurae vel distinctiones que graeci vocant: distinctio, subsdistinctio, media distinctio. Distinctio est ubi finitur plena sententia; hujus punctum ad summam litteram ponimus. Subdistinctio est, ubi non multum superest de sententia, quod tamen necessario separatum mox inferendum sit. Hujus punctum ad iman litteram ponimus. Media est, ubi fere tamtum de sententia superest, quantumm jam diximus, cum tamen respirandum sit. Hujus punctum ad mediam litteram ponimus. In lectione tota sententia periodus dicitur, cujus partes sunt cola et commata'. And according to Garruccium, 541: 'Punctum ad mediam litteram in inscripcionibus usurpobatur'.
12. Ferreira (1938).
13. He was the father of Sebastião José de Carvalho e Melo, on whom King Joseph bestowed the titles of Count of Oeiras and Marquis of Pombal. He expropriated many properties and condemned their legitimate owners and lawyers to death or exile. See Castelo Branco (1882); Costa (1960).
14. In 1674, the function of the Bailiff of the Inquisition was the equivalent of an informer of the political police. According to Marco António Azevedo Coutinho (1737), 'The Holy Office . . . by private and secret paths, through the attorneys of the people, tried to recruit men that they called Bailiffs of their Court', transcription of Saraiva (1918), p. 50.
15. Torres (1929).
16. All the Hebrew words found by the investigator are biblical. The dictionary used was J. Buxford, *Lexicon hebraicum et chaldaicum* 6th ed. (Basle, 1655).
17. See pp. 500–1.

19   The Prophecy of the Sibyl

It was the Jews who developed the graphic arts all over Europe and introduced them into Portugal.[1] But even before this, they had a flourishing industry in illuminated manuscripts, which germinated from a *'scriptorium'* in Lisbon that was granted to them by King Alphonse V and then protected by Queen Leonor, wife of King John II.

As far as is known, the first book published in Portugal was the *Pentateuce*, printed by Samuel Gacon in 1487,[2] in an edition coordinated by Elisier Toledano. A few years later, the works of Valentim Fernandes came off the press, immediately followed by the publications of Samuel d'Ortas and his sons, one of whom, Abraão, became famous.

VALENTIM FERNANDES

The printer that interests us most is Valentim Fernandes, a graphic artist of unusual intellectual ability. He was a historian and geographer and voyaged with the Portuguese to Senegal, Guinea and Cape Verde. He worked as a printer and engraver in Portugal from 1495 to 1516. On the orders of Queen Leonor and in conjuction with Nicholas of Saxony, he printed the Portuguese translation of Rudolph of Saxony's *Vita Christi*. He was also known as Valentim of Moravia, where he was born. As a stockbroker on the Lisbon market for spices being exported to Germany, he was appointed as the public scribe of Germanic merchants in Lisbon by King Manuel I, through a royal charter of 21 February, 1503.[3] In the work of the New Christian Duarte Pacheco Pereira,[4] *Esmeraldo de Situ Orbis*, printed by Valentim Fernandes, Saúl dos Santos Ferreira discovered that the word 'Esmeraldo' formed an anagram EMSLODRAE which, in turn, transliterated into Hebrew phonetic sounds formed a message: 'with Salvador of Madeira'. Santos Ferreira suggested a probable collaboration of Columbus in the preparation of the book, just as he had collaborated in the drafting of the report of the rounding of the Cape of Good Hope with Bartholomeu Dias.

Valentim Fernandes was the son-in-law of the German Wesler, whose firm was the first to set up an agency in Lisbon after the discovery of the seaway to India. He was a companion of the famous German printer Stevan Gabler. He was born in Moravia and settled in Lisbon in 1494, although it is thought he had already been working in the graphic arts for some twelve years. In 1488, he struck up a friendship with Konrad Peutinger, who was 'Secretary for the City of Augsburg' and counsellor of Emperor Maximilian of the Holy Roman Empire for Oriental trade (i.e., in Portugal). The Emperor frequently received Konrad Peutinger in the court. It is thus easy to see why Valentim Fernandes settled in Lisbon. He enjoyed many privileges and was appointed as a squire in the service of Queen Leonor, besides being of use in supplying information to Peutinger regarding the situation in Portugal.

The chronicler Duarte Pacheco Pereira, who was of Jewish descent, was a knight of the House of King John II and commanded the nau 'Espírito Santo' (Holy Ghost) in the fleet of Dom Afonso de Albuquerque that sailed for India in 1503; he demonstrated such heroism on this and other expeditions that he organised against the foe that he was dubbed the 'Lusitanian Achilles'. After being governor of India, his enemies intrigued against him and King Manuel I ordered him to be sent back to Portugal in chains in the hold of a ship. In spite of proving his innocence and being an illustrious writer and soldier, he died in misery.

We know that Pacheco Pereira was a friend of Bartolomeu Dias and that he was on the Island of Principe (Archipelago of S.Tomé) on parallel 0 degrees, taking astronomical measurements. He returned to Portugal with Dias on the latter's voyage home after rounding the Cape of Good Hope. We also know that Bartolomeu Colón took part in the expedition to the Cape of Good Hope and that Columbus himself sailed the Atlantic with Dias. Even if they did not sail on the same ship, we can admit, as Santos Ferreira did, that the two knew each other extremely well.

And we are now sure that Valentim Fernandes, who wrote a letter to Konrad Peutinger in which was included the *Profecia da Sibila* ('The Sibyl's Prophecy', see p. 251), also counted Columbus among his friends. There is one particular detail in the letter that deserves close attention. Valentim Fernandes signed his engraving on the last page of the *Book of Marco Polo* with the letters 'V FRZ'. . The 'FRZ' is an abbreviation of Fernandez or Fernandes. This led me to think that Colón-Zarco had constructed an abbreviated anagram of Fernandes when he signed (*Xpo*) *Ferens*. As Fernandes is a family name, Salvador Fernandes Zarco would be, according to the norm, the son of a Fernando and as everything points to the fact that he was the grandson of the discoverer of Madeira, it could be either Fernandes Zarco or Gonçalves Zarco.

19.1 The official signature of Valentim Fernandes.
As the notary of Germanic merchants established in Lisbon (dated 21 February 1503).

(Photographed by the author from a sixteenth-century text)

Valentim Fernandes' signature contains a triangle (or equilateral set square) and the symbolic interlacement (of the strap) of the cabalistic ternary (group of three elements or letters) (see Figure 19.1). The engravings in the *Book of Marco Polo* that Fernandes printed and made are also of great interest (see Figure 19.2). The bird that, in the upper frieze of the engraving of the *Book of Marco Polo* (Figure 19.2), seems to be pecking grapes, appears to confirm that the printer belonged to a hermetic group of cultists of the alchemic cabala. Some investigators have mistakenly interpreted this bird as a pigeon and thus related it to the name 'Colombo', but it is a phoenix (see Figure 19.3), which was so widely represented in churches and monasteries from the twelfth to the sixteenth centuries. It can be found in both the Monasteries of the Hieronymites in Lisbon and the Order of Christ in Tomar. Although the phoenix is usually depicted rising from the flames in the act of resurrection, it can also be found, already resurrected, on the 'Vine of the Lord'.

Other cabalistic figures are represented in the engravings of the *Book of Marco Polo*: dragons (a symbol that was adopted as the crest of King Manuel I); apes riding hounds; a mantichora (an esoteric, winged proboscidean that was never seen, as it inhabited inaccessible regions); a lion, which devours this monster in another frieze; and five-petalled flowers, such as the red rose.

THE MEANING OF THE PROPHECY

Was Valentim Fernandes imbued with the spirit of the Templars, which had also raised its head in Central Europe? He at least seemed to know

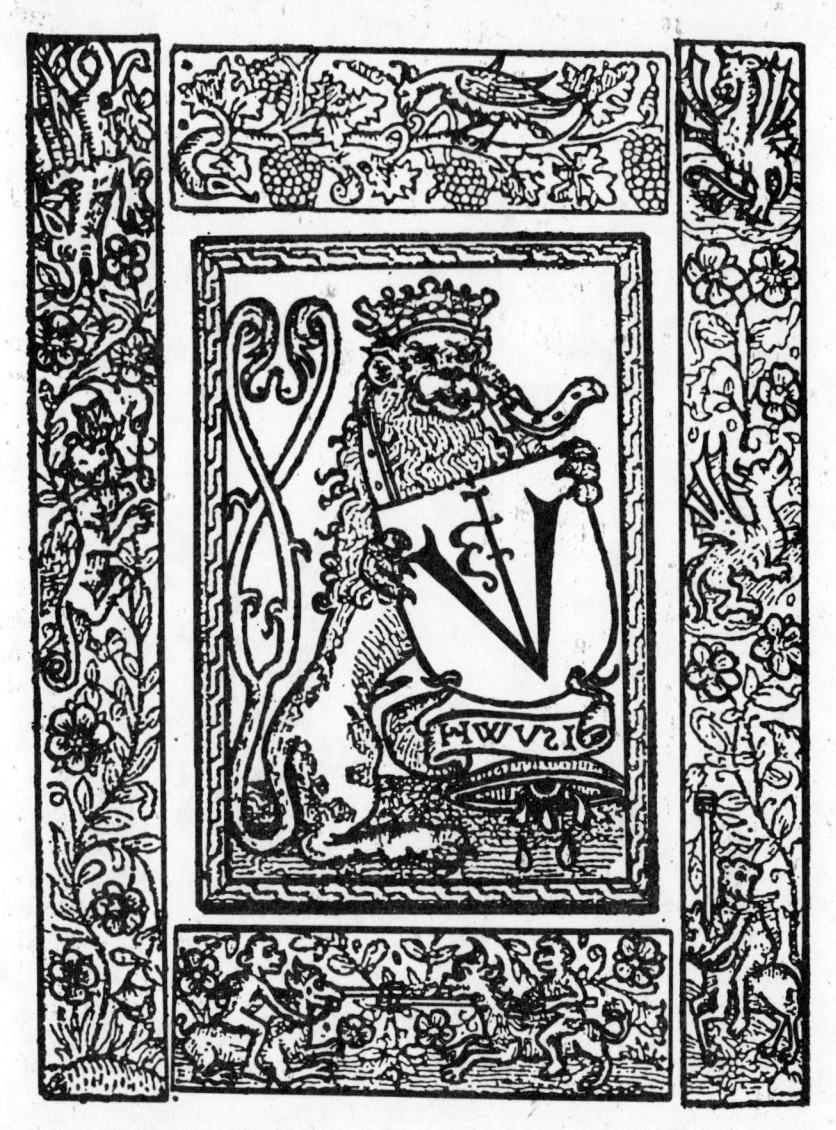

19.2 The *Book of Marco Polo* (1497).
An engraving by Valentim Fernandes, on the last page; his emblem appears on all the works printed by him from 1496 onwards, some of which (as here) bear an ornamental border.

the Templar alchemic cabala. It was this fact that led me to study a text by Fontoura da Costa in *Cartas das Ilhas de Cabo Verde de Valentim Fernandes* ('Letters of Valentim Fernandes of Cape Verde Islands').[5] This includes a letter that the printer presumably wrote to his friend Konrad Peutinger, in which he told an unusual story, which I have summarised:

19.3 Phoenixes and eagles.

    a The phoenix in an engraving from the *Book of Marco Polo* (see Figure 19.2). According to legend, the phoenix inhabited the Arabian deserts and lived for many centuries. When it felt death approaching, it made a nest smeared with fragrant resins, exposed it to the sun and burned itself. From the ashes came an egg which produced another phoenix. This myth, considered as one of the symbols of Hermes, was interpreted from the beginning by Christianity as a symbol of Christ's resurrection.

(*Engraving taken from Conradus Lycosthenes*, Prodigorum ad Ostentorum Chronicon, Basle, 1557)

    b Head of a phoenix that surmounted the 'Great Seal' of the USA from 1782 to 1902.
    c The head of the eagle that replaced the Hellenic–Hebraic–Masonic phoenix in 1902 on the 'Great Seal'.
    d The 'Great Seal' of official American documents since independence, having a circle of clouds and not ten stars in splendour, because as there were thirteen original States the number of stars would be at variance with the cabalistic decenary; they were therefore grouped geometrically in the splendour.

On a beach of the Promontory of the Moon[6] which people call the Rock of Sintra, three stone columns with Latin inscriptions were found, one being taller than the other two. They had been buried and covered with stones, so that they had to be pulled up with chains. Due to the erosion of the sea and rain, the inscription was nearly illegible.

Transcribing the Latin inscription, Fernandes told his friend that it had been deciphered in the presence of King Manuel I by his secretary. As it is highly unlikely that the King would have gone to Cape Roca to see three columns, or that they were taken to the court, Fernandes probably meant that he had been informed of the inscription, which read:

> Prophecy of the Western Sibyl decreed by the Eternal Gods. The stones will return to their natural disposition and the numbers to their correct order, when the West sees the riches of The East and the Ganges and the Indus see the Tagus, the admiration will be great, exchange their merchandise. To the Eternal Sun and the Consecrated Moon (I am).

This is obviously the plan of a temuric cabala, with the three columns or 'pillars of Intelligence' or of 'Wisdom' in which the decenary (group of ten elements or letters) is written.[7]

1. The 'Prophecy of the Western Sybil' suggests the application of the method of the Templar (or western) alchemic cabala instead of the traditional Hebraic cabala.
2. The 'Eternal Gods' specified the duality of Good and Evil, of the *Sephera Prima* of the 'Sephirothic system'.
3. 'The stones will return to their natural disposition and letters [numbers] to their correct order' indicates the clockwise movement, '*dextratio*' in the numeration of the sephirothic senary (group of six elements or letters), which is typical of the western cabala and replaced the zig-zag Hebraic numeration (see p. 302).
4. 'When the West sees the riches of The East' seems to indicate some alteration in the application of the normal method in which numeration is begun in the *Sephira 1* (in the position of North of the sephirothic senary). It may be said that this new rotation from the right to the left would not begin at the top of the senary, but from the 'Light' of the 'Candelabrum of the Seven Arms' to the 'Shade' of the 'Table of Atonement of the Tabernacle' (i.e., from east to west, beginning from the *Sephira 2*, see p. 313).
5. '[When] the Ganges and the Indus [see] the Tagus' would indicate

the three *channels* of each sephirothic ternary, joining the two external angles to the internal one, the latter being formed by one of the umbilical letters of the same system, as is usual; it could also mean 'when the Portuguese reach India'.
6. 'Exchange their goods' would mean the permutation of letters.
7. 'The admiration will be great' would correspond to the unexpected decodification of the cabalistic message.
8. 'Sun and Moon' are clearly an invocation of the elementary and occult sciences and of the two complementary powers, masculine and feminine (i.e., the 'Father' and the 'Mother').

I am of the opinion that Fernandes sent Peutinger some cabalistic message and was now indicating the specific temuric methods for its decipherment by means of 'Prophecy'. But what message could it have been? Did it concern the siglum that Colón-Zarco used to sign his name? Even if it were not this, I could in any case try those methods; and if it were, the Moravian printer, by communicating a secret to his friend, would be making things easier for him (and for me) in our research.

Columbus however, had died three months before, in 1506, eight years after the discovery of the seaway to India. The *Prophecy*, therefore, was out of date. But Fernandes knew this. And he had not written the work, he merely transmitted it to his friend.

Notes and References

1. After Gutemberg's German Jewish creditors had disposed of all the printing machines that he had so laboriously set up, they were able to launch the printing press all over the Europe to the advantage of the Jewish community. It was only in 1465, blind and ill and three years before his death, that Gutemberg saw the merit of his work recognised by Archbishop Adolf von Nassau, the Elector of Mainz, who appointed him Counsellor of the State.
2. Gusburg (1897) pp. 815–20.
3. *Torre de Tombo, Chancelaria de Dom Manuel I*, Book 35, p. 53.
4. Carvalho, Joaquim Barradas de (1982).
5. 16 August 1506.
6. Esoteric name of Sintra Mountain Range, Cape Roca.
7. All the nomenclature used here will be explained in Chapter 20.

20 Astrology and Alchemy

Astrology, or the science of the stars, dealt not only with the position of the stars in relation to the Earth, but also their influence on man's destiny. But after the science had ceased to be used to forecast the future in the sixteenth century, it became known as astronomy. The astrologers who claimed to foresee the future were frequently accused of charlatanism and of exploiting the gullibility of laymen. About 200 years after the closure of the 'School of Athens', astronomy, an ancient science of Egypto–Hellenic origin, continued under the auspices of the Arabs. Alfragan (9th century) participated in the revision of *Ptolemy's Tables* and wrote a primer called *Celestial Movements*. Albumasar – Abu Massar Gáfar-ben-Mohammed-ben Omar al Balhi (776–865) – wrote *Introduction to the Science of Astral Legislation*, which was translated into Latin by Hermanus Secundus in 1140 with the title of *Introductorium in Astronomiam*. The father of Prince Henry the Navigator, King John I, possessed a copy of this manuscript,[1] which was to launch the western nations into the science of astrology, until then the preserve of the Jews.

Ptolemy's *Almagest* had been translated into Arabic, but Hassan-ben-Haiten (died 1038) criticised it from a scientific point of view. About 1080, a watchmaker by the name of Arzaquel had invented the astrolabe and created the canons for the *Toledan Tables*[2] which, with the Tables of Albaten, were to serve as the basis for the *Alphonsine Tables* of Alphonse X, who founded the 'Academia del Alcazar Galeana' (which was the equivalent of the Junta of Mathematicians that King John II of Portugal later founded to replace the 'School of Sagres') and appointed, among others, the Jews Jehuda (known as 'the Coneso') and the Royal Astrologer) of Toledo, Rabiçag-aben-Cayut, Juda fi-de-Mosé and Don Abraham (who was also a Royal Astrologer) to reorganise the science of astronomy. Several studies were compiled in the form of a treatise under the title *Book of Knowledge*.

The work of the Sevillian Moor, Djaber-ben-Afflah (known as 'the Gerbe'), *De Astronomia*, which corroborated and divulged Arzaquel's

Toledan Tables, was translated into Latin by Gerard of Cremona only in 1534, under the title of *Gebri filii Afflah Hispanalensis De astronomia libri IX*, and published in Nuremberg by Peter Appian. By this time, the Portuguese and Spanish experts had already read the work in Arabic. The late publication of these Tables beyond the Pyrenees may explain the scientific backwardness of Martin Behaim, who based his calculations on those of Ptolemy, and saw them rejected by the Portuguese Junta.

The Castilian Junta of Mathematicians, the successor of the 'Academia del Alcazar Galeana', paradoxically stagnated, or even took a step backwards, in the field of cosmographic science. This was due to the fact that an 'adventurer' had bewildered them with the inaccurate maps of Behaim and Toscanelli which placed Japan on the longitude of the Antilles. King Ferdinand later censured the pilots of the 'Casa de la Contratacion' for their ignorance in placing the Antilles on the wrong latitude.

An attempt had been made to separate astronomy from astrology by St Isidor, Bishop of Seville (560–630), who purged the study of the stars from any use to predict man's destiny. He maintained that astronomic studies should simply be applied to the movement of the heavenly bodies, the seasons, etc. and that astrology was superstition, as it attempted to foresee the future in the stars and forecast the birth of men.[3] (see Figure 20.1).

The Visigoths and the Suevi looked for auguries in the entrails of animals and prisoners and had certain magical practices, but they assimilated little or nothing of the Graeco–Roman culture. Who at that time could then have known enough about cosmology to be able to use it to make predictions?

Between the tenth and seventh centuries BC, the Jews had founded various settlements in the Iberian Peninsula, such as Escalone, Toledo, Maqueda and Yepes.[4] They penetrated into Celtiberian territory and there are tombstones at Castelo de Moviedo that lead some archaeologists to believe that the first Jewish settlement in Iberia took place at the time of King Solomon. People who had a knowledge of the Egypto–Hellenic science affronted St Isidor because astrology and astronomy were so entwined with magic that cosmographic knowledge took second place to the practice of witchcraft, especially as prediction was lucrative and cosmography beyond the understanding of the people in general and lacked material interest. The fruits of the separation of the two sciences were reaped only nine centuries later, at the time of the Discoveries.

The Iberian Jews were more inclined to devote themselves to mystical speculation, such as alchemy and the Cabala, than to the study of astronomy; and despite the fact that astronomy and astrology had been condemned by St Augustine and prescribed as separate by St Isidor, all

20.1 Alchemy and esoteric alchemy.
Fifteenth-century woodcut, an allegory of the Sciences on which esoteric astrology was then based. Alchemy was placed in the same field as scientific astrology, which entered into decline owing to concurrent practices of false auguries in prediction, diabolism and witchcraft and the falsification of coins through the plating of base metals with silver or gold.

(*Sixteenth-century engraving*)

the Portuguese Kings up to John II were attended by Jewish physicians and astrologers who predicted the future with a blend of science and superstition.

THE ORIGIN OF THE PERESTRELO FAMILY

The Jew David Gabriel Pallastelli (known as 'the Mantle of the Stars') was in 1395 the astrologer of Giovanni Galeazzo, Lord of the Duchy of Milan. Pallastelli became a Christian and adopted a crowned lion surrounded by ten stars for his coat-of-arms, which revealed his Jewish origin (the Lion of Judah and the decenary of the cabala). When Sicily and Aragon were united under the same crown in 1409, another astrologer and mathematician, Pallastrelli (possibly Pallastelli's son) moved from Milan to Aragon. When Princess Leonor, daughter of Ferdinand I of Aragon came to Portugal in 1428 to marry Prince Duarte (future King and brother of Henry the Navigator), there was one Micer Filipe Pallastrelli in her retinue who also possessed a coat-of-arms with the same lion, but with only a ternary of stars and a senary of five-petalled roses. These heraldic symbols suggest not the Hebraic but the western Christian and Templar alchemic hermeticism.

The newly-arrived astrologer and mathematician took the name of Palastrelo and King John I (father of Prince Henry the Navigator) confirmed his Lombardian coat-of-arms in 1434, but with an uncrowned lion. It is possible to ascertain only that from his first marriage to Catarina Vicente Palastrelo begot two sons, Rafael and Bartolomeu Perestrelo. The latter was in the service of Prince João[5] and married Brites Furtado de Mendoça, daughter of Dom Afonso Furtado de Mendoça (see p. 285). Among the several children of his second marriage to Isabel Moniz was another Bartolomeu Perestrelo, a navigator of the Order of Christ who married Guiomar Teixeira of the family of the navigators Vaz Teixeira. From this marriage issued another Bartolomeu Perestrelo, who wed Aldonça Delgado, granddaughter of João Gonçalves Zarco. From his second marriage to Isabel Moniz, the first Bartolomeu Perestrelo had a daughter, Filipa Moniz, who married Columbus.

All these marriages clearly took place in the circle of the navigators of the Order of Christ who were in the service of the crown or the princes. The marriage of Filipa Moniz to a Genoese wool-carder would have been an aberrant exception, even more so since he would not yet have made a name for himself and was still far from becoming the Admiral of the Indies in Spain. But if he were Colón-Zarco, grandson of João Gonçalves Zarco, it would have been the norm. And both Columbus and Filipa had Jewish ascendants.

On the day after the death of King John I in 1433, the Jewish physician and astrologer, Master Guedelha (Guêdêliah) requested an audience with the new King, Duarte,[6] to warn him of impending disaster if he did not postpone his coronation. The Sovereign turned a deaf ear to the warning. After a year of peace came disaster at the siege of Tangier; four years later, the King succumbed to an outbreak of plague that decimated the population of the kingdom.

The practice of astrology and alchemy continued in Europe for many years; bishops, cardinals and monarchs devoted themselves to these pseudo-sciences, but it was from these false sciences that many scholars reached conclusions of scientific accuracy.

THE ALCHEMISTS

Alchemy had its deepest roots in the Orient. It is known that it was practised in China around 4500 BC, but it was Taoism (a doctrine attributed to Lao-tse, c. 604–531) that gave rise to the research carried out from the third century AD. Alchemy was one of the occult disciplines of Hindu and Buddhist Tantrism; it spread to Egypt, Chaldea and Persia, and the Jews and the Crusaders, especially the Templars, were paramount in its introduction into Europe in the eleventh century.

The oldest alchemic writings are attributed to writers such as Plato, Aristotle, Zoroaster, Pythagoras and Moses. Alchemy was considered as a sacred art in Alexandria. It flourished in the Byzantine empire in the sixth century and a hundred years later, Chalid-ibn-Jazid, (who reigned in Egypt in the first half of the century) brought the doctrine into the Arab world.[7]

Alchemy took root in the Iberian peninsula after the Muslim invasion in the eighth century, although it was the Jews, as students of the Cabala, who did most to cultivate it. As already mentioned, some Crusaders devoted themselves to alchemical studies and, on returning from the Middle East, played an important role in the dissemination of such learning in Europe in the twelfth and thirteenth centuries, adapting the alchemic doctrines to the gnosis of Christianity.

Alchemy was also called the 'science of Hermes' or 'Hermetic philosophy'. It was basically dedicated to the study of matter and the possibility of transmuting the molecular constitution of metals,[8] into gold (*chrysopoeia*) or into silver (*argyopoeia*). Alchemy had implications in medicine in the search for the 'elixir of Life' (*panacea*) and also tried to produce the 'philosopher's stone', which the alchemists called the 'great Work', engendered in the 'philosophical egg' – a closed crystal globe containing a secret mixture and acting as a womb, which was called a chamber – '*Câmara*' in Portuguese, the name we know João Gonçalves Zarco chose when he was raised to the peerage.

PALAEO-CHRISTIAN AND ALCHEMICAL SYMBOLS

Owing to the many frauds perpetrated by alchemists and the fact that its practices deviated from Christian doctrine, the Church condemned alchemy, although Pope John XXII (1316–34) showed a certain leniency towards its practitioners. The Templars founded a 'College of Alchemy' in the twelfth century.

Alchemy was in many ways an offshoot of astrology and like the Hebraic Cabala was divided into theoretical (mystical) and practical aspects. Several treatises were written on the subject, mostly by Jewish authors. In the thirteenth and fourteenth centuries, alchemical writings became cryptographic and cabalistic, replete with symbols, some of which were inspired in palaeo-Christian symbology. Here are some examples:

- *Christ* and the *pelican* (which represented Him): they corresponded to the 'philosopher's stone' for the transmutation of base metals into precious metals.
- The *dragon* and the *sword*: they represented the fire of creation that resuscitated the phoenix.
- The *phoenix*, reborn, symbolised the breaking of a new day, the vigilance of the Christians in order to prevent Peter denying Christ a second time.
- The *anchor* was hope in God, that one would find a safe harbour to shelter from storms (it was associated with the dolphin).
- An *open or concave fishing net* and a *ship's cable* corresponded to Divine mercy.
- The *vine* represented the image of Paradise and the Eucharist.
- The *pigeon* was simplicity, innocence, baptism and peace.
- The *moon* symbolised the silver of the '*argyopoeia*' and the Mother.
- The *sun* symbolised the gold of the '*chrysopoeia*' and the Father.
- The *crown* represented the '*chrysopeia*' already achieved.
- The *rose* represented the heart, love in all its forms (it was a red five-petalled rose, this number sometimes being doubled by the sepals to a ten-petalled flower).
- The *chamber* was the vault, the ceiling and the 'philosophical egg' (or the nuptial chamber or sepulchre).
- The *leather strap* symbolised secret fraternity, the hermetic link.[9]
- The *wild artichoke in blossom*, which after being burned re-blossoms in purple, was a fusion of the phoenix and the rose.

Twelfth-century hermeticism laid special emphasis on the *Emerald Tablet* (*Tabula Smaragdina*), from which the following text is taken:

> It is true, without lies, certain and very true. That which is below is like that which is above and that which is above is like that which is below, for the fulfillment of the miracles of one sole thing. And in the same way that all things come from the One, thus are all things born from this one thing, through adaptation. The Sun is its father and the Moon is its mother, the wind carried it in its womb, the Earth suckles it; the *Telema* [perfection] of the whole world is here.[10]

St Albert the Great (1193–1280) and St Thomas Aquinas (1226–74) were interested in alchemy, and the latter thought it a valid science, so long as it did not stray into the field of magic.[11] Roger Bacon (1214–94) carried out experiments in the transmutation of metals and enjoyed the protection of King Edward I of England. Arnauld de Villeneuve (1245–1313), who was a friend of Pope Clement V, developed the theory (taken from the Cabala) that there was a *'Spiritus'* that could act as a vehicle between the stars and man.[12] Raymond Lully (1235–1313) of Majorca was a disciple of Villeneuve, but modern scholarship now regards his works of astrology as apocryphal.

According to the defenders of the Catalan thesis of Columbus's origins, the cosmological knowledge of their navigators came from the works of Lully, which aroused great interest in the fourteenth century but which were out of date and of little value a hundred years later. Guillaume de Lorris, who was a contemporary of Lully's and wrote several celebrated treatises on Alchemy, claimed that Divine Grace and the philosopher's stone had the same meaning as the five-petalled rose. Esoteric speculation influenced Christian literature, as in the case of Dante's *Divine Comedy*.

Alchemy was revealed to be an enlightened doctrine in the fifteenth century, but as there was a proliferation of heresies at this time, which the Church condemned because they were usually linked to witchcraft, the alchemists, fearing involvement in potential heresy, made their practice even more secretive.

Several notable alchemists appeared in this period, among them being Basile Valentin, who wrote *Twelve Keys*; he may have lived in a Benedictine monastery around 1413, although his writings were published only in 1602. Some authors, however, consider him to be a mythical figure. A similarly mythical figure was William of Baskerville.[13]

One of the alchemic theories that most shocked theologians was that of 'sexual dualism', which was widely proclaimed by hermetic authors who believed that all sympathy and antipathy in a philosophical sense came from two complementary principles, one masculine and active and another feminine and passive. These theories were based on millenary Oriental concepts: God was a hermaphrodite before the creation; He then split into two beings of the opposite sex and created the world through their coitus. The Sun was thus masculine and the Earth feminine, although the main figure of femininity was the Moon, which was the Mother, the ever-fecund Goddess, but always virgin, represented by a woman crowned with ten stars and bearing a crescent moon on her body (see Figure 20.2).

The union of man and woman corresponded to the opposition of the principle of creation to the principle of fecundity. All these graphic symbols of sexual origin,[14] shown by primitive man in his artistic representations, were expressed in an infinitive variety of forms and gradually gave rise to the various alphabets.

20.2 Sexual dualism.
Fifteenth-century xilograph depicting the union of a man/Sun with woman/Earth and Moon, both holding lilies of purity which, with the third one held by the dove of the Holy Ghost, make the symbol of the Son/Christ.

(*Photographed by the author from an eighteenth-century text*)

Alchemy was closely linked to the Cabala as well as to astrology and its cultists used the *temurah* for cryptographic communication. The mystical Cabala influenced notable people, such as Pico de Mirandola (1463–94), for instance, who in 1486 published several theses against the philosophy and theology of his time, daring even to allege that the Cabala contained the soundest proofs in favour of the divinity of Christ. Mirandelo, however, had been initiated in the Cabala by the Jew Johanan Aleman,[15] and Hebrew thinkers certainly tried to provoke defections from the Christian doctrine in an attempt to dismember the Church into other reformist factions. The notable Portuguese writer Dom Francisco Manuel de Mello (1611–66) wrote the *Treatise on the Science of the Cabala*, which was later 'corrected' by the 'Censor of the Court'. De Mello was very interested in the occult sciences, but his work was extremely superficial.

Most of the monasteries, cathedrals and churches built between the twelfth and sixteenth centuries in Europe are replete with palaeo-Christian symbols that showed alchemic and cabalistic influences,[16] especially in Portugal, where the presence of the Templars was felt: the Monastery of Christ in Tomar is an outstanding example of this. These symbols are also to be seen in the images of some saints and even of the Virgin Mary, in old vestments of priests and in objects used in the Holy Mass.

Notes and References

1. See Luciano Freire, *Lusitania*, Vol. II, fasc. 1 (September 1924). Erhard Ratdolt copied this work and published it in Augsburg in 1489, under the auspices of Emperor Frederick III (married to the Portuguese Princess Leonor, sister of Prince Fernando, Master of the Order of Christ). It was republished in Venice in 1506.
2. The criticism of Hassan-ben-Haiten and the study of the *Toledan Tables* led Djaber-ben-Afflah of Seville to reject Ptolemy's *Almagest* as well.
3. In '*Differentia astrologicae et astronomicae*', *Elogiarum*, Chap. 27, Book III. 'Inter astronomiam et astrologiam aliquid differt. Nam astronomia conversionem coeli, artus, obitus, motusque siderum, certosque temporum staciones. Superstitiosa vero est illa, quam matematici sequuntur, que stelio augurantur, quique etiam duodecim signe per singula animae vel corporis membra disponunt, iderumque cursu nativitates hominum et mores proedicere conantur'.
4. Acosta (1719); the original place names are Escalon, Toledoth, Maquedah and Yappe.
5. Son of King John I.
6. Father of King Alphonse V (father of King John II), who possessed the *Book of Magic* written by Juan Gil Burgo.
7. Bacon (1960); Mély (1895); Bertheler (1938).
8. Marchard (1959); Read (1936); Stillman (1960).
9. Teste (1950); Schneider (1962); Burkhardt (1960).
10. The cabalistic method of the 'mirror' is based on this concept in relation to the existence of a Macrocosm and a Microcosm, diametrically opposed but complementary: 'God can be understood only through the image of His creation': see Poisson (n.d.), pp. 2–3; Thorndike (1934), Vol. II, Chap. XLV; Ganzenmüller (1975).
11. *Suma Teologica*, Vol. II, Chap. LXXVII, Art. 2.
12. Haven (1896); Sansonetti (1972); Arnold (1955).
13. Ouy (1972); Baskerville was the main character of Umberto Eco's *The Name of the Rose* (Milan, 1980), and Sean Connery played the role of Baskerville in the film of the same name.
14. Villeneuve (1970); Saint-Martin (1979).
15. Tryon-Montalabert (1979), p. 230.
16. Fuscanelli (1926) and (1970).

    Other sources: Paracelse (1933); Alveydre (1950); Alleau (1945), (1953) and (1976); Chenu (1975); Cohen (1958); Hutin and Cavon (1971); Tresmontant (1969); Figuier (1970); Lulle (1953); Chevalier and Gherrboaut (1970); Plancy (1970); Garçoh (1971); Delumeau (1971); Gazenmüller (1975). Bacon (1977).

21 The Labyrinth of the Cabala

The oldest books that make up the doctrinal sources of the *Qabalah* or *Kabbala* date from the second century AD – the *Yetzirah*, which is thought to have been written by Rabi-Akiba, and the *Zohar*, attributed to his disciple Rabi-Semeon-ben-Yochai. (There are, however, doubts about the authorship of the latter work, as it came to light only in the thirteenth century.[1])

The Hebraic Cabala is based on Oriental writings that came down through the ancient Egyptians. The writings of the Rabbis divided the study of the Cabala into two branches – theoretic (*Iymith*) and practical (*Maasith*). The theoretic Cabala refers to philosophy and mysticism, also known as 'theology'; the practical Cabala deals with the secret science that teaches 'the art of making superior powers act, in any circumstance, over the inferior world'.

In the theoretic Cabala, which is speculative and metaphysical, God is said to be 'All Eternity' and the Hebrews call Him '*Ain*' ('Nothing') or '*Ain-Soph*' ('Limitless' or 'Infinite'), considering God to be an abstraction that can be understood only through His actions: 'The world was created from nothing'. According to the Bible, David said: 'I raise my eyes to the mountains from whence the Nothing will come to my aid'.[2]

Ain-Soph, which is the beginning of everything of everything (*Ilath-Ha-Iloth*), is divided into four 'emanations' or 'worlds' that the cabalists called *Atziluth, Briah, Yetzirah* and *Assiah*.[3] *Atziluth* is divided into ten 'essential qualities' that make up the 'Holy Decad' or 'Decenary' of the *Sephiroth*.[4] *Briah* (Creation) is the beginning of everything. *Yetzirah* (Formation) envelops the 'Angels', incorporeal and individual beings that are surrounded by a luminous involucre or aura. *Assiah* (Production) submits matter to continuous variations that are born and die, are composed or divided.

The Cabala was based on the 'Holy Decad' of ten spheres, represented by circles that are known as *sephiroth*.[5] The *Sephiroth* (the sephirothic system of the 'Holy Decad') was formed by two groups of spheres – the first contained three that were known as 'intellectuals' and the other seven were designated as 'attributes'.

In the sephirothic group of 'intellectual' spheres, the first S corresponds to *Kether* (the Crown), the second S to *Chokhmah* (Wisdom) and the third to *Binah* (Intelligence). In the group of the 'attributes', the seven remaining Ss correspond respectively to *Chesed* (Mercy), *Geburah* (Severity), *Tiphareth* (Beauty), *Netzah* (Victory), *Hod* (Glory), *Yesod* (the Foundation) and *Malkuth* (the Kingdom) (see Figures 21.1–21.4).

The *sephiroth* are interconnected by 'channels' or 'rivers' of communication, which are paths of reciprocal influence, like those mentioned in the *Prophecy of the Sibyl* when referring to the 'Ganges, Indus and Tagus' (see p. 251).

The whole tree of the sephirothic system is surmounted by another sphere called the *Sephira Prima*, also known as *'The Horizon of Eternity'*, followed by the *Suma Corona*. This, in turn, is divided in two hemispheres, the upper one, representing the 'Light' of Good, and the lower one the 'Darkness' of Evil.

THE WESTERN CABALA AND ITS SYMBOLOGY

The Western Cabala (also known as the Templar or alchemic Cabala as it was the one preferred by the alchemists) started spreading in the thirteenth century and reached its apogee at the end of the fourteenth and throughout the fifteenth centuries. It first appeared in the Iberian peninsula and soon spread all over Europe, particularly England and Germany, where the philosophers of the Order of the 'Rose Collar'[6] attributed it to the Templars, whose Order was by that time extinct.

This Cabala was impregnated by the divine personality of Jesus Christ, by the Trinity, by all the symbolism of Bethlehem and Calvary, by the three Magi who were later called Kings. And among its symbols there appeared the cabalistic image that corresponded to the Hebraic *Chesed* (Mercy) – the female pelican tearing out its own breast in order to feed its young, which represented Christ's sacrifice for the salvation of Mankind (see Figure 21.7a).

This symbol was adopted by King John II and, initially, by his wife Leonor, the founder of the institution of 'Mercy', which had the pelican as its emblem. The alchemist designated the sephirothic system as the 'Tree of Life' or the 'Door of Life', which was represented by the three pillars of 'Intelligence' or 'Wisdom', in which the 'Channels of Wisdom' passed through each of the Sephiroth, indicating the descending path of power (see Figure 21.3). They sometimes drew a diagram of the same system which showed three superior emanations. *Ain*, *Ain-Soph* and *Ain-Soph-Aur*, instead of the *Sephira Prima* (see Figure 21.4).

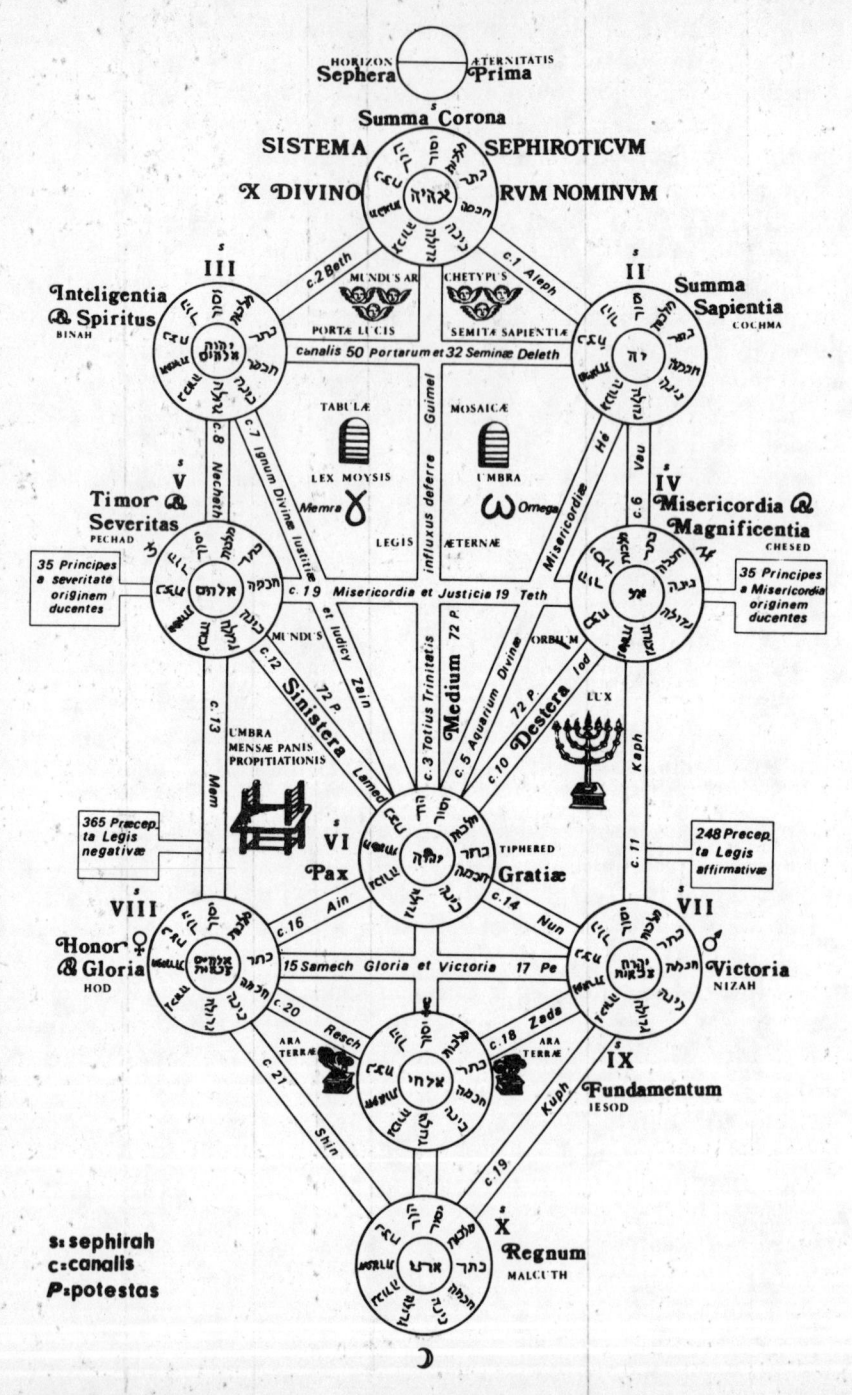

21.1 The *Sephirothic* system.
(Copied by the author from an eighteenth-century text)

THE CABALA 265

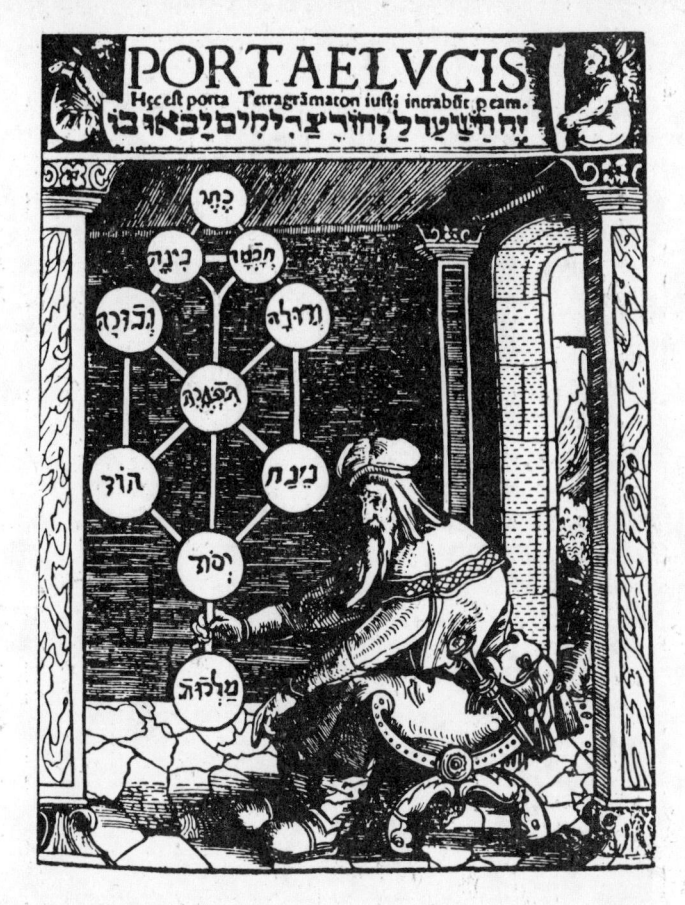

21.2 The *Sephirothic* decenary.
On the frontispiece of *The Gates of Light*, meaning 'the gates of heaven'.

(*Photographed by the author from Touchaix, 1978*)

In the western (Templar or alchemic) Cabala, clearly Christian in essence,[7] the two groups of the three angels that appeared in the sephirothic senary were placed above the crown and the divine triangle of the Trinity was inserted between the crown and the *Sephira Prima – Xpo* (Christ) in His triple essence (the Father, the Son and the Holy Ghost), linked to God and to man. I shall consider cabalism further only where it is necessary to interpret the cabalistic emblems of the time that were used by royalty or the nobility or – above all – to decipher Colón-Zarco's siglum.

In the divine Cabala, the fusion of the 'emanations' of more than one sphere of the system can be carried out through the 'channels'. The 'emanations' are, up to a point, hermaphroditic, none of them being strictly masculine or feminine. This trait is noticeable in Helleno–Roman

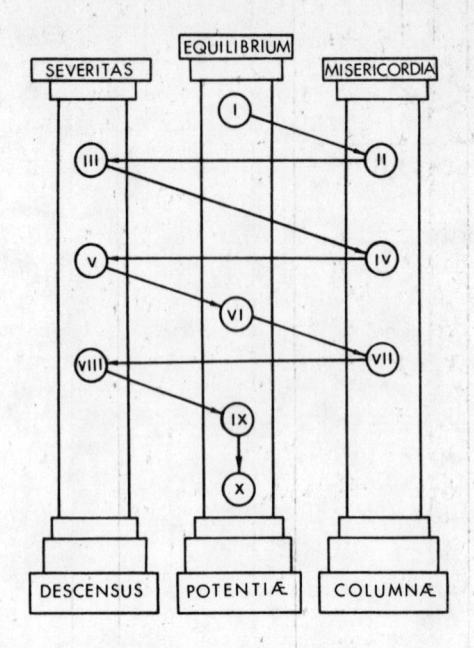

21.3 The three Pillars (or Columns) of the Descent of Power.
In the decenary of the *Sephirothic* system, showing the Hebrew 'zig-zag' sequence.

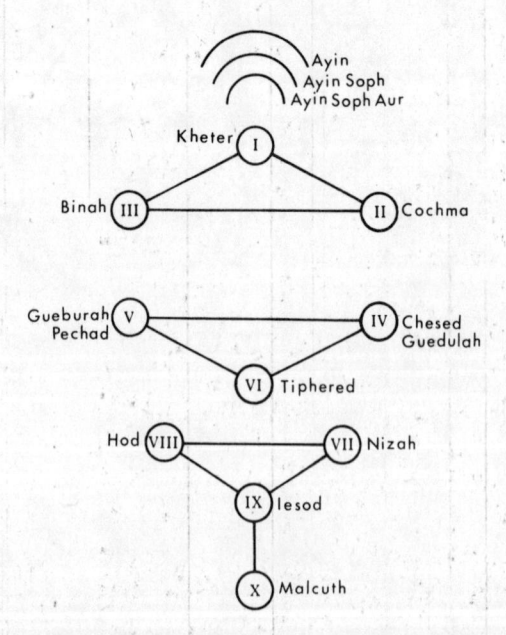

21.4 The *Sephirothic* Ternaries and the Auras.
Diagram of the Ternaries and the three Auras, retaining the Hebraic 'zig-zag' sequence.

1. The Sagres School.

   A circle divided into 60 sections in the fortress at Sagres, where Prince Henry gathered the mariners of the Order of Christ. This circle was used for practical training in orientation by the sun and the Polar Star with the quadrant and the astrolabe, the navigational instruments of the day.

2. The altars at Tomar and Newport.
   a The round altar in the Monastery of Christ in Tomar, headquarters of the Order of Christ in Portugal, showing 'initiation' symbols.
   b The altar in Newport, USA, presumed to have been built by the Portuguese who stayed in America or were shipwrecked there on the return trip of Gaspar Corte-Real (1500), brother of Miguel Corte-Real, author of the inscription on the Dighton Rock (see Figure 13.1).

3. The twelfth-century church at Borgund. A Viking church, built of wood, as were most Viking houses until at least the sixteenth century, at Borgund in Norway.

4. A window in the Monastery of Christ at Tomar. The Manueline style incorporated a profusion of symbolic decorations such as the 'ropes' of Mercy, the five-petalled rose (see Chapters 21 and 34), the artichoke in blossom and the leather strap.

5. Henry of Bolingbroke.

   (1367–1413), son of John of Gaunt, uncle of Prince Henry the Navigator and future King Henry IV of England; he wears a 'Livery Collar' of 18 *S*s.

6. Sir Thomas More.

   (1478–1535); he wears a 'Rose Collar' of 24 *S*s, a senary sequence of the Templars' motto, and the five-petalled rose (see Chapter 21) with five sepals, which the English called the Tudor rose.

7. A sixteenth-century English cannon.

   The five-petalled rose was adopted by the order of the Rose-Cross Brotherhood when it was founded in the seventeenth century, and by the Franco–Masonic Brotherhood in the eighteenth century. From 1400 to 1700 pieces of ordnance were decorated with esoteric symbols, and some of these pieces were known as 'dragons', 'serpents' and 'colubrines' (i.e. a grass snake). Some of these pieces may be seen in the Portuguese Military Museum.

8. A sixteenth-century Portuguese cannon.

   This bronze cannon bears rings in the form of serpents and the arms of Portugal surmounting a 'Wheel of Knives' (see Chapter 34), the hagiographic symbol of St Catherine (see Plate 9).

9. St Catherine.

   St Catherine with the wheel on which she was tortured; Portuguese Queens and Princesses particularly venerated this saint. This may be seen in the Portuguese Military Museum.

10. Colón-Zarco's portrait (?1507)

    Although posthumous, this is considered to be the only true portrait. It may be seen at the Madrid National Library (see Plate 15).

11. The Polyptych of Nuno Gonçalves.

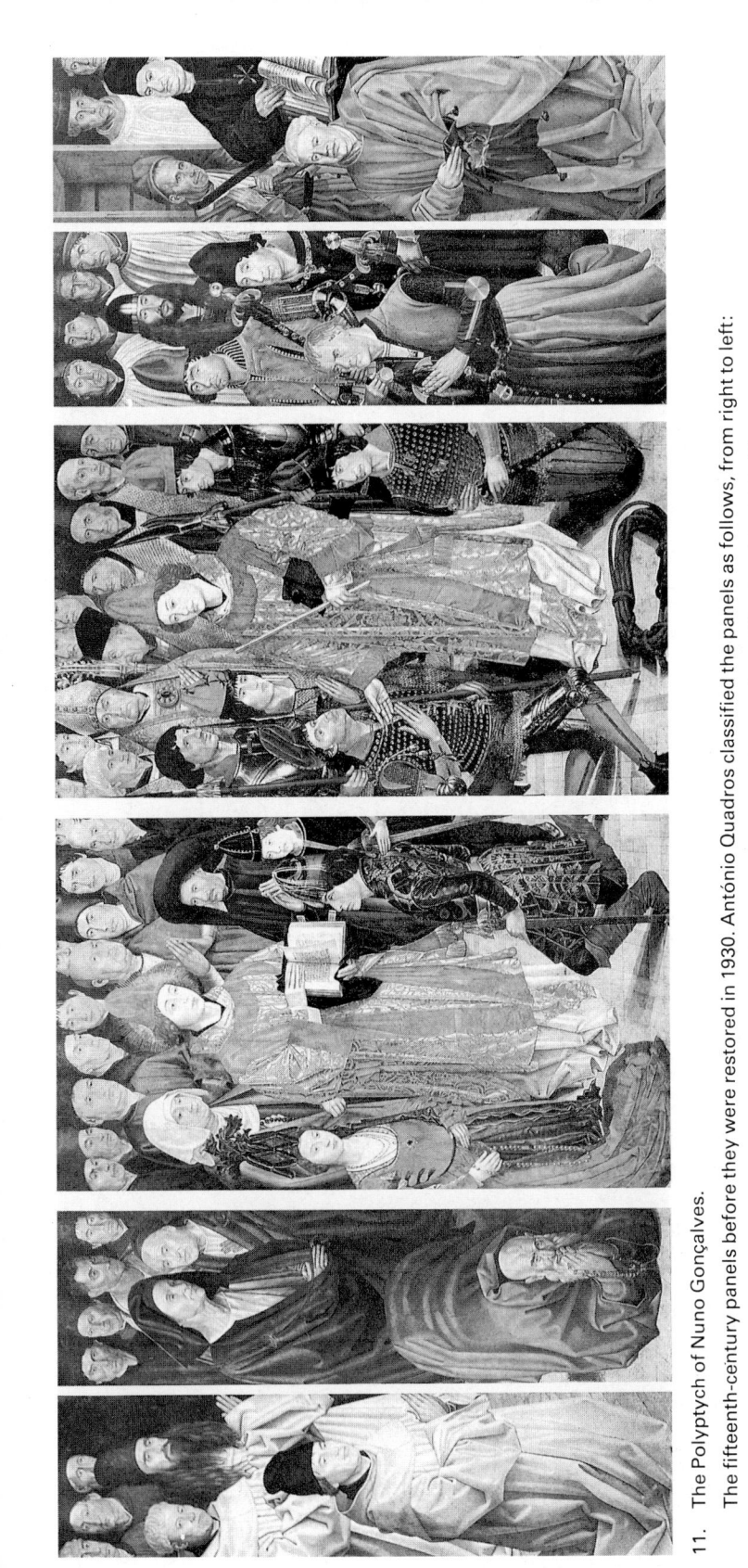

The fifteenth-century panels before they were restored in 1930. António Quadros classified the panels as follows, from right to left:

I Panel of Cistercian Monks;   II Panel of the Brotherhood of the Holy Ghost;   III Panel of the Alliance of the Holy Ghost;   IV Panel of the Mission of the Order of Christ and Avis;   V Panel of the Knights;   VI Panel of the Relic.   *See next page for identification key.*

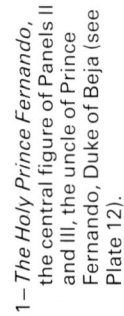

1– *The Holy Prince Fernando*, the central figure of Panels II and III, the uncle of Prince Fernando, Duke of Beja (see Plate 12).

2– *King Alphonse V*, uncle of Colón-Zarco.

3– *The Holy Princess Joana*, daughter of King Alphonse V (see Plate 16).

4– *King John II*, before ascending the throne.

5– *Prince Fernando*, Duke of Beja, father of Colón-Zarco (see Plate 12).

6– Prince Fernando's eldest son, *Dom João* (see Plate 13).

7– Prince Fernando's second son, *Dom Diogo* (see Plate 14).

8– *Dom Fernando II*, Duke of Braganza.

9– Dom Fernando II's brother, *Dom João*, Marquis of Montemor-o-Novo.

10– *Charles* (known as 'the Bold') *of Burgundy*, first cousin to Prince Fernando, Duke of Beja. António Quadros believes, I think correctly, that this figure has previously always been wrongly identified as Prince Henry the Navigator.

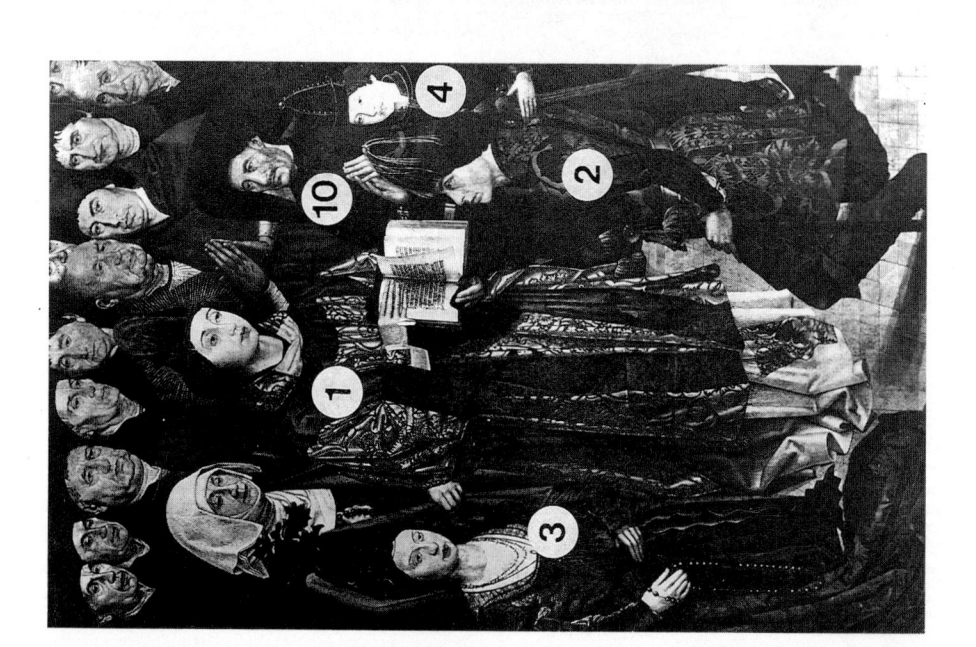

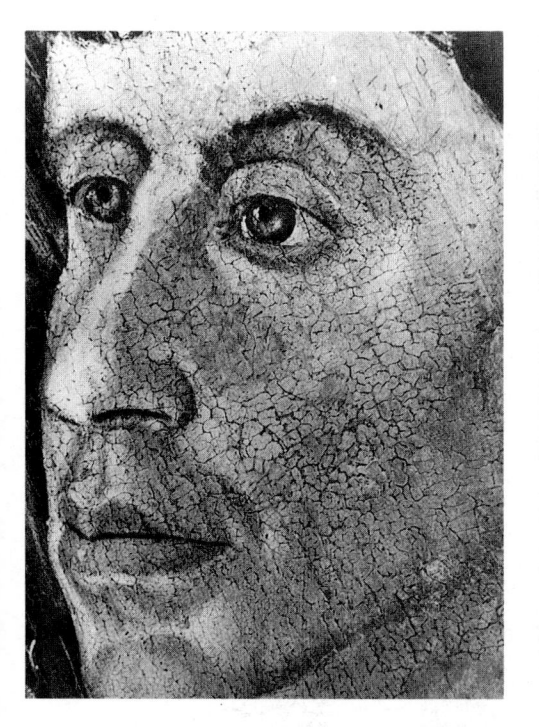

12. Prince Fernando, Duke of Viseu and Beja, Master of the Order of Christ and father of Colón-Zarco.

13. Prince Fernando's eldest son, Dom João; as he died young the Prince's positions and titles were inherited by his second son, Diogo (see Plate 14).

14. Prince Fernando's second son, Dom Diogo; having conspired against John II, his cousin and brother-in-law, the King stabbed him to death.

15. Colón-Zarco's portrait (?1507): an enlarged reproduction.

16. The Holy Princess Joana, daughter of King Alphonse V and first cousin to Colón-Zarco.

mythology, in which Venus may appear with a beard and Hercules is represented dressed in female attire. Hebraic characters that have a specific interpretation may appear in a western cabalistic cryptogram.[8]

For a better understanding of cabalistic symbology, we need to examine four key symbols.

1. The five- (or sometimes ten-) petalled *rose* that was considered as the symbol of the Round Table represents the height of spiritual energy (see Figure 21.5). It also depicted the heart and love, for alchemic cabalists, its colour being the blood of the Saviour of Mankind.

   The pelican (Figure 21.6) not only symbolised the 'philosopher's stone', but also Jesus sacrificing Himself for Mankind, as we saw above. For the western cabalists, this was located in the *Netzah* (Triumph) emanation, the last level of the pillar of Mercy.

21.5 The five- or ten-petalled rose.

21.6 The pelican.

In hermetic symbolism, the rose and the pelican together represented the highest expression of divine and human affection. Although of Oriental origin, the *compass* came into history as an invention of Talus, the nephew of Daedalus (Figure 21.6). For astrologers and alchemists, it was the triangle that was under the hands of the Centaurus (coat-of-arms) of the Andrades of *Arco*, (see p. 61 and 241) of the southern constellation and represented the 'measurements' and 'proportions' of the earth in contrast to the Infinity of God (see also the Dighton Rock inscriptions, p. 150).

2. The *phoenix* (see Figure 21.7a), reborn from the flames in a mystery of resurrection, meant the indispensable 'vigilance so that Christians will never again deny Jesus Christ' for the western cabalists. King John II, who adopted the pelican as his emblem with the motto 'For Law and for the People', replaced it with the phoenix, as can be seen in Valentim Fernandes's engraving in *Vita Christi* (Figure 21.7b). Knowing the severity and perspicacity of the King, the engraver would not have made such an alteration without the royal authority.

Although the Cabala has a mystical origin, its symbols can be connected with human beings in cryptographic messages. Let us speculate on the reasons that led John II to adopt the phoenix. He made the substitution after the attempts on his life. If the pelican symbolised the sacrifice of God for the benefit of Mankind, it could equally represent the efforts of a King for the good of his subjects as can be read in his motto. And if the phoenix invoked the 'vigilance of Christians so that Christians will never again deny Jesus Christ', it could also refer to the vigilance over his subjects so that a Duke would not molest the person of the King again. But the King granted his pardon to all of them under the sign of 'Mercy'.

3. A *Rosicrucian emblem* (see Figure 21.5) attributed to one Johann Valentim Andreae, supposedly the founder of the Order of the Rose Cross, but who was probably no more than a mythical personage created by Sir Francis Bacon (1561–1626), philosopher and Lord Chancellor of King James I of England. The emblem appeared much later than the rose of the western Cabala and must not be confused with the ones of the Templars, the Tudors and the Order of Christ.

4. Queen Leonor initially had the same pelican as her husband, John II, as her personal emblem and also attributed it to the institution of 'Mercy' that she founded. She replaced this, however, by the emblem reproduced in Figure 21.8b from Valentim Fernandes's *Vita Christi*, which includes the *rope*, a *pair of compasses* and a *fishing net*. The rope and the fishing net are both symbols of the emanation *Chesed* (Mercy). Some investigators wrongly supposed that the central arm suspended from the ring was the arm of the *anchor* ('the hope that God would lead people's souls to a safe harbour'), but in the emblem

21.7 The phoenix instead of the pelican.

  a Illustration in the *Gramatica de Pasgrana*, printed by Valentim Fernandes in 1497.
  b Ornamental frieze from the *Vita Christi*. The emblem of King John II, symbolising 'Mercy', can be seen at the top, but with the phoenix replacing the pelican, with a five-petalled (doubled) rose, which represents the Love of Christ.

(Photographed by the author from a sixteenth-century text)

it is merely a sephirothic 'channel', the ring of which represents the *Sephira Yesod* (the Foundation). The *rope* of 'Mercy' is entwined in circles that form the sephirothic senary and the respective representations of Light (seven-armed Candelabrum) and Darkness (the Table of the Bread of Atonement of the Tabernacle (see Figure 21.8a)). The Sephiroth *Netzach* (Victory) and *Hod* (Glory) are intentionally marked with two small shields, each one on the arm of a pair of compasses. This is one of the symbols of the emanation *Chokhmah* (Wisdom) at the top of the pillar of Mercy and is closely related to the armillary sphere that was adopted by King Manuel I as his emblem. The Sephirah

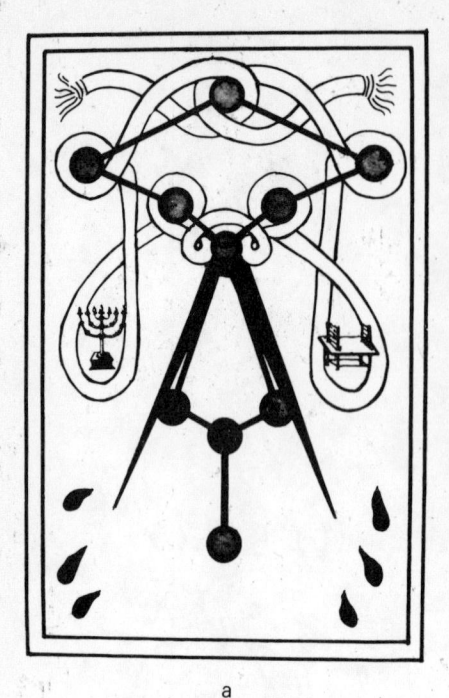

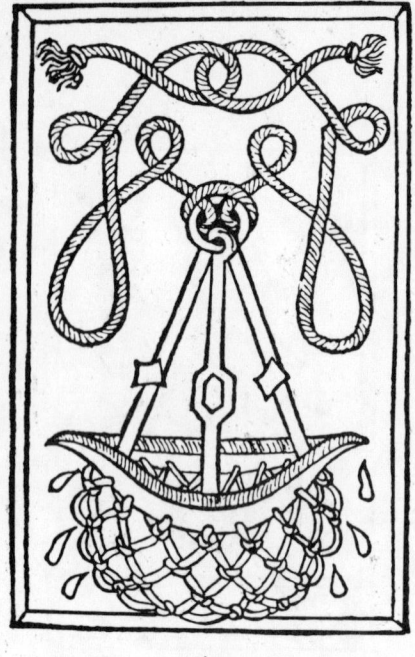

21.8 The rope and the fishing net.
   a *Sephirothic* system shown in the emblem of Queen Leonor.
   b The emblem of Queen Leonor.

Malkuth (the Kingdom) is represented by the *fishing net* (of the concave frame) (see Figure 21.8a), which signifies, 'My Kingdom is Mercy'. The six drops of water (or tears or blood) indicate the sephirothic senary of sacrifice.

In order to conceal all that cabalistic symbolism, it was said that the Queen had chosen the new emblem when her son, Afonso, was killed in a riding accident in 1491. His body was supposedly pulled out of Tagus in a fisherman's net, but this story is a fiction. The new emblem was probably adopted by Queen Leonor fifteen years later and published in 1505 in *Os Autos dos Apóstolos* (The Play of the Apostles), the author of which was patronised by the Queen.

It is obvious that the Catholic Church could not accept the concepts of the Hebraic Cabala or, even more so, those of the western Cabala which established a veiled analogy between the Trinity of Christianity and the trilogy of Osiris (the first God of a human nature who, after being killed by Seth, the God of Evil, was resuscitated in his son), Isis and Horus. These concepts may have contributed to the arrangement of certain

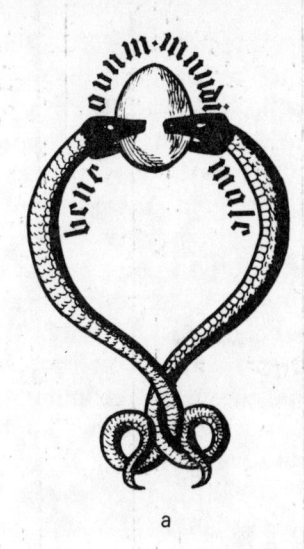

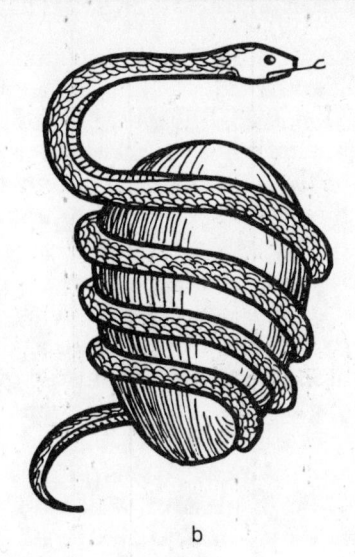

21.9 The two principles of Persian Zoroastrian dualism.
  a Fighting for possession of 'The egg of the world'.
  b A victorious serpent enveloping the egg.

elements in the deciphering of Colón-Zarco's siglum. As this was a cryptographic signature, it was possible that it contained some clue to the identification of his family – the *Kether* emanation would correspond to the father, the *Binah* to the mother and the *Malkuth* to the author of the siglum.

The practical Cabala, of course, is linked to the theoretical Cabala. The latter is based on the existence of two opposing principles – *Ormudz* (Good) and *Ahriman* (Evil). The good angels (of Light) are under the command of *Metraton*, the 'archangel with the divine face' or Bright Angel, first minister of the 'Heavenly Court'. The evil angels (of Darkness) are led by Satan, who is the archangel of Hell. They are represented by a benign serpent and a malign serpent who struggle to dominate the world (see Figures 21.9a and 21.9b).

THE RITE OF MITHRAS

It is thought that the Greeks of Asia Minor spread Zoroastrianism in Europe after the conquests of Alexander the Great, although Plutarch attributed its expansion to the hordes of Pompey, who introduced it into Italy in 68 BC following the war against *Mithridates*. Mithras was an Iranian divinity, created but then dethroned by Ormudz, thus giving rise to Mithraism, whose symbol was a six-pointed star that represented Light and Truth. A bull was sacrificed in a sacred cave and if the peoples that followed this rite did not possess cattle, they sacrificed a lamb instead, exactly as at the Christian Easter rite. In the initiation ceremony, the priest, dressed in white, told the neophyte that the Word was the main creator, with power to transform iron into gold and arms into the prosperity of the human race. The sanctuary where the initiation ceremony was carried out had a vaulted roof painted in blue. The priest blew on the neophyte's forehead and sprinkled water on his head to purify him. He then pointed a sword at his heart and said that he must always be aware of the misery of Mankind and shouted 'Mercy'! After placing a gold cup in the neophyte's hands, the priest recommended: 'despise wealth, for it is infernal' and, in a vow of poverty, the neophyte returned the cup to the priest. Following this, another priest gave the neophyte a glass of wine and a piece of bread and recommended: 'Drink, my son, and give liquid to whomsoever is thirsty and nourishment to whomsoever is hungry'. Charity was considered to be a 'magic fire' *par excellence*, able to transform the most selfish souls into founts of generosity that approached that of Mithras, the 'balsam of human tears'.[9]

The Cathars and the Templars could obviously have found certain affinities between the Christianity that they professed and the ancient cult of Mithra, particularly in the symbolism of the bread and the wine, the purifying water and the blowing on the neophyte's forehead, the six-pointed star that was similar to that of David and the symbol of the tears represented by drops of water.

In some schemes of the sephirothic system of the western Cabala, there are angels in the *Yetzirah* emanation between the *Sephira Prima* and the Crown, their task being to carry out the commandments of the Divinity. The sphere of *Malkuth* stretches, in the opposite direction to the Infinite, to the Hells, the dwelling of malign angles (*Olipoth*) that are sometimes represented by apes or satyrs, but never actually appear in the sephirothic system. Two other angels hover below the hemisphere of Darkness in the *Sephira Prima*. But they are incarnate (the *genii*) of the two sexes and are placed between spiritual angels and man. The masculine *genii* appear frequently in Arabian literature.

The dualist cabalistic philosophy was adopted by Mohammed, either through the influence of the Persian or Hebrew Rabbis, many concepts of whom he echoed. It was also a pretext for one of the accusations made against the Templars, who were suspected of observing this Zoroastrian belief in which the forces of Good and Evil are in permanent conflict for power: the black and white Templar flag.

In the alchemic or Templar Cabala, the *Briah* and *Assiah* emanations can also be depicted in the *Atziluth* decenary as an 'umbilical centre' represented by the 'Tablets of Moses', with the symbol M (*Mêm* in

Hebrew) and the Greek *omega* as the Hebrew alphabet does not contain vowels. M, which explicitly refers to the *Maschiah* (Messiah), can also expound the idea of 'eternal darkness', death, the end. It is generally said that *Mêm* and *omega* which are so common in artistic representations in churches – in Portugal also in glazed tile panels – mean 'the Beginning and the End'.

THE TEMURAH METHOD AND THE NAMES OF 'BABEL' AND 'TOMAR'

The practical Cabala is divided into the Temurah, Gematria and Notariqon. The Temurah consists in 'arbitrarily transferring the letters of a word or, according to certain rules, replacing them with the other letters of the alphabet, in order to form a word that is different from the one that is in the message'. Using the 'permutation' (*ath bash*)[10] for instance, that last letter of the Hebraic alphabet (*Tàw* = T) is replaced by the first (*'Aleph* = A); the penultimate (*Sïn* = S) by the second (*Bêth* = B) and so on, so that the word SCHESSCHAC (name of an unknown country), made up of the letters *Sïn*, *Sïn* and *Kaph* (K), is transformed into BABEL (the Hebraic name of the temple of *Bel*-Babylon and the tower which was never finished), composed by the letters *Bêth*, *Bêth* and *Lamedh* (L).[11]

The Hebraic cabalists also used the 'inversion' of the letters of the words. The Portuguese encyclopaedia, as we have seen says that the toponymy of the town of TOMAR derives from the Saxon etymon 'tomian' (p. 241), but Isaak Bitton has shown it to be a cabalistic temurah name given to the hill where the town is located and where a Jewish colony was settled before the foundation of the Portuguese Kingdom. The Zarco family took the name from the synagogue of the 'Arco' (arch) and the Templars built their castle and monastery on that hill.

THE GEMATRIC METHOD

The Gematria, the second section of the practical Cabala, is based on the numerical value of the letters of the Hebrew alphabet (see Appendix 4 p. 514); I have chosen an example that, within the parameters of the Gematria, demonstrates the power of God and the Devil, according to the Zoroastrian concepts: the word 'Naschasch' (serpent in the sense of the Devil) is composed of the letters *Nûn* = 50, *Heth* = 8 and *Sîn* = 300, which makes 358. The word 'Maschia' is made up of *Mêm* = 40, *Sîn* = 300, *Yôdh* = 10 and *Heth* = 8, which also totals 358. But this is a case of mathematical fallibility, since God has always had the power to overcome the Devil in Hebraic doctrine.[12]

The fourth part of the Hebrew *Pentateuch*[13] is entitled *The Book of Numbers*, and the Hebrews gave great importance to the intimate relationship between letters and numbers. St Augustine wrote: 'The understanding of numbers leads to a better understanding of the Holy Scriptures'. Yet showing a certain scepticism regarding the intentions of the Cabalists he added: 'the good Christian should be on his guard against all those who devote themselves to impious predictions, above all, when these people, in agreement with the Devil, do deceive their soul and do entangle their person in the web of a pact with the Devil'.[14]

Father António Vieira, a notable Portuguese Jesuit writer (1608–97), who was persecuted by the Inquisition, dedicated one of his sermons to the value of numbers.

The Jews had a special predilection for the decenary, the senary and, above all, for the tenary. The Hebrew language itself manifests a constant triplicity. The word 'soul', for instance, varies according to the 'emanation' it belongs to – in *Briah, nephesh* = soul, in the *Yetzirah, ruah* = spirit and in *Assiah, neschamah* = life.

The nautical cryptography used by the Order of Christ came from the Temurah, as did that of military and espionage services, although different methods are followed today. The Jewish cabalists preferred 'arrangements' (in the mathematical sense) of three letters (tenary) or of six (senary), as well as their respective geometrical representation: the triangle and the hexagon.

THE EXCELLENCE OF THE TERNARY

In his *Timaeus*, Plato states: 'It is impossible to combine two things perfectly without a third, as something is needed to link the two'. This is why one of the symbolic letters M or ω may be the link between two spheres in the alchemic Cabala in order to form a ternary (see p. 287).

The Egyptians had already established a divine ternary – Osiris, Isis and Horus. Zoroaster considered the number 3 as sovereign of the universe, but attributed the beginning of everything to the number 1 – the Verb or Word that the Jews called *Memra*, which is the same M (*Mêm*) that I have already ascribed to Mashiah, which means the 'Beginning' in the 'Holy Decad'.

Jean Servier[15] says:

> In primitive tradition, the number ONE (1) symbolises masculinity and number TWO (V) femininity. Put together they form the siglum (⋎), i.e. 'shin', one of the first that man engraved in stone [which like the first two are patent in the ideographic engravings of all pre-historic peoples]. It may evoke a spring or a stick to make a fire and is undoubtedly the first religious sign, from which comes

the heraldic symbol of the Fleur-de-lys. Ethnologists and palaeohistoriographers have always called it the 'mark [claw-print] of the bustard' which represents fecundation and conception, . . . because this migratory bird . . . appears in the Spring when the corn grows and departs after the harvest (see Figures 21.10a–d, 21.11a and b).

And Claude Peygnot, in a palaeographic study of the symbols of the ternary, in his book *Le Nombre – le Language de Dieu*[16], points out: 'The bird has always been the sign of the Spirit. Instead of the bustard, Christianity adopted the Dove as the symbol of the Holy Ghost because it took an olive branch to Noah and the Third Person of the Trinity is represented by this bird at the baptism of Christ'. It is also present in the Conception of the Virgin.

Don Fernando de Colón refers to the dove on the first chapter of the *Historia del Almirante*, when he denied that his father was called Colombo.

Although the sign 'Sîn' cannot be mistaken for the head of an arrow or a lance in wall engravings, as the central arm of the latter is longer, the early Christians adopted the bustard clawprint to represent the head of a lance. As Peygnot says: 'During the Passion of Christ, the 'shin' of the trinity is represented by the wound made by the soldier's lance. On penetrating the Saviour's flesh, the point left the mark of the sacred letter . . . Christ received his wound unquestioningly. The lance is a symbol of the Father and designates the son as the victim, being manifested in the Spirit'. This is, for example, 'what the altar stone in the Sault Museum [Vaucluse] reveals. The date [1554] is framed by sacrificial axes and the sign of the trinity is represented by the letters SHI, i.e. IHS [Jesus Hominus Salvator] . . . , this inversion being due to the fact that Hebrew is read from right to left'.

Peygnot continues: 'For the Jews, the sign could be to recall that Christ did not come to abolish the Law but to fulfill it and that the Saviour is identified with 'shin', the beginning of life'.

For me, it is an initiation symbol of the conversion of Jews into New Christians, perhaps through the Templars.

THE HEXAGON AND THE *DEXTRATIO* PROCESSES

The senary 'traditionally represents terrestrial harmony, recognisable in love and beauty only when conjugated with the binary or ternary'. The sephirothic decenary represents the process through cycles: 'Nothing is definitely acquired in this lower world [Earth]. Everything is transformed, evolves and revolves in cycles. The wheel turns and the decenary reflects its perpetual change'.[17]

276 COLUMBUS REVEALED

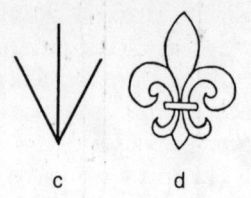

21.10 Number and symbol.

    a The numbers 1 and 2.
    b Claw-print of a bustard (see p. 275).
    c The tip of the lance that wounded Jesus on the Cross.
    d The heraldic fleur-de-lys.

21.11 The sacred lance and the sacrificial lamb.

    a In ancient times, the bustard was sacrificed at the time of the winter solstice. After the discovery of America the symbol was replaced by the Christmas turkey or goose, but without any religious significance. The symbol of the lance with its point in the shape of a fleur-de-lys appeared in sacred representations up to the end of the eighteenth century.
    b The lamb was for the ritual sacrifice at Easter.

(Both reproductions of a glazed tile panel, the Chapel of Ease (1777), Monaghan, Ireland, photographed by the author from an eighteenth-century text)

The senary may be represented by a hexagon or two crossed equilateral triangles forming the 'Shield of David'. Pythagorean philosophers considered the regular hexagon to be 'perfect beauty', because the radius of a circle in which it is inscribed is equal to any one of its sides. St Augustine also considered the number six to be 'perfect' in its esoteric sense. The rotation of the senary is also one of the processes used in the alchemic Cabala of the Temurah, in order to enable the movement of the spheres in the sephirothic system, always clockwise: '*Semper dextratio, nunquam retrocedere*' (Always turn to the right, never retrocede).

The movement of the hands of a mechanical clock[18] (of Oriental semitic origin, see Figure 21.12) has a metaphysical function, which is based on the relationship between infinite Space and Time and between Good and Evil. It is the movement of the earth's rotation on its north–south axis and its motion around the Sun. The sephirothic world of 'eternal darkness' or Evil was the '*Occasum*', which corresponded to the sunset, which is to the left of the earth, while Light comes from the East and, therefore, to the right. We thus have the words *dextera* and *sinistra*, the former meaning 'right, favourable, propitious, clever, ready', originating the adverb *dextere*, 'happily', the latter meaning 'left, pernicious, fateful, ill-boding, disorganised'.[19] Moving clockwise, one achieves the aim of '*dextratio conjugere*', which signifies 'to make peace or an alliance between people or things'[20] (see Figure 21.12).

21.12 The '*dextratio*'.
'From Light to Sunset: Always turn to the right and never retrocede' (see p. 313). This image is also known as the 'Wheel of Fortune', with six spokes.

(*Redrawn by the author from a nineteenth-century text*)

278 COLUMBUS REVEALED

In the alchemic Cabala of the Temurah, this sense of *'dextratio'* is also transmitted within the integral concept of Good itself, emanating from the *Menorah* (*Candelabrum of the Seven Arms*), here symbolising *Aur* (Light, the East) moving round to the protective shade of the Tabernacle (*Misham*), where the *mensa panis propitii 'the table of the bread of Atonement'* (*Tebah*) is to be found. Both the 'Table' and the 'Candelabrum' are represented in the sephirothic senary, each of them possessing specific cabalistic functions. Flanking the ninth Sephirah are the 'Altars of the Earth', the two sacrificial altars of the 'holocaust' (immolation by fire) and the 'perfumes' (incense).

THE MIRROR PROCESS

Another fundamental process of the alchemic Cabala is that of the *mirror*, based on the entwined double triangle of the 'Shield of David' (or the 'double triangle of Solomon'), which represents the two principles of the Cabala – the *Macroprosopos* (Macrocosm) and the *Microprosopos* (Microcosm) – i.e., the God of Light (the White God) and the God of Reflection (the Black God) (see Figure 21.13).

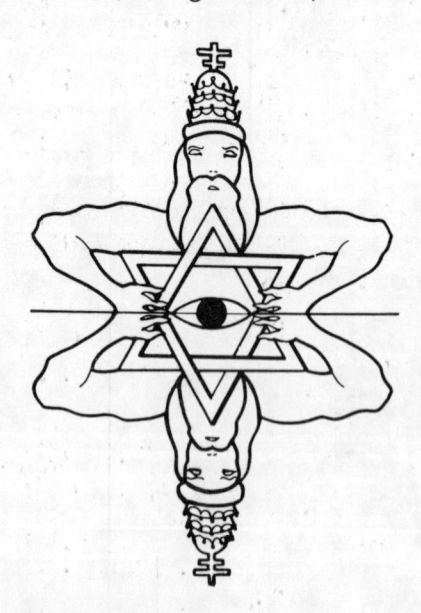

21.13 The Macrocosm and the Microcosm.
An engraving based on the texts in Eliphas Lévi's *Great Seal of Solomon*. The mirror of the waters reflects the divine work, on the principle that God can be understood only through His own creation. It also reflects the Eye inside the Shield of David (or the Seal of Solomon for Western alchemists).

(*Redrawn, with amendments, by the author from Eliphas Lévi, Great Seal of Solomon, London, 1922*)

In cabalistic hermeticism, the image of the *Ain* (the Nothing, the Divine) is reflected in the 'Sea of Eternity', which is black due to its infinite depth and which means that God, who is immaterial, is visible only in the material nature that He created. It also symbolises 'Mercy' and 'Revenge'.[21]

The alchemic cabalists represented the Mirror with an Eye that symbolised *Ain Soph* when closed and 'Intellectual Power' when open.[22] The Latin word *occulum* means 'eye, sight', *speculum* mirror and *specularis* 'see, observe, examine'. From this corelation comes the concept *'Occuli speculum animae'* (The eyes are the mirror of the soul). The temuric process of the Mirror[23] gives us *'Realitas alter imago'* (The other image of reality), with the concept of *'Per speculum veritas'* (The truth through the mirror).

THE NOTARIQON METHOD AND A HEBRAIC TEMURIC EXAMPLE

The third section of the practical Cabala, the *Notariqon*, uses the junction of the last letters of several words to form a single word. An example of this is from Genesis: *'Bra elohim lasoth'* (i.e., *'Creavit Deus ut facere'*,[24] God created to fulfil), in the sense of 'that which reigns in the world is that which is real, the Truth'. This transliterates into the word *'Emeth'*, which means *'Truth'*.[25]

Of these three forms of the practical Cabala, the Temurah, the Gematria and the Notariqon, the first seems to be the most suitable to decipher Colón-Zarco's siglum, as in the 'mirror' it depicts the senary of the Temurah with its six Ss and the umbilical words *Mêm* and *omega* (see Figure 21.14).

21.14 Using the Temurah to decipher Colón-Zarco's siglum.

In order to explain the reason for choosing the temuric Cabala more clearly, we need to consider briefly the nature of the process. It is based on the *permutation of letters or groups of letters in the same word*. I have chosen the first word of the verse of Genesis noted above as an example, but this time taken from the Gospel of St John. The word is BERESHITH:[26]

בראשית

The root of this word is formed by the three central letters:

ראש

which also form a word that means 'head', 'chief', the spiritual principle that acts'.

This new word also has a root, formed by:

אש

which, in turn, means 'fire' in the sense of 'the Spirit that gives a reason to everything'. It symbolises the first step for the materialisation of everything that exists in a universal sense.

Let us now examine the first three letters of the initial word:

ברא

They form a word that means 'He created' in the sense that He moved from 'Principle' to 'Essence'.

Finally, let us look at the last three letters of BERESHITH:

שית

The cabalists maintain that it is the contraction of

שש or שיש

which means six, from which we can draw the conclusion 'He created six', an idea frequently expressed in cabalistic texts.

Genesis tells us, in fact: 'God created the Universe in six days and rested on the seventh. He created Man on the sixth day'.

This led the cabalists to the following conclusion:

'God created the Macrocosm – the Universe';
'God created the Microcosm – Man'.[27]

It must be remembered that each letter of the Hebrew alphabet corresponds to a number and has its own meaning (see Appendix 4).

The only thing left now is to decipher Colón-Zarco's siglum.

Notes and References

1. Moses of Leon compiled the *Zohar, Book of Splendour* published in 1304; he later published *Sephira Yetzirah*.
2. Verse 1, Psalm 121 of the Hebrew Bible; God is not known as *Nothing* or *Infinite* in the Catholic Bible, but as the Lord (I raised my eyes unto the hills/from whence cometh my succour/my succour cometh from the Lord).
3. There is considered to be the third dimension following *Ain* and *Ain-Soph* in the alchemic Cabala: *Ain-Soph-Aur*, where the angels of the *Yetzirah* are to be found.
4. 'Sephira' in the singular.
5. It is difficult to explain the meaning of *Sephira*'; it is similar to both *sephar* (numeration) and *sepher* (writing). It cannot really be translated as 'sphere', even though this is its graphic representation.
6. Apart from the mythical 'Round Table' of King Arthur, the English never had military orders like those of the Temple and the Hospital, but they did set up an institution of the hermetic fraternity that was called the 'Rose Collar' (which, according to the alchemic Cabala, symbolised the *heart* and *love*). The Order of the Garter had the same symbol, but it was not a military order.
7. It was later adopted by the Order of the Rose Cross and by the original Masonry, whose mysticism was inspired in Hebraism and in a hermetic doctrine attributed to the Templars; see Fortune (1965); Frank (1967); Grad (1968) and (1974); Knight (1976); Hall (1977); and Papus (1978).
8. Touchais (1978), Vol. I, pp. 11–15.
9. Saunier (n.d.).
10. The *ath bash* is applied not only to letters but also to words within the sephirothic system.
11. The Hebrew script contains only consonants that are vowelised.
12. In the Hebrew script, the letters are used with the function of numbers: the name of a number corresponds to the letter that designates it. The part of the text that refers to the theoretical and practical Cabalas is a synthesis of

Cabala ou Caballa in *Dictionnaire de la Conversation et de la Lecture* (1833).
13. As already mentioned, it was the first book printed in Portugal; see Anselmo, Artur (1981). The 'Pentateuch' is depicted, although with an imaginary script, in the polyptych of Nuno Gonçalves in the National Art Museum in Lisbon. The open book is in the hands of a Jew on whose breast a six-pointed star was later painted in order to conceal a Margarita (Marigold), an emblem of the decenary that was used by high-class Jews instead of the 'Shield of David'. (See Fig. 35.2, p. 477).
14. St Augustine, *Patrologia*, vol. XXXIV, 279.
15. Servier (1961), p. 74.
16. Peygnot (1987).
17. Foster (1983); Warustel (1946); Gobert (1962).
18. The most famous mechanical clock with a round face (*Menzanah*) was one of the 'three marvels' of the Palace of Mechuar of the ancient Kings of Tlemcen.
19. All the Latin expressions in the book are taken from the *Magnum Lexicum* (see Cabral and Ramalho (1867)).
20. See Reuchlin (1615).
21. See Hogan (1922), cited by Lévi, *Great Seal of Solomon*, p. lxxx.
22. Eliphas Lévi , pp. cxvii, lxxiv.
23. Used by Ribeiro, Patrocínio (1927) in his deciphering of Colón-Zarco's siglum and by Pestana Junior (1928) (see p. 242).
24. In Latin characters, this sentence does not, of course, correspond phonetically to the terminal letters transliterated, as they are in Hebrew.
25. Verse 3, Chap. II of Genesis, Vulgate Bible – 'God created to fulfil'.
26. Hebrew is written from right to left. *Bereshith*: 'At first it was a verb'; Verse I, chap. I in *The Holy Gospel of Jesus Christ according to St John* of *The Holy Bible containing the Old and New Testaments*, according to the Latin Vulgate (António Pereira de Figueiredo, Lisbon, 1819, an edition sponsored by Queen Mary II of Portugal in 1842).
27. See Touchais (1978), from which this explanation of *Bereshith* is taken.

22 On the Other Side of the Mirror

Now that we know something about the elements of the Cabala and of the sephirothic rules of the Temurah, let us begin a methodical decipherment of the siglum conceived by Colón-Zarco to express his crypto-identity – a signature that is much more than a mere name and explains the personality of the man himself.

In view of the tolerance of the temuric rules, which allow one a wide choice of permutations, movement of symbols, fusion of 'worlds of emanation' and rotation of the numeric mechanism, it is relatively easy to compose a cabalistic siglum, just as it is for the secret services to arrange a code today. On the other hand, its decodification is a long, exhaustive and complex process.

The deductive method demands laborious, systematic research and an indefatigable persistence. I was, however, also able to use the inductive method, by following up the results of investigations carried out by earlier Portuguese researchers – and, above all, by being convinced that the thesis that I have produced is the truth.

As it is unlikely that Colón-Zarco was the natural son of Prince Henry the Navigator (1394–1460), I have defended the hypothesis that he was the bastard son of Prince Fernando (1433–70), Duke of Beja, Master of the Order of Christ and nephew of Prince Henry. The first step was to find the masculine names of Salvador and Fernando in the siglum and a feminine name that could give a clue to the identity of the navigator's mother, possibly a daughter of João Gonçalves Zarco. The most common names of the ladies of the Portuguese royal family and of the navigators in the service of the Order of Christ (since these two lines intermarried) were Beatriz (or Brites), Branca, Catarina, Constança, Elena (or Helena), Filipa, Inês, Isabel, Joana, Leonor, Maria and Mécia.

As far as as Colón-Zarco's birthplace was concerned, it would have to be found in the names of nations, cities, towns and villages – Portuguese if possible, but never ruling out the possibility of foreign place names if it became necessary.

The siglum was made up of a ternary of Ss and the letters A, X, M and Y. By using the mirror method already described in pp. 242–78, the ternary became a senary and the letters O and L were acquired.

This seemed to rule out all the three-letter abbreviation formed by B, D, E, F, G, H, K, N, P, Q, R and T, which would eliminate Italy, France, England, Germany, Aragon, Castile, Catalonia, Genoa, Barcelona, Vigo and even Portugal and most of its provinces and cities – Alentejo, Algarve, the Beiras, Trás-os-Montes, Minho, Beja, Évora, Guarda, Oporto, Viana, Viseu and many others.

At this stage, Lisbon seemed to be high on the list of possibilities. In a siglum, however, the name of a place could be in Latin, which would increase the chances of finding an adequate solution. It would also be convenient to look for some indication of a date of birth – the day, month or year. In accordance with the senary, both Genoa (*Genua* or *Janua* in Latin) and Doménico and Susana Fontanarossa were immediately excluded, as the city has a G in its name the 'Genoese father' a D and the 'Genoese mother' an F.

There are two known forms of Colón-Zarco's siglum. One which appears on all the suspect documents and does not contain the colon that preceded Xpo, and the genuine one that definitely does bear the colon (see Figure 22.1).

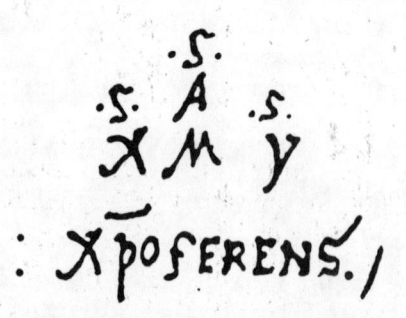

22.1 Colón-Zarco's siglum, showing the 'colon'.

It was clear from the beginning that Xpo FERENS had no place in the sephirothic senary and would have to be studied separately. Let us, therefore, look at the seven letters. We shall start with the Ss (Figure 22.2a and b).

S (*Sîn*) is the symbol of *Naschassch* (serpent) and God, indicating Him as the *Saviour of the World* (*Salvador* in Portuguese). The S of *Samael* (Devil) had been dethroned by Christianity. In the Sacred Catholic Mass '*Sanctus*' is evoked three times before the Prayer to the Eucharist and this is also represented by three Ss. In the cult of the Paraclete (Pentecost), moreover, both the Cathars and the Templars and their successors of the

```
         ·S·
    ·S·  A  ·S·              ·S·
      x m y            ·S·        ·S·
         a                   b
```

22.2 The Ss in the siglum.
  a The six letters.
  b The ternary.

Order of Christ invoked the *'Sanctus Spiritus Salvator'*, which they inscribed on fillets (the horizontal divisions on an heraldic shield) by using three Ss.

The *Serpe*, or the serpent that was so common in Portuguese heraldry, was also used to represent 'science'. It was adopted as the emblem of medicine and pharmacology, used by the Templars in the illumination of manuscripts and in architecture and was especially dear to alchemists.

In writing, whether engraved on seals or printed, it is used to separate words, especially when it is necessary to occupy a space in an inscription.[1]

THE MEANING OF THE 'S'

The Plantagenet Prince John of Gaunt,[2] grandfather of Prince Henry the Navigator, introduced into Portugal the 'Livery Collar', which was made up of 18 Ss in a sequence of the senary with the initials of the Templar motto *'Sanctus Spiritus Salvator'*. The first Portuguese to receive the 'Order of the Livery Collar' was Dom Afonso Furtado de Mendoça,[3] who had commanded an English fleet that sailed from Plymouth to Galicia, where the Duke of Lancaster's troops disembarked on their way to fight in Catalonia. When the Duke came to visit King John I, the sovereign sent Mendoça to receive him officially when his ship sailed into the River Tagus. His coat-of-arms bore a crest of a golden wing inscribed with an S and an SS in each quarter.[4]

I have already analysed some of the various interpretations of the siglum. All the investigators have tried to find the words that suited them, but in vain, because Colón-Zarco never made any attempt to transmit a simple and descriptive message through his signature. Only Santos Ferreira made a positive contribution of undoubted merit – the sign (./) that stands for 'colon' in Greek and corresponds to *'zarga'* in Hebrew, thus identifying 'Zarco'. But Ferreira did not know the symbolism of S for *Salvador* and fabricated a *Gonçalves* using an imaginary junction of S to which he attributed the meaning of *'Salve – Salve – Salve'*. He did, however, interpret *Xpo* as *'Salvador'*, as had already been pointed out in Manuel de Carvalho's *Theatro Genealógico*. He also knew of the existence of the 'Salvadores' as the names of the descendants of João Gonçalves Zarco, which was repeated down the generations in the family of Corrêa de Sá, Viscounts of Asseca.

Patrocínio Ribeiro, in turn, had the brilliant idea of using the 'Mirror' method, but did so only experimentally and did not understand the clear indication expressed in the siglum that this is the method that should be adopted.

Let us now make a temuric analysis of the siglum. The Ss in the ternary may represent the *sigilum* of the motto of the knights of Languedoc, the Temple and the Order of Christ, as well as *'Spiritus Sanctus Salvator'*, or even 'Saviour', seen in palaeo-Christian inscriptions and fifteenth century Portuguese seals and medallions (see Figures 22.3 and 22.4).

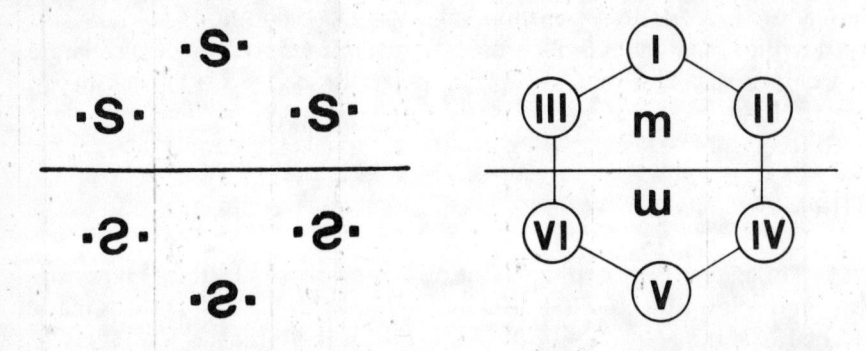

22.3 Temuric analysis of the siglum: the ternary and the mirror.

22.4 Temuric analysis of the siglum: the ternary, the umbilical centre and the mirror.

THE UMBILICAL CENTRE AND THE TRIANGLES

The umbilical quality of *Mêm* and *omega* indicates that their function is to form words or abbreviations with another two letters of the senary in a sephirothic triangle. As *Xpo* is a Greek abbreviation, this would suggest that there is a place for new Greek characters resulting from the reflection of the 'mirror' (see p. 274):

X (*Chi*)=C; (*omega*)=O; (*lambda*)=L; (*alpha*)=A (see Figure 22.5a and b).

$$\frac{\begin{array}{c}A\\x\ m\ y\end{array}}{\begin{array}{c}x\ ω\ λ\\∀\end{array}}$$

a

$$\frac{\begin{array}{c}A\\x\ m\ y\end{array}}{\begin{array}{c}C\ O\ L\\A\end{array}}$$

b

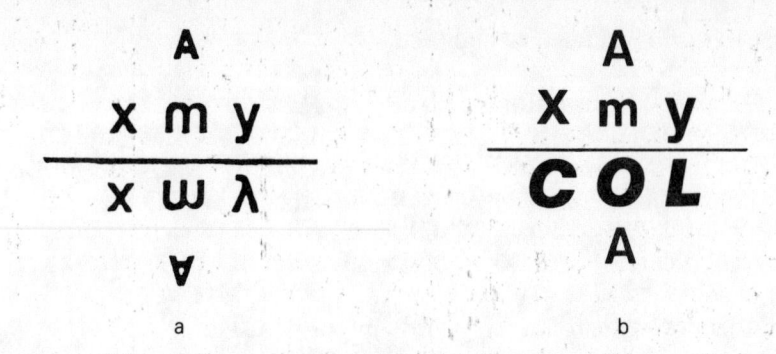

22.5 Temuric analysis of the siglum: using *sephirothic* triangles and Greek characters.
  a Sephirothic triangles.
  b Greek characters.

Three consonants (L, M and X) and three vowels (A, I and O) were obtained in this way, which continued to fit into the Cabala of the ternary and had the aim of forming six words or abbreviations, as had been suggested by the six Ss flanked by points that were the result of the reflection of the 'mirror'. Each ternary should be formed by two peripheral and one umbilical letter. As a vowel, O would obviously be used more often than M. Therefore, six triangles were isolated, according to the alchemic Cabala, producing the following order (see Figure 22.6):

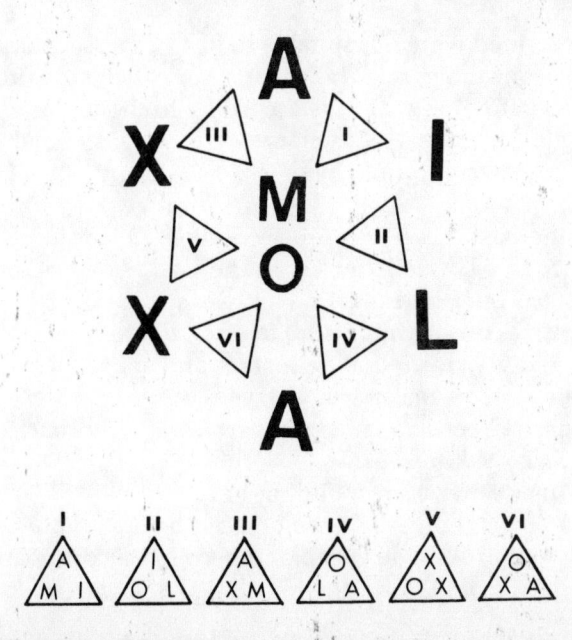

```
         A
    X ◁III  I▷ I
       M
    ◁V    II▷
       O
    X ◁VI  IV▷ L
         A
```

```
  I      II     III    IV      V      VI
 △       △      △      ▽       ▽      ▽
 A       I      A      O      X       O
 M I     O L    X M    L A    O X     X A
```

22.6 Order of *sephirothic* triangles.

The following order was obtained:

I	AMI	AIM	IAM	IMA	MAI	MIA
II	IOL	ILO	LOI	LIO	OIL[5]	OLI
III	AXM	AMX	XAM	XMA	MXA	MAX
IV	OLA	OAL	AOL	ALO	LAO	LOA
V	XOX	XXO	OXX	–	–	–
VI	OXA	OAX	XOA	XAO	AOX	AXO

From this order the following words might be used:

AMI   *Amicus* = friend; *Amita* = aunt; *Amitini* = cousins; *Amittere* = to lose, to forgive.

IMA   *Imago* = image, face; *Im(m)aculata*[6] = immaculate; *Im(m)anis* = cruel.

MAI   *Maius* = May.

MIA   (accepting the barbarism for *MEA*) = my, mine.

ILO   *Il(l)o* = over there; *Il(l)ocabilis* = indisposable, unmarriageable.

ALO   *Al(l)ocution* = language, speech.

AXO   *Axon* = central line (of the sun dial).

OXA   *Oxalis* = sorrel dock, acid plants.

MAX   *Maxila* = chin; *Maximus* = maximum.

The words obtained were not suitable to form a message unless it had a very subjective meaning, for which many more than six words would be necessary. Therefore, and always according to the temuric rules, other values for the letters of the senary were sought. So the following correspondences were found:

O   In the past the O was replaced by U. O was also written U in place of AU (see letter U); in the past *Hecoba, nostrix* was said instead of *Hecuba, nuxtri*.

U   Is also a consonant, V (*uaco* for *vaco*, *uox* for *vox*, etc.); according to Plautus, Lucretius Sallustius and others, *voltis* instead of *vultis*; *volgus* instead of *vulgus*, etc. . . . Those who tried to abolish Greek letters in Latin wrote *Sulla, Sullanus*, etc., instead of *Sylla, Syllanus*, etc.

Y   a Greek letter corresponding to I.

I   Is a vowel; however, when placed before another vowel making with it a single syllable, it becomes a consonant. In voices derived from Greek, the I is always a vowel (*iambus*). In those derived from Hebrew it is a consonant (*Iudex, Iupiter* instead of *Judex, Jupiter*, etc.), with the value of a J.

X Is a double semi-vowel. It correspond to CS and in the past *Pacs, Grecs* were written *Pax* and *Grex*.

C (also with an S sound) *Caesar* with the value of *Cesar*.

These different values of the letters, partially used by Patrocínio Ribeiro, permitted other possibilities in the formation of words or in three-letter abbreviations:

(O = V = AU = U) (V = U = Y = I) (I = J) (X = C = S)

If the U has the value of an I, AU might be regarded as AI.

Returning to the cabalistic siglum, I tried to find new arrangements that could form words that had a valid meaning in a message from Colón-Zarco. In order to avoid being accused of spurning any plausible probability, I appraised each word at its face value, even when I felt it was incongruous. But all the words that had no meaning in a message of Colón-Zarco were excluded – such as *Maialis*, the castrated pig that was sacrificed to the Goddess Maia – and I did not repeat any etymon (if the verb *Immaculare* had already been chosen, for example, words from the same root such as *Immaculata, Immaculabilis*, etc. were excluded, although the suitablity of adjusting them to the abbreviation IM (M) (A) was considered). This gave me Triangle I:

AMI (O = AU = V = U) (U = Y = I = J)

AMI	AIM	IAM	IMA	MAI	MIA
AMJ	AJM	JAM	JMA	MAJ	MJA
AMO	AOM	OAM	OMA	MAO	MOA
AMU	AUM	UAM	UMA	MAU	MUA
AMV	AVM	VAM	VMA	MAV	MVA
AMAU	AAUM	AUAM	AUMA	MAAU	MAUA
AMAI	AAIM	AIAM	AIMA	MAAI	MAIA

Only the following words or abbreviations have an acceptable Latin equivalent:

1	2	3	4	5	6	7	8	9	10	11
AMI	IMA	MAI	MIA	JAM	MAJ	AMO	OMA	AMU	MAU	MAIA

(see Appendix 6, Triangle I).

290　COLUMBUS REVEALED

Triangle II is as follows:

IOL (O = AU = V = U) (U = Y = I = J)

IOL	ILO	LOI	LIO	OIL	OLI
JOL	JLO	LOJ	LJO	OJL	OLJ
IUL	ILU	*LUI*	LIU	UIL	*ULI*
JUL	JLU	LUJ	LJU	UJL	ULJ
IAUL	*ILAU*	LAUI	LIAU	AUIL	*AULI*
JAUL	JLAU	LJAU	LJUA	AUJL	AULJ
IVL	ILV	LVI	*LIV*	*VIL*	VLI
OVL	OLV	LVO	LOV	*VOL*	VLO
OUL	OLU	LUO	LOU	UOL	ULO
VUL	ULV	LUV	LVU	VLU	UVL
OOL	*OLO*	LOO	(repetition)		
IIL	ILI	LII	(repetition)		
UUL	ULU	LUU	(repetition)		
IAIL	ILAI	LAII	LIAI	AIIL	AILI
JAIL	JLAI	LAIJ	IJAI	AIJL	AILJ

Only the following have an equivalent:

1	2	3	4	5	6	7	8	9	10	11	12
OLI	*LUI*	*UL(L)I*	*JUL*	*IL(L)AU*	*AULI*	*LIV*	*VIL*	*VOL*	*VUL*	*OLO*	*IL(L)I*

(see Appendix 6, Triangle II).

Triangle III: (X = C = S = G)

AXM	AMX	XAM	XMA	MXA	*MAX*
ACM	AMC	*CAM*	CMA	MCA	MAC
ASM	AMS	*SAM*	*SMA*	MSA	*MAS*
AGM	AMG	*GAM*	GMA	MGA	*MAG*

They correspond to:

1	2	3	4	5	6	7	8
MAX	*CAM*	*SAM*	*SMA*	*MAS*	*AGM*	*GAM*	*MAG*

(see Appendix 6, Triangle III).

Triangle IV: (O = AU = V = U) (U = Y = I = J)

OLA	OAL	AOL	ALO	LAO	LOA
ULA	UAL	AUL	AL(L)U	LAU	LUA

THE OTHER SIDE OF THE MIRROR 291

VLA	VAL	AVL	ALV	LAV	LVA
ILA	*IAL*	*AIL*	*ALI*	LAI	LIA
JLA	JAL	AJL	ALJ	LAJ	LJA
AULA	AUAL	AAUL	ALAU	LAAU	LAUA
AILA	AIAL	AAIL	ALAI	LAAI	LAIA

They correspond to:

| 1 | 2 | 3 | 4 | 5 | 6 | 7 | 8 |
| ALO | AL(L)U | LAU | ALV | LAV | IL(L)A | ALI | AULA |

(see Appendix 6, Triangle IV).

Triangle V: (X = C = S = G) (O = AU = V = U) (U = Y = I = J)[6]
In this series – XOX – the repetition of the letter X can be seen. The intervening letter used to form a word or abbreviation, therefore, cannot be a consonant, so that arrangements with the semi-vowels V and J become superfluous.

XOX	XXO	OXX	COX	CXO	OCX	OXC	XOC	XCO
XUX	XXU	UXX	CUX	CXU	UCX	UXC	XUC	XCU
XIX	XXI	IXX	CIX	CXI	ICX	IXC	XIC	XCI
COC	CCO	OCC	COS	CSO	OCS	*OSC*	SOC	SCO
CUC	CCU	UCC	CUS	CSU	UCS	USC	*SUC*	SCU
CIC	CCI	ICC	*CIS*	CSI	ICS	ISC	*SIC*	SCI
SOS	SSO	*OSS*	SOX	SXO	OSX	OXS	XOS	XSO
SUS	SSU	USS	SUX	SXU	USX	UXS	XUS	XSU
SIS	SSI	ISS	SIX	SXI	ISX	IXS	XIS	SXI
GOG	GGO	OGG	*COG*	CGO	OCG	OGC	GOC	GCO
GUG	GGU	UGG	CUG	CGU	UCG	UGC	GUC	GCU
GIG	GGI	IGG	CIG	CGI	ICG	IGC	GIC	GCI
XAUX	XXAU	AUXX	CAUX	CXAU	AUCX	AUXC	XAUC	XCAU
CAUC	CCAU	AUCC	SAUX	SXAU	AUSX	AUXS	XAUS	XSAU
SAUS	SSAU	AUSS	SAUC	SCAU	*AUSC*	AUCS	*CAUS*	CSAU
XAIX	XXAI	AIXX	CAIX	CXAI	AICX	AIXC	XAIC	XCAI
CAIC	CCAI	AICC	SAIX	SXAI	AISX	AIXS	XAIS	XSAI
SAIS	SSAI	AISS	SAIC	SCAI	AISC	AICS	CAIS	CSAI
GAIG	GGAI	AIGG	SAIG	SGAI	AISG	AIGS	GAIS	GSAI

They correspond to:

| 1 | 2 | 3 | 4 | 5 | 6 | 7 | 8 | 9 | 10 | 11 |
| COC | OCC | CIC | SOS | OSS | SUS | SIS | COX | COS | OSC | SOC |

| 12 | 13 | 14 | 15 | 16 | 17 | 18 | 19 | 20 | 21 | 22 |
| SCO | CUS | SUC | SCU | CIS | SIC | SCI | COG | GIG | AUSC | CAUS |

(see Appendix 6, Triangle V).
Triangle VI: (X = C = S = G) (O = AU = V = U) (U = Y = I = J)

OXA	OAX	XOA	XAO	AOX	AXO
OC(C)A	OAC	COA	CAO	AOC	ACO
OSA	OAS	SOA	SAO	AOS	ASO
UXA	UAX	XUA	XAU	AUX	AXU
UCA	UAC	CUA	CAU	AUC	ACU
USA	UAS	SUA	SAU	AUS	ASU
VXA	VAX	XVA	XAV	AVX	AXV
VCA	VAC	CVA	CAV	AVC	ACV
VSA	VAS	SVA	SAV	AVS	ASV
IXA	IAX	XIA	XAI	AIX	AXI
ICA	IAC	CIA	CAI	AIC	ACI
ISA	IAS	SIA	SAI	AIS	ASI
JXA	JAX	XJA	XAJ	AJX	AXJ
JCA	JAC	CJA	CAJ	AJC	ACJ
JSA	JAS	SJA	SAJ	AJS	ASJ
OGA	OAG	GOA	GAO	AOG	AGO
UGA	UAG	GUA	GAU	AUG	AGU
VGA	VAG	GVA	GAV	AVG	AGV
IGA	IAG	GIA	GAI	AIG	AGI
JGA	JAG	GJA	GAJ	AJG	AGJ
AUXA	AUAX	XAUA	XAAU	AAUX	AXAU
AUCA	AUAC	CAUA	CAAU	AAUC	ACAU
AUSA	AUAS	SAUA	SAAU	AAUS	ASAU
AUGA	AUAG	GAUA	GAAU	AAUG	AGAU
AIXA	AIAX	XAIA	XAAI	AAIX	AXAI
AICA	AIAC	CAIA	CAAI	AAIC	ACAI
AISA	AIAS	SAIA	SAAI	AAIS	ASAI
AIGA	AIAG	GAIA	GAAI	AAIG	AGAI

They correspond to:

     1   2   3   4   5   6   7   8   9   10
OC(C)A COA ACO CAU ACU SUA AUS VAC CAV ACI
    11   12   13   14   15   16   17   18   19
   ISA ASI JAC GAU AUG VAG AGI CAIA GAIA

(see Appendix 6, Triangle VI).

I must confess that I am a novice in the field of the Cabala, so I think it is essential that I bring all the arrangements and combinations with the letters used to the notice of the reader. I also hope that experienced cultists of the Cabala will make a closer scrutiny of my research.

I made up 144 phrases with the six Latin words used, but none of them were found to be suitable for any message that would identify Colón-Zarco as 'Cristóbal Colón' or 'Cristóforo Colombo'. Then – always respecting the sephirothic order – I gathered the selected words in a three-columned chart, as it seemed to make the combinations more coherent (see Table 22.1).

From Table 22.1 (which includes 43 verbs, 30 nouns, 13 adjectives and 15 link words) a new selection was made eliminating synonyms and experimenting with the verbs in the past tense (see Table 22.2).

Only the following elements seemed to be essential:

1. *Feminine names*:

   JUL – Julia; *LIV* – Branca; *SUS* – Susan; *AUG* – Augusta; *ISA* – Isabel

2. *Masculine names*:

   JUL – Júlio; *CAM* – Câmara; *SAM* – Samuel; *AUG* – Augusto

3. *Relatives and personal relationships*:

   AMI – aunt; *IL(L)A* – linked to; *COG* – relatives; *COA* – grow up together

4. *Birthplace*:

   MAI – Maia; *OLI* – Lisbon; *MAG* – Minorca; *AUS* – Italian *CAI* – Caia; *GAI* – Gaia; *AC(C)I* – Azores

5. *Months*:

   MAI – May; *JUL* – July; *AUG* – August

6. *Link expressions*:
   MAI – my/mine; *IL(L)A* – on the part of/by means of; *SOS* – his, her/hers, your/yours, their/theirs; *S(C)IA* – known as/said; *CAU* (cause) – origin (linked with 'my/mine' in the sense of 'generation').

The large number of verbs were conjugated in the present and past participles, making the combinations a long slow task. For a long period

Table 22.1 Deciphering Colón-Zarco's siglum: the three-columned chart

	Action		Qualification		Connection	
I	*Amiciare*	= to cover	*Amita*	= aunt	*Mia (Mea)*	= my/mine
	Imaginare	= to imagine	*Maia*	= Maia	*Jam*	= now
	Amitere	= to lose	*Maius*	= May	*Jampidem*	= a long time ago
	Amovare	= To move, to withdraw	*Majestas*	= nobility		
II	*Luire*	= to redeem	*Olisipo*	= Lisbon	*Oli (Illis)*	= that place
	Vilipendere	= to despise	*Julius*	= July/Julia	*Illic*	= in that place
	Volere	= to want	*Aulicus*	= courtier		
	Volvere	= to stir	*Livida*	= livid, white		
	Vulnerare	= to wound	*Volubilis*	= inconstant		
	Illicere	= to seduce	*Voluntas*	= will, wish		
	Illigare	= to connect to	*Vulsus*	= ruined		
III	*Cambire*	= to change, to leave	*Camara*	= chamber, Madeira Island		
			Campus	= field		
			Samuel	= Samuel		
			Mascarpio	= dishonest		
			Magnarius	= dealer		
			Mago	= harbour in Minorca		
IV	*Alluctare*	= to fight	*Alogia*	= madness	*Illac*	= by someone's order, in another way
	Laudare	= to praise	*Illaqueatus*	= connected to, embarrassed with, entangle	*Alias*	= another time, other place
	Illabor	= to fall, to insinuate oneself, to slip	*Illaudibis*	= reprehensible	*Alius*	= some one else
	Illabore	= to work			*Alisusmodo*	= in another way

Alienare = to separate *Aliciare* = to entice *Alligare* = to charm		
V	*Occedere* = to arrive *Occidere* = to kill, to wound *Occipiere* = to begin *Occubare* = to hide *Occurrere* = to come, to help *Sistere* = to restrain *Suscitare* = to resurrect *Sucingere* = to involve *Succurrere* = to help *Cognoscere* = to know	*Sosper* = safe *Sussurrans* = denouncer *Socius* = companion, partner *Scopus* = finality *Custos* = spy, perceptor *Scordia* = madness, cowardice *Susana* = Susan *Cognatio* = relatives	*Sos (Suos)* = his, her, its, their *Sus* = up and down *Sic* = in such a way
VI	*Occalescere* = to be patient *Coaccedere* = to join *Coactare* = to force *Coalere* = to grow together *Coagimentare* = to unite *Causificare* = to forgive *Suadere* = to persuade *Auscultare* = to listen, to obey *Cavere* = to avoid, to prevent *Accidere* = to cut, to ruin *Accingere* = to undertake *Accipiere* = to accept	*Occasus* = West, sunset *Coevus* = contemporary *Acolantus* = prodigal *Causa* = cause, origin *Ausonius* = Italian *Ausem* = difficult task *Accipiteris* = Azores (islands) *Isabella* = Elisabeth *Asia* = Asia *Augustus* = August (month) *Vagabundus* = wandering *Caia* = Caia	*Aconiti* = without effort *Vacanter* = in vain

Table 22.2 Deciphering Colón-Zarco's siglum: eliminating synonyms and experimenting with verbs

I	II	III	IV	V	VI
Amitiare covered	*Luire* redeemed	*Cambire* swopped/left	*Alutare* fought	*Occedere* arrived	*Coacedere* joined to
Imaginare imagined	*Viliapendere* despised	*Camara* chamber/(Madeira Island)	*Illabor* fell/insinuated myself	*Occidere* killed/wounded	*Coactare* forced
Amovere put away	*Volere* wanted	*Campus* field/camp	*Illaborare* worked	*Occipiere* started	*Coalere* grew up together with
Amita aunt	*Volvere* stirred	*Samuel* Samuel	*Illatebrare* hid	*Occubare* hid	*Coagimentare* joined
Maia Maia (village)	*Vulnerare* wounded	*Mascarpio* dishonest	*Aliciare* enticed	*Occurrere* came/helped	*Causificare* forgave
Maius May	*Illicere* seduced	*Magnarius* merchant	*Alienare* separated	*Suscitere* resurrected	*Suadere* persuaded
Majestas nobility	*Illigare* connected to	*Mago* Minorca (Balearic Island)	*Alligare* charmed	*Cognoscere* knew	*Auscutare* heard/obeyed
Mia (Mea) my/mine	*Olisipo* Lisbon		*Alogia* madness	*Sosper* safe	*Cavere* avoided
Jam now	*Julius* July/Julius		*Lavernio*[(1)] thief	*Sussurrans* denouncer	*Accidere* cut/ruined
	Alius courtier		*Illaqueatus* connected to	*Socius* partner	*Accigere* undertook

Japidem a long time ago	*Livida* extremely white	*Illaudibilis* reproachful	*Scopus* purpose	*Accipiere* accepted
	Volubilis inconstant	*Illac* by someone's order/in another way	*Custos* spy	*Ocasus* sunset/west/darkness
	Voluntas will/volition		*Scordia* cowardice	*Causa* origin
	Vulsus ruined	*Alias* other time/other place	*Susana* Susan	*Ausonius* Italian
	Oli (Illis) that place	*Alius* someone else	*Cognatio* relatives	*Ausem* difficult task
	Illic in that place	*Aliosmodo* in another way	*Sos (Suos)* his/her/its/their	*Accipiteris* Azores
			Sic in such a way	*Isabella* Elisabeth
			Scia known as/said	*Augustus* August/Augustus
				Vagabundus wanderer
				Aconiti easily
				Vacantes in vain

1. A choice inspired in the Gematric version in Hebrew, by Santos Ferreira; see p. 244 'defrauded his Prince; ... he that has robbed will disappear'.

and demonstrating a distinct lack of experience, I tried to form phrases of six words that would produce a coherent message. I drew a complete blank. On conceiving a siglum as his personal signature, Colón-Zarco would certainly not have included unnecessary information: besides his name, he would merely indicate his parent's names and his place and date of birth.

We now know that Columbus (see p. 195) was born in 1448. However, we should also analyse other hypotheses.

Columbus died in 1506 at an age which could be put at between 55 and 75 (i.e., he was born between 1431 and 1451). We can use only three letters. Suppressing the initial unnecessary Roman numeral *MCD* (1400), we are left with (*MCD*) XLI = 1441 and (*MCD*) XLV = 1445. Even extending the limits of his age to 54 and 76, we would get (*MCD*) XXX = 1430 and (*MCD*) LII = 1452. To indicate his date of birth, it would be possible only by coincidence to find three letters that would correspond and, what is more, none of the combinations mentioned above appear in any sephirothic arrangement.

IRRELEVANT MESSAGES

Colón-Zarco, it seems, did not bother to mention the year of his birth, in which case the month would also be of no importance. The incognito continued, but I could now eliminate the months of May, July and August. As far as names are concerned, 'Augusto' does not appear among the fifteenth-century navigators and nobility, who almost always baptised their children with names that already existed in the family. On the other hand, 'Samuel' was too Hebraic to be adopted by New Christians of this period, and even more so for the Old Christians. The feminine name 'Augusta' can be eliminated for the same reason as its masculine counterpart, just as 'Isaura' had been, even though it had the same abbreviation of *ISA*. It also seemed necessary to exclude Branca from *Livida*.

As I could not count on the expressions 'I was born' or 'I come from', I temporarily rejected Mago (Minorca), Azores, Lisbon, Maia, Gaia and Caia. I was compelled to do this because of the limit of six words. And I had already verified that the use of any of the selected verbs would demand a long composition of phrases, from which I could only hope for a personal identification. For this, the present indicative tense of *esse*: *sum, est or sunt* would suffice. But none of these words would fit, as the siglum does not have an E or an N, besides the fact that the rules of the temurah do not allow the formation of a ternary of letters that includes two umbilicals (M and U). I also had no T in the senary.

Table 22.3 Deciphering Colón-Zarco's siglum: initial combinations

I	II	III	IV	V	VI
MIA (MEA) my/mine	JULIUS/A Júlio/a	CAMARA Câmara	IL (L/AC) on behalf of	COGNATIO relatives	AUSONIUS/A Italian
AMITA aunt			IL (L) AQUEATUS/A married to/ being the lover/ the mistress of/ having an affair with	SOS (SUOS) his, etc	CAUSA origin
				S (C) la known as; said	COALEO grew up together
				SUSANA Susan	ISABELLA Isabel

There are cases in Latin inscriptions on statues, tombs and fillets in which the predicate of the verb *esse* is omitted. This led me to shed the predicate in order to experiment with new arrangements. The verbs selected can remain in abeyance for future use. Meanwhile, I tried out the following arrangements with the selected elements (see Table 22.3).

I will save the reader from a complete explanation of the various arrangements obtained and give just one free version:

JULIUS AUSONIUS MEAE SUSANAE ILLAQUEATAE CAMARA (sum).

(I am) an Italian Júlio of my Susan[7], having an affair with a Câmara.

Notes and References

1. See Anselmo (1981) (The Figures were erased from the book).
2. Duke of Lancaster, 4th son of King Henry III of England and father of Philipa of Lancaster, sister of Henry IV of England, who married John I of Portugal.
3. See the reference on p. 463; the first Furtado de Mendoça (Hurtado de Mendoza), descendant of the Lords of Biscay, came to Portugal in the retinue of Queen Beatriz, wife of King Alphonse III, and married Guiomar, of the noble family of Resende. It was the surname of Perestrelo's first wife.
4. See *Armorial Lusitano*, pp. 358–60.
5. In triangle II *IOL* or *MIL* could be used, but the ternary *IOL* was preferred because *MIL* supplied only abbreviations that were either meaningless or

unadaptable to an acronym of personal identification, except the word *Militare* (to serve in the army or to be a soldier); the others are: *Milium* (a thousand); *Milus* (collar); *Milvus* (hawk), and other similar or derived words. In another order: *Lima* (file); *Limarius* (mud); *Limitatio* (demarcation); *Limpidus* (limpid) and *Limus* (wet mud).
6. In triangle V, containing 3 consonants, it is impossible to form any Latin abbreviation.
7. Susana Fontanarossa, wife of Domenico Colombo and Cristoforo Colombo's mother.

23 *From Light to Darkness*

The umbilical *Mêm* and *omega* designated 'creation and formation'. The location of *Xpo Ferens* had not yet been studied within the sephirothic system, but everything indicated that *Xpo* (the abbreviation for Christ) should be placed below the *Sephira Prima*, since it is the link between God and man.

I had not, meanwhile, attributed words to *sephiroth* VII, VIII, IX and X outside the senary. The X, in principle, should include the signature 'Zarco' (./). In accordance with the western Cabala of the temurah, both the senary and its umbilical nucleus should express 'symbolic' and 'intellectual' concepts that embraced the Crown, Wisdom, Mercy, Courage and Beauty (see p. 308).

The words that I proposed were not suitable for the concepts of the sephirothic order, but there was nothing to oblige Colón-Zarco to stick to the classical rules, as he could make use of the arbitrary permutation of the *ath bash* method. If he ignored those rules, it would make it more difficult for us to decipher the Cabala; on the other hand, if he stuck to the norms of the mystic meaning, we would have to look for suitable words that could be adapted to a sephirothic order.

I was out to unravel a personal identification. I hoped to find some indication of a man's origins and ancestry. I therefore gave up following the classical Hebraic sephirothic order of the senary and decided to experiment with the western numerical disposition, which is clockwise and not zig-zag, especially as it was this method that Valentim Fernandes had referred to in his letter to Konrad Peutinger (see Figure 23.1).

This clockwise movement was, as we know, called the *'Dextratio'*. After many attempts, I decided to experiment with a message made up of the following abbreviations:

    MIA   JUL   CAM   IL(L)A   SCI   ISA

The two Shields of David in Figure 23.2 show the alteration in the order of the sephirothic triangles.

302   COLUMBUS REVEALED

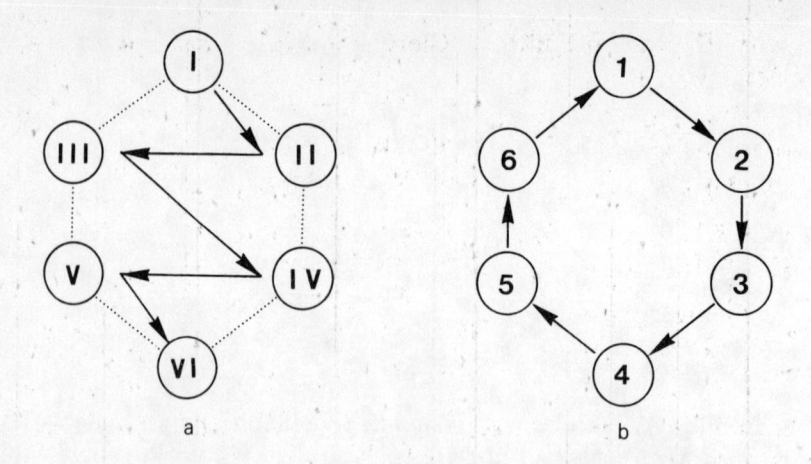

23.1 Deciphering Colón-Zarco's siglum: using the Western Cabala.
   a   Classical/Hebraic.
   b   Western/Alchemic.

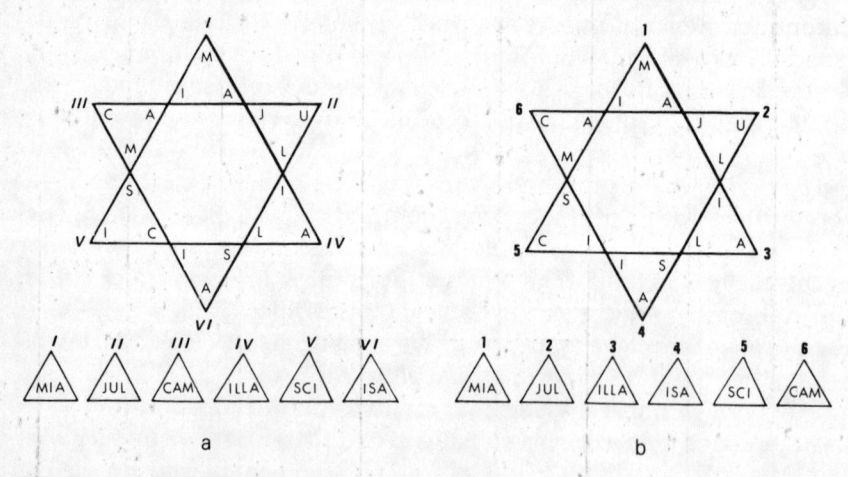

23.2 The two Shields of David.
   a   Classical Hebraic 'zig-zag' method.
   b   Western Alchemic/Templar *'dextratio'*.

The message now obtained was still rather confusing, but seemed to be closer to Colón-Zarco's identity:

Mia Julii Illaqueata Isabella Scia Camara

My Isabel, said Câmara, concubine of Júlio

In order to improve the sense of the words now obtained, but still without a definite link, I decided to resort to the *sephiroth* VII, VIII and

FROM LIGHT TO DARKNESS

IX, which represented Victory, Glory or Majesty and the Foundation respectively.

At the same time, I noticed that the siglum presented another ternary: *Xpo*, which was surmounted by the *roof* (see Figure 23.3).

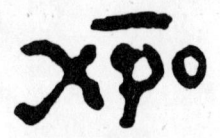

$$\overline{Xpo}$$

23.3 The 'roof' of *Xpo*.

As an abbreviation with two consonants (X and P) and one vowel (O), there was little room for manoeuvre.

I then realised that all the letters of the senary and the word FERENS were in capital letters and the only word of the siglum that deserved to be written in capital letters, seeing that it meant 'Christ', Colón-Zarco had written in small letters. This could mean that it would be necessary to continue with the 'mirror' method in order to increase the chances, especially as a small p would produce a b when inverted, while a capital P would yield nothing (see Figure 23.4).

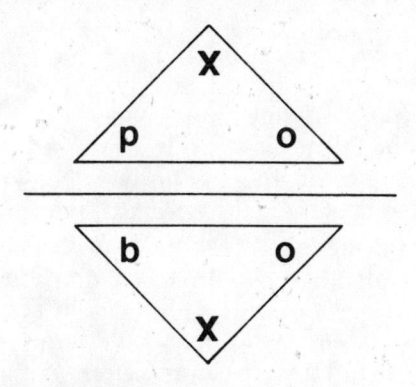

23.4 The mirror method of deciphering *Xpo*.

I then formed the following from *Xpo*:
Upper Triangle: (X = C = S = G) (O = V = U = AU) (AU = AI) (U = I = Y = J)[1]

XPO	XOP	POX	PXO	OXP	OPX
CPO	COP	POC	PCO	OCP	OPC
SPO	*SOP*	*POS*	*PSO*	*OSP*	*OPS*
GPO	GOP	POG	PGO	OGP	OPG
XPU	XUP	PUG	PGU	UGP	UPG

CPU	*CUP*	PUC	PCU	UCP	UPC
SPU	*SUP*	*PUS*	PSU	USP	UPS
GPU	GUP	*PUG*	PGU	UGP	UPG
XPI	XIP	PIX	PXI	IXP	IPX
CPI	*CIP*	PIC	PCI	ICP	IPC
SPI	SIP	PIS	PSI	ISP	IPS
GPI	GIP	PIG	PGI	IGP	IPG
XPAU	XAUP	*PAUX*	PXAU	AUXP	AUPX
CPAU	CAUP	*PAUC*	PCAU	AUCP	AUPC
SPAU	SAUP	*PAUS*	PSAU	*AUSP*	AUPS
GPAU	GAUP	PAUG	PGAU	AUGP	AUPG
XPAI	XAIP	*PAIX*	PXAI	AIXP	AIPX
CPAI	CAIP	PAIC	PCAI	AICP	AIPC
SPAI	SAIP	PAIS	PSAI	AISP	AIPS
GPAI	GAIP	PAIG	PGAI	AIGP	AIPG

They correspond to:

1	2	3	4	5	6	7	8	9
COP	SPO	POS	CUP	SPU	SUP	PUS	PUG	CIP

10	11	12	13	14	15	16	17	18	19
PIC	SPI	SIP	PIS	PIG	PAUX	PAUC	PAUS	AUSP	PAIX

(see Appendix 6, Upper Triangle).

As it happens, the name *Salvador* is expressed only in the Greek abbreviation *Xpo*. The Ss in the siglum have the specific function of indicating that each one of them is a *sephirah*. They intentionally point to the necessity of using the sephirothic system.

Each S is flanked by two points and Patrocínio Ribeiro, without interpreting them as *Sephira*, felt that he should adopt the ternary, in which each point would be a letter that could form a word or abbreviation in conjunction with another central letter. With great lucidity, he based his idea on a Latin rule that I transcribed in an earlier chapter (p. 245, note 11). These two points also have another function in the temuric Cabala – they indicate *repetition*. Each S between two points must be repeated. This is an explicit reference to the 'mirror method', which is to duplicate the pentagram.

This sign that denotes repetition was also adopted for musical scores in remote times. It is known today by the Italian expression *'Dal Segno'* (see Figure 23.5).

Repetition through the 'mirror', therefore, was essential, not as a fortuitous attempt, but because the points that flank the Ss are determining factors. And as they formed a triangle, I was faced with a

FROM LIGHT TO DARKNESS 305

23.5 Repetition and the Ss.

disposition of the sephirothic senary plus two traditional umbilical letters – Mêm (M) and omega (ω) (see page 286).

Here are the arrangements from the lower 'mirror' triangle (b, o and x) (see Figure 23.6):

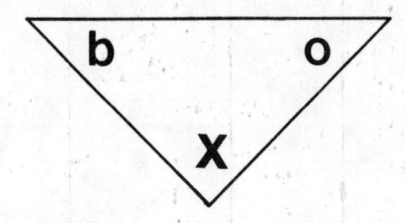

23.6 The lower 'mirror' triangle.

Lower Triangle: (X = C = S = G) (O = V = U = AU) (AU = AI) (U = I = Y = J)

XBO	XOB	BOX	BXO	OXB	OBX
CBO	COB	BOC	BCO	OCB	OBC
SBO	SOB	BOS[3]	BSO	OSB	OBS
GBO	GOB[4]	BOG	BGO	OGB	OBG
XBU	XUB	BUX	BXU	UXB	UBX
CBU	CUB	BUC	BCU	UCB	UBC
SBU	SUB	BUS	BSU	USB	UBS
GBU	GUB	BUG	BGU	UGB	UBG
XBI	XIB	BIX	BXI	IXB	IBX
CBI	CIB	BIC	BCI	ICB	IBC
SBI	SIB	BIS	BSI	ISB	IBS
GBI	GIB	BIG	BGI	IGB	IBG
XBAU	XAUB	BAUX	BXAU	AUXB	AUBX
CBAU	CAUB	BAUC	BCAU	AUCB	AUBC
SBAU	SAUB	BAUS	BSAU	AUSB	AUBS
GBAU	GAUB	BAUG	BGAU	AUGB	AUBG
XBAI	XAIB	BAIX	BXAI	AIXB	AIBX
CBAI	CAIB	BAIC	BCAI	AICB	AIBC
SBAI	SAIB	BAIS	BSAI	AISB	AIBS
GBAI	GAIB	BAIG	BGAI	AIGB	AIBG

306   COLUMBUS REVEALED

They correspond to:

1	2	3	4	5	6	7	8	9	10	11
SOB	CUB	BUC	SUB	GUB	CIB	BIC	SIB	BIS	GIB	BIG

(see Appendix 6, Lower Triangle).

Two triangles had thus been formed – one with its vertex pointed upwards and the other pointed downwards; the latter was perfectly adjusted to the Sephiroth VII, VIII and IX (see Figure 23.7a and b).

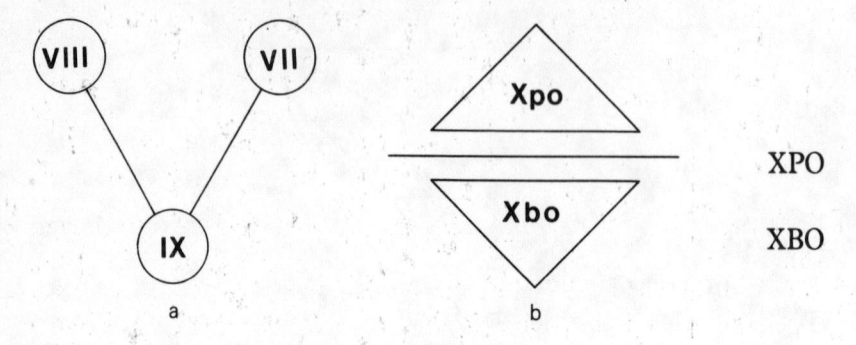

23.7 The upward- and downward-pointing vertices of the triangles.
   a The *Sephiroth* 7, 8 and 9 of the siglum subject to the mirror method.
   b *Xpo* subject to the mirror method.

As I have already explained, only three letters can be inserted into the sephirothic table, referring to sephiroth VII, VIII and IX respectively, and the downward-pointed triangle seemed to be appropriate for the ternary *Xbo*. The idea of 'The Foundation, Basis,' led to a search for a word that expressed that sense and I selected *Soboles* = stock, lineage, race, generation, family.

The next problem was where to move the surplus upper triangle *Xpo*. It could not be placed between Sephiroth IX and X as it would interrupt the numerical sequence of the system. Its destination was indicated in the siglum. I recalled that the sign *roof*, which designated *Xpo* as *under the roof*, referred to the *Sephira Prima* that represented the 'Nothing, Beginning, Verb, God'. *Xpo* (Christ) should, of course, be placed above the 'world' of *Yetzirah*,[5] immediately below the *Sephira Prima* and, therefore, above the Crown.

Xpo was thus moved to the top of the sephirothic senary and, being above the whole of the decenary, carried out its cabalistic function of 'the link between God and Man'. The 'Emanation of the Angels' is related to the *ibbur* (Impregnation), which is 'the meeting of several souls in one body, which happens when the human soul needs outside

FROM LIGHT TO DARKNESS 307

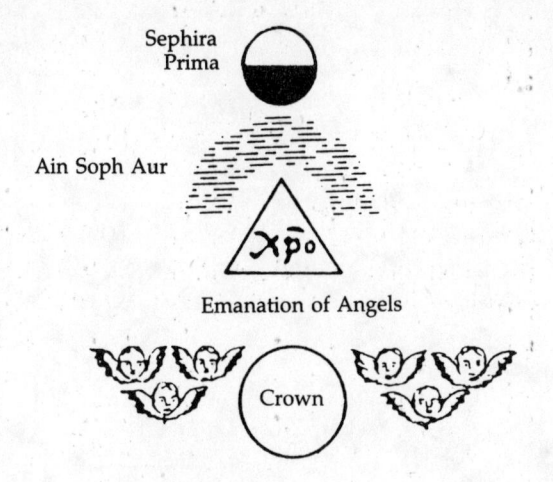

23.8 *Xpo* as a link between God and man.

help in order to achieve its aims; Pardon and eternal Peace'[6] (see Figure 23.8).

I looked for a Latin word in the upper triangle and found *Suppetere* (help), but then immediately found another one that was more in keeping with the cabalistic sense: *Paix*[7], which means 'Peace' and 'Pardon' or, in the genitive 'of Peace, of Pardon'.

'FERENS' AND FRUSTRATING MESSAGES

The only thing left to do now was place the linear word FERENS in the sephirothic system (see Figures 23.9a and b). I had already considered this to be an abbreviation of Fernandes, as it was similar to the one used by Valentim Fernandes. As FERENS is made up of six successive letters, it did not allow the use of a normal triangle and the 'mirror' method did not supply any inverted letters (see Figure 23.10).

For this reason, I decided to divide the word into two three-letter fractions: FER and ENS. This would correspond to the two hemispheres of the *Sephira Prima* (see Figure 23.11).

In this way, and with the whole sephirothic system seemingly complete, its configuration was compared to the words attributed to the siglum and produced the following result:

FER ENS PAXIS MEAE. JULIUS ILLAQUEATUS (cum) ISABELLA SCIA CAMARA SOBOLES (sunt).

FER ENS of my peace. Júlio lover of Isabel, said Câmara, are my generators (parents).

308 COLUMBUS REVEALED

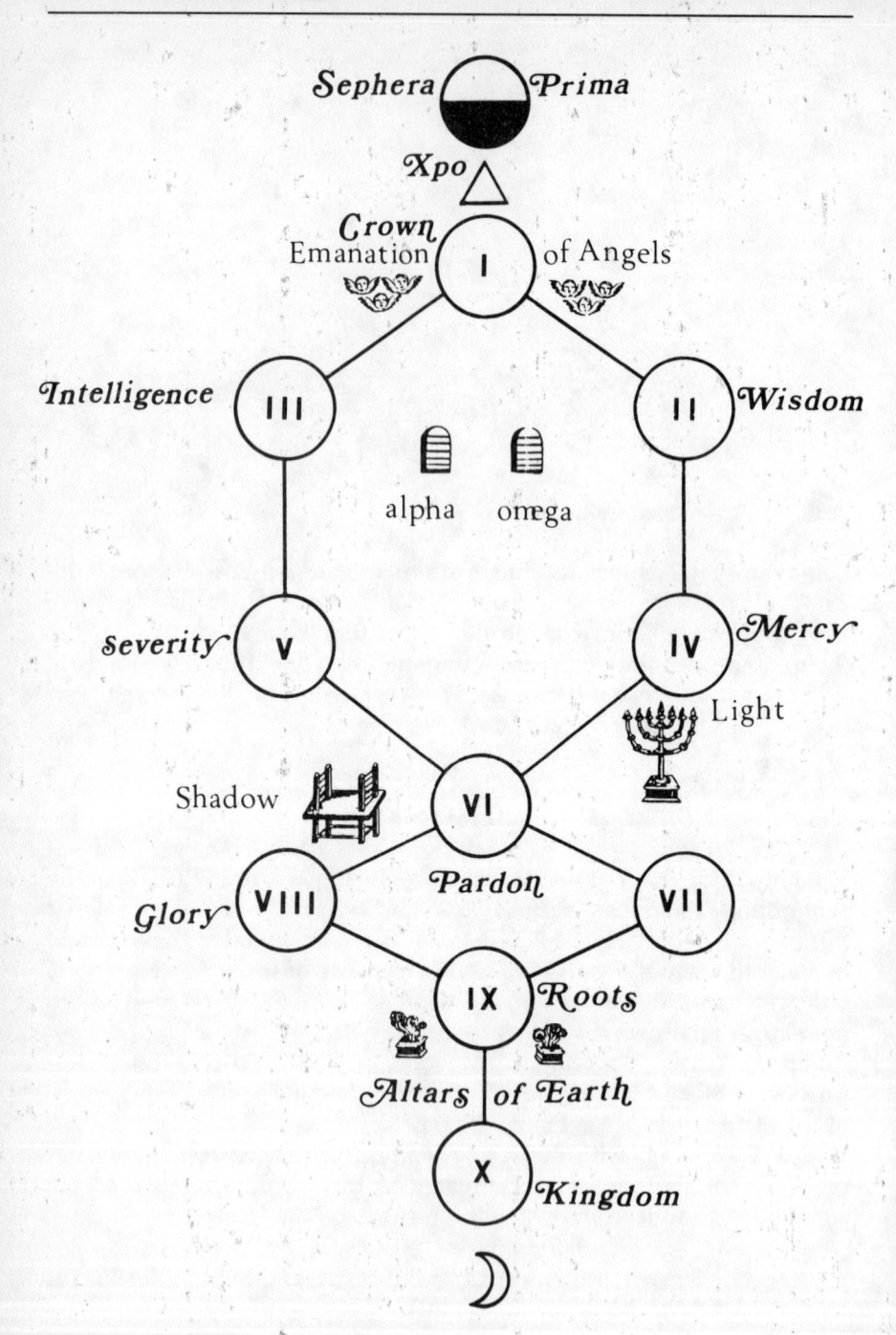

Sephera Prima

Xpo

Crown of Angels
Emanation
I

Intelligence III   II Wisdom

alpha   omega

Severity V   IV Mercy

Light

Shadow
VI
Pardon

Glory VIII   VII

IX Roots

Altars of Earth

X Kingdom

23.9 Placing FERENS in the *Sephirothic* system.
 a Western/Templar *Sephirothic* system.

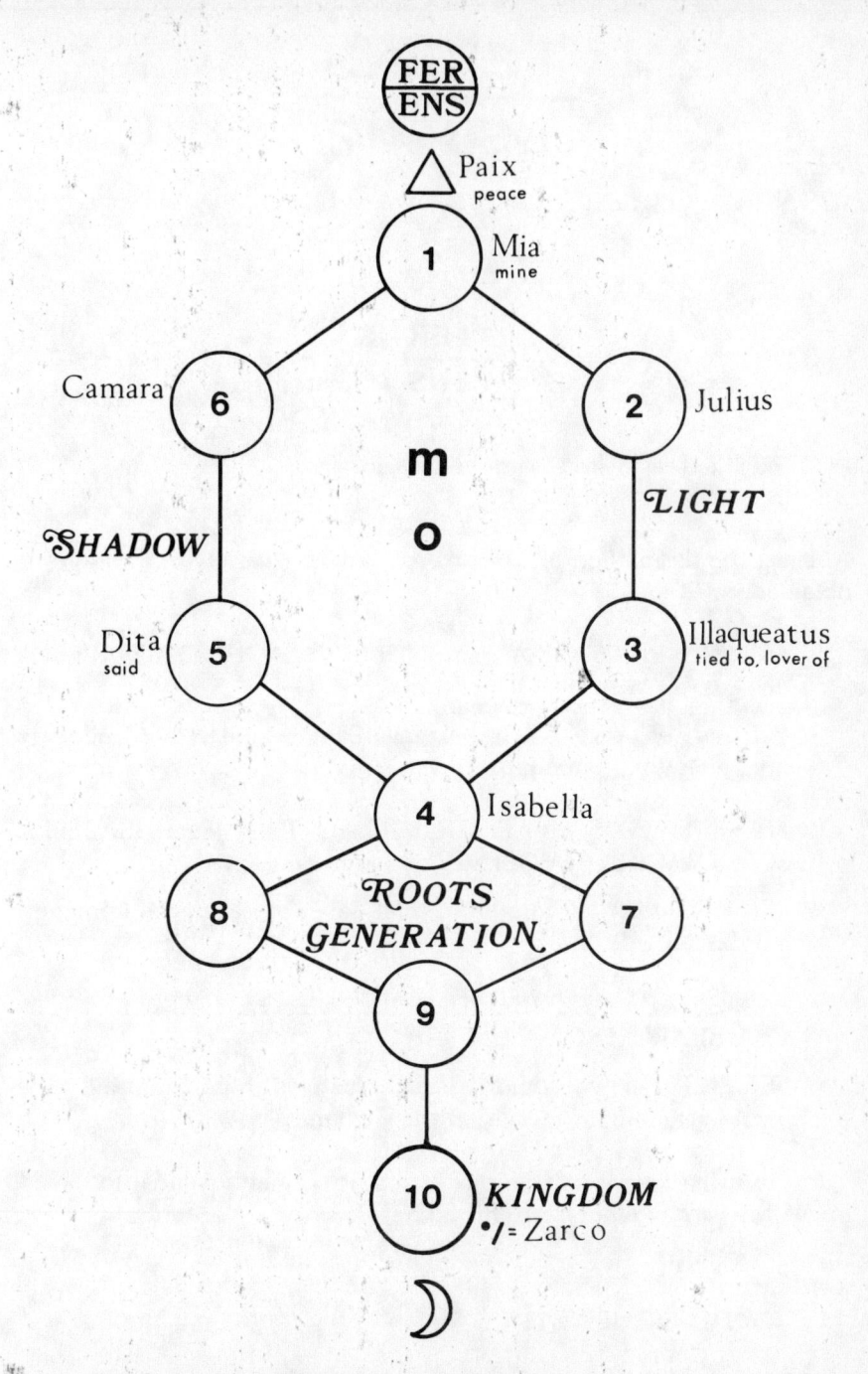

```
          ┌─────┐
          │ FER │
          │ ENS │
          └─────┘
            △ Paix
               peace
          ╱─────╲
         ( 1 )  Mia
          ╲───╱  mine

 Camara  ( 6 )        ( 2 )  Julius

              m            LIGHT
              o

𝒮HADOW

 Dita   ( 5 )        ( 3 )  Illaqueatus
 said                        tied to, lover of

          ( 4 )  Isabella
               ℛOOTS
      ( 8 )  GENERATION  ( 7 )
          ( 9 )

          ( 10 ) KINGDOM
                 •/= Zarco
              ☾
```

b The author's first attempt to decipher the Cabala.

FERENS
ꟻEЯEИS

23.10 Using the mirror method for FERENS.

```
  FER
  ───
  ENS
```

23.11 FERENS in the two hemispheres of the *Sephira Prima*.

Altering the declination of Latin words, a wider variety of messages is obtained, such as:

FER ENS PAXIS. MEA (cum) JULIO ILLAQUEATA ISABELLA SCIA CAMARA SOBOLES (sunt).

FER ENS of peace. My Isabel, said Câmara, mistress of Júlio, are my generators (parents).

FER ENS PAXIS MEAE. JULII ILLAQUEATI (cum) ISABELLA SCIA CAMARA SOBOLESCENS (sum).

FER ENS of my pardon. From Júlio, lover of Isabel said Câmara, I am descended.

FER ENS PAXIS. MEA JULII ILAQUEATA ISABELLA SCIA CAMARARUM SOBOLUM (est).

FER ENS of my pardon. My Isabel, mistress of Júlio, (is) said to be of the generation (lineage) of the Câmaras.

In the face of such frustrating results, it became essential to experiment with arrangements of FER ENS.

UPPER HEMISPHERE: *FER*

FER   FRE   EF(F)R   ERF   REF   RFE

They correspond to:

1	2	3	4
FER	FRE	EF(F)R	REF

(see Appendix 6, Upper Hemisphere).

LOWER HEMISPHERE: ENS (S = C = X = G)

ENS	ESN	NES	NSE	SEN	SNE
ENC	ECN	NEC	NCE	CEN	CNE
ENX	EXN	NEX	NXE	XEN	XNE
ENG	EGN	NEG	NGE	GEN	GNE

They correspond to:

1	2	3	4	5	6	7	8	9	10
ENS	NES	SEN	ENC	NEC	CEN	NEX	ENG	NEG	GEN

(see Appendix 6, Lower Hemisphere).

This last series (GEN), with its words 'birth', 'father', 'mother', 'family', 'origin', 'relationship', etc. seemed truly promising, but at the same time disturbing for the 'Zarco' thesis because of the appearance of the words 'Genoa' or 'Genoese'. Looking for a masculine family name that could be used in the upper hemisphere, I obviously selected *'Fernandus'*; I chose three chances for the lower hemisphere: *genuus* = Genoese; *Genua* = Genoa; *Ensifer* = who carries the sword, who commands. And I constructed several phrases, among which were:

A   *Fernandus genuus paxis* . . .

    Ferdinand, Genoese of peace . . .

B   *Fernandus Genuae paxis* . . .

    Ferdinand of peace of Genoa . . .

C   *Fernandus ensifer paxis* . . .

    Ferdinand who carries the sword of peace . . .

There was no logical sequence, besides the fact that the words found were outside the concepts of sephirothic norms.

Had Colón-Zarco branched off on a different path? This would have obliterated a precious cabalistic trail.

THE SIBYL'S ADVICE

I thought of Valentim Fernandes and tried to imagine his intentions when he sent that strange letter to Peutinger. Had he enclosed Colón-Zarco's siglum so as to challenge the latter's art of cabalistic deciphering? And why did they talk about Columbus in that particular year of 1506? Was Valentim Fernandes's letter a reply to a consultation made by Peutinger? It is dated 16 August – i.e., exactly three months after Columbus's death on 19 May. The post was slow in those days and I can imagine that the correspondence was exchanged immediately after 19 May. Peutinger would have received news of Columbus's death about six weeks after it had occurred; Fernandes would have received his consultation a month and a half later and would have replied at once.

The German certainly did not spend so much time or suffer so much as I have done. The letter with the 'Sibyl's Prophecy' would have helped him. So why should it not help me to find my way through the mist in which I was enveloped? Besides suggesting the *'Dextratio'* rotation in regard to the numerical sequences of the Sephiroth of the senary, it had specified 'from the light to darkness', a movement frequently referred to by western cabalists prior to the thirteenth century. Re-examining the sephirothic schema, I came across two symbols that I had not yet taken into consideration – 'The Candelabrum of the Seven Arms' and the 'Table of the Bread of Atonement', which respectively represented *Light* (from the East) and *Darkness* (the refuge of the Tabernacle).

According to the Templar (western) Cabala, Peace and Pardon were emanated by *Christ*, but I might not forget the original Jewish Cabala in which the same emanation of Peace and Pardon were between the 'Candelabrum of Seven Arms' and the 'Table of the Bread of Atonement' (see Figure 23.12).

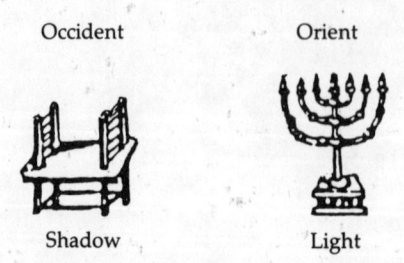

Occident   Orient

Shadow   Light

23.12 The 'candelabrum of Seven Arms' and the 'Table of the Bread of Atonement'.

In fact, as was written in Chapter I of the *Historia* of Fernando Colón, his father had been given the mission of 'taking the Light of Christ to the Lands of Darkness' that were lacking the word of God. It fell to him to lead an expedition over the 'deep waters' from the Iberian Peninsula

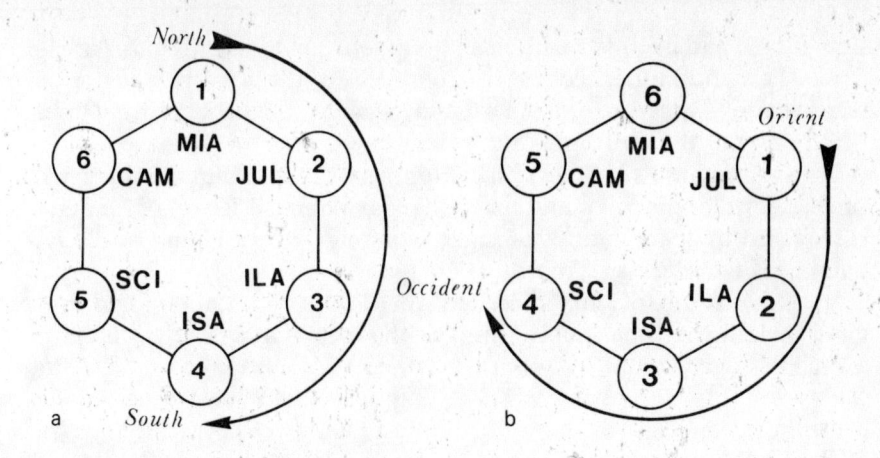

23.13 Reading the *Sephirothic* system 'in the tracks of the sun'.
   a North–South.
   b East–West.

(east–northeast) to the Antilles (west–southwest). In this case, I should try a new reading of the system, beginning the rotation from East to West and not from North to South, in the tracks of the Sun (see Figure 23.13).

Within the norms of the sephirothic system, the disposition of the words now read:

JULIUS ILLAQUEATUS SCIA CAMARA MEA

Júlio lover of Isabel said Câmara mine

Fiat lux! I had at last found a suitable solution for Colón-Zarco's identity. It was enough to use the feminine *Júlia* instead of the masculine *Júlio*.

FER.	ENS.	PAIX	JUL.	IL(L)A.	ISA.
Fernandus	Ensifer	Paxis	Juliae	Illaqueatus	Isabella
SCI.	CAM.	MIA	SOB.	./	
Scia	Camara	Mia	Soboles	(sunt)./	

Fernando, who holds the sword of *Pax Júlia*, lover of Isabel, said Câmara, are my lineage. Zarco.

Pax Julia was the Roman name of Beja, and was always named in official documents.[8] Prince Fernando, who lived in the city for several years, was the nephew of Prince Henry the Navigator, who adopted him in 1436 and bequeathed him all his assets. On the death of his uncle, Prince Fernando inherited the Duchy of Viseu, the Archipelagos of the Azores and Cape Verde, the lordship of many towns and the Grand-Mastership of the Order of Christ.

Prince Fernando's brother, King Alphonse V, made him the first Duke of Beja, Lord of the town and the castle, gave him the Island of Madeira, discovered by João Gonçalves Zarco, squire of Prince Henry and navigator of the Order of Christ.

In 1450, Prince Henry donated the Island of Madeira[9] to João Gonçalves Zarco and raised him to the peerage with the name of Câmara.[10]

'CÂMARA'

The origin of the name Câmara is as mysterious as that of Zarco.[11] It was said that when he 'refound' Madeira, João Gonçalves Zarco had seen a lot of wolves 'in a big black cave that sheltered many of these animals, which the sailors hunted. João Gonçalves Zarco and his descendants took the name Câmara from that place'.[12] The legend is without foundation as these animals do not exist on the island. The animals that were seen on the coast were sea-lions, which are more common in the cold northern waters but which migrate as far as the Canaries in winter. The crest of the navigator's arms was a seal, with another two flanking a keep-tower (see Figure 23.14a). The arms were later altered and the seals were replaced by wolves, one standing, as the crest, and two others rampant supporting the tower (see Figure 23.14b).

We know that nearly all the symbolism of the old coats-of-arms was based on esoterism. The *tower*, for instance, the symbol of *Chokmah*, represents the phallus, which gave rise to menhirs, obelisks, triumphal columns, minarets, pillories, bell-towers, etc. Another symbol is the *mermaid*,[13] a mythological cabalistic being with a remote origin in the legend of Persephone and which means that 'the soul survives the body'. The evolution of its image was complex, as it appears with wings among the Greeks. But Christians, who always showed the pagan gods as devils, transformed it into a woman with her two legs in the form of fish-tails, and this was how it appeared in wall paintings, church capitals and other sculptures, sometimes holding the two tails in its hands so as to be represented as a female figure with two vaginas. It can also be seen with masculine body and face – a bearded marine being that frequently decorates fifteenth- and sixteenth-century cartography and engravings of esoteric books. It underwent another morphological metamorphosis at this time, with the bottom part of the body being fused into one fish-tail. Yet a being of this type had already existed in Helleno–Roman mythology and can be seen, for example, in a mosaic in the Roman ruins

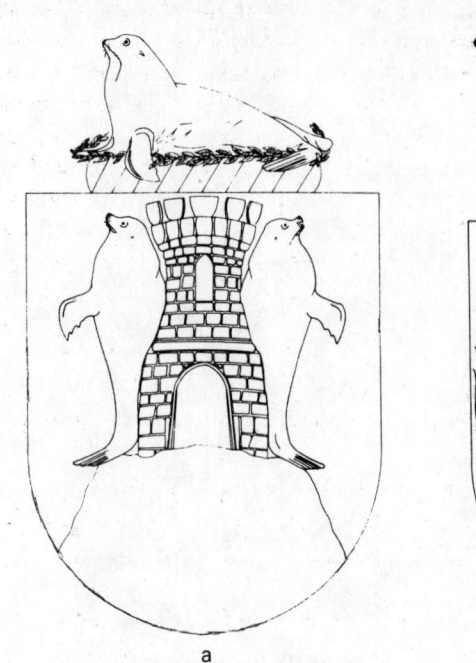

23.14 The Câmara (Zarco) coat-of-arms.
  a Initially presenting sea-lions (or seals).
  b Later altered to wolves, which did not exist in the archipelago of Madeira, replaced to avoid amphibians which could have been interpreted symbolically as esoteric mermaids.

of Conimbriga in Portugal. And inside the cloister of St Peter of Galingnams in Gerona.

The only coats-of-arms known in Portuguese heraldry that bore a mermaid are that of the Marinhos, a Portuguese family of remote origin, and that of the D'Ornelas (or Dornelas), a branch of which is descended from a brother of João Gonçalves Zarco (see Figure 23.15a and b).

Although '*Camara*' means 'a vaulted roof', it may also represent 'the uterus of the gestation of the philosopher's stone' (i.e., the 'philosophical egg').

The immediate descendants of João Gonçalves Zarco took the precaution of changing the *dos-à-dos* seals for wolves rampant and, although they kept the title of Câmara, tried to discover if it could be construed as 'cave' or 'overhanging rock' and not 'arch' or 'arc' (or, even less, as the Templar alchemic or Hebraic cabalistic 'Egg'). Everything was conjugated to decipher Colón-Zarco's siglum, identifying him as the bastard son of Prince Fernando, Duke of Beja and Lord of Madeira, and Isabel da Câmara, whose father had been raised to the peerage with the title of

23.15 The Marinhos and Dornelas coat-of-arms.

    a The coat-of-arms of the Marinhos family, with a mono-tailed mermaid as a crest and a shield of five fleur-de-lys on a plain field. Another branch of the family had a shield of wavy stripes representing the waves of the sea.

    b The arms of the Dornelas family, with a bi-tailed mermaid as a crest and a field containing two identical mermaids and a band with three fleur-de-lys. One branch of the Dornelas family descends from João Gonçalves Zarco's brother.

Câmara. Although born out of wedlock, Colón-Zarco was the grandson of King Duarte, nephew of King Alphonse V, half-brother of Queen Leonor and therefore cousin and brother-in-law of King John II. He was also half-brother of the future King Manuel I.

THE FINAL SOLUTION

There was still one sign in the siglum to decipher, however. My attention was now focused on the colon (:) that precedes the abbreviation *Xpo* (see Figure 23.16).

: X͠p͠oſERENS./

23.16 The first colon in the base of Colón-Zarco's siglum.

I have already emphasised the value of the colon as a sign of repetition, as adopted in musical scores. Returning to the rules of the western Cabala, however, I saw that the abbreviation *Xpo* was in the 'emanation of the angels' of *Yetzirah*, and that this is depicted by two groups of three angels in the sephirothic system, each group in a triangle, a clear indication to explore two ternaries of the Cabala.

Using the 'mirror' method, I had already obtained the word *SOBOLES* (generation, stock, lineage, family) from PAIX. I now had to discover the two complementary words demanded by the sign of repetition, which was clearly indicated by the presence of the two 'Image of the Earth' altars that flank *sephira* IX (the Foundation) (see Figure 23.17).

Altars – Image of the Earth

The Root – Creation

23.17 The Altars and the Root.

It was now essential to obtain another arrangement for each of the ternaries – a complementary word for the space between FERENS and PAIX and another as a sequence of *SOBOLES*, at the end of the decenary, immediately before *Malkuth* (the Kingdom) (see Figure 23.18).

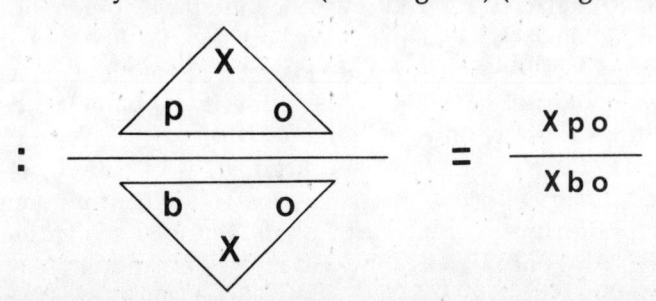

$$: \frac{\triangle\ {}^{X}_{p\ o}}{\triangledown\ {}_{b\ o}^{\ \ \ X}} = \frac{X\,p\,o}{X\,b\,o}$$

23.18 Deciphering 'Cuba'.

Following a logical sequence, as a continuation of the previous arrangements, it was easy to select the words:

$$\frac{Xop}{Xob} = \frac{Cop}{Cob} = \frac{Cop}{Cub} = \frac{COPIA}{CUBA}$$

Everything seemed to have fallen into place. The message completed by Salvador Fernandes Zarco in the second and final form of his siglum was complete:

FERNANDUS ENSIFER COPIAE PACIS JULIAE ILLAQUEATUS (cum) ISABELLA SCIA[14] CAMARA MEA SOBOLES CUBAE (sunt). Zarco

FERNANDO, WHO HOLDS THE SWORD OF POWER (*dux* or Duke) OF BEJA, LOVER OF ISABEL, SAID CÂMARA, (are) MY GENERATION OF CUBA.

And he signed '*Xpo Ferens./*', meaning *Salvador Fernandes Zarco.*

It is clear that as the Letter D does not appear in the siglum Colón-Zarco had to substitute the word *DUX* (he who commands, Duke) with the metaphor 'who holds the sword of power'. Prince Fernando was certainly the Constable of the Realm and the sword was the symbol of this post.

'*Copia*' not only meant power, but also army, the Commander of which was the Constable. As an analogy, he was the Master of the Order of Christ. It also means 'crowd', 'population' and the Duchy of Beja was under his Governorship. In order to mean 'army' he would have had to have written 'copiarum'.

Today a town in the District of Beja, Cuba was then no more than a small village huddled around a 'big house' (a villa or a '*monte*'). During the reconquest, which was concluded by King Alphonse III, many villages of the Alentejo were the result of the agglomeration of the defeated Berber population around the 'big house' built by the squire to whom the King had entrusted the defence and development of the area.

Even if Colón-Zarco had not been born exactly in Cuba, but in a neighbouring locality of this area of red earth – '*terra rubra*' – such as Vila Ruiva, he would not have been able to specify its name as the Cabala allowed him to supply only the ternary *CUB*.[15]

As far as Zarco's signature is concerned, it is found, as already mentioned, in the *Sephira* that corresponds to the Kingdom in the sequence of the 'Image of the Earth' altars. But which kingdom? When his brother, Alphonse V, was crowned, Prince Fernando was recognised as second in line to the throne if the king died without issue. But his illegitimate son, Colón-Zarco, had nothing but the memory of the place where he was born. He was not even able to keep his own personality: 'My Kingdom is merely my real name'.

THE MYSTERIOUS MONOGRAM

Since the publication of my book in Portugal, Dr Manuel Luciano da Silva, whose investigations regarding the discovery of North America by the Portuguese have already been cited, has achieved the extraordinary feat of decodifying the monogram which appears to the left of the siglum in fifteen documents signed by Christopher Columbus (see Figure 23.19) and to which, negligently, I did not pay much attention.

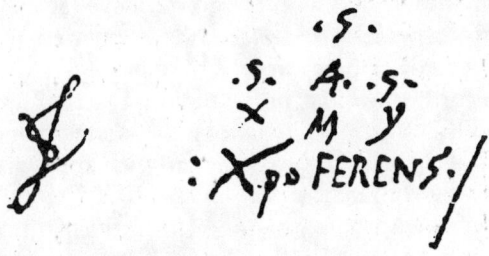

23.19 Colón-Zarco's monogram and siglum.
(Dr Luciano da Silva)

Monogram   Siglum

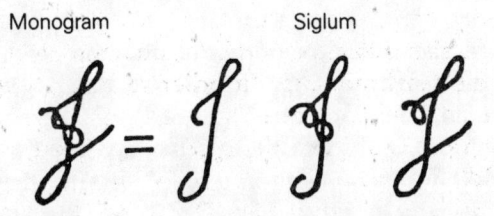

Monogram   'S', 'F' and 'Z' in handwriting style of
the time, meaning *Salvador Fernandes Zarco*.

23.20 Deciphering Colón-Zarco's monogram.
(Dr Luciano da Silva)

If the Monogram is broken down (see Figure 23.20), the three letters S, F, Z are obtained, the initials of Salvador Fernandes Zarco. The decodification of the monogram confirms my decipherment of the siglum of the sephirothic Cabala of the Temurah through which Columbus transmitted his own identity and his place of birth.

THE DISCOVERER PRINCE

The historian Agostinho Ferreira Gambetta provided me with some valuable materials for my thesis, when he presented a paper at the Lisbon Naval Academy and referred to the work of the historian Demétrio Ramos, *Why did Colón have to offer his services to Spain?*,[16] in which it was claimed that Colón was a Prince (Infante) of Portugal. Professor Gambetta also supplied me with precious details extracted from the writings of Las Casas, Colón's most faithful chronicler (apart from Fernando).

1. Pero Vasquez de la Frontera, 'who was a very wise man in the ways of the sea . . . he had once gone on a voyage of discovery [to terra firma] *with the Prince of Portugal*'.[17]

    This Pero Vasquez de La Frontera was born in Galicia and in 1451 or 1452 had already sailed with Diogo de Teive (see p. 58) and on his second voyage in which the latter had discovered the westernmost island of the Azores, Flores, from where he had sailed further west. There is no record, however, of his having reached terra firma. But it must be noted that Vasquez does not mention Diogo de Teive, so that it can be supposed that the expedition with the Portuguese Prince took place at a later date.

2. Martín Alonso Pinzón (see pp. 160, 161 and 218) 'was informed by Pero Vasquez de la Frontera that he had gone to discover this land with a Prince of Portugal and said that they had almost succeeded, and that they were prevented from doing so because of the algae in the sea' (the Sargasso Sea – see pp. 155–66).[18]

3. — 'because the said Prince of Portugal, because he did not cross it failed to find the said land'.[19] Although not directly, he thus identified the Prince as Salvador Zarco, son of Prince Fernando, Duke of Beja and Master of the Order of Christ, as it could be no other.

    It is obvious that this expedition took place after the one of Diogo de Teive and after 1474, when Salvador Zarco had already adopted the name Cristóvam Colom, as can be seen in the letters he exchanged with Toscanelli. Besides this, Pinzón testified in the *'Pleyto'* that he had sailed to the Western Atlantic with a map of the zone that had been drawn by Columbus and handed over to him by Friar Gaspar Gorricio.

    According to the documents of the *'Pleyto'* (vol. 5, p. 119, 15 December 1535), the navigator Alonso Gallego 'testified that he had heard Pero Vasquez de la Frontera say . . . that the said Colón had obtained information from Pero Vasquez, as a person that had been in the service of the King of Portugal and knew of the said lands and islands'.

It is also important to stress that Pero Vasquez never mentioned the name of Cypangu, but merely a western land, certainly the one that Teive called Antilia. Cypangu was simply used by Columbus to deceive the Catholic Monarchs.

No other Portuguese Prince was a mariner.

Notes and References

1. The series (O = V) and (U = J) were not formed because in the arrangements with B and P it would have resulted in the joining of three consonants.
2. The series (O = V = J) was not formed because it would have resulted in three consonants.
3. It was not considered as it presented only two unsuitable Latin words: *Bos* = ox *Bosphurus* = Bosphorus.
4. As above; *Gobius* = gudgeon.
5. A Divine Emanation which was represented by the *Ain Soaph-Aur* in the western Cabala. It is designated as *Emanation of the Angels* in Hebraic Cabala, but appears around the Crown, generally within the senary.
6. In *Cabala ou Caballa, Dictionnaire de la Conversation et de la Lecture* (1833).
7. In Vulgar Latin; it was written like this in old Provençal and is still written so in French.
8. Colón-Zarco could not have indicated the abbreviation of Beja in the siglum as he did not have the letters B and E in the sephirothic senary. But he clearly expressed the name of the city as it was then written in documents of the time.
9. Confirmed by letters patent of King Duarte of 15 November 1445.
10. Confirmed by a letter of King Alphonse V of 9 July 1460.
11. See pp. 237–41.
12. In Maximiano Lemos, *Encyclopedia Portuguesa Illustrada* (Oporto, 1901).
13. Generally known as a Triton when masculine.
14. The decodifying of a cabalistic message can be univocal and apodictic only when the key is known and the inductive method is used. As I was unaware of many historical elements, I sometimes had to opt for the deductive method in this decipherment, which is always insecure and imprecise. In this case, in the face of words that corresponded to the ternary S(C)IA (see Triangle V, p. 524), I could select only the word *scia*. Although the solution seems to be somewhat deficient, it can be justified by the fact that none of the direct descendants of João Gonçalves Zarco, apart from Salvador Fernandes himself, used this surname of Jewish origin.
15. The historiographer João Paulo Estêvens Taborda of Vila Alva (quoting João Tello de Magalhães from his *Cadastro da População do Reino*, 1527), states that Cuba had only 172 families in 1532; 63 of them living in a suburb

three miles from the town. He adds that Vila Ruiva was raised to the rank of a town by Queen Mary I in 1782.
16. Madrid (1971).
17. Las Casas (1875–6), Book I, Chap. XXIX, Vol. I, p. 110 (*Pleyto Colombino*).
18. Las Casas (1875–6), Book I, Chap. XXXI, Vol. I, p. 118.
19. Las Casas (1875–6), Book I, Vol. VIII, p. 301.

Part VI
Check and Mate

24 *After the Heat of the Battle*

Las Casas[1] related that when Bartolomeu Colón was in London in 1490, he showed a map to King Henry VII. This map was found among Columbus's papers but later disappeared.

'TERRA RUBRA'

Las Casas stated that he had seen this map, captioned with two stanzas of poetry. While the first stanza was an erudite composition in fairly acceptable Latin, the second was made up of two lines of deplorable writing, in which Bartolomeu Colón wished to unveil his birthright: *'Janua cui patria est nomen cui Bartholomeus Columbus de Terra rubra opus edidit'*.

This 'pidgin' Latin can be translated as 'Genoa whose name is the homeland where Bartolomeu Colón of Terra Rubra published the work'.

There is no need to stress the anomaly of the reasonable writer of the first stanza becoming the Latinist of the second. Immediately afterwards, however, the composition improves again. It must be noted that *'Terra Rubra'* is not in Latin, which would have been the genetive *'Terrae Rubrae'* without the *'de'*. Was Bartolomeu Colón such a poor Latinist?

The chronicler considered that the phrase had been written in 'very bad and corrupt hand', because there were words superimposed on erasures: to introduce the word *'Janua'* and alter the surname *'Colonus'* to *'Columbus'* other letters would have to be omitted.

Fernando Colón[2] also declared that he had seen many of his father's signatures, including those before he had acquired the titles granted him by the Catholic Monarchs, when he was signing 'Columbus of Terra Rubra', or *'Terra Rubia'* in Spanish, together with his siglum. This was confirmed by Las Casas: it is said that Columbus came from some place in the province of Genoa; wherever he was born or whatever name is given to the place, the only thing known is that before he reached the position he reached he called himself Cristóbal Colón of Terra Rubia, as did his brother Bartolomeu Colón (Book I, Chap 2).

One should not be surprised at the surname *'Columbus'*, even though Fernando insisted, in Chapter I of the *Historia* that his father's surname was 'Colonus'. Las Casas mainly copied the distorted Fornari Italian version, as it was the only one then published. One should not be surprised either by the fact that the originals of the Colón brothers also disappeared from the archives.

The 'Genoists' interpreted 'Terra Rubra' as 'Terrarossa', but this is only *one* word and not two. This would be like translating *Ulixbona*[3] as *Ulix* plus the adjective *bona* (good). Correctly in Latin would be *Terrarossae* (Genitive). The writer was undoubtedly someone who wanted Bartolomeu to be seen as a Genoese Colombo. And he did not erase *'Terra rubra'* as it could be used to translate into Terrarossa. The argument is based on a suspect notarial deed that states that in 1471 *Bartolomeo Colombo* 'left Genoa for Terrarossa in order to join his father [who had meanwhile moved to that place] and to be closer to the wool industry'. But, nobody knows if this document is genuine.

In the fifteenth century, moreover, Terrarossa was not the town it is today, 60 km from Genoa by modern highways. Did it possess a more prosperous and advanced wool industry than the huge port of Genoa, where Doménico followed his trade? Was this document created in order to reinforce the inscription of Bartolomeu Colón's *mappa mundi* on which he claims he was born at Terra Rubra? But if Doménico Colombo moved to Terrarossa only in 1471, how is it that his son, who according to the Genoese thesis was fifteen at the time, was born in Terrarossa?

If Bartolomeu was still a wool-carder in that year, how could he have been his brother Cristóforo's master if Cristóforo (already with the name of Cristóvam Colom) was corresponding in Latin with Toscanelli in Florence in his status as a scholar of cosmography and cartography only three years later? If he belonged to the fifteenth-century cultural environment of wool-carders and tavern-keepers, where would Bartolomeu have learned to compose erudite poetical verses and cite classical personages in Latin? It does not seem to be a thing that someone would have learned while making maps in order 'to earn his living'.

What was life like in those days? Except in rare cases, the demands of daily life bound most people to the trade of the family. Boys worked with their fathers from a very early age and had no chance of studying like their present-day counterparts. Students belonged to the upper classes, as the poor could not dispense with the family's labour and schooling was expensive. Even amongst the Jews, who were so inclined to study, only the sons of the most illustrious families had access to the exclusive School of Sciences.

On the other hand, to learn cosmography, cartography and scientific navigation was truly transcendental. As I have already mentioned, the knowledge of the Italian and Spanish mariners of that time was way behind that of the Portuguese. This was undoubtedly proved by the difficulty Columbus encountered in making two Castilian Juntas of Experts understand what he was trying to explain before they accepted his proposals. Navigating the Ocean Sea in the fifteenth century was still a highly complex operation, much different from coastal navigation or cruising the Mediterranean. Far from land, without the radio and other modern technology with which ships of today are equipped, the navigator was on his own. Only his own skill could steer his ship to where he wanted to go. Cosmographic-nautical science was a field for a chosen few and was learned through a long apprenticeship at sea. The epic of the discoveries was a very different thing from the coastal piracy and commerce of the time. If Doménico Colombo had been a nobleman or rich bourgeois with influential connections, he could have given Bartolomeu a fine education in the field of letters and sciences, but he was merely a rustic tradesman with five children to support and up to his eyes in debt.

THE FULLING AND WEAVING INDUSTRY OF BEJA

The following passage is taken from Afonso Zúquete's book of genealogy.[4]

> Prince Fernando [future Duke of Beja] had married his cousin Beatriz, daughter of Prince João [son of John I] and Isabel [daughter of the first Duke of Braganza] in 1447 . . . In Beja, Beatriz protected the wool industry; at her request, King John II granted her privileges for the fulling-mills that she had built for the fullers of the city [1490].

The mills were used to full the wool after it had been carded. The wool industry had not only been developed in Beja, but also in Alvito. Sheep were raised in the fluvial areas of the small rivers Odearce and Odivelas. The former supplied the villages of Cuba and Faro (not the capital of Algarve but a village of the Duchy of Beja) and the latter Alvito and Vila Ruiva. 'Ruiva' is the Portuguese word for the Castilian 'Rubia', and 'Rubra' is also a Portuguese synonym of 'Ruiva', meaning 'red'.

The colour of the Alentejo soil gave rise to the names of communities. Some examples of this are Campo Branco (White Field), which comes from the light-coloured earth of the area, and Castro Verde (Green Fortification), which did not receive its name from the local vegetation but from the greenish schistose soil. Several other places took the name of 'Ruiva', such as Alcaria Ruiva, because of the reddish argillaceous soil of the surrounding area. Vila Ruiva stands only 11 km from Cuba by the old Roman road, and less than seven from Alvito.

As we already know that Cristóbal Colón was, in fact, Salvador Fernandes Zarco, born in Cuba, in the area of the 'red lands', it is now possible to accept that Bartolomeu had been born at Vila Ruiva. Does this mean that they were born on neighbouring *'montes'*,[5] and were either blood brothers whose mother had moved house or foster-brothers brought up in the same family?

On analysing the siglum, we can see that as Colón-Zarco had only three letters at his disposal (see the layout of the lower triangle on p. 317), he could never have indicated the two words *'Terra Rubra'*. Was this the reason that he decided to indicate the name of the neighbouring town Cuba as his place of birth? Or had he really been born in Cuba and then taken to the *'monte'* at Vila Ruiva that belonged to the Duke of Braganza, his close relative? It is just a detail, but later I shall try to find an answer to this problem.

It must be stressed that if Bartolomeu Colón had written *'Villa Rubra'* in Latin, the word 'villa' would not have meant village, but merely a country house; it could not have indicated anything more than a 'farm house with red walls'. So we have two wool industries, one is Genoa/Terrarossa and another in Beja/Terra Ruiva.

BARTOLOMEU COLÓN AT THE ENGLISH AND FRENCH COURTS

If we admit that Colón was not a wool-carder and tavern-keeper but came from a higher stratum of society, it is possible to accept his claim that he wrote to the Duchess of Bourbon and the Kings of France and England with the proposal to discover India by the west in an effort to speed up the decision of the Catholic Monarchs. He said that those foreign Kings had answered his letters.

According to Las Casas, Bartolomeu Colón had been the bearer of his brother's letters, but that on the way to England his ship had been boarded and sacked by French pirates and he had lost everything. This was the reason that the King of France later gave him 100 ducats so that he could return to Spain. Accepting as true the facts related by chroniclers that were Columbus's contemporaries, we may assume the following:

1. To deal with such an out-of-the-ordinary and important proposal to the English and French sovereigns – the possibility of the spice trade and other riches of India, China and Japan – Columbus did not wish to go himself and sent his brother Bartolomeu.
2. After seven years of being rebuffed, he remained in Spain in order to continue putting pressure on the sovereigns.

3. When Bartolomeu returned from France, Columbus was coming to the end of his first voyage in the service of Spain.
4. On his second voyage in 1494, Columbus met his brother for the first time for a long time and appointed him as the military and civil governor of Cuba.

It can thus be concluded:

(a) Columbus was not the slightest bit interested in sailing in the service of the Kings of England or France. His only aim was to divert the attention of the Catholic Monarchs to the west and his brother's journeys to the courts of Spain's rivals were purely to encourage the Spanish sovereigns to come to a positive decision.
(b) If Bartolomeu was a wool-carder's son, he would not have been received in the courts of England and France and would not have received 100 ducats for having been robbed at sea. Kings were not in the habit of indemnifying pirates' victims, unless they were worthy of esteem and possessed the right credentials.
(c) If Columbus appointed his brother as civil and military governor of Cuba it was obviously because he deserved the position and possessed the necessary qualities to carry out his duties. This Bartolomeu proved in the pacification of San Domingo. They were to sail together again in 1502, in the discovery of the coast of Verágua (Panama).

How can it be explained that not only the Kings of England and France but also the Duchess of Bourbon received Bartolomeu Colón with such unusual compliance? It must be remembered that in 1491 Columbus was not yet in a position to present those monarchs with worthy credentials as an out-of-the-ordinary navigator, which means they must have known him better than modern historians do. If we resort to genealogy we shall see that:

1. Princess Catarina, Colón-Zarco's half-sister, was betrothed to Edward IV of England in 1463 but died the following year; and in 1491, Henry VII of England and John II of Portugal were allies.
2. Charles VIII of France was the son of Louis XI, who had been betrothed to the Holy Princess Joana of Portugal, first cousin to Colón-Zarco.
3. And the Duchess of Bourbon, Anne de Beaujeu, was the daughter of Louis XI and was Regent of France until her half-brother, Charles VIII became of age.
4. The Duchess's brother, the Count of Savoy, in turn, had been an intimate friend and brother-in-law of Charles (known as 'the Bold') of Burgundy, because this Prince was married to Anne's sister, Isabel.

5. Charles 'the Bold' was son of Philip (known as 'the Good') of Burgundy and Princess Isabel of Portugal, Colón-Zarco's great aunt; he was father of Mary of Burgundy, who married Emperor Maximilian of the Holy Roman Empire, who – as the son of Emperor Frederick III and Princess Leonor of Portugal, Colón-Zarco's aunt – was Colón-Zarco's first cousin.
6. The Duchess of Bourbon (sister-in-law of Charles 'the Bold') was a cousin of Columbus.

It can thus be understood why Bartolomeu Colón was so well received at the English and French Courts.

Colón-Zarco later wrote to the Catholic Monarchs:[6]

> In order to serve Your Highnesses I did not wish to come to an understanding with France, nor with England, nor with Portugal, and Your Highnesses have seen letters from the Kings of these countries from the hands of doctor Villabon.

In the letter that Colón-Zarco wrote to the Royal Council in 1500, when he returned from his third voyage from the Antilles in chains with his brothers, he said:

> Sirs, I have served these Monarchs in the enterprise of the Indies for the last 17 years . . . I have done so out of love and I told France, England and Portugal that those lands and seigniories were for my Lords the King and Queen. The promises [of the foreign Monarchs] were neither few nor false . . . [but I gave my Sovereigns] more than one thousand seven hundred islands besides Hispaniola, which is vaster than the whole of Spain.

A COINCIDENCE OF TOPONYMY

It is important to point out that Salvador Fernandes Zarco gave the name of St Salvador to the first island at which he anchored. It is plausible that he called it this after Christ (Saviour), but we now know that he baptised it with his own name: Salvador, the same as one of the two only parishes of Beja. It was usual to baptise a child with the patronal name of his parish. Later he translated it 'Christóvam' = Saviour.

To the second island he gave the name of Fernandina, and not Fernanda, apparently in honour of King Ferdinand (Fernando) of Aragon. But why would he name it after a King who had shown little or no interest in his undertaking, instead of honouring in first place Queen Isabella, who had supplied him with the means to make his voyage?

Colón-Zarco may not have been over-enamoured with his father, Prince Fernando, Duke of Beja, with whom he must have had very little contact, but he must have respected him as the Master of the Order of Christ. It therefore seems possible that he baptised the island with his father's name. However, Salvador Zarco was 'Fernandes', a surname that means 'son of Fernando'. Everything indicates that the name 'Fernandina' is more suited to Fernandes than to Fernando. This diminutive points to the fact that he named the first two islands after himself – Salvador Fernandes.

Only to the third island did he give the name 'Isabella' (and not 'Isabelina') which, although seemingly honouring the Queen of Spain, he named after his mother, Isabel da Câmara. He also baptised a settlement – today a town – on the coast of Puerto Rico with the name of 'Isabella'.

He named the fourth island 'Juana' (and not 'Juanina') in honor of his cousin and brother-in-law, King John II of Portugal. Colón-Zarco, however, gave the impression that he had named the fourth island after Prince Juan, the son of the Catholic Monarchs. But when he visited his King on his return to Europe, he certainly explained his true intentions to him. It was because the King thought that the name seemed too obvious and could raise suspicions that Colón-Zarco began writing 'Cuba' in his logbook instead of Juanina or Juana.

We now know that he was born at Cuba in the zone of Beja. But he took care to explain that the name was the one used by the local people. This was not exactly true. The local population, the Tainos, who belong to the Arawak linguistic group, called the land 'Cóba', which is much more resonant than Cuba. But Colón-Zarco wished it to remain like this for both geography and history. If necessary, he would be able to allege that metaphorically the bay resembled a 'cuba', which is a cask for wine in Spanish or a bed (*cubilis*, from *cubo*) in Latin. There was nowhere else in the world with that name.[7]

The name 'Cuba' in Portuguese comes from the Arab word *côba*, which means 'tower'. João Gonçalves Zarco's coat-of-arms bore a 'côba', which with the prefix of the Arab article 'al' (the) formed 'alcoba', meaning vaulted tower. This, in turn, became 'alcova', which also meant 'camara' (chamber or room), the name Zarco adopted on being raised to the nobility (see Figure 24.1).

More in fear that someone would suspect the true motive of the name Fernandina – there were suspicions of his true aims – rather than humiliate the King himself, especially after Colon had been rewarded, he renamed the island 'Ilha Longa' ('Long Island').

I have made a close study of the names that Colón-Zarco gave to the places he discovered as well as to those that were discovered by his navigators and inserted into his cartography. Many of them seem to be

24.1 The heraldic symbol of the tower.
  a The *côba*.
  b The castle.

named after villages that are to be found around Beja. Was it mere chance that Columbus gave Portuguese names translated into Spanish – and not always correctly – to these newly-discovered lands? Many are closely related to Salvador Fernandes Zarco, the Duchy of Beja, Madeira, Azores and Africa. As the Admiral of the Indies, he altered several place-names that had been given by his navigators.

- *Alpha* and *Omega* – names, which cannot be found anywhere in the world except on the eastern tip of Hispaniola, are Hebraic and palaeo-Christian symbols that were very common in the artistic representations of the Orders of the Temple and Christ and were the umbilical centre of Salvador Fernandes's cabalistic siglum.
- *Assumpcion* – sea channel (see Santarém Channel below).
- *Belém* (Bethlehem in English) – there is no Bélem anywhere in Spain or in any other part of the world apart from the Bethlehem in Palestine,[9] the Bélem that the Portuguese baptised in the sixteenth century in Brazil and the Bélem in Panama. The two in the Americas had the same origin – the beach and the Monastery of Jeronimos at Bélem in Lisbon, from where many navigators set sail on their voyages of discoveries, among them Vasco da Gama.

It is not surprising that the Catholic seamen named so many places after the saints they worshipped and venerated. Some of these names, however, deserve attention.

- *Buena Vista* – a settlement in Cuba and the Island of Boa Vista in the Archipelago of Cape Verde.
- *Brasil* – a port on the southern coast of Hispaniola. One of the islands of Azores was called Brazil before they were 'refound' and the name

was later given to the land that was officially 'rediscovered' by Pedro Álvares Cabral in 1500. The Portuguese had been there many years before, and Zarco knew this well.
- *Concepcion* – the name of the bay where Colón-Zarco anchored in Hispaniola. There is a place with this name (*Conceiçao*) in the *Duchy of Beja* and Columbus's stepmother is buried in the Church of St Mary of the *Conception* (in Beja) with four of her children who died in infancy. Colón had the Church of Concepcion built in San Domingo so that he could be burried there. There is also a place in Madeira called Conceição.

No Italian word appears in any of the place-names adopted. According to the thesis that defends a non-Portuguese Columbus he spent more time in Spain than in Portugal and travelled with Castilians, whose language he was obliged to speak.
- *Faro* – near Barraquilla, on the Atlantic coast of Colombia, is a place called Punta Faro, which must not be confused with Punta del Farol (Morant) in Eastern Jamaica, which was named long after Columbus's fourth voyage. Faro is the capital of the Algarve and there is a place of this name near Columbus's birthplace, Cuba. The Genoese did their utmost (as can be seen by the colour prints in the *Raccolta*, which are covered by tracing-paper on which the lighthouses of Genoa and Savona are outlined in black) to emphasise the coincidence. Yet Colón-Zarco did not call that Colombian point Punta del Faro or Punta del Farol, but purely and simply 'Faro'. There has never been a lighthouse in Faro on the Algarve coast and there certainly has never been one at the other Faro in the middle of the Alentejo plains (Duchy of Beja).

Faro comes from the Arabic name 'Harum'. The Emir Ibn Harum built new walls around the present capital of the Algarve and as there was an image of the Virgin on the battlements the Christian Mozarabs called the 'town' 'Santa Maria de Harum'.

Since there is no aspirated 'H' in Portuguese, however, *Harum* became *Farum*. In his *Cronica do Conde Dom Duarte de Menezes*, the chronicler Azurara spelt it *Faraão* and Alão de Morais wrote *Farão* in his *'Pedatura'*, under the title of 'Monizes' (p. 706).

In Castilian, *Harum* produced Haro, Haran, Harana and Araña (spider). The name has nothing to do with this arachnid, even though the Aranas's coat-of-arms bore fleur-de-lys with a shield in the centre that contained a border with three spiders.
- *Galera* – there are two 'Punta de Galera' in Columbian toponymy. '*Galera*' is 'galley' in English and was a ship that Colón-Zarco never used for oceanic navigation. On his first Atlantic crossing he rigged his ships in the Portuguese manner and transformed them into a type of caravel. But he had sailed the Guinea coast, where the Portuguese had named a prominent rock that jutted out into the sea 'Ponta da

Gale' (Galley Point). In Portuguese *'gale'* and *'galera'* are synonyms; and they have given the name 'Ponta da Galera' to a cape on the Island of São Miguel in the Azores, whose Captain-Governor was Colón-Zarco's uncle, Rui Gonçalves da Câmara (see p. 445, Tree XII brother of 43). There are two 'Punta da Galera', one on the Atlantic coast of Colombia, one to the north of Cartagena, and another in the northeast of Trinidad.

- *Graciosa* – an island in the Bahamas and another in the Azores.
- *Guadalupe* (Guadelupe) – island and port. The Spanish can, of course, rejoice that they have the highly-venerated Our Lady of Guadelupe. It happens, however, that the Arab word *'guade'* (river) also exists in Portuguese toponymy. Serra de Guadalupe is the name of a mountain range in Portugal. There was, and still is, a hermitage dedicated to Our Lady of Guadelupe on the Island of Terceira in the Azores and there is another hermitage of the same name at Serpa, near Beja, 1km from Monte Guadalupe, which has been a place of pilgrimage for the local people since the fourteenth century. Salvador Zarco certainly visited the hermitage in his childhood. And there is still another fourteenth century Hermitage of Guadalupe near Sagres (today Vila do Bispo) where the sailors of Prince Henry 'the Navigator' used to pray.
- *Guinchos* – an island near Cuba. In Castilian, *'guincho'* means 'pointed stick' or 'hook',[10] and Spanish dictionaries indicate that *'guincho'* is a bird of the falcon family in Cuba. The word is certainly Portuguese; there is a beach fifteen miles from Lisbon called Guincho and the fort that stands there (it is a hotel today) has a swift as its heraldic shield. *'Guincho'* in Portuguese is the shriek that is emitted by that bird. There is also an islet by the name of Guincho in front of Madeira island, where Colón-Zarco lived for some time. It seems certain that when Colón-Zarco saw those birds in the Bahamas he gave the island the Portuguese name of their cries that rent the air.
- *Puerto Santo* (the port of Barracoa in Cuba) – Porto Santo was a place between Alvito and Grândola in the former Duchy of Beja which later became known as Porto d'el-Rei. Porto Santo is also one of the islands of the Archipelago of Madeira and was the birthplace of Colón-Zarco's wife Filipa Moniz Perestrelo.
- *Rojo* – a cape in Porto Rico and also the name of the first cape the Portuguese saw on the Guinea Coast, Cape Roxo, where Colón-Zarco claimed to have navigated.
- *San Antonio* – a name given to several places in the Archipelago of the Antilles, among which is the westernmost cape of Cuba. St Anthony was a Portuguese thaumaturge and considered to be one of the leading figures of his time. There are many places in Continental Portugal and the Archipelagos of the Azores and Madeira with his

name, including the easternmost point of the Algarve. It is also the name of the church at Serpa, in the Duchy of Beja, that Colón-Zarco certainly visited in his youth during the pilgrimages to the nearby hermitage of Our Lady of Guadeloupe. (The Italians always mystify the tourists saying that the saint was from Padua.)

When Colón-Zarco was returning to Lisbon after being received by King John II at Azambuja following his first voyage to the Antilles, he went to the Convent of St Anthony to kiss the hand of his half-sister, Queen Leonor.

- *San Bartolomé* – another common name in the Antilles. This could have been in honour of Colón-Zarco's brother, but there are two places of this name in the Duchy of Beja, one near Viana do Alentejo and another near Santiago.
- *Sanctus Spiritus* – to the northeast of Trindad in Cuba. Like all the members of the Order of Christ, Colón-Zarco was a devotee of the Holy Ghost. There is a village by the name of Espirito Santo in the Duchy of Beja, to the south of Mértola.
- *San Jorge* – St George was always the patron of Portuguese knights and it is the name of the castle that dominates Lisbon, an island of the Azores, a city in the Cape Verde Islands and a port of the island of Madeira. In the frequent struggles between Portugal and Castile in the Middle Ages, the Portuguese went into battle under the protection of St George, while their adversaries invoked St James. In the *Raccolta*, the Genoese used a colour print, as they did with their lighthouses, to prove that Genoa had a St George's Bank. This is common knowledge, but it is doubtful that Colón-Zarco was interested in advertising a bank. He obviously used the toponym St George because he was Portuguese.
- *San Juan* – a name found in Puerto Rico and several places in the Antilles and very common in Spain. But there is also a São João near Colón-Zarco's birthplace, Cuba.
- *San Luis* – a place in Cuba and to the north of Odemira in the Duchy of Beja.
- *San Miguel* – a cape on the island of Hispaniola. It is the name of one of the main islands of the Azores and is to be found on the coast of the Algarve. There is a village of this name near Mértola in the Duchy of Beja.
- *San Nicolas* – a Caribbean channel (it was spelt S. Nicoli on Las Casas's map) and a cape on Hispaniola. São Nicolau is a parish of Lisbon and the name was given to one of the islands of the Archipelago of Cape Verde (see Santarém Channel below).
- *Santa Catarina* – a common name in Columbian toponymy. It is a place in Curaçao. It is also to be found on the coasts of the Algarve, the Azores, Madeira and Mina in Africa, where Colón-Zarco claimed

to have sailed with Bartolomeu Dias. It is the name of a village near Alcácer do Sal in the Duchy of Beja. This saint was particularly venerated by Queen Isabel (wife of King Alphonse V) and her sister-in-law, Princess Catarina (sister of Prince Fernando, Duke of Beja), who were aunts of Salvador Fernandes Zarco. The Portuguese raised the standard of the 'Wheel of Knives', emblem of St Catherine, in battle during the fifteenth century.

- *Santa Clara* – a place in Cuba, as well as in Madeira and about 1km from Beja. The Convent of Beja belonged to the nuns of St Clare.
- *Santa Cruz* – a name left everywhere by the Portuguese, such as Madeira and the Azores (on the islands of Flores and Graciosa). There is a village of this name near Santiago in the Duchy of Beja. The first name of Brazil was 'Terra de Santa Cruz'.
- *Santa Lucia* (or Luzia) – a Caribbean island. This name is to be found on the coast of the Algarve and in the Archipelagos of the Azores and Cape Verde, as well as near Ourique in the Duchy of Beja.
- *Sant'Ana* – near the port of Guadelupe. The name is to be found on the island of Madeira and in the Duchy of Beja between Portel and Vidigueira (the title that Vasco da Gama took when he was made a Count after discovering the seaway to India), as well as near the Monastery of Arrábida near Setúbal.
- *Santarém* – Columbus's cartography includes a channel between the Bahamas and Cuba that was explored by his navigators. To the south is the Channel of St Nicholas and to the southeast, as a continuation of the two, the Channel of Assumption, later designated by Portuguese cartographers as the 'Old Channel of the Bahamas'. I do not know of any other place in the world, whether it be town, port or cape, that has the name of St Irene or, even less, Santarém – apart from this channel and a town in Brazil. But it is the name of one of Portugal's most important cities. It was conquered from the Moors by the first King of Portugal, Alphonse Henriques, who gave the command of the garrison to Mem Moniz. At the time of the Roman occupation of the Iberian Peninsula, Lusitania was divided into three 'provinces' – *Scalabis* (Santarém), *Pax Julia* (Beja) and *Caceres*, which became Spanish.

When Colón-Zarco dropped anchor in the River Tagus at the end of his first voyage to the Antilles, he had to go to the Vale do Paraíso estate near Azambuja in order to be received by King John II. He then journeyed another 30 km to Santarém to spend the night at the palace of Prior do Crato, Dom João de Menezes. Chapter 29, which includes the genealogical trees of the families directly related to Colón-Zarco, will give the reader a clearer idea of the relationship that linked the host to his guest (see pp. 393 and 399 and Trees IV, V, VI, and IX, 42).

The second wife of King Alphonse V, an aunt by marriage of Colón-Zarco, lived in Santarém, after being widowed, from 1478 to

1530, having professed in the Convent of St Clare in 1480. Pedro Álvares Cabral, the 'official discoverer' of Brazil, is buried in Santarém. This dissertation on Santarém is to emphasise the names given to the three channels between the Bahamas and Cuba. The city of Santarém had two parishes at that time – St Nicholas and Our Lady of the Assumption.

- *Santiago* – After the tragic death of Dom Diogo, the Duke of Beja and Master of the Order of Christ, King John II 'inherited' not only this Order but also the Order of Santiago (St James). Santiago was the name of a town of the Duchy of Beja and of one of two only parishes of the town (the other was São Salvador). And one of the islands of Cape Verde, the discovery of which I have already related, was also called Santiago.
- *Santo Domingo* – a place on the island of Hispaniola. It was also the name of a cape on the Guinea coast (São Domingos) where Colón-Zarco had sailed, besides two villages in the Duchy of Beja, one to the northeast of Mértola and another to the southeast of Santiago, which is also the name of a port on the island of Cuba.
- *Trinidad* – a Caribbean island and a town in Cuba, to the south of Sanctus Spiritus. There is a Trindade four miles from the city of Beja. Although such hagiological toponymy is common to both Portuguese and Spanish geography and to nearly all the Christian peoples of the Romance languages, I am not demoralised by the coincidences, especially when I see the large number of place-names from the Duchy of Beja scattered all over the islands and the South American coast discovered by Salvador Fernandes Zarco/Cristóbal Colón. I should also like to point out another coincidence.
- *San Vicente* – a place on the island of San Salvador. St Vincent was one of the most venerated saints in Portugal and is the patron saint of Lisbon. Born in Spain at the end of the third century, he was tortured to death on the orders of the Emperor Diocletian and his body abandoned to the beasts and birds of prey.

According to legend, however, a raven protected his body from the beasts. On learning of this, his torturers decided to throw his body into the sea with a millstone tied round his neck. But the cords became loosened and it was washed up on to the beach, where it was found by some Christians and hidden. He was much later translated to the Basilica of Valencia. Up to this point, his martyrdom is related to Hispania under the Roman occupation. From this point on, his story is a part of Portuguese history. When Valencia was taken by the Saracens four hundred years later, some Christians took the Saint's remains to the Promontorio Sacro.[11] In the eleventh century, the Muslims allowed the Christians to build a chapel on the cape to keep the saint's mortal remains. According to the Arab historian and geographer Edrisi, there were always nine ravens on the roof of the chapel and the Muslims called it 'The Chapel of the Ravens'. The priests claimed that the ravens were direct descendants of the one that had defended the body after the martyr's death. In about 1170, King Afonso I sent a ship to have the saint's remains transferred to Lisbon and had the Church of St Vincent built to house them, the first church he had constructed in the city.[12] Even today, ravens are known as *'vicentes'* in Lisbon. It is not surprising, therefore, that Salvador Zarco gave the name of San Vicente to several places in the Antilles, the first of them on San Salvador.

One of the Cape Verde islands and villages on Madeira and in the Duchy of Beja, near Alvito and Terra Ruiva, have this name. In the fifteenth century Cuba had only one parish: São Vicente.

- *Vale do Paraíso* the name of the place where Colón was received by King John II.

Other names in the Antilles – Santo Anton, San Martin, Concepcion, Pinos, San Gonzalo, Punta del Sol, San Lorenzo and Puerto de La Cruz – have their equivalents on Madeira.

Much more could be written on this subject, but I should like to point out that Salvador Zarco called his son Fernando after his grandfather, the Duke of Beja. It was an opportune 'coincidence' that the King of Castile had the same name. Figure 24.2a and b show place names in the Beja region and in the Indies.

ZARCO IN MADEIRA

Salvador Zarco's stays on the Island of Madeira also deserve consideration. He married Filipa Moniz Perestrelo in 1479 or 1480 and, according to Las Casas, his son Diogo was born in 1481. The *Historia del Almirante* relates that Colón-Zarco's mother-in-law had given him 'the writings and sea charts that had belonged to her husband'; I found this strange, since it shows a complete disrespect for the 'rule of secrecy' by Perestrelo in leaving the documents to his widow. But the text has been amended by Fornari.[13] All the names of the Portuguese navigators had been corrupted ('Pero Vasques' transformed into 'Pietro Velasco', for instance), and other mistakes made.

Notes and References

1. Las Casas (1875–6), Book I, Chap. XXV.
2. Colón, Don Fernando (1985), Book I, Chap. 2.
3. The name of Lisbon derives from the old Phoenician *Alis* (pleasant) plus *Ubo* (bay), but when translated into Latin, *Ulixbona*, the word cannot be divided.
4. Zúquete (1961), Vol. I, p. 312.
5. This name, meaning 'mount', is given in Alentejo to the house of a big landowner. It stands in the middle of the property, surrounded by the houses of the farm labourers.
6. Cólon, Don Fernando (1985), Chap. XIII, final para.
7. It was only in the seventeenth century that the name of Cuba was given to a village in the interior of Mexico (today a town in New Mexico, USA) and in the nineteenth century to a small city in New York State.
8. Meaning a house like the Town Hall, and a large room.
9. A town in the Union of South Africa was called *Bethlehem* in the eighteenth century, and there are a few *Belems* in Latin America.
10. Owing to the influence of Portugal's British allies at the time of the Napoleonic invasions in the nineteenth century, the word 'winch' (used for lifting cannon and other heavy equipment) also gave rise to the word '*guincho*' in Portuguese. But '*guincho*' (or the cry of the swift) has been used in Portuguese since the Middle Ages.
11. '*Promontorium Sacrum*' was the Graeco–Roman name for the present-day Cape St Vincent, or Sagres, where Prince Henry set up his School of navigation. According to Guyango's Castilian version of the *Cronica del Moro Rasis* (Arrazi), the legend was confirmed by Abolacine, a knight of Fez (Morocco), who had been at Sagres at that time.
12. When the body of St Vincent was translated to Lisbon, the ship was escorted the whole way by ravens. The emblem of Lisbon is a '*caravela*', with ravens on the prow and the stem; see Virgilio Correia, *obras*, Vol. II, p. 221.
13. Colón, Don Fernando (1985) pp. 60–2 and 183.

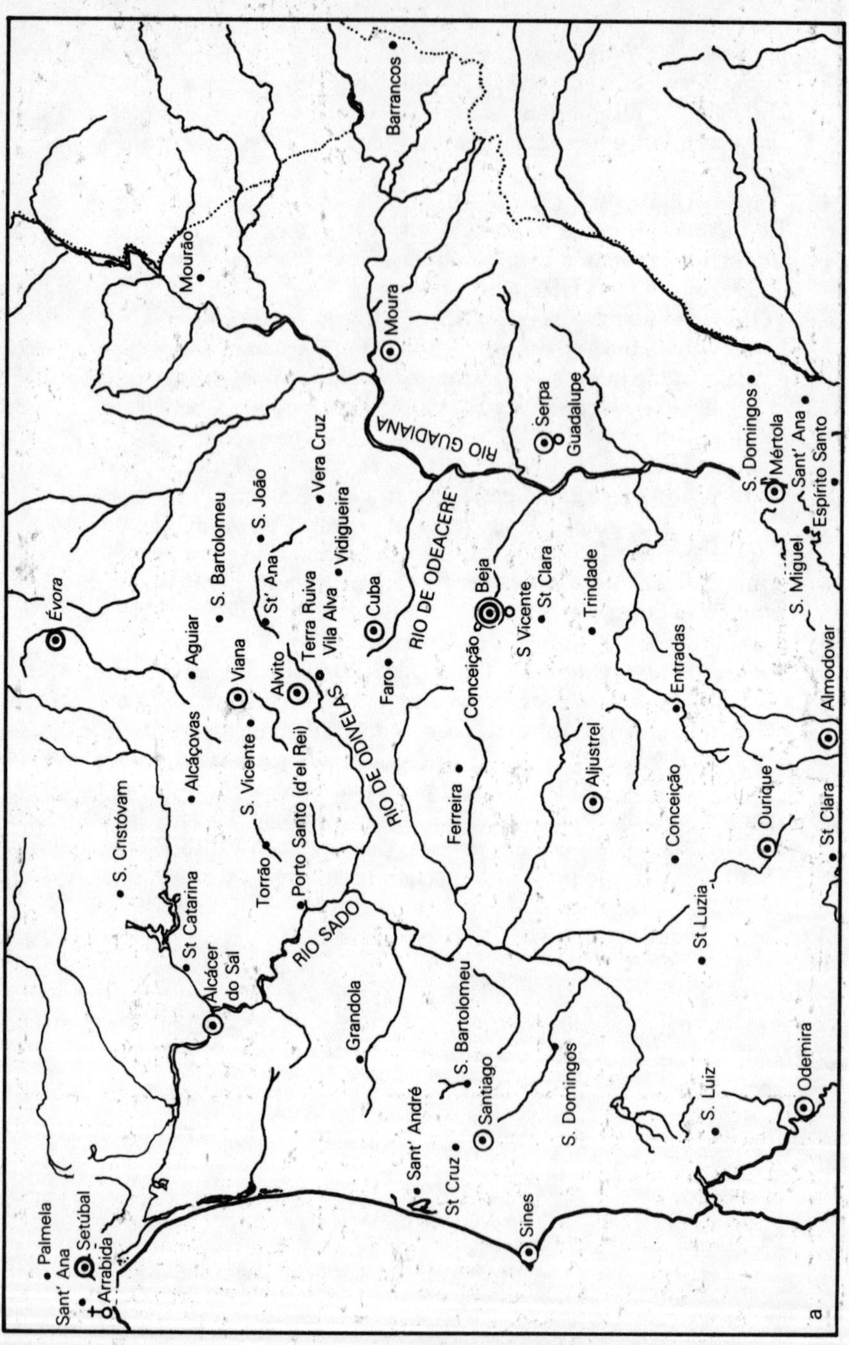
Palmela
Sant' Ana
Setúbal
Arrabida
S. Cristóvam
St Catarina
Alcácer do Sal
Alcáçovas
Torrão
S. Vicente
Porto Santo (d'el Rei)
RIO SADO
Grandola
Sant' André
St Cruz
Sines
S. Bartolomeu
Santiago
S. Domingos
S. Luiz
Odemira
St Luzia
Conceição
Ourique
St Clara
Almodovar
Entradas
Aljustrel
Ferreira
RIO DE ODIVELAS
Conceição
Faro
Cuba
Alvito
Terra Ruiva
Vila Alva
Viana
Aguiar
S. Bartolomeu
St' Ana
S. João
Vera Cruz
Vidigueira
Évora
Mourão
Moura
Barrancos
RIO GUADIANA
Serpa
Guadalupe
RIO DE ODEACERE
Beja
S Vicente
St Clara
Trindade
S. Miguel
Espírito Santo
Sant' Ana
Mértola
S. Domingos
a

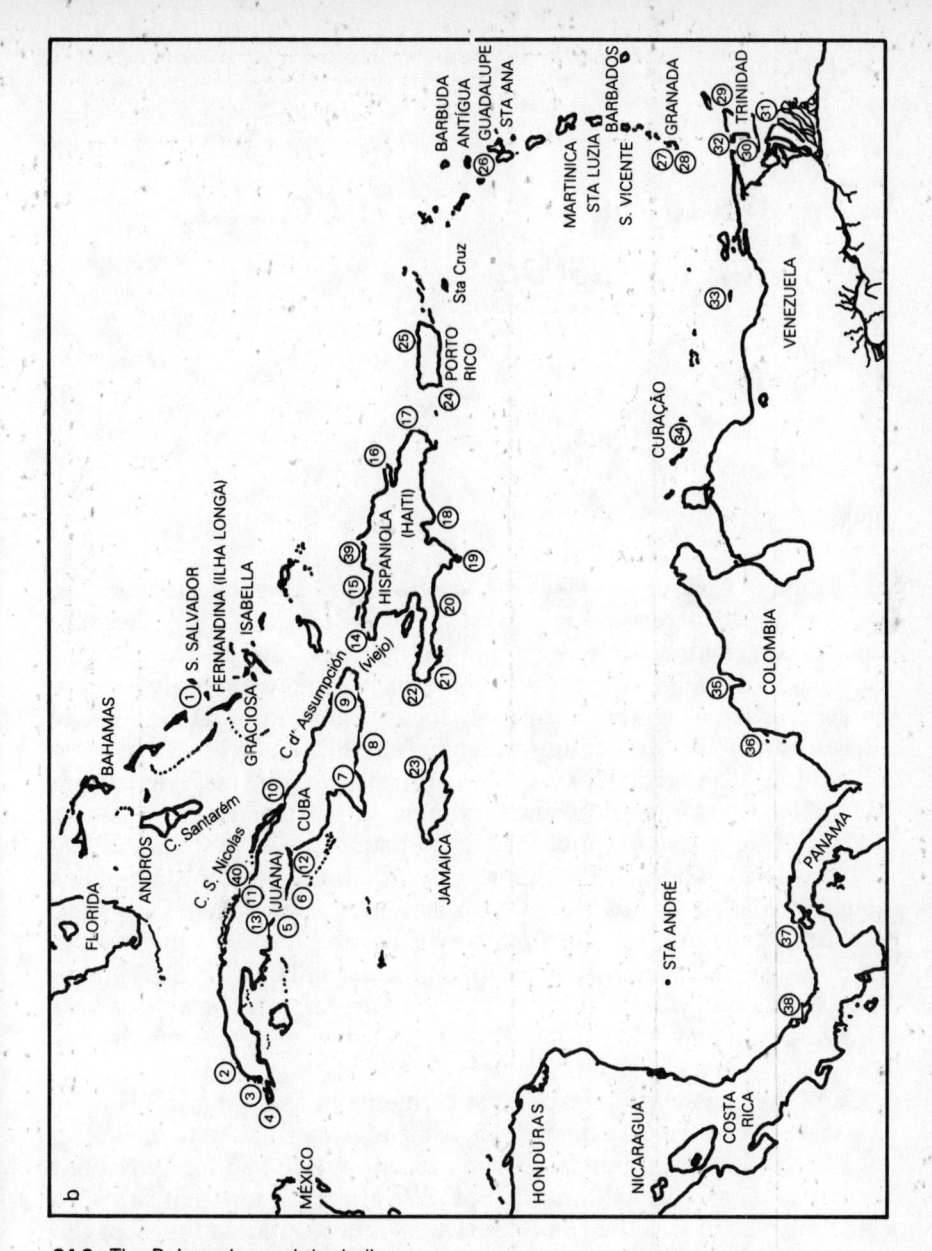

24.2 The Beja region and the Indies.
  a Place names in Beja.
  b Place names in the Indies: 1 S. Vicente; 2 Sta Luzia; 3 Guadiana; 4 Ponta de Santo António; 5 S. João Baptista; 6 Sta Cruz; 7 S. Luis; 8 Santiago; 9 Porto Santo; 10 Mourão; 11 Isabel; 12 Sanctus Spiritus; 13 Sta Clara; 14 S. Nicolau; 15 Conceição; 16 Cabo de S. João; 17 Cabo Alfa; 18 S. Domingos; 19 Cabo Roxo; 20 Brasil; 21 S. Miguel; 22 Cabo Ómega; 23 Sto António; 24 Cabo Roxo; 25 S. João Baptista; 26 S. João; 27 Sta Catarina; 28 S. Jorge; 29 Ponta da Galera; 30 S. Bernado; 31 Boca das Serpentes; 32 Boca do Dragão; 33 Margarita; 34 Sta Catarina; 35 Ponta Faro; 36 Ponta da Galera; 37 Belém; 38 Boca de Touro; 39 Cabo Isabel; 40 Ilha dos Guinchos.

25 A House on Madeira and a Convent in Lisbon

'ESMERALDO'

In relation to Columbus's stays in Madeira, a new personage appears on the scene – João Esmeraldo – who has been inexplicably ignored by Columbian historians. It is on record that João Esmeraldo came to Portugal in the reign of King Manuel I, but we know that only because he asked that King for permission to use the arms that he already possessed. He received authorisation in May 1520. João Esmeraldo was a lord of Flemish origin. He was born at Artois and was descended from the D'Alvargne families of Fimes and Noduchel. The studies by Álvaro Rodrigues de Azevedo (published in a bulletin issued by Funchal City Hall archives), who was the first to give details about João Esmeraldo's house in Madeira, and Afonso Dornelas have shown that Esmeraldo may have gone to Madeira in 1480, which probably means that he had been in Lisbon before this.[1] He had his house enlarged in 1494, there being a window with that date on it, and four years later he baptised his son with the name of Cristóvam Esmeraldo, which suggests that the boy was named after Columbus.

One other interesting detail must be mentioned. Owing to his deep devotion to the Holy Ghost, João Esmeraldo gave that name to an Entailment that King John II granted him and which had formerly been known as 'Lombada da Porta do Sul'. We already know that Salvador Fernandes Zarco was the son of the Duke of Beja and had been born in Cuba. What link did he have with João Esmeraldo that would allow him to lodge in the latter's house?

The surname 'Esmeraldo' clearly has an esoteric origin, from the word 'Emerald'. There is nothing Flemish about it and it is not a first name of baptism. It sounds like an 'alchemic-astrological' surname, similar to that of Pallasteli (the blanket of stars) that gave rise to the Perestrelos (see p. 255). And what was the relationship between Esmeraldo and Duarte Pacheco Pereira, who entitled his work *Esmeraldo De Situ Orbis*, and Valentim Fernandes, who inserted the inscription 'with Salvador of Madeira' in the same work in a Hebraic–gematric anagram?

This mystery led me to explore some earlier history (see Figure 25.1). Princess Isabel, daughter of King John I and sister of King Duarte (grandfather of Salvador Zarco), married Philip (known as 'the Good'), Duke of Burgundy and Count of Flanders.[2] An embassy arrived in Lisbon in the autumn of 1428 to arrange the marriage and stayed there until May 1429 to attend the wedding festivities. The wedding was held by proxy in Lisbon in June 1429 and was attended by many Flemish lords, who came in a fleet of thirty-nine ships in order to escort the bride to her husband.

Three sons were born of this marriage – Anthony and Joseph, who both died in infancy, and Charles (known as 'the Bold'), one of the most notable princes of his day. He married Princess Isabel of Bourbon and begot Mary, who married Emperor Maximilian I of the Holy Roman Empire (see p. 444, Tree XI). Maximilian's adviser for Portuguese affairs was the Konrad Peutinger, a friend of Valentim Fernandes. It was Peutinger who signed the *Imperial Letters Patent* which proved that João Esmeraldo had a coat-of-arms of Brabant so that King Manuel I could confirm his Portuguese arms (see p. 463).

A brief glance at all these links is enough to see how hard it would have been for a Genoese wool-carder to maintain such intimate relations with João Esmeraldo that he could have been a guest in his house (as proved) before becoming the illustrious Admiral of the Indies. Salvador Fernandes Zarco, on the other hand, would have been more than welcome in Esmeraldo's house, since the Emperor that had signed the latter's *Letters Patent* was his cousin.

THE CONVENT OF SANTOS (SAINTS)

People got married very early in life in those days, sometimes before the age of fourteen. Life expectancy was little more than twenty-five years though some people did reach the age of ninety, like Dom Pedro de Sousa, first Count of Prado (see Tree III, 29, p. 366), who took Joana de Mello as his third wife; or even 100, like Dom Henrique Henriques, Commander of São Salvador and Lord of Alcáçovas (see Trees XII and XIII, 52, pp. 445 and 447) great-grandson of João Gonçalves Zarco.

The bride and groom were usually from families that had close links and always of the same social stratum. Extra-marital unions and the subsequent illegitimate offspring were common and accepted as normal.

Some defenders of the Portuguese thesis have believed that the mother of 'Colón-Zarco' or 'Colmo-Palha' ran away to Genoa to have her child, thus explaining how Columbus came to be Genoese, but they

CHECK AND MATE

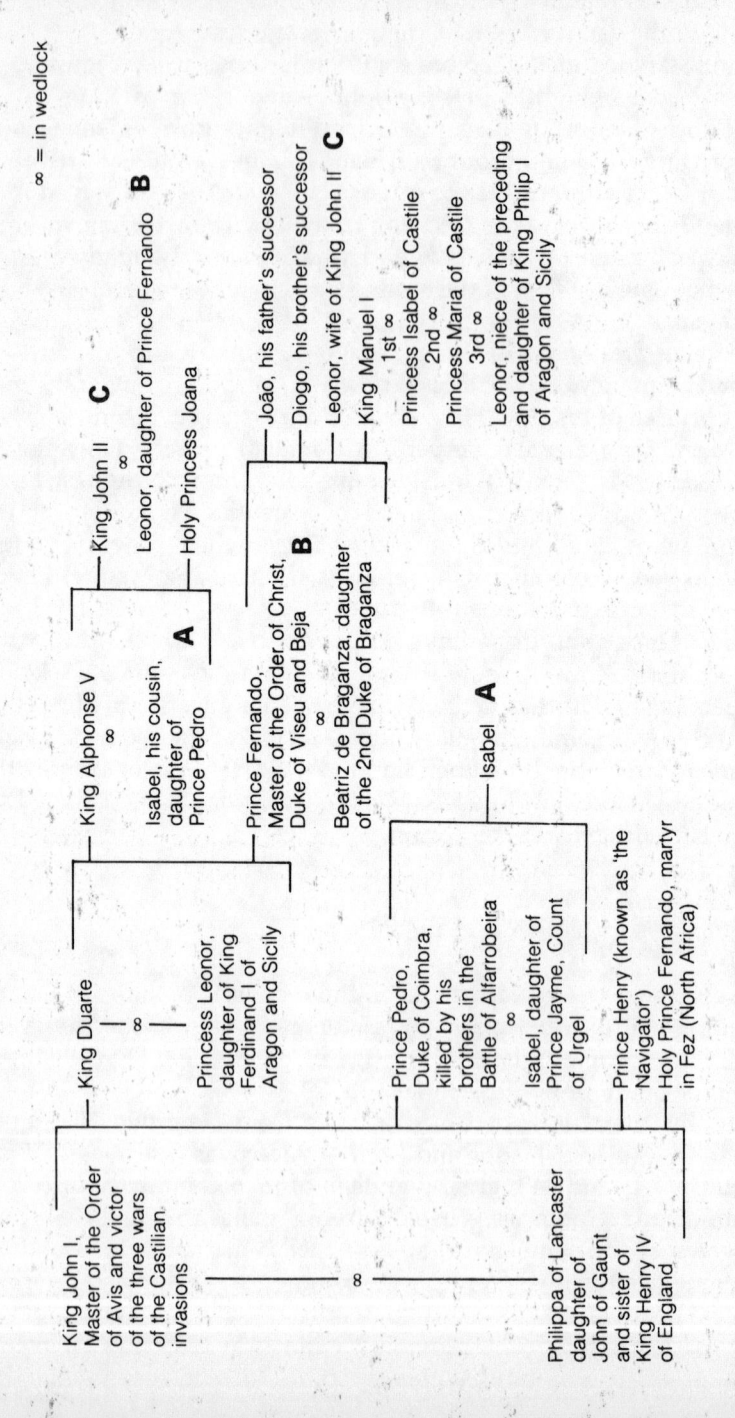

∞ = in wedlock

King John I, Master of the Order of Avis and victor of the three wars of the Castilian invasions
∞
Philippa of Lancaster, daughter of John of Gaunt and sister of King Henry IV of England

├── King Duarte ∞ Princess Leonor, daughter of King Ferdinand I of Aragon and Sicily
│   ├── King Alphonse V ∞ Isabel, his cousin, daughter of Prince Pedro **A**
│   ├── King John II ∞ Leonor, daughter of Prince Fernando **B**
│   └── Holy Princess Joana
├── Prince Pedro, Duke of Coimbra, killed by his brothers in the Battle of Alfarrobeira
│   └── Isabel, daughter of Prince Jayme, Count of Urgel
│   └── Isabel **A**
├── Prince Henry (known as 'the Navigator')
├── Holy Prince Fernando, martyr in Fez (North Africa)
└── Prince Fernando, Master of the Order of Christ, Duke of Viseu and Beja **B**
      ∞ Beatriz de Braganza, daughter of the 2nd Duke of Braganza
      ├── João, his father's successor
      ├── Diogo, his brother's successor
      └── Leonor, wife of King John II
         King Manuel I **C**
           1st ∞ Princess Isabel of Castile
           2nd ∞ Princess Maria of Castile
           3rd ∞ Leonor, niece of the preceding and daughter of King Philip I of Aragon and Sicily

25.1 The most notable descendants of King John I of Portugal.

were not aware how suspect the Italian documents are.

Young girls who gave birth to offspring of an affair could still easily arrange a legitimate marriage, above all if they had rich and influential parents who could supply them with a dowry. If this were not the case, they could retire to a convent. Some, after terminating their noviciate, sought ecclesiastical authorisation to return to temporal life and marry. Were there young girls in this situation in the Convent of Santos (of the Order of St James) to which the young Filipa Moniz Perestrelo retired, not as a novice but as a paying boarder in order to refine her education? There is no documentary proof whatsoever, but I have mentioned the fact as some Genoese supporters have referred to it, among them Heers,[3] who did not cite any of his sources.

The Mother Superior of the convent of Santos,[4] 'Madre' Violante Nogueira, was the daughter of the Governor of Lisbon, Afonso Nogueira, and Constanca Vaz de Almada (see pp. 375 and 386), and was considered to be a 'nun of the highest moral virtue'. (Some writers have made the mistake of confusing her with Ana de Mendoça, daughter of Dom Nuno Furtado de Mendoça, Alphonse V's Lord Chamberlain of the royal Household, who was a niece of 'Madre' Violante.)

At the age of sixteen Ana gave birth to the bastard son of King John II (Duke of Coimbra and Master of the Order of St James (see page 372). The King had the Convent of Santos-o-Novo (the New) built for her in 1492 and made her its beneficiary. He gave her son into the care of his sister, the Holy Princess Joana, who was a cousin of Colón-Zarco. It is interesting to note that the affair that King John II began with Ana when she was fifteen was encouraged by Dom Diogo de Almeida, Prior of Crato, who was the grandson of the Archbishop of Lisbon, Dom Pedro de Noronha, and Isabel (or Branca) Moniz Perestrelo, sister-in-law of Colón-Zarco.

A son of Dom Diogo, Dom Estêvão de Almeida, who went to the service of Empress Isabel, wife of Charles V, was Bishop of Astorga and Cartagena (Spain) and was one of the patrons of Fernando Colon.

We must not forget that the Empress Isabel, known as 'Princess of Portugal', was grand-daughter of Maximilian who was married to Princess Maria de Borgonha, both cousins of Colón-Zarco, and granddaughter of Queen Isabel (known as 'The Catholic'). And this Queen was granddaughter of Princess Isabel of Bragança, sister of King Duarte of Portugal (grand-father of Colón-Zarco). There is no news, therefore, of any immoral boarders at the Convent of Santos-o-Velho (the Old) and Ana became Mother Superior of Santos-o-Novo (the New) only 14 years after Filipa Moniz Perestrelo had left to marry Salvador Fernandez Zarco (see p. 491).

And that is all as far as Cristóforo Colombo, the Genoese, wool-carding, amorous wooer is concerned.

WAS FILIPA PROMISED IN MARRIAGE TO COLÓN-ZARCO?

Many matrimonial contracts were signed while the future bride and groom were still in their childhood. Had Salvador Fernandez Zarco been promised to Filipa Moniz, who entered the convent of Santos-o-Velho to wait his return from his expeditions to Africa and other places? It is also said that Filipa Moniz Perestrelo had family in the Azores. Such was the lack of high-class ladies in those recently peopled islands that those that lived there sometimes married more than once. The high-class colonists had brides sent out from continental Portugal or even the Canaries or took a black slave as a concubine. For instance, Rui Gonçalves da Câmara (Colón-Zarco's uncle), second Donatary-Captain of the Island of São Miguel and successor of João Soares de Albergaria (see pp. 392 and 445, Tree XII), who after his wife's death (Maria de Bettencourt) lived with a native of the Canaries, Elvira Gonçalves, and begot a daughter, Beatriz da Camara. This half-caste lady married Dom Francisco Coutinho, son of Dom Pedro de Albuquerque (see p. 420).

On the death of Bartolomeu Perestrelo (see p. 256, Trees I and V, 3), Isabel Moniz had no need to leave Madeira. On the contrary, she could have stayed there to marry off her daughters. It must be remembered that many people of the higher classes left the continent to settle in the Archipelagos. Even foreign nobility with coats-of-arms to prove their title requested royal authorisation to settle there and so get rich.

So if the widow of Bartolomeu Perestrelo and sister of the notable Monizes left Madeira to live in Lisbon, boarding her daughter Filipa in the Convent of Santos, foregoing the protection of the family, giving up her properties and disdaining an excellent opportunity of arranging a good marriage for her daughter, she did not do so owing to a sense of adventure. She rented the Captaincy of Porto Santo in 1458 for an annuity which (as I showed in detail in the Portuguese edition of this book) was the equivalent of a general's pay today. So we may conclude that (contrary to what is expounded in the Genoese thesis) Isabel Moniz was far from being poor. The annuity realised on the lease of the Captaincy of Porto Santo would have been a substantial dowry for Filipa Moniz, who must have been three or four years old at the time. But she had two younger sisters, Isabel and Catarina, the latter being born in the year her father died.

In the deed of the lease of the Captaincy of Porto Santo to Pero Correa[5] it states that 'Isabel Moniz, her mother, and Diogo Moniz, her brother, tutors of the said Bartolomeu Perestrelo . . . were present' (see p. 256 and Tree I, B, p. 362).

'Diogo Moniz, her brother' means Isabel's brother and, as administrators of Bartolomeu Perestrelo's estate, his mother and brother took responsibility for the sale, the deeds of which were signed in 1458.

Finding herself a widow with five young children, one of which was still being breast-fed, Isabel Moniz had no time to devote to the administration of the Captaincy. The presence of Bartolomeu's uncle and tutor, Diogo Moniz, suggests that the annuity was not only valid for the lifetime of Isabel, but until the coming of age of her son and heir of the Captaincy. Isabel Moniz remained on the Island of Porto Santo for twenty years after the lease of the Captaincy. She moved to Lisbon with her daughters Filipa, Isabel and Catarina after the marriage of her son Bartolomeu to Aldonça Delgado, a grand-daughter of João Gonçalves Zarco. And her eldest daughter, Violante (or Brigulada), who married Miguel Moliarte, moved to Huelva only in 1484, at the time of Colón-Zarco's arrival in the same town – when the future Admiral was still a 'poor coloured prints seller'.

Isabel Moniz had rented a precious asset and had moved from the place where she had been born and where she had friends and family to lose herself in a big city where life was more expensive and nobody knew her. Why? The voyage between Madeira and Lisbon was long and uncomfortable in the small carracks and caravels of the time.[6] Why had she brought her marriageable daughters to the capital of the kingdom, where there were far more women than men due to the constant warfare? And what whim led her to board Filipa in a convent while her two other daughters stayed with her? Did Isabel and Catarina not deserve the same education? I think that Filipa Moniz Perestrelo was already promised in marriage to Salvador Fernandez Zarco. She was twenty-tree or twenty-four years of age and he had already undergone the metamorphosis into 'Christóvam Colom' (see p. 361). Was it King John II who used his influence so that the fiancés should meet in Lisbon? A 'secret agent' who was married would raise less suspicion than an adventurer in the pay of a third party. It would be better if nobody could recall the Salvador Fernandes who had served his apprenticeship as a navigator. After many voyages, however, he turns up again in 1477, an adult and married. He remained in Lisbon for a year or so and only following the birth of his son Diogo did he venture to return to Madeira. He was not now the young Salvador Fernandes, but a grown man by the name of Cristóvam Colom who had met and married Filipa Moniz in Lisbon. The only people who knew him now as Salvador were his 'brothers' of the Order of Christ who had sailed with him, but they had taken an oath of secrecy and would not denounce his new identity.

Among those who were in on the secret were Diogo and Bartolomeu Colón, but they would appear on the scene only much later. They were his 'brothers' in the Order of Christ, besides being his step-brothers, as I shall demonstrate later. Once in Madeira, Colón-Zarco lived in the house of João Esmeraldo, far from any township.

As far as Filipa's sisters are concerned, Isabel became the mistress of

Dom Pedro de Noronha, Archbishop of Lisbon, Catarina married Ayres Annes of Beja, King John II's private secretary, and Violante married the German Miguel Moliarte, as already mentioned.

Notes and References

1. See Zagallo (1945); Visconde de Porto da Cruz (1936).
2. To celebrate this occasion, the Duke of Burgundy and Count of Flanders, Philip III, instituted the *Order of the Golden Fleece*, which would be limited to 39 knights – one for each ship of the Flemish fleet – and would be conferred only on reigning sovereigns, princes and the highest nobility. See Bachman (1884–1894).
3. See Heers (1981).
4. This 'Convento de Santos' became known as 'Santos-o-Velho' (the Old) after the construction of a second convent of the same name and which became known as 'Santos-o-Novo' (the New).
5. *Alguns documentos . . ., Chancelaria de Dom Afonso V,* Tombo National Archive.
6. Garcia de Mello, son of Ruz Gonçalves da Câmara (son of João Roiz da Câmara and 2nd degree cousin of Colón-Zarco) died in a shipwreck with his wife and daughters between Lisbon and Madeira. Such shipwrecks were frequent.

26 The Genealogical Web

To understand the vast web of families in which Salvador Fernandes Zarco was included, it is necessary to look into the Portuguese genealogy of the fifteenth century, especially the nobility linked to the Discoveries. One of the startling phenomena that we come across is the prolificacy of bastardy; here I shall cite only the number of bastard offspring of royalty, from Count Henry of Burgundy to King John II, that were recognised as legitimate: Count Henry – two; King Alphonse Henriques – four; King Sancho I – eight; King Alphonse II – one; King Alphonse III – five; King Dennis – seven; King Peter I – several, only one of whom was legitimised; King Ferdinand I – one; John I (bastard son of King Peter) – two; King Duarte – one; King John II – one.

Some of the highest nobility of Portugal came from these bastard offspring.

Princes were even more prolific, and the legitimation of these bastards was not always registered, as was the case of Friar João Manuel, who was the Bishop of Ceuta, son of Prince Fernando, Duke of Beja, and Joana Manuel. The affair between Prince Fernando and Isabel da Câmara was relegated to obscurity, certainly not because she later married Diogo Afonso de Aguiar, son of João de Aguiar, but because she was the mother of Cristóbal Colón/Salvador Zarco. Some cases like this remained unknown, but none as momentous as this.

Besides this, there were the numerous cases of the bastard children of priors, bishops and cardinals.

One of the most notable figures of Portuguese history, for instance, Nuno Alvares Pereira, who led the Portuguese army in three consecutive victories over the Castilians and played a decisive part in placing King John I on the throne of Portugal, was one of the thirty-two bastards of prior Álvaro Gonçalves Pereira, master of the Order of the Hospital, his mother being a nursemaid of the court, Iria Gonçalves do Carvalhal.[1]

The illegitimate son of King John I and Inês Pires Esteves was Afonso,

349

first Duke of Braganza, who married Beatriz Pereira Alvim, daughter of the Constable, Nuno Alvares Pereira. This marriage gave rise to the greatest noble house of Portugal and it was a Braganza who led the uprising against the Spanish domination that restored Portugal's independence in 1640 and ascended the throne as John IV, beginning a dynasty that would rule the country until 1910.

One of the sons of the first Duke of Braganza was Dom Afonso, Count of Ourem and the first Marquis of Valença, who had three bastard children. One of these was Dom Francisco de Portugal, first Count of Vimioso, who in turn was an ancestor of Dom Álvaro de Portugal, descendant of and heir to all the assets and titles of 'Cristóbal Colón'.

Many children and mothers died in childbirth, as happened to Queen Isabel, the daughter of the Catholic Monarchs (who first married Prince Afonso, son of King John II, and then King Manuel I) in giving birth to her only son, Prince Miguel da Paz, who was recognised as heir to both Iberian thrones. In a way, adultery was a means of defending a man's legitimate wife, in an era when medicine was extremely primitive. On the other hand, it could have been the result of the fact that couples did not marry out of love. Marriages were arrangements of convenience, often related to the acquisition of wealth and land. In these cases, the fiancés had the privilege of knowing each other before their betrothal, but among the nobility the couple often met for the first time only at the wedding. This gave rise to illicit passions, often linking men to the women of their birthplace, as if they were searching for long-lost roots, in a sensual union with telluric forces.

There were certainly extreme cases, like the one of Dom Fernando d'Eça or Deza, son of Prince João,[2] who married several women in Galicia. We know the identity only of his last wife, Isabel de Avallos, daughter of Don Juan Lopez de Avallos, Constable of Castile.

Some ladies also had extramarital relations and their husbands were usually tolerant if the lover was a person of high rank. One of these ladies was Brites (Beatriz) Gonçalves, who became the mistress of Rui Gonçalves da Câmara, Donatary-Captain of the island of São Miguel in the Azores (see pp. 334 and 393), who already had three bastard children. A daughter was born from this affair and King Manuel I authorised the father legally to adopt his bastard offspring. Rui Gonçalves da Câmara was the son of João Gonçalves Zarco and, therefore, brother of Isabel da Câmara, the mother of Salvador Fernandes. But cuckoldry was not always passively accepted and many adulteresses paid with their lives. This was the case of Beatriz Henriques, who was killed by her husband Nuno Pereira, Governor of Fronteira (Alentejo), when he discovered that she had given birth to a lover's child. In accordance with the *Laws of the Realm*, King John II forgave him and he later remarried.

Beatriz Henriques was the sister of Dom Henrique Henriques, later

Huntsman of King Manuel I, who gave him the town of Alcáçovas in 1505 and made him Lord Chamberlain two years later. Dom Henrique married Filipa de Noronha, daughter of João Gonçalves da Câmara from his second marriage to Maria de Noronha and, therefore, niece of Colón-Zarco's mother.

I have used these examples merely to explain the socio-family structure of the society of the time, in which bastardy was benevolently accepted and adultery not always so. They may also lead us to ask the following questions: did Isabel da Câmara leave Madeira to become a lady-in-waiting to the future Queen Leonor (Colón-Zarco's half-sister), or one of the princesses, or even to Beatriz, future wife of Prince Fernando, Duke of Beja and Lord of that Archipelago? Under the feudal system then in force it is possible that Prince Fernando installed Isabel de Câmara in a 'big house' of the region of Cuba, where Colón-Zarco may have been born.

The second Duke of Braganza, Dom Fernando, married, in 1429, Joana de Castro, daughter of João de Castro, Lord of Cadaval. Dom Fernando of Braganza was the proprietor of a castle at Vila Viçosa (Alentejo), but hunted in the forests between the streams of Odéacere and Odivelas and in the 'Seixal Wood' near Cuba. The hunting party was lodged either in this town or at Vila Ruiva. The third Duke of Braganza, Dom Fernando, was at Vila Ruiva in 1472 when, while his father was still alive, he married Isabel, daughter of Dom Fernando, Duke of Beja.

This Dom Fernando, the third Duke of Braganza, and his brother-in-law Dom Diogo were both accused of being involved in the conspiracy against King John II and executed. One was the Lord of Vila Viçosa and Vila Ruiva,[3] the other was Lord of Beja and Cuba. They were respectively half-brother-in-law and half-brother of Colón-Zarco.

Why did Colón-Zarco serve the monarch who had ordered the deaths of his kinsmen? Was it because the king was also his half-brother-in-law, married to his half-sister? Was it because Dom Diogo despised the future Admiral as much as he despised the Order of Christ? Was it because it was King John II himself that took over the reins of control of the Discoveries after the death of Prince Fernando or because Salvador Fernandes – who had been at sea since he was fourteen years old – had respected the oath of the Order of Christ? Fortunately, I was able to count on the research of three genealogists of my own family who served Kings Alphonse VI, Peter II and John V.[5] They have given me precious information regarding the connections between certain families, besides bequeathing me easily readable genealogical trees.[6]

Many names used in the fifteenth century were altered in the two following centuries. This is the case of 'Brites', which became 'Beatriz', and both names can be found in treatises, sometimes referring to the same person. This also occurs in relation to the surname 'Roiz' and 'Rodrigues'. Abbreviations have undergone an even greater evolution and sometimes names are spelt differently in the same text, thus making their interpretation ambiguous. As many children were given the name of their parents and some parents baptised two of their offspring with the same name, it becomes very difficult to place people chronologically due to the fact that there is no record of the year in which they were born, died and married and only occasionally is there any mention of the reign in which they lived. There is also the added difficulty of family names being replaced by new surnames that, in turn, derived from nicknames and cognomens. Another phenomenon of genealogical books is that while some families that remained in relative obscurity are mentioned in great detail, some even keeping their foreign surnames, others that achieved fame and were raised to the peerage and granted coats-of-arms receive scant attention, some of the descendants being mentioned only when they married into other families.

In their genealogical analyses, for instance, Alão de Morais and others omit the trees of the Perestrelo and Esmeraldo families as if they had been uprooted by the storm of the Inquisition, even though their arms are registered in the Portuguese College of Heralds. Even the family name of Zarco was replaced by that of Câmara, adopted when the family was raised to the peerage, while the holders of other peerages also retained their natural surnames. It is interesting to point out that there is a genealogical reference to an Esmeraldo only in the eighteenth century: Cristovam Esmeraldo Atouguia da *Câmara*,[4] son of Luis Atouguia Esmeraldo de Aguiar and Isabel Esmeraldo. It is even more interesting to recall that, 300 years before, Diogo Afonso de Aguiar married Isabel Gonçalves da Câmara, mother of Salvador Fernandes Zarco, in Madeira.

THE OBSCURITY OF NEW CHRISTIAN SURNAMES AND THE INQUISITION

My lack of experience of palaeography and the abbreviations of first and family names and surnames frequently led me to hesitate in the interpretation of genealogical texts. Even in works printed in the seventeenth and eighteenth centuries, there are many doubts regarding the sequence of generations, as there are many lacunae, either through erasures or deliberate omissions.

The New Christians were able to use their 'Lusitanised' Hebrew names during the fifteenth century, but they were quietly dropped after the Inquisition had been instituted in Portugal, especially during the reigns of the Spanish Kings Philip I and II. There are records of the intervention of the Holy Office in cases of apostasy and relapse. One of these cases is that of Henrique Moniz Telles, son of Egas Moniz (brother of Isabel Moniz Perestrelo) and Maria Roiz and nephew of Filipa Moniz Perestrelo. He went to Brazil as a young man and got married to a white lady by the name of Leonor – the genealogists have suppressed her surname – whose sister married Sebastião de Faria. The sisters were New Christians and devoted themselves to Jewish practices, which led to them being burned at the stake. This was worrying to their respective husbands, as Henrique Moniz Telles was the grandson of a lady with the surname of Cabral (her first name is omitted), daughter of Diogo Cabral (cousin of Pedro Álvares Cabral, 'official rediscoverer of Brazil'), who was married to Brites Gonçalves, daughter of João Gonçalves Zarco and aunt of Colón-Zarco.

The Holy Office in Portugal was aware that the first Câmaras and Perestrelos had been New Christians and cosmographers, possibly astrologers and cabalists. The Holy Office in Spain always kept a close eye on Cristóbal Colón, as it suspected that he dabbled in those sciences that were condemned as quasi-magic. The descendants of these families thus found it necessary to behave with the utmost caution.

The war against heresy brought great changes to the kingdom during the reign of King John III, even more so as King John I – great-grandfather of Colón-Zarco – had possessed a *Book of Magic* that, in his *Book of Hunting*, he attributed to the astrologer João Gil.

To what extent could the suspicions of the Dominican Inquisitors be justified? Certain families linked to the Order of Christ included members of the nobility who, despite the high posts they occupied and fidelity to the Crown, even during the reigns of Kings Philip I and II, continued to observe the cult of the Templars and cabalistic practices. 146 years after the death of Colón-Zarco, Dom Rodrigo da Câmara, seventh Captain of the Island of São Miguel (Azores), third Count of Vila Franca and a descendant of João Gonçalves Zarco, was stripped of his post and title by the Holy Office. He spent a year in the prison of the Inquisition and was burned in 1652 for the 'nefarious crime' of cabalistic and alchemical practices, which were then classified as magic, although they had not been considered as such at the time of King John II.

I have already mentioned that several people had exactly the same name and I should like to point out some cases. On delving into the origin of Dom Diogo Furtado de Mendoça (brother of Dom Afonso – see pp. 285 and 386), grandfather of the Archbishop of Lisbon, Dom Pedro de Noronha, whose mistress was Colón-Zarco's sister-in-law, Isabel (or Branca) Perestrelo, I came across the marriage of Leonor de Noronha and Martim Vaz Mascarenhas. This could be one of the three following men:

(a) Commander of Santiago of the Order of Christ, in the duchy of Beja, who married Isabel Correa of Madeira.
(b) Commander of the Order of St James at Aljustrel, who married Maria de Noronha (daughter of Henrique Henriques, Lord of Alcáçovas, and Filipa de Noronha, Colón-Zarco's aunt).
(c) The above-mentioned Commander of Aljustrel, of the Order of St James, married to Leonor de Noronha, daughter of Dom Fernando de Noronha and Constança de Castro e Albuquerque.

We have no dates regarding the birth of these three; we know only they served the Crown during the reigns of John II and Manuel I.

THE TOMBS OF ISABEL AND FILIPA MONIZ

The will of Dom Diogo Colón, legitimate son of Salvador Fernandes Zarco, states that he wished to translate the mortal remains of his mother to his tomb:

> and bring the body of Felipa Moniz, his legitimate wife, my mother, who is in the Monastery of Carmo, in Lisbon, in the Chapel of Pity, which belongs to the Moniz family.

In an effort to reinforce his thesis of a Portuguese Colon, Pestana Júnior studied a document written by Friar Gaspar de Sant'Ana which is to be found in the records of the *Chronica dos Carmelitas* and which clearly states that Filipa's mother, Isabel Moniz, was buried in the Monastery of Carmo. But in trying to attribute Colón-Zarco with the identity of Simão Palha, he included a couple in the genealogy of the Monizes that were not a part of the family. (I dealt with this fact in detail in the Portuguese edition, but I think it is enough here to present an extract of the genealogical tree of Filipa Moniz Perestrelo.)

Prior Alvaro Gonçalves Pereira, Master of the Order of the Hospital (see p. 36), was son of Dom Gonçalo Pereira, Archbishop of Braga. He begot, among thirty-two children, Dom Nuno Alvares Pereira, Constable of the Realm, already mentioned, and Dom Payo Gonçalves Pereira (both sons of Elvira Gonçalves). He was the father of Vasco Martins Moniz (I), who married Beatriz Pereira, his cousin, and fathered Henrique Martins Moniz.

I shall deal with the descent of this Henrique, who married twice: his first wife was Grimaneza Pereira, his cousin and widow of Dom Afonso Telles Barreto, Governor of Faro; his second wife was Inês de Menezes, daughter of Gonçalvo Nunes Barreto (brother of the above Dom Afonso and his successor as Governor of Faro). The only one of these five children that is of interest to us is Vasco Moniz Barreto (II), grandson of Vasco Martins' Moniz (I).

Vasco Moniz Barreto was married three times and begot nine children (see p. 496, Trees XIV and XV, 5B, p. 452). His first marriage was to Isabel Cabral, daughter of Diogo Cabral (cousin of Pedro Álvares Cabral). His second wife was Brites (or Beatriz) de Goís, daughter of her namesake, widow of Lançarote Teixeira. He took Joana Teixeira, daughter of Lançarote, as his third bride.

Let us now take a look at a synthetised genealogical tree of Isabel Moniz (Barreto), daughter of Joana Teixeira and wife of Bartolomeu Perestrelo (see Figure 26.1).

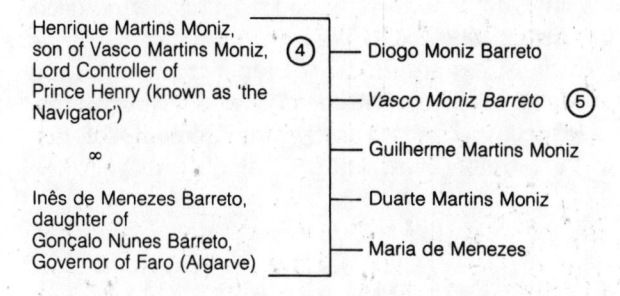

```
Henrique Martins Moniz,
son of Vasco Martins Moniz,     (4)  ├── Diogo Moniz Barreto
Lord Controller of
Prince Henry (known as 'the          ├── Vasco Moniz Barreto    (5)
Navigator')
                                     ├── Guilherme Martins Moniz
           ∞
                                     ├── Duarte Martins Moniz
Inês de Menezes Barreto,
daughter of                          └── Maria de Menezes
Gonçalo Nunes Barreto,
Governor of Faro (Algarve)
```

References:
5 (Tree I and III); see 5 and 66 (Tree XVI); 4 and 66 (Tree XV); 68 (Tree XVI) and 5 (Tree XIX).

26.1 Descendants of Henrique Moniz.

There is no doubt, therefore, that Isabel Moniz could well be buried in the Monastery of Carmo, which was founded by Nuno Alvares Pereira, who was her uncle and a comrade in arms of her great-grandfather, Vasco Martins Moniz. If her daughter Filipa was buried there, it was through the female line of the Monizes and not through the Perestrelos. Isabel would not even need the influence of the Archbishop of Lisbon, who had already given her three grandchildren, to be buried there. And as we already know, through the decipherment of his siglum, that 'Cristóbal Colón' was a Zarco, the bastard son of Prince Fernando, my argument gathers even more strength. Friar Dom João Manuel, bastard son of King Duarte and half-brother of the legitimate Prince Fernando, was educated under the wing of the Constable Nuno Alvares Pereira. Pope Eugene IV appointed him Provincial of the Order of Carmo in 1441. King Alphonse V gave him the post of Royal Chaplain in 1450. Although he was made Bishop of Guarda, he never left the court and begot two bastard sons, Dom João Manuel and Dom Nuno Manuel, by one Justa Rodrigues. He died in 1476 and was interred in his monastery.

∞ = in wedlock
ɸ = out of wedlock

Isabel Moniz (Barreto)
∞
Bartolomeu Perestrelo, discoverer and first Donatary-Captain of Porto Santo

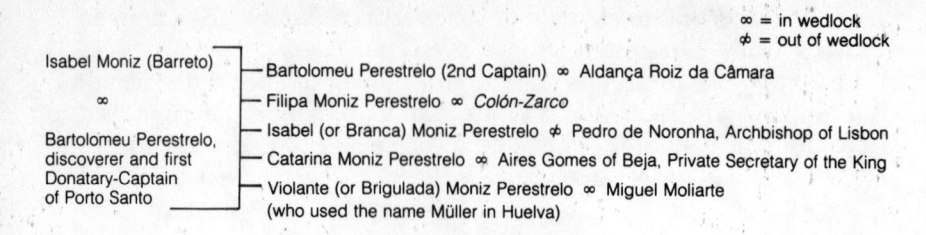

— Bartolomeu Perestrelo (2nd Captain) ∞ Aldança Roiz da Câmara
— Filipa Moniz Perestrelo ∞ *Colón-Zarco*
— Isabel (or Branca) Moniz Perestrelo ɸ Pedro de Noronha, Archbishop of Lisbon
— Catarina Moniz Perestrelo ∞ Aires Gomes of Beja, Private Secretary of the King
— Violante (or Brigulada) Moniz Perestrelo ∞ Miguel Moliarte
 (who used the name Müller in Huelva)

26.2 Descendants of Isabel Moniz and Bartolomeu Perestrelo.

Isabel Moniz may have still been alive to attend her daughter Filipa's wedding in 1478 and in a city where the nobility had been greatly reduced in numbers, she would surely have been well acquainted with Friar Dom João Manuel, the uncle of her son-in-law Colón-Zarco. As family relations exercised a powerful influence in those days, we can be sure that Isabel Moniz was buried in the Monastery of Carmo, like her daughter Filipa Moniz Perestrelo, Dom Diogo Colón's mother (see Figure 26.2).

THE NAVIGATORS CONVERGE ON BEJA

The scientific coordination of the so-called 'School of Sagres' of the Order of Christ was not always planned at Sagres itself. The members of the Order followed Prince Henry 'the Navigator' to wherever he periodically settled. On the Prince's death, all the assets, posts and titles were inherited by his nephew and adopted son, Prince Fernando, who thus became the Duke of Viseu and Administrator of the Order of Christ. Fernando's brother, King Alphonse V, made him the Duke of Beja and gave him the castle of that town, where he already lived.

The cosmographers, cartographers, geographers and pilots of the Discoveries began to converge on Beja, even though they belonged to other parts of the kingdom. Many of them came from old noble families, others had been raised to the peerage at the time of King John I through their deeds on the battlefield in the wars against Castile, in the campaigns of the Western Crusade in North Africa and in the maritime explorations. To be a squire of the Master of the Order of Christ at the time of Prince Henry was almost the equivalent of being a knight – a 'knight of the sea'.

Besides Lisbon, the Algarve, Madeira and the Azores were departure points for the Discoveries and the members of the Order of Christ settled in all of these places. As they frequently had to confer with Prince Fernando, however, many of them set up house in the Duchy of Beja when they reached middle age and were appointed to administrative posts and received Entailments and governorships. This helped to develop the vast, sparsely populated region of the Alentejo, which is nearly a third of the country. Even today the area is very thinly populated. This is why many noblemen from the north of the country and the neighbouring Algarve, but above all from the Azores and Madeira, settled in this area in the fifteenth century.

JUVENILE MARRIAGES

We know that primitive peoples of Africa, Asia and Oceania generally carry out an 'initiation' ceremony consisting in circumcision or defloration of children at the age of ten moons (about nine years), thus enabling them to maintain sexual relations. People got married very early in fifteenth-century Europe, not only as a result of war but also epidemics. Some cases of early marriages involving Portuguese royalty can be mentioned: Princess Beatriz was married to Prince Fernando in 1447 when she was thirteen and he was fourteen, but the marriage was consummated later; their daughter, Leonor, was also married at thirteen, to the future King John II, who was sixteen. This King had a mistress of fifteen, Ana de Mendonça, a lady-in-waiting to Princess Joana, who gave birth to the King's son, Jorge, first Duke of Coimbra, at the age of sixteen. Prince Fernando's sister, Leonor, was also thirteen when she was given in marriage to Frederick III, Emperor of the Holy Roman Empire. The same prince's niece, Joana, was betrothed to King Louis XI of France at the same age, although the marriage did not take place. Fernando, third Duke of Braganza (who would become Prince Fernando's brother-in-law on taking Princess Isabel as his second wife), married, at the age of seventeen, Leonor de Menezes, who was fourteen. King Afonso V (Prince Fernando's brother) married Princess Isabel, daughter of Prince Pedro, at the age of ten. Some time before, Prince Fernando's daughter, Isabel, was given in marriage to Dom João de Menezes, who died at a 'tender age' when she was eight.

In this age of primitive medicine, many babies were stillborn and many mothers died in childbirth. Innumerable royal princes died in the first years of their lives. It is not known whether all of these juvenile marriages were consummated immediately following the wedding, but there is evidence that some adolescents of the nobility maintained sexual relations with maidens in the service of the ladies or with the girls of their serf's families at the onset of puberty.

'Cristóbal Colón' was born in 1448; if Prince Fernando was married in 1447, he could, therefore, have sired a bastard at the age of fifteen.

Marriages were arranged between the children of these members of the Order of Christ, sometimes between branches of families that had the same names. The most common surnames to be found during the fourteenth and fifteenth centuries at the time of the Discoveries, the colonisation of Brazil and the building of the Portuguese Empire in Asia and Oceania, were Albuquerque, Almeida, Andrade, Ataíde, Barreto, Cabral, Câmara, Castelo Branco, Castro, Correa, Corte-Real, Costa, Cunha, Furtado de Mendoça, Gama, Henriques, Mascarenhas, Mello, Moniz, Noronha, Perestrelo, Sylveira, etc. All these surnames gave rise to place-names in the Alentejo, Algarve, Madeira, the Azores and Lisbon. Many more names had been added to this list by the end of the eighteenth century, either of families closely linked to the Crown or of gentlemen who were gifted or moved by the missionary spirit (or, above all, impelled by an ambition to get rich quickly). As usual, there were courtiers in the palace that ingratiated themselves into the King's favour and lived an opulent life at the cost of the 'doers'. And to paraphrase Camões 'at times there were traitors'.

We already know the identity, the names of the parents and the birthplace of Salvador Fernandes Zarco/'Cristóbal Colón' through the deciphering of his siglum. And nearly everything was explained in this respect by what he himself, his son Fernando and Las Casas wrote. The only thing left to do now is identify his 'brothers' Diogo and Bartolomeu and the 'maiden of Córdoba' who bore his bastard son Fernando. As the families of the navigators were so intertwined, I thought it would be valuable to delve into the genealogical trees of the families connected to the Discoveries. I based my study on several genealogies,[7] which were not always in agreement for the reasons I have already explained. But I was careful to make a note of the doubts and stick to facts.

In the synthesised genealogical trees presented[8] (Tree I, p. 362) emphasis is laid on the interconnection of some families whose eldest sons were either seamen in the service of the Order of Christ in the exploration of the seas or 'Western Crusaders' in North Africa and those that opened up new continents and discovered distant islands.

For more than a century, the descendants of these families rose to the highest positions in the realm – Viceroys, Governors of overseas colonies, Admirals of the fleet and Commanders of the strongholds that the Portuguese had set up all over their newly-discovered and newly-won Empire.

Notes and References

1. See Martins, Oliveira (1917).
2. Dom João d'Eça, son of King Peter I of Portugal and Inês de Castro, 'who became Queen after her death'. He married Maria Telles de Menezes, sister

of Queen Leonor Telles. Being convinced by the latter that her sister had been unfaithful to him and at the same time being promised the hand of his niece, Princess Beatriz, in marriage, Dom João murdered his wife. He was obliged to flee to Castile, where King John I, who also had his eye on the Portuguese throne, imprisoned him for the rest of his life. Another Dom João, Master of the Order of Avis and bastard son of Peter I of Portugal, finally ascended to the throne (see p. 25).

3. The 'monte' of Vila Ruiva was transformed into the 'Cadaval Castle' after Dom Nuno Alvares Pereira de Mello, 1st Duke of Cadaval (1638–1727), inherited the estate.
4. Married Helena Thereza Luiza de Castro e Sylveira in Madeira. Prior to this, there is only a record of the first João Esmeraldo, married to Agueda de Abreu. This surname in the XI century was *Hebreu* (Hebrew).
5. Dom Afonso Manuel de Menezes, Belchior Andrade Leitão and José Freire Monterroio Mascarenhas.
6. Working together, one of my great-great-grandfathers and his son-in-law, my great-grandfather, elaborated genealogical trees of their ancestors Andrade, Barreto, Correa, Mascarenhas and Sylveira, a work that has been extremely useful in my research.
7. Sousa, Dom António Caetano de (1739); Antonio de Lima and H.H. de Noronha, *Nobilarios* (1948); Count Dom Pedro, *Nobilario* (n.d.); Morais (1688); Castelo Branco (1864); Gayo (1875).
8. Some men begot children before they were 20, some after they were 50; there were many offspring, the eldest sometimes being over 20 years older than the youngest; some ladies were widowed very early and remarried; unions between uncles/nieces and aunts/nephews made it impossible perfectly to coordinate the chrono-genealogical columns of the trees.

27 'Terra Rubra' and La Rábida

Knowing that the families of the navigators of the Discoveries, members of the Order of Christ or serving in its orbit, intermarried amongst themselves, and knowing the most common surnames, I delved into the respective trees and studied the common genealogical links, eliminating the people who seemed dispensible. Believing now that Bartolomeu Colón was telling the truth when he wrote that he had been born at *'Terra rubra'*, I went in search of his family origins. As the three Colóns called themselves brothers, Salvador having been born at Cuba and Bartolomeu at Vila Ruiva, Diogo must have come into the world nearby.

Based on this assumption, I paid particular attention to the families from Alentejo, the Algarve and Madeira that were, in some way or another, connected to the region of Beja, under the Master of the Order of Christ. The Order of St James was more widespread in the Alentejo, but its members had already begun to participate in the Discoveries. The first genealogical tree presented (Tree I, p. 362) shows that there was a distinct movement of members of the Order of Christ to the region of Beja and that the families of the navigators frequently intermarried. It is known that Bartolomeu Perestrelo was the grandfather of Filipa, Colón-Zarco's wife. I then discovered that:

(a) *Pedro Álvares Cabral*, 'official discoverer' of Brazil, married Isabel de Castro, granddaughter of Isabel Perestrelo (Filipa's sister); he was, therefore, a cousin of Colón-Zarco by marriage (see relation 9A on Tree I).

(b) *Fernão Vaz Corte-Real* (relation 6), married to Judith de Goís (sister of Filipa, Colón-Zarco's wife), was great-uncle of Columbus by marriage; and he was the son of João Vaz and brother of Miguel and Gaspar Corte-Real, the explorers of the western seas.

(c) *Luis de Camões* (relation 22), author of the *The Lusiads* and who served in Africa, India and China, was the great-grandson of João Vaz de Camões, grandfather of the husband of Inês Dias da Câmara; as this

lady was a half-sister of Colón-Zarco, this means that the poet was a kinsman of Columbus by marriage.
(d) *Vasco da Gama* (relation 24) was the great-grandson of Estêvão da Gama, uncle of Guiomar Vaz da Gama; that means he was the great-great-uncle of Luis de Camões, who made him the main character of his epic poem. As the same Guiomar was the aunt of Lopo Vaz de Camões, husband of Inês Dias da Câmara (half sister of Colón-Zarco), Vasco da Gama and Colón-Zarco were *kinsmen*;
(e) *Bartolomeu Perestrelo*, who discovered Porto Santo with João Gonçalves Zarco, was father of his namesake Bartolomeu Perestrelo, who married Aldonça da Câmara, first cousin of Colón-Zarco. This means that Bartolomeu Perestrelo was the cousin of Colón-Zarco's wife Filipa Moniz Perestrelo.

All the Discoverers were thus kinsmen among themselves.

I gradually whittled down my area of research to a zone around the towns of Alvito and Cuba, in the middle of the geological area of 'red lands'. It must be pointed out that Cuba is encircled by other little towns, among which Vila Alva and Vila Ruiva, the latter being nearer Alvito than Cuba. Who were their Lords (see Tree II, p. 365)?

VILA RUIVA

João Afonso de Albuquerque, Master of the Order of St James of Castille, was the lord of many lands, among which was Vila Ruiva. On his death, he bequeathed them to his grandson, Pedro de Albuquerque, who was married to Isabel de Mello (daughter of João de Mello, Lord Steward of King Alphonse V) and an intimate friend of Dom Fernando, Duke of Braganza. This friendship linked them in the organisation of the hunts that were held in that area and in the conspiracy plotted against King John II (see p. 164). The two men were executed when the plot was discovered, and following this in 1484, Vila Ruiva became the property of the crown.

Dom Pedro's grandfather had been the father-in-law of Dom Fernando Coutinho, son of Dom Vasco, in whose house João Gonçalves Zarco had lived in his youth in Matosinhos. Dom Fernando Coutinho was the brother of Dom Gonçalo Coutinho, and a nephew of theirs, Dom João Coutinho, married Catarina de Castro, sister of the second Duke of Braganza, who lived in the Alentejo town of Vila Viçosa.

Dom Pedro de Menezes, a relative of the Duke of Braganza, and Lord of Viana do Alentejo, married four times: his first wife was Margarida de Miranda, daughter of the Archbishop of Braga, his second was Filipa Coutinho, daughter of the above-mentioned Dom Gonçalo, his third was

∞ = in wedlock
ɸ = out of wedlock

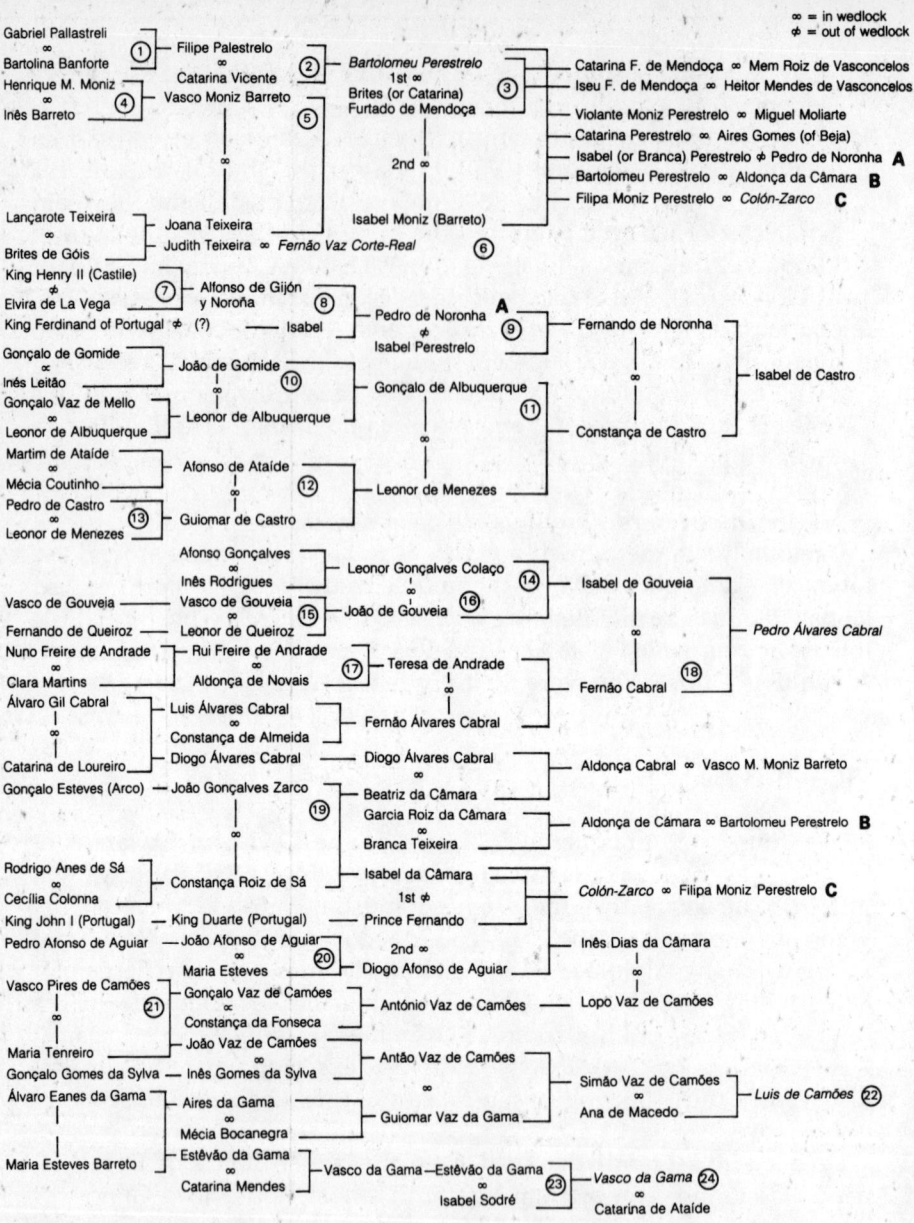

Genealogical Tree

- Gabriel Pallastreli ∞ Bartolina Banforte — Filipe Palestrelo (1) ∞ Catarina Vicente — Bartolomeu Perestrelo (2)
- Henrique M. Moniz ∞ Inês Barreto (4) — Vasco Moniz Barreto (5)

Bartolomeu Perestrelo (3)
- 1st ∞ Brites (or Catarina) Furtado de Mendoça:
  - Catarina F. de Mendoça ∞ Mem Roiz de Vasconcelos
  - Iseu F. de Mendoça ∞ Heitor Mendes de Vasconcelos
  - Violante Moniz Perestrelo ∞ Miguel Moliarte
  - Catarina Perestrelo ∞ Aires Gomes (of Beja)
  - Isabel (or Branca) Perestrelo ɸ Pedro de Noronha **A**
  - Bartolomeu Perestrelo ∞ Aldonça da Câmara **B**
  - Filipa Moniz Perestrelo ∞ Colón-Zarco **C**
- 2nd ∞ Isabel Moniz (Barreto) (6)

- Lançarote Teixeira ∞ Brites de Góis — Joana Teixeira — Judith Teixeira ∞ Fernão Vaz Corte-Real
- King Henry II (Castile) ɸ Elvira de La Vega (7) — Alfonso de Gijón y Noroña (8)
- King Ferdinand of Portugal ɸ (?) — Isabel

Pedro de Noronha A ɸ **Isabel Perestrelo** (9)
  - Fernando de Noronha ∞ Isabel de Castro
  - Gonçalo de Albuquerque (11) ∞ Constança de Castro

- Gonçalo de Gomide ∞ Inês Leitão — João de Gomide (10)
- Gonçalo Vaz de Mello ∞ Leonor de Albuquerque — Leonor de Albuquerque
- Martim de Ataíde ∞ Mécia Coutinho — Afonso de Ataíde (12) — Leonor de Menezes
- Pedro de Castro ∞ Leonor de Menezes (13) — Guiomar de Castro

- Afonso Gonçalves ∞ Inês Rodrigues — Leonor Gonçalves Colaço (14) ∞ João de Gouveia (16) — Isabel de Gouveia
- Vasco de Gouveia ∞ Fernando de Queiroz — Vasco de Gouveia (15) — Leonor de Queiroz
- Nuno Freire de Andrade ∞ Clara Martins — Rui Freire de Andrade — Aldonça de Novais (17) — Teresa de Andrade — Fernão Cabral (18) ∞ → *Pedro Álvares Cabral*
- Álvaro Gil Cabral ∞ — Luis Álvares Cabral ∞ Constança de Almeida — Fernão Álvares Cabral
- Catarina de Loureiro — Diogo Álvares Cabral — Diogo Álvares Cabral — Aldonça Cabral ∞ Vasco M. Moniz Barreto
- Gonçalo Esteves (Arco) — João Gonçalves Zarco (19) — Beatriz da Câmara — Garcia Roiz da Câmara — Aldonça de Câmara ∞ Bartolomeu Perestrelo **B**
  - Branca Teixeira
  - Isabel da Câmara 1st ɸ — *Colón-Zarco* ∞ Filipa Moniz Perestrelo **C**
- Rodrigo Anes de Sá ∞ Cecilia Colonna — Constança Roiz de Sá — Prince Fernando
- King John I (Portugal) ∞ King Duarte (Portugal) — João Afonso de Aguiar (20) 2nd ∞
- Pedro Afonso de Aguiar — Maria Esteves — Diogo Afonso de Aguiar — Inês Dias da Câmara ∞ Lopo Vaz de Camões
- Vasco Pires de Camões ∞ — Gonçalo Vaz de Camões (21) ∞ Constança da Fonseca — António Vaz de Camões
- Maria Tenreiro — João Vaz de Camões — Antão Vaz de Camões ∞ Guiomar Vaz da Gama — Simão Vaz de Camões ∞ Ana de Macedo — *Luis de Camões* (22)
- Gonçalo Gomes da Sylva ∞ Inês Gomes da Sylva
- Álvaro Eanes da Gama — Aires da Gama ∞ Mécia Bocanegra
- Maria Esteves Barreto — Estêvão da Gama ∞ Catarina Mendes — Vasco da Gama — Estêvão da Gama (23) — *Vasco da Gama* (24) ∞ Catarina de Ataíde
- Isabel Sodré

References:

1 and 2 – Italian New Christians, knights and astrologers; between all entries; 3 – Discoverer, with Zarco (19), of Porto Santo and its 1st Donatary- Captain; 4 – Knight and Administrator of Prince Henry's Household; 5 – who succeeded his father-in-law as Governor of Faro; 6 – Navigator like his father and brothers to the West of Azores; 7 – Bastard son of King Alphonse XI of Castile; 8 – Count of Gijón and Noroña; 9 – Archbishop of Lisbon; 10 – Lord of Vila Verde; 11 – Lord of Castanheira; 12 – Count of Atouguia; 13 – Lord of Cadaval; 14 – Daughter of the King's nurse Inês Rodrigues; 'Colaço' means two children nursed together as brothers; 15 – Lord of Colmeal and Almendra; 16 – Lord of Colmeal and Almendra and of Valhelhas, Governor of Castel-Rodrigo; 17 – Knight-Commander of the Order of Sant'Iago (St James); 18 – Governor of Belmonte; 19 – Discoverer, with Teixeira (see Tree XIII – I), of Madeira and it's 1st Donatary-Captain – he was elevated to the peerage and took the name Câmara; 20 – Purveyor of Évora, grandson of the Count of Montemor; 21 – Lord of Estremoz and Avis, Governor of Portalegre; 22 – Knight in Africa, India and China, author of *The Lusíadas*; 23 – Commander of the Fortress of Sines; 24 – Discoverer of the seaway to India and 1st Count of Vidigueira.

I Relation: Cabral/Corte Real/Camões/Gama/Colón-Zarco

Beatriz Coutinho, niece of Dom Gonçalo and daughter of Dom Fernando Coutinho, and his fourth was Genebra Pessanha, daughter of Carlos Pessanha, a descendent of the Italian admiral, Pezagno, who had come to Portugal at the time of King Dennis.[1]

Although he had legitimate children, Dom Pedro de Menezes, Admiral of the Kingdom, also sired a bastard son by a New Christian by the name of Isabel Domingues. This was Dom Duarte de Menezes, who was one of the most notable soldiers of his time, Governor of Ceuta and who, after having taken part in the conquest of Alcazar Ceguer, died at Tangier when saving the life of King Afonso V. Dom Duarte de Menezes kept a Jewish school in a building that belonged to the chapel of his father.

Dom Duarte was married twice. His first wife was Isabel de Mello, daughter of Martim Afonso de Mello, Governor of Évora, Olivença, Campo Maior and Castelo de Vide, all towns of the Alentejo, and his second was Isabel de Castro, cousin of Joana de Castro, wife of the second Duke of Braganza.

After the execution of the third Duke of Braganza, Dom Fernando, his successor, Dom Jaime, grandson of Joana de Castro, was exiled to Spain. Eight generations of surnamed Portuguese descended from him and up to 1739 they successively married descendants of 'Cristóbal Colón'. These marriages did not take place by chance. While one branch came from the Dukes of Braganza, the other branch descended from the Dukes of Beja, and not from a Genoese wool-carder.

Following King John II's death, King Manuel I authorised Dom Jaime to return to Portugal and gave him Cuba. Dom Jaime, however, decided to exchange it for Azurara, Cascais and Setúbal, which belonged to Dom Rodrigo de Mello.

A descendent of Dom Rodrigo's, Dom Nuno Álvares Pereira de Mello, inherited Vila Ruiva. Through the Castros, he became Lord of Cadaval and the first Duke of that name. From the ruins of his forebears' mansion near Vila Ruiva, he constructed an opulent building that became known as Cadaval Castle.

ALVITO

Rui Dias Lobo, Lord of Alvito, married Margarida de Vilhena, daughter of Martim Afonso de Mello. After the death of her husband, this lady married João Rodrigues de Sá, to whom I shall refer on pp. 498–9 when I broach the subject of the origins of the Portuguese Colonnas. As Rui Dias Lobo left no heirs, the possession of his lands reverted to his sister, Maria Lobo, second wife of Dom João da Sylveira, the first Baron of Alvito and Lord of Aguiar, about whom I shall speak later when mentioning the birthplace of Friar Juan Peres de Marchena, Colón-Zarco's patron in Spain.

A descendant of the Sylveiras, Dom Pedro de Sousa, Lord of Prado and Governor of Beja, was the grandson of Dom Lopo de Sousa, who was Dom Jaime of Braganza's preceptor when he was exiled in Spain. Dom Pedro de Sousa, who died in the Battle of Alcazar-Kibir with most of the Portuguese nobility, inherited the town of Alvito and its castle. Some days before leaving for the war in Morocco, Dom Pedro de Sousa was one of the noblemen who received King Sebastian at Cuba.

Starting with Dom Nuno Álvares Pereira de Mello and Dom Pedro de Sousa and going back in time, I have elaborated genealogical Trees II and III (pp. 362 and 365), thus discovering the first Lords of Vila Ruiva and Alvito, Colón-Zarco's contemporaries.

In Tree III, it can be seen that the Câmaras were closely linked to the Henriques and the Mellos, as well as to a new family, the Aguiars. The only person of this name that had appeared in Tree I was Diogo Afonso de Aguiar, who had married Isabel da Câmara, Salvador/Colón's mother.

Concentrating on the reigns of Kings Alphonse V, John II and Manuel I, the period of Salvador Zarco's life, and excluding those that, for one reason or another, had spent very little time in the area of Cuba, Vila Ruiva and Alvito, I elaborated another genealogical tree beginning with Dom Martim Afonso de Mello. The main aim continued to be to find out who had been in charge of the manor of Vila Ruiva during the absence of the owners. I thus elaborated Tree IV (p. 367) through a process of selection from Trees I–III.

Tree IV shows that Dom Duarte de Menezes was the brother-in-law by marriage of Dom Fernando Henriques, cousin of Dom João Henriques, who in turn was the father-in-law of Isabel Zarco's brother, João Gonçalves da Câmara, and so uncle of Colón-Zarco. This means that Dom Duarte's son, Dom João de Menezes, was a second-degree cousin of Isabel and a third-degree cousin of Colón-Zarco.

THE FRIARS MARCHENA

I shall now briefly deviate from the genealogical trees and mention the conquest of Alcazar Ceguer in North Africa. After the conquest of the city by Dom Duarte de Menezes, it was besieged by the forces of the King of Fez, who planned to reconquer the fortress. This plan, however, reached the ears of a Portuguese, one Pêro de Marcham, who had been a prisoner in Fez and had made the acquaintance of a certain Moor who, in turn, had been captive in Tavira (in the Algarve) and spoke Portuguese. After being promised that he would be well rewarded, the Moor

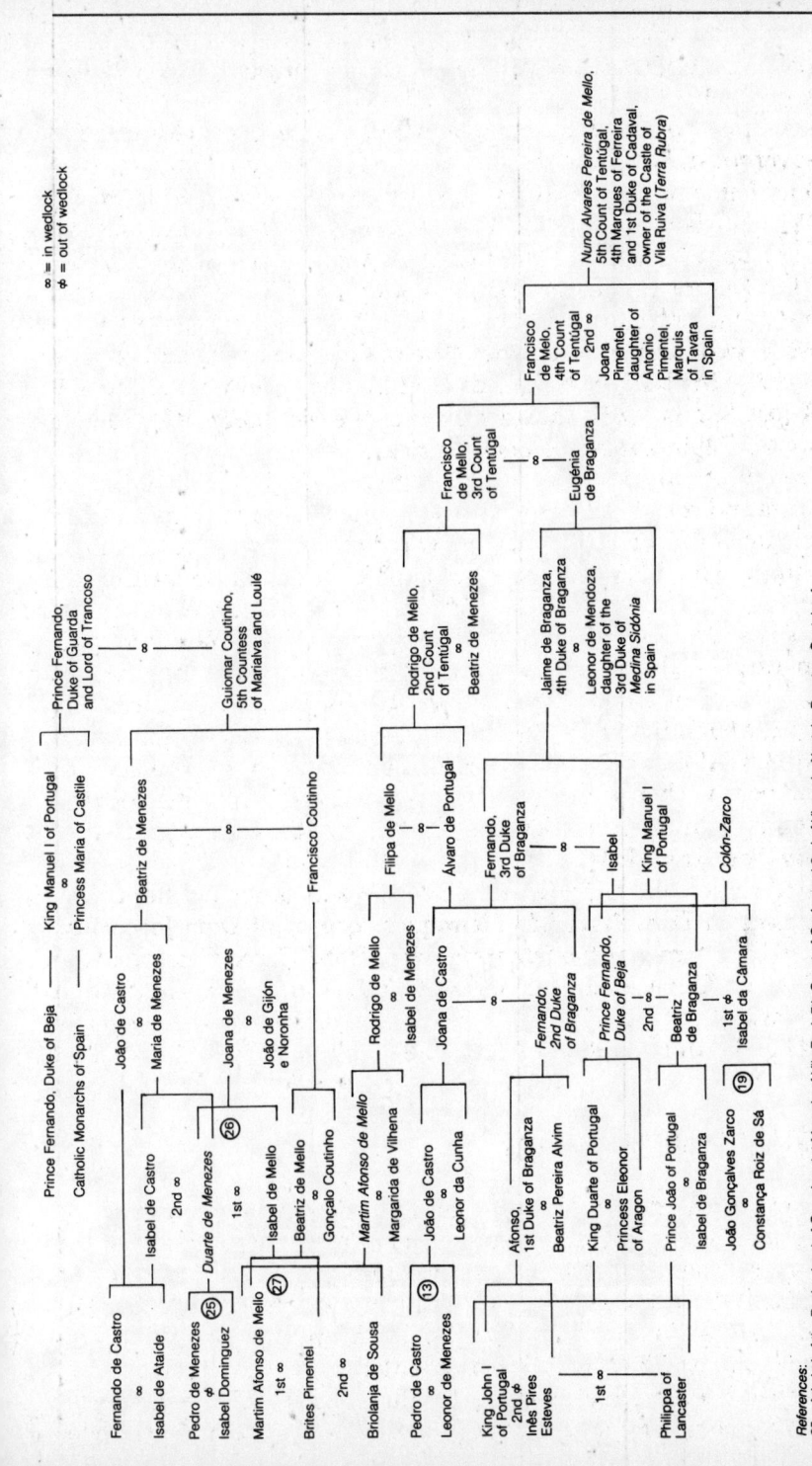

'TERRA RUBRA' AND LA RÁBIDA   365

∞ = in wedlock
⌀ = out of wedlock

■ Vila Ruiva/*Terra Rubra*: ascendants and relatives of Nuno de Mello

References:
25 – Admiral of the Realm, Lord of Canaries, 1st Marquis of Vila Real, 7th Count of Ourém and 2nd of Viana do Alentejo, Donatary-Captain of Ceuta (North Africa); 26 – 3rd Count of Viana do Alentejo, Governor of Ceuta and Alcazar-Ceguer (North Africa); 27 – Royal Huntsman and Commander of the King's Guard; 13 and 19 (see Tree I).

III Alvito: ascendants and relatives of Rodrigo da Sylveira

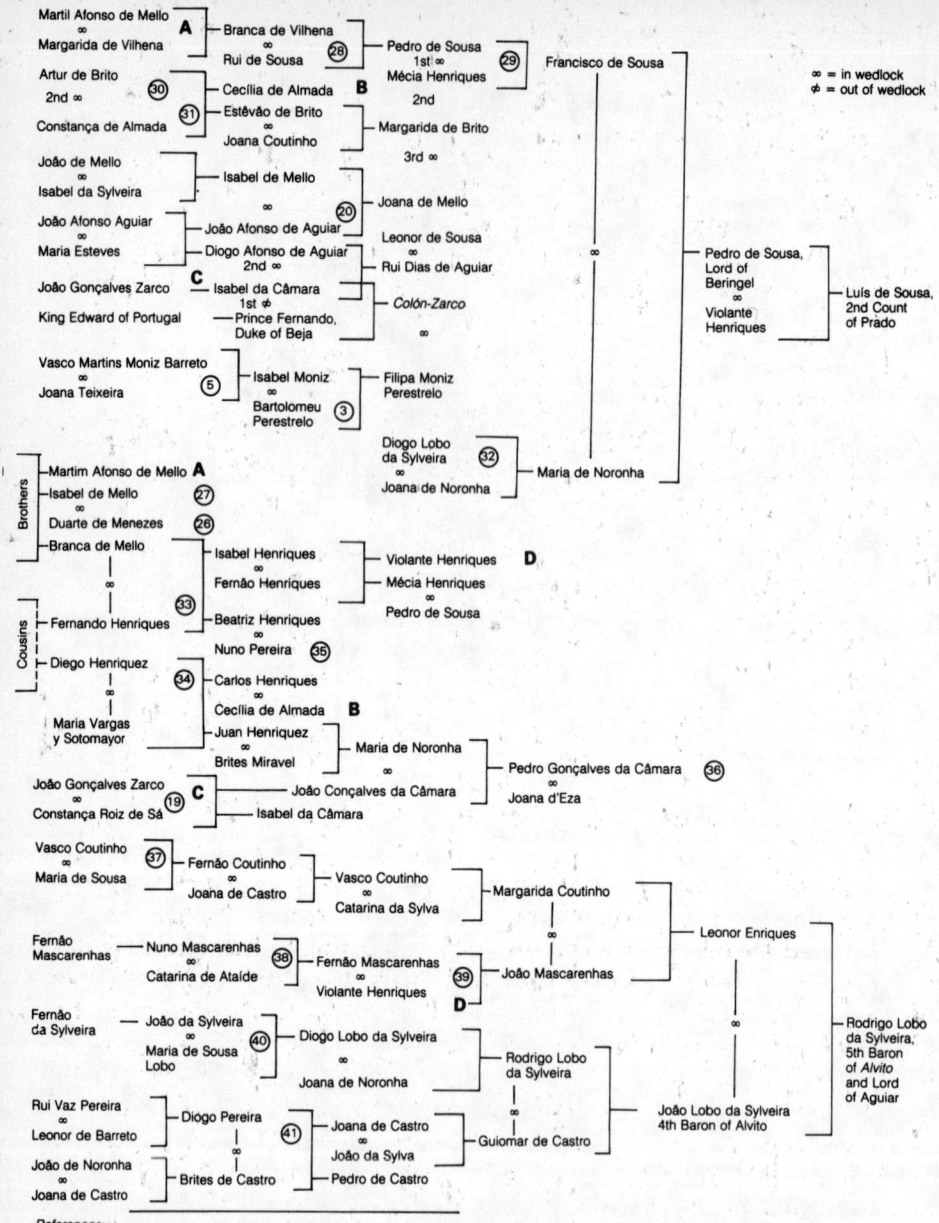

Legend: ∞ = in wedlock; φ = out of wedlock

References:
3, 5, 19 and 20 (Tree I); 26 and 27 (Tree II); 28–1st Lord of Sagres and Beringel; 29 – 1st Count of Prado, Lord of Beringel, Governor of Alcácer-do-Sal and Beja; Captain of Azamor and Alcazar-Ceguer (North Africa); 30 – Governor of Beja, and Royal Huntsman; 31 – Sister of Álvaro Vaz de Almada, Admiral of the Realm, 1st Count of Avranches (Normandy) by concession of King Henry IV of England; 32 – 2nd Baron of Alvito, Lord of Oriola, Niza and Aguiar, Lord High Steward of King John II and Administrator of King Manuel's proprietorship; 33/34 – grandsons of King Henry II of Castile; 35 – Governor of Fronteira, who killed his wife; 36 – Royal Huntsman, son-in-law of João Fogaça de Albergaria, Administrator of King John II's proprietorship; 37 – Marshal of the Realm, Governor of Santarém, 1st Count of Marialva, Lord of the Couto (hunting reserve) of Leomil and owner of the mansion of Matozinhos, where João Gonçalves Zarco lived when young; 38 – Knight-Commander of Almodovar (order of Christ); 39 – General-Commander of the Cavalry, Governor of Montemor-o-Novo – he was the closest companion of King John II, who died in his arms; 40 – 1st Baron of Alvito, Lord of Niza and Aguiar; Royal Huntsman; – 41 – 2nd Count and Governor of Vila da Feira.

'TERRA RUBRA' AND LA RÁBIDA 367

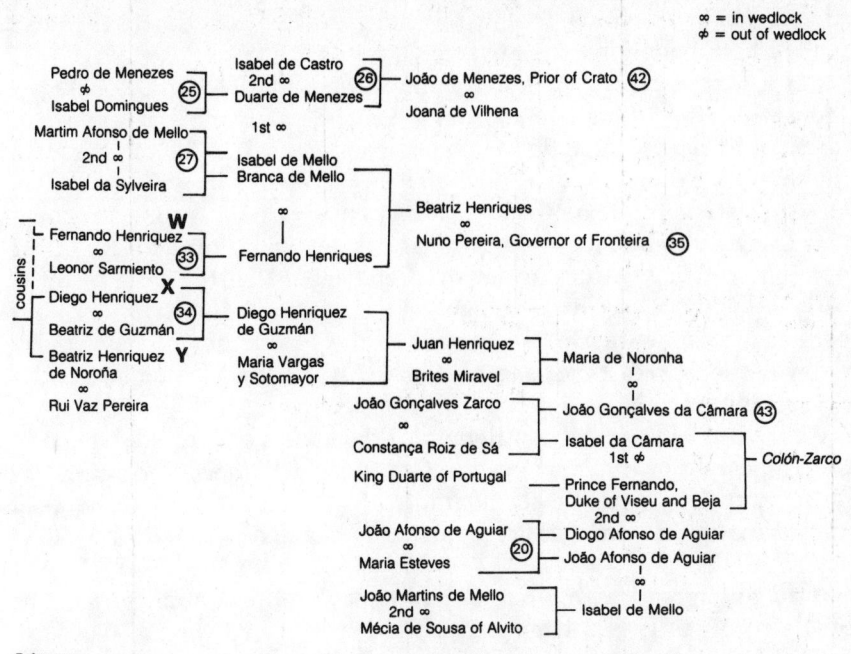

∞ = in wedlock
φ = out of wedlock

```
Pedro de Menezes        Isabel de Castro
      φ            ─25─  2nd ∞          ─26─ João de Menezes, Prior of Crato ─42
Isabel Domingues         Duarte de Menezes      ∞
Martim Afonso de Mello                          Joana de Vilhena
      │                  1st ∞
    2nd ∞          ─27─  Isabel de Mello
      │                  Branca de Mello
Isabel da Sylveira
                W                               Beatriz Henriques
   ┌ Fernando Henriquez                              ∞
   │       ∞          ─33─ Fernando Henriques─  Nuno Pereira, Governor of Fronteira ─35
   │ Leonor Sarmiento
cousins
   │ Diego Henriquez X
   │       ∞          ─34─ Diego Henriquez
   │ Beatriz de Guzmán      de Guzmán         ─ Juan Henriquez
   │                        ∞                      ∞           ─ Maria de Noronha
   └ Beatriz Henriquez Y   Maria Vargas          Brites Miravel    │
     de Noroña              y Sotomayor                             ∞
          ∞                                    João Gonçalves Zarco ─ João Gonçalves da Câmara ─43
     Rui Vaz Pereira                               ∞                     │
                                               Constança Roiz de Sá    Isabel da Câmara
                                                                         1st φ              ─ Colón-Zarco
                                               King Duarte de Portugal ─ Prince Fernando,
                                                                         Duke of Viseu and Beja
                                                                         2nd ∞
                                               João Afonso de Aguiar  ─ Diogo Afonso de Aguiar
                                                    ∞           ─20
                                               Maria Esteves         ─ João Afonso de Aguiar
                                                                         │
                                               João Martins de Mello    ∞
                                                 2nd ∞                 Isabel de Mello
                                               Mécia de Sousa of Alvito
```

References:
20 (Tree I and III); 25 and 27 (Tree II); 33, 34 and 35 (Tree III); 42 – Prior of Crato, who was granted dispensation to marry and became Commander of Arzila and Tangier and Count of Tarouca; 43 – 2nd Captain of Madeira (see p. 445 and Tree XII).

IV Menezes/Henriques/Aguiar/Colón-Zarco

decided to warn Dom Duarte of the imminent attack. The Portuguese were thus able to pre-empt the enemy and a surprise cavalry charge shattered the beseigers when they were preparing for the final assault.[2]

I felt impelled to discover who this Pêro de Marcham was. Assuming that he was of the nobility, I resorted to heraldry[3] and found only one 'Marcham', of Andaluzian origin (Huelva): Christóvam de Marcham, married to Constança de Ataíde, daughter of Dom Pedro de Ataíde. Although the Ataídes (through Nuno de Mascarenhas – Tree III, 38) were closely related (through Violante Henriques – Tree III, 33) to the Sylveiras of Alvito, I gave up the search and focused my attention once more on Dom Duarte de Menezes.

When the Portuguese were disembarking to attack Alcazar Ceguer in December 1459, Dom Duarte de Menezes lost an infantry captain that he had taken with him from Beja. This captain (today the equivalent of a colonel), Rui Gonçalves de Marchena, descended from a family of Huelva. On following up the captain's genealogical trail, I found out that he was the father of Antonio de Marchena, 'novice' of the Order of St Francis. This impelled me to conclude that the 'novice' Antonio de Marchena could have been transferred – already as a friar – from the

Monastery of Arrábida to the Monastery of La Rábida (its Spanish name), both belonging to the same Order of St Francis.

Las Casas wrote: 'By reading some letters of Cristóbal Colón to the Monarchs, which I have seen, from the island of Española, it seems that it was one Friar Antonio de Marchena who helped him a lot in persuading the Queen to accept his petition'.[5]

A document from the Catholic Monarchs to Colón reads as follows: 'We think you should take a good astrologer with you and are of the opinion that Friar Antonio de Marchena would be the best because, besides being a good astrologer, he has always supported your petitions'.[6] Columbus himself wrote to the Sovereigns: 'I have never received the help of anyone, except Friar Antonio de Marchena, besides that of Eternal God'.[7]

According to Friar Alonso Remón[8] 'Friar Juan Peres de Marchena of the Monastery of La Rábida of the Order of St. Francis, Portuguese by birth, knowledgeable in cosmography, convinced Colón to make it known that King John II of Portugal had not agreed with what he had to say.

'Had not agreed with what he had to say' is the central prop of the whole ruse. As I shall later prove, the meeting that John II had with Colón-Zarco and the Junta of mathematicians was a superbly woven intrigue to mystify the Spanish Monarchs.

Let us see another fringe of the story. In 1480, the year following the signing of the *Treaty of Terçarias de Moura*, the Portuguese Ambassador Dom João de Sylveira, first Baron of Alvito (see p. 363) and Lord of Aguiar, went to Córdoba. On the list of the retinue's attendants, there is a page, João Peres, who, I presume, was still a boy. The aim of the embassy was to annul the combined marriages of Prince Juan of Castile to Princess Joana of Portugal and of Prince Afonso (son of King John II) to Princess Isabel (eldest daughter of the Catholic Monarchs), all of whom were still children.

I believe that the incognito João Peres was Marchena, because, for some unknown reason, he did not return to Portugal. We know, however, that a young Juan Peres de Marchena, after having later professed, became confessor of the Catholic Queen Isabella. According to Friar Remón, Juan Peres had asked the Queen that he be allowed to leave the Court and devote himself to his studies (cosmography). It is opportune to remember that when Colón arrived at the Monastery of La Rábida, he spoke Portuguese to Friar Juan Peres de Marchena (see p. 94).

Peres meant 'son of Pero (or Pedro)'. This leads me to ask the following question: 'if the Portuguese captive in Fez that warned Dom Duarte de Menezes of the enemies' plans was called Pero de Marcham, could he not have been the father of the future confessor of Queen Isabella'. The spelling of 'Marcham' (or 'Marchão') could be a phonetic corruption in the evolution of 'Marchem' instead of 'Marchena', which is the name of an Andaluzian village. It must be remembered that 'Harum', which gave origin to Farão and Faro, became 'Haro' in Castilian, which in turn led to 'Haran', 'Harana', 'Arana' and 'Araña', from whence came the name of the Portuguese family Aranha, later linked to the Menezes.

Narrating the Battle of Trancoso at the time of King John I, the chronicler Fernão Lopes de Castanheda referred to the death of one Pêro de Marcham. As this King was the grandfather of Prince Fernando, Duke of Beja (commander of Dom Duarte de Menezes in the campaign of Alcazar Ceguer), it is quite possible that the above-mentioned Pêro de Marcham was the grandfather of his namesake and of Rui de Marchena. And could these two not have been cousins, despite the distortion of the former's surname? I was unable to determine this kinship. I did discover one surprising thing, however. I knew only that Cristóvam de Marcham, of Huelva, married Constança de Ataíde, daughter of Dom Pedro de Ataíde. Later I found out that this Dom Pedro was cousin of Dom Martinho de Ataíde, second Count of Atouguia, who in turn was first cousin of his wife, Constança de Castro.

But this Constança de Castro was the sister of Isabel de Castro (already mentioned), daughter of Dom Fernando de Castro and Isabel de Ataíde (sister of Dom Pedro), who was married to Dom Duarte de Menezes. Consequently, Dom Duarte was a cousin by marriage of Cristóvam de Marcham. What degree of kinship could link him to Rui de Marchena, also of Beja and belonging to a family from Huelva? Everything points to the fact that the surnames 'Marcham' and 'Marchena' are one and the same (by distortion of 'Marchem'), from a place situated between Córdoba and Huelva.

Following up the lineage of Fernando de Castro, Governor of the House of Prince Henry 'the Navigator' and Governor of Covilhã, I found out that he was the son of Dom Pedro de Castro,[9] who married Maria Ponce de Marchena in Huelva. Pedro had fled to Huelva after betraying King John I. He was pardoned, returned to Portugal and took part in the conquest of Ceuta in 1415 with the same Prince Henry.

And then the genealogy gave me an unexpected surprise: Dom Duarte de Menezes, third Count of Viana, was a bastard son of Dom Pedro de Menezes, first Count of Vila Real, who was married to Maria Ponce de Marchena (namesake of the above, her aunt). Therefore, Duarte de Menezes, cousin of Colón-Zarco, was a stepson of Maria Ponce de Marchena, daughter of Pedro Ponce de Leon, Lord of Marchena, and Maria de Ayala y Guzmán of Córdoba, living in Huelva (see p. 411, Tree IX, 26).

As at that time the population was very reduced, there were no other families surnamed 'de Marchena' between Córdoba, Huelva and Beja.

This impelled me to believe that the two friars 'de Marchena' could only be kinsmen of the Lord of Marchena, especially because they were familiar to the Court and well-educated cosmographers. Their kinship with Menezes explains why Colón-Zarco sought the friars' advice and requested them to put his case to Queen Isabella 'the Catholic', so that she would back his proposal to discover the seaway to India by the west.

The time has come to abandon all hypotheses and return to simple facts. It was up to Colón-Zarco to contact the two friars of La Rábida. According to Friar Remón, they were both Portuguese, irrespective of the fact that they were cousins or were kinsmen of Dom Duarte de Menezes and of Colón-Zarco himself.

THE COUTINHOS

Through the genealogical trees presented, it can be seen that the future Admiral of the West Indies was surrounded by relatives who were members of the Orders of Christ and St James, as well as Huntsmen and Falconers of the King and Princes. A great part of the Portuguese noble houses sprang from these kinsmen of his. In my research of Vila Ruiva, I came across Dom Vasco Coutinho (p. 361, Tree III, 37) and delved into the origin of the name. One of the oldest families of the Realm, the Coutinhos descended from Dom Garcia Roiz da Fonseca.[10] The name came from the diminutive of the word '*couto*', a hunting reserve, because the 'Couto de Leomil' that the family owned was very small. Other lands came into possession of the family through marriages, such as that of the Marialvas and the Santiagos in the Duchy of Beja. The eldest son of Dom Vasco, Gonçalo Coutinho, was appointed Marshal of the Realm by King John I and his grandson and namesake, the above-mentioned Vasco (Tree III, 37) was made the 1st Count of Marialva by King Alphonse V. It was Dom Vasco, the grandson, who warned King John II of the planned attempt on his life, thus enabling the King to pre-empt his would-be assassins. And it was Dom Vasco (the grandfather), member of the Order of Christ and with a manor in Matosinhos that married Maria de Sousa, illegitimate daughter (later legitimised) of Dom Lopo Dias de Sousa, Master of the Order of Christ, direct antecedent of Prince Henry 'the Navigator'.

A descendent of Dom Vasco's, Leonor Henriques, married Dom João Lobo da Silveira, third Baron of Alvito (Tree III, 40), the first sons of whose family – beginning with Dom Fernão da Sylveira – were Masters of the Order of Christ. They owned lands at Ribeira de Niza and at Aguiar in the future Duchy of Beja and became the Lords of Alvito through marriage. In referring to Dom Vasco Coutinho, it is interesting to recall that João Gonçalves Zarco lived (was maybe even born) in the 'big house' at Matosinhos from the time of Gonçalo de Leomil, first Marshal of the Realm. It is also important to mention the fact that Simão Gonçalves da Câmara, grandson of João Gonçalves Zarco and first cousin of Salvador Fernandes Zarco, left Madeira when he felt old age coming on and settled in the house of the Coutinhos in Matosinhos, where he spent the rest of his life.

The Order of St James in Portugal had its headquarters in the Monastery of Santos-o-Velho in Lisbon up to the time of King Alphonse I. It was moved to Alcácer do Sal in the reign of Sancho I and to Mértola in the reign of Sancho II. Both towns are in the Alentejo and both were in the Duchy of Beja at the time of Prince Fernando, who was the Master of the Order of Christ and, temporarily, of St James. It is impossible to delimit the territory of the Duchy of Beja, but it seemed to cover what is today called the Lower Alentejo. Évora and its adjacent lands belonged to the Bishopric of that city.

King Alphonse V ordered the construction of a new headquarters in Palmela, which was already an important base of the order, in 1447, confiding the Mastership *in nomine* to the son of King John II, Prince Afonso.

THE ORDER OF ST JAMES AND THE CONVENT OF SANTOS

The Order of St James was founded in Castile and its Masters were Castilian. King Dennis, who transformed the Order of the Temple into the Order of Christ, managed to separate the Portuguese Order of St James from its Spanish counterpart through a Bull of Pope Nicholas IV in 1288, the first Portuguese Master being Dom João Fernandes. At the insistence of the Castilian Sovereigns, however, Popes Celestine V and Boniface VIII reintegrated the Portuguese Order into the Spanish one, a situation that continued to the death of Pope Clement V. Only then did the Portuguese venture to elect Dom Lourenço Annes as the Master and, despite fierce opposition, the two parts of the Order were separated once and for all. The last Master of the Order in Portugal was Dom Jorge, Duke of Coimbra, bastard son of John II. On his death, the Mastership passed into the hands of King John III, who also took the reins of the Orders of Christ and Avis. I mention this in an attempt to justify the inclusion of members of the Order of St James in the undertaking of the Discoveries of the Order of Christ and the argument that King Dennis already foresaw the extension of the Western Crusade beyond the seas. The frontiers of Portugal had been established since the time of his father, King Alphonse III and the Portuguese Crusade would be conducted in North Africa. The Castilian Crusade was to drag on in the Peninsula until 1492 and the conquest of Granada – the same year that Salvador Fernandes Zarco/'Cristóbal Colón' reached the Antilles.

One other interesting element appeared in regard to the Order of St James. It also had a convent at Arruda, which was transferred to Santos-o-Velho when the knights moved to Alcácer do Sal. The role of this institution was 'to receive the wives and daughters of the Commanders and dignitaries of the Order when they were away at war'.[11] It also received, as paying boarders, 'daughters of the nobility when their betrothed were absent'. King John II transferred this institution to the Convent of Santos-o-Novo in 1492 (see p. 345). It can be confirmed, therefore, that this was the Convent where Filipa Moniz Perestrelo was boarded while waiting for her betrothed, Salvador Fernandes Zarco, to whom John II had entrusted a difficult and mysterious mission – the alteration of the coordinates that divided the world, with the Treaty of Tordesillas.

Notes and References

1. See p. 384.
2. Azurara (1841); Pina (1790); Leitão (1630), Book VI, p. 23; Leão, *Primeira parte das Chronicas dos reis de Portugal reformadas* (1600).
3. *Armorial Lusitano* (Faria, 1961).
4. King Sebastian meditated in this monastery before his meeting in Cuba and departure to the Battle of Alcazar Kibir.
5. Las Casas (1875–6), Pt I, chap. 32.
6. *Carta mensagera* from the Catholic Monarchs to Columbus (22 September 1493), see Navarrete, *Documentos diplomaticos* (1915).
7. Letter from Columbus to the Catholic Monarchs, see *Codice diplomatico-americano* (Havana, 1867).
8. Remón (1530).
9. Son of Dom Álvaro Pires de Castro, Count of Arraiolos, 1st Constable of Portugal, Governor of Lisbon.
10. Garcia Rodrigues da Fonseca took part in the conquest of Lamego, during the formation of the Kingdom of Portugal. Dom Vasco Fernandes Coutinho (grandfather of his namesake), 6th Lord of Leomil, was the Royal Bailiff of King Ferdinand I. He married Brites Gonçalves de Moura, Lady of the Bedchamber of Queen Philipa of Lancaster, wife of King John I.
11. Lima, Dom Luiz Caetano de (1734), pp. 360–540.

28 Huntsmen and Admirals

I think it is essential to go deeper into the origin of the 'post' of 'Huntsman', which seemed to have been an open door to higher positions and titles.

The Aguiar family (pp. 349 and 352, Trees I and IV, 20), which was so closely linked to the Mellos (pp. 363 and 378, Trees II, 27; IV under 20; IV, 47), the Zarco–Câmaras and the Henriques (pp. 392 and 413, Trees II, 43; III, 34 and 36), particularly aroused my interest.

The vast area between the rivers of Odeácere and Odivelas (today reduced to streams) was what was known as an 'open reserve' at this time. Although it was not walled, it was reserved for royal hunts and, of course, the big landlords of the Duchy of Beja, especially the Dukes of Braganza. Only later did Dom Jaime, fourth Duke, start the construction of the ducal palace at Vila Viçosa and set up a 'closed reserve'. Tournaments started to be less frequent in the Iberian Peninsula from the end of the fourteenth century and, as training for warfare, were replaced by bull-baiting[1] and hunting with hounds[2] and falcons,[3] arts that went hand in hand with the post of Huntsman in the fifteenth century.

Through the words of the chronicler Fernão Lopes and of King Duarte,[4] we know that King John I wrote the *Livro da Montaria* (Book of Hunting), which has since disappeared, 'written and revised with the agreement of very good huntsmen'. These huntsmen were lords that guarded the open reserves, many of which were their own lands. But some of them had to take charge of royal reserves and organise royal chases.

On a trip to Estremoz, in the company of Pisani Burnay, in search of weapons made prior to the seventeenth century, at the time of the Council of Europe Exhibition in Lisbon in 1983, we visited the History and Ethnography Museum. There we found a collection of funerary stelae which proved the ecumenism of the Templars, with Christians and Jews buried in the same Holy ground. And we were able to study the evolution of the Portuguese crosses of the Order of the Temple, some of them already showing signs of the cross of the Order of Christ, even before its foundation.

Pisani Burnay then called my attention to a tall, bare tower – a true *côba* (see Figure 24) – the interior of which had iron bars fixed in its decaying walls. We gradually realised that it was a 'falcon tower', and that the bars were perches for the hunting birds. The door hinges were pre-sixteenth century. The royal arms above the door were of the second half of the fifteenth century. The tower showed how much the Kings of that time appreciated falconry, and the discovery led us to read Dom Luiz Caetano de Lima, who wrote:[5]

> The post of 'Caçador Mor' was concerned with hunting with falcons, while that of 'Monteiro Mor' was connected with hunting with the hounds. The two were merged in the post of 'Monteiro Mor'.
>
> The post of 'Caçador Mor' seems to be the same as that of the 'Falcoeiro Mor' that the Chronicle of King Ferdinand refers to, mentioning João Gonçalves as the holder of the post up to the year of 1370.

I shall now note the Huntsmen that were contemporaries of Salvador Fernandes Zarco/'Cristóbal Colón':

(a) Fernando Afonso Pereira, cousin of Diogo Pereira (p. 366, Tree III, 41; see p. 493, Tree XVIII, 20 and 58), married to Brites de Castro, and of Nuno Pereira, who killed his wife Beatriz Henriques, daughter of Fernando Henriques (Tree III, 35).
(b) Afonso Vaz de Brito, cousin of Estêvão de Brito, father-in-law of Dom Pedro de Sousa, Count of Prado, who killed his second wife Margarida de Brito (Tree III, 29).
(c) Pedro de Castro, son of Diogo Pereira and Brites de Castro (Tree III, 41).
(d) Henrique Henriques, married to Filipa de Noronha, grand-daughter of João Gonçalves Zarco and uncle by marriage of the above-mentioned Dom Pedro de Sousa (see also p. 421 and 443, Trees XII and XIII, 52).

THE AGUIARS

As can be seen from Tree III (p. 366), these four Huntsmen were all closely related to the Aguiars and to Colón-Zarco. Before the creation of this post by King Alphonse V, however, Huntsmen already enjoyed special privileges from Kings and Princes, as well as from the Dukes of Braganza, who were almost as powerful as the Kings. After the reign of

Alphonse V, some of the Huntsmen who were not given the position of 'royal' received other rewards and privileged posts, sometimes being raised to the peerage.

These are to be found in the Aguiar, Albuquerque, Alcoforado,[6] Almada, Almeida, Dias, Furtado de Mendoça, Mello and other families. Either directly or through marriage, they all appear in the above genealogical trees and many were navigators and some were Admirals of the Realm.

I explained the origin of the Aguiar family in the Portuguese edition of the book and referred to some of its most illustrious members. Among these were Dom Estêvão de Aguiar, Abbot of Alcobaça, and his son (or nephew) and heir, Dom António Esteves de Aguiar (see p. 493, Tree XVIII, 78), who was Bishop of Ceuta at the time of King John I and who used the arms of the Colonnas, the family of the Archbishop of Lisbon, Dom Agapito Colonna, father of Pope Martin V. The genealogists[7] do not agree in regard to the kinship of Dom Estêvão and Dom Pedro Afonso de Aguiar, Governor of Montemor-o-Novo. Some claim they were brothers and others say they were cousins.

Alão de Morais[7] claims that the Governor was the brother of João Afonso de Aguiar, who married Maria Esteves in the Algarve and begot two sons, Pedro Afonso and Diogo Afonso, besides another one, João Afonso (known as 'the Hawker'), who was born in Viana do Alentejo in the future Duchy of Beja. They lived in the reign of Alphonse V and the eldest, Pedro Afonso, married Mécia de Sequeira, Lady-in-Waiting of Queen Isabel and daughter of Francisco de Torres (the Lord Chamberlain) and Violante Nogueira, who was a kinswoman of her namesake, the Mother Superior of the Convent of Santos where Filipa Moniz was boarded before marrying Salvador Fernandes Zarco. This Pedro Afonso de Aguiar was the father of Jorge de Aguiar, Admiral of the Home Fleet and of an India nau in the reign of Manuel I.

Figure 28.1 is from vol. III, 1st part, p. 229, of Alão de Morais's work, in which he refers to the two other sons of João Afonso de Aguiar: João Afonso and Diogo Afonso.

As already mentioned, fifteenth- and sixteenth-century genealogies are replete with omissions and erasures owing to the Inquisition, particularly in cases of people of Jewish origin or who descended from incestuous or ignominious bastardy. Besides this, information is fairly exact, although sometimes names have been omitted or merely registered as follows:

– 'no children as far as is known'
– 'did not marry'
– 'died young' or 'at birth'
– 'nun'
– 'not known whether he/she had offspring'
– 'had B' (bastard, unnamed) or 'had BB' (bastards).

§

2   João A.º de Aguiar f.º 2.º ou 3.º de I.º A.º de Aguiar n. 1 Foi Prov.ᵒʳ de Evora. Casou cõ D. Isabel de Mello f.ª de J.º de Mello Copeiro mór delRei D. A.º 5.º E de sua 2.ª m.ᵉʳ D. Melicia de Sousa e teve

        3   ... ...
        3   ... ...
        3   D. Joana de Mello m.ᵉʳ de D. P.º de Sousa 1.º Conde de Prado
        3   D. Leonor de Mello m.ᵉʳ de Gomes de Fig.ᵈᵒ Prov.ᵒʳ d'Evora.

§

2   *Diogo* (λ) (¹) Affonso de Aguiar hirmão de P.º A.º de Aguiar n. 1. (²) casou na Ilha da Madeira cõ Izabel Gls. *da Camera* filha de Joam Gonçalves Zarco e teve

        3   Ruy Diaz de Aguiar
        3   P.º A.º de Aguiar
        3   D.º A.º de Aguiar o moço
        3   Inez Dias da Camera m.ᵉʳ de Lopo Vaz de Camões cõ g.

3   Ruy Diaz de Aguiar filho 1.º deste faz delle memoria Andrade chron. delRei D. J.º o 3.º 2. p. c. 82. fls. 118. Casou com Leonor Homẽ de Sousa sua prima cõ hirmã q tàbem foi m.ᵉʳ de Duarte Pestana filha de Garcia Homẽ de Sousa E de sua m.ᵉʳ Cn.ª Gls. da Camera s. g.

(λ) — Deste procedem os Aguiares das Ilhas outro livro lhe chama Diogo A.º de Aguiar E he certo.

(¹) — N. E. — Escrito sobre uma palavra que parece: *Damião*.

(²) — N. E. — aliás n. 2.

28.1   Alão de Morais's *Pedatura*.
     a   1st part, vol. III, p. 229.

/2/ João A° de Aguiar, 2nd or 3rd son of the 1st A° de Aguiar. He was Purveyor of Évora. Married Isabel de Mello, daughter of João de Mello, Lord High Steward of the Royal Household of King Alphonse V. His second wife was Melicia de Sousa, who bore him

/3/ . . . .
/3/ . . . .
/3/ Joana de Mello, wife of Dom Pedro de Sousa, 1st Count of Prado
/3/ Leonor de Mello, wife of Gomes de Figueiredo, Purveyor of Évora

§

/2/ Diogo (A)[1] Afonso de Aguiar, brother of Pedro A° de Aguiar N° 1.[2] married in Madeira to Isabel Gonçalves da Câmara, daughter of João Gonçalves Zarco and begot

/3/ Ruy Diaz de Aguiar
/3/ Pedro A° de Aguiar
/3/ Diogo António de Aguiar the younger
/3/ Inez Dias da Câmara, wife of Lopo Vaz de Camões, had children

Andrade mentions Ruy Diaz de Aguiar in the *Chronicle of King John III*, part 2, chap. 82, page 118. He married Leonor Homem de Sousa, his first cousin, who was also the wife of Duarte Pestana and daughter of Garcia Homem de Sousa and Constança Gonçalves da Câmara. No children.

(A) The Aguiars of the islands descend from him. Another book calls him Diogo A° de Aguiar, which is correct.
(1) *Note*: Written over a word which seems to be: *Damião*
(2) *Note*: otherwise n° 2.

b Translation of 1st part, vol. III, p. 229.

Omissions are rare when referring to legitimate marriages between fairly important people and are non-existent concerning people of high position and the nobility – except, as already mentioned, in cases of bastardy or Jewish blood. We must therefore explain the very strange omission of the names of the first two sons of João Afonso de Aguiar (known as 'the Hawker' and son of his namesake) who was the 'huntsman' of the Duke of Braganza and married Isabel de Mello. He later became the Purveyor of Évora (see p. 362, 366, 367 and 493, Trees I, III, IV and XVIII, 20).

THE ELIMINATED BROTHERS

Let us now see who the father-in-law of this João Afonso de Aguiar was.

The Mello family first appear in the genealogical trees with Mem Soares de Mello in the reign of King Sancho II. In the fourth generation, Vasco Martins de Mello begot three children, the eldest being Martim Afonso de Mello (p. 453, Tree XVI, 27).

1. Martim Afonso de Mello – Captain of the Royal Guard of King Duarte and Governor of Olivença. He married Margarida de Vilhena (daughter of the Royal Bailiff). He would be the grandfather of Filipa de Vilhena, who was to marry Dom Álvaro de Portugal (see p. 453, Tree XVI, 71), son of Dom Fernando II, Duke of Braganza. Through this marriage, Filipa became the sister-in-law of Dom Jorge de Portugal (married to Isabel de Colón), from whom descended seven lines of heirs of Colón-Zarco's titles and assets.

Martim Afonso de Mello took Briolanja de Sousa as his second wife and begot six more children.

1. Isabel de Mello – married in digamy to Dom Duarte de Menezes (p. 365 and 366, Trees II and III, 26).
2. Martim de Mello – Governor of Castelo de Vide. Married Brites de Azevedo, daughter of the Lord of Aguiar.
3. João de Mello – Governor of Évora and Lord High Steward of the Household of King Alphonse V. Married Isabel da Sylveira (p. 367, Tree IV, 27), daughter of the King's Private Secretary, Nuno da Sylveira.
4. Diogo de Mello – died young.
5. Branca de Mello – married Dom Fernando Henriques, Lord of Alcáçovas (Tree III, 33).
6. Brites (or Beatriz) de Mello (Tree II, daughter of 27) – married Dom Gonçalo Coutinho, second Count of Marialva (see p. 453, Tree XVI, 74).

Nearly all of the descendants of these six were the lords of many lands, became mayors of many towns, occupied high positions in the court or were raised to the peerage.

Our attention will focus only on the fourth child, João de Mello, first married to Isabel da Sylveira and then to Mécia de Sousa in digamy.

The following nine children were born of his first marriage:

1. Martim Afonso de Mello – heir of his father's assets, but did not succeed him as Governor of Serpa. Married Leonor Barreto, daughter of Gonçalo Nunes Barreto, Mayor of Faro. One of his sons, Jorge de Mello, Governor of Redondo and Pavia (both places in the Alentejo), went to India with Jorge de Aguiar as an Admiral of the Fleet.
2. Garcia de Mello – Governor of Serpa and Lord High Steward of the Household of King Alphonse V. Married Filipa da Sylva, daughter of Henrique Pereira, Commander of Santiago and Governor of the Household of Prince Fernando, Duke of Beja.
3. Henrique de Mello – lived at Serpa. Married Brites Pereira Barreto, daughter of Nuno Pereira de La Cerda (see p. 425).
4. Briolanja de Mello – married Diogo de Sampayo, Lord of Ansiães and Vilarinho.
5. Leonor de Mello – married Nuno Barreto, Governor of Faro.
6. Filipa de Mello (or da Sylveira) – married Pedro de Moura, Governor of Marvão.
7. Constança de Mello – did not marry.
8. Branca de Mello – married Rui Dias Pereira of Serpa.
9. Isabel de Mello – married João Afonso de Aguiar, Purveyor of Évora.

From his second marriage, to Mécia de Sousa, daughter of Diogo Lopes Lobo, Lord of Alvito, João Afonso de Aguiar had one daughter:

1. Brites de Mello – married Fernão da Sylveira.

These ten brothers and sisters had cousins (see p. 453, Tree XVI) who were raised to the peerage or given high posts, such as Dom Rodrigo de Mello, who became the first Count of Olivença, Dom Gonçalo Coutinho, second Count of Marialva, Dom Leonel de Lima, first Viscount of Vila Nova de Cerveira, besides the Governor of several towns throughout Portugal (see p. 402, Tree VI, father of 45).

Neither the Mellos or the Aguiars nor the ladies they married had any Jewish blood. They were all cultured, good Catholics and were never involved in any injurious disputes. Yet how then can the absence of João Afonso Aguiar and Isabel de Mello's first two children from the genealogical trees be explained? Their parents, grandparents and closest relatives occupied high positions in the administration of the Kingdom.

It is not, therefore, the result of provincial illiteracy. And as they were not New Christians, there was no reason for the genealogists to fear the Inquisition.

We know that the Portuguese Dom Gonçalo Annes de Aguiar, great-grandson of Dom Pedro Mendes de Aguiar, moved from Portugal to Castile in the reign of King Ferdinand II (known as 'the Saint') and was a favourite of King Alphonse X (known as 'the Wise') (1252–84). He served in the Sevillian War and the King granted him some lands in Andaluzia to which Dom Gonçalo gave his family name, but in the Castilian form of Aguillar. Owing to the high esteem in which he was held, he was buried in the Royal Chapel in Córdoba. He married Berenguela of Cardona, daughter of the Count of Cardona, and started the Aguillar families in Portugal and in Córdoba. The Portuguese branch settled in Elvas (Alentejo).

The Portuguese surname Aguiar that had been transformed into Aguillar in Cordoba later returned to its original form in Portugal. This mutation of surnames is still frequent today.

The Aguiars did not live on the *'monte'* of that name – which had been given by one of their ancestors – as it belonged to the Azevedo family at that time and soon after was passed on to the Sylveiras, Lords of Alvito.

Even if João and Diogo had been born at Aguiar, which is unlikely, it is only five miles from Viana do Alentejo, which belonged to the Menezes, and Pedro Afonso de Aguiar and his wife could easily have baptised their children there.[9] The proof that they were born is to be found in the genealogical register. Only the names were suppressed. Why? They were both legitimate sons. The two daughters that were born later were both baptised in Beja. The two sons remained unknown to history.

In view of the close connections of kinship among the families of Beja, Vila Ruiva and Alvito and taking into consideration the 'open reserve' between the Odeácere and the Odivelas rivers and the high qualifications of the Huntsmen, is it not possible that João Afonso de Aguiar had been the 'guardian' of the mansion of Vila Ruiva before he was raised to the position of the Purveyor of Évora, a post that was received later in life? Or had Diogo de Aguiar occupied the mansion as its guardian? Or had both brothers been there?

In his siglum, Salvador Fernandes Zarco states that he had been born at Cuba. Naturally his father, the Duke, could not have kept a mistress in the town of Beja, where he lived with his legitimate wife, as a matter of discretion. On the *London Map* Bartolomeu said that he had been born at *'Terra Rubra'* (i.e., Vila Ruiva).

It is essential now to relate an episode in the life of Prince Fernando, Colón-Zarco's father. He abandoned Beja and his wife Beatriz, in 1452, and secretly left the country. He intended to join his uncle Afonso, King of Naples, in the hope that, as the King had no heirs, he would adopt him as such. Immediately his father, King Alphonse V, heard of his departure, he ordered the Count of Odemira, Dom Sancho de Noronha, son of Dom Afonso de Noronha, Count of Gijón (who was a bastard son of Henry II of Castile and who started the Noronha family in Portugal), to take a fleet to block the Straits of Gilbraltar and intercept the Duke's vessel. This he did and brought the fugitive back to Lisbon to face his father's ire. Colón-Zarco would have been four years old at this time.

I cannot invent the date in which Prince Fernando abandoned his mistress Isabel da Câmara and his son, but I can advance a conjecture. If Isabel had returned to Madeira, Prince Fernando would have had no reason to continue visiting Cuba or Vila Ruiva. He would certainly have remained in the Castle of Beja with his wife and surrounded by his 'Court' of knights of the Order of Christ, when he was not fighting the Moors in North Africa.

After the death of Dom Afonso, first Duke of Braganza, his son and heir, Dom Fernando and his wife Joana de Castro went to live in the Braganza's castle at Vila Viçosa. From then on, Vila Ruiva was nothing more than a chase.

There remains one mystery to explain. What led Isabel da Câmara to go to the Alentejo? Even if she had been a Lady-in-Waiting of Princess Beatriz, who lived in Beja, or the Duchess of Braganza, who resided at Vila Viçosa, how can her departure from Madeira be explained? Young ladies of those days did not travel around looking for jobs in the Court or Ducal Houses. The fact is that her brother, João Rodrigues da Câmara (second Captain of Madeira) was at Serpa, a few kilometres from Beja and Cuba. Like many gentlemen from Madeira and the Azores of those days, he had gone there in search of a bride. He married Isabel de Mello, daughter of Rui Pereira de Serpa. His brother Rui had acquired the Captaincy of São Miguel from João Soares de Albergaria (see p. 392), who had also gone to Beja looking for a bride; João naturally went to take possession of his captaincy, and took his bride with him. His sister, meanwhile, whom he had taken to Beja with him in order to find her a good future, had settled in Beja or Vila Viçosa. This, as I have said, is only conjecture, but it is based on sound logic.

Being unable to obtain documentary proof regarding Isabel, I can do no more than resort to logical reasoning in an attempt to reconstitute her story.

1. On being abandoned by Prince Fernando, she could no longer remain at the mansion of Cuba or perhaps now in Vila Ruiva.
2. On leaving the country, she could have returned to Madeira with her illegitimate child. But she preferred to leave him in the care of his powerful father; soon he would be not only Duke of Beja but also

Lord of Madeira. Salvador Fernandes would receive a much better education in continental Portugal than on one of the islands.
3. Diogo Afonso de Aguiar had fallen in love with Isabel Zarco, and after her love affair with Prince Fernando had come to an end, he asked for her hand in marriage.
4. Did Prince Fernando break with Isabel on the insistence of his wife or did he suspect her love for Diogo Afonso de Aguiar? If the two of them had planned to marry and start a new life far from Beja, Madeira would have been the place to go. But without a bastard son.
5. Diogo Afonso de Aguiar was neither a common ploughman nor an experienced seaman – as his descendants would be. He would certainly not have wished to leave a prosperous region where he could count on influential family ties and a rosy future and emigrate to Madeira unless he had a good reason – either his love for Isabel Zarco or fear of reprisals on the part of Prince Fernando, who may have thought himself betrayed. This is mere speculation, but a distinct possibility.
6. Isabel da Câmara returned to Madeira and Diogo followed her in order to settle down with her. Salvador Fernandes stayed in the area of Beja under the protection of his father, but not alone at the house where he was born. His mother would naturally have been a friend of her neighbour Isabel de Mello, who was married to João Afonso de Aguiar (see p. 367, Tree IV, son of 20) who, I presume, had the house of Vila Ruiva under his guardianship, and who later had two children. Salvador Fernandes could have been entrusted to her care. He would go to Madeira to serve his apprenticeship before the mast – the future of the youth of those days in Portugal.
7. According to my sub-thesis, the two sons of João Afonso de Aguiar and Isabel de Mello were Diogo and Bartolomeu.

BARTOLOMEU AND DIOGO COLOM

Although I believe that Bartolomeu was older than Diogo, the omission from the Aguiar's genealogical tree make it impossible to prove this. We cannot even tell if they really were the 'Colón brothers'. However, I shall continue.

8. Salvador Fernandes must have been five or six years older than Bartolomeu and ten or twelve years older than Diogo. The difference of ages between Colón-Zarco and Diogo may explain why Columbus did not pay the same attention to him than to his brother Bartolomeu. There are some letters written to Bartolomeu but none written to Diogo.

Diogo decided to enter the Church, which he did after taking part in the third voyage to the Antilles with his brothers. He did not sail on the fourth voyage. Bartolomeu was to become a cartographer. Salvador, grandson of the King, would become the Admiral of the West Indies.

Salvador Zarco = *Cristóvam Colom*
Bartolomeu de Aguiar = *Bartolomeu Colom*
Diogo de Aguiar = *Diogo Colom*

9. When did this metamorphosis take place? I can see Diogo studying religion in Beja or Évora; Salvador/Cristóvam and Bartolomeu in the Order of Christ, but, in view of the five or six years' difference in age, the former had gone to Madeira long before the latter. Both knew the art of seamanship and of mapmaking – each one followed his own profession. And in Porto Santo, Salvador Fernandes Zarco met his relative, the young Filipa Moniz Perestrelo.

The alleged metamorphosis, the crucial point of my sub-thesis, happened before King John II killed Dom Diogo, Duke of Viseu and Beja, (a half-brother of Colón-Zarco) and took over the reins of the Order of Christ, in view of the lack of interest shown by the Master, Dom Diogo.

'Cristóvam Colom' came into being only in order to pressure the Castilians into signing the Treaty of Tordesillas. His brothers were his back-up in 'Operation Castile'. Before this, nobody had ever heard of them.

10. It is an enigma that historians have never tried to explain: three Colombos, navigators, cosmographers, cartographers, leaders of men, experienced in warfare, versed in Latin – and from a fifteenth-century family of wool-carders, tavern-keepers and cheese-makers.

It can be assumed that Bartolomeu and Diogo became 'Colom' because Salvador Fernandes had done the same thing: Colom = Arco = Zarco. They could thus be passed off as brothers. This also makes it easier to understand the propensity of Salvador Fernandes/'Cristóbal Colon' for theology. It was surely not the friars of La Rábida (the Marchena cousins) alone that made him familiar with the intricacies of religion. In their intimate contacts on board ship and in the Antilles, Diogo de Aguiar had probably made his contribution to enlightening him in the Divine Mysteries. Salvador Fernandes Zarco's family in the Duchy of Beja, particularly on the paternal side, was deeply imbued with Christian faith.

There is no doubt that Salvador Fernandes Zarco received an education that was exclusively reserved for princes, noblemen and prelates in the fifteenth century. In no way could he have come from the Genoese family that has been assigned to him, and even less could he have been a pirate.

11. On studying the genealogical trees, it is not difficult to deduce who suppressed the names of Bartolomeu and Diogo de Aguiar and relegated them to obscurity. Fernando da Sylveira, a Doctor of Law of Alvito, was Chief Justice and Lord Chancellor at the beginning of John II's reign. Aires Annes of Beja, married to Catarina Moniz Perestrelo, sister-in-law of Salvador Fernandes Zarco, succeeded him as the King's Private Secretary. And thus the three 'brothers' disappeared from the genealogical registers. Either one of these two men could have received orders from the King to consign one Zarco and two Aguiars to anonymity. And there they remained – until a siglum was deciphered five centuries after it had been conceived.

ADMIRAL KINSMEN OF COLÓN-ZARCO

'I am not the first Admiral in my family' said 'Cristóbal Colón'. The Genoese concluded that he was boasting and being vain. I shall prove that he was not.

Dom Caetano de Lima[10] tells us:

> Dom Fuas Roupinho[11] was appointed the first Admiral of Portugal in the reign of King Alphonse I; the royal chronicler, however, Friar António Brandão, gave him no more than the title of *'Captain'*[12] when he described the victory he achieved over the Moors off Cape Espichel in 1180.

It was only in the time of King Dennis that the title of 'Admiral' began to be used, on a par with that of 'Captain of the Fleet', and the 'position' was defined by Manuel Severim de Faria (1583–1655) in the *Notícias de Portugal* (1555). According to Friar Francisco Brandão,[13] there was a difference in this reign between 'Admiral of the Fleet' and 'Admiral of the Galleys', who led the struggle against the Moors that preyed on the Portuguese coast. It so happens that João Gonçalves Zarco, Salvador Fernandes Zarco's grandfather, was the 'Admiral of the Galleys', so Colón-Zarco was not the first Admiral in his family.

Not satisfied with this, however, I continued my search in chronicles and genealogical trees, and discovered that the first Admiral that really bore the title was Nuno Fernandes Cogominho. In 1317, King Dennis appointed a Genoese, Manoel Pezagno, to the post. He had already achieved a certain success against the Moors, but in the waters of the Mediterranean. His three sons, Carlos, Bartolomeu and Lanzarote, who adopted the Portuguese name Pessanha, succeeded the father and each other in the position. But they proved to be a dismal failure and King Peter I sacked the last one. Later, the new appointee was Dom João Afonso Tello (p. 365, Tree II, 5), brother-in-law of King Ferdinand I and married to Guiomar de Portocarrero a surname that would later be linked to that of Colón. He was the grandfather of Dom Duarte de Menezes (pp. 365, 366 and 398, Trees II, III and V, 26) who, as we have already seen, was a cousin by marriage of Colón-Zarco.

Dom João was succeeded by Dom Rui Freire de Andrade, son of Dom Nuno Freire, Master of the Order of Christ and great-grandfather of Pedro Álvares Cabral who, in turn, was the father-in-law of Fernão Álvares Cabral (p. 362, Tree I, father of 18). The last-named was a first cousin of Diogo Cabral, married to Brites Gonçalves, said Câmara and thus brother-in-law of Isabel Gonçalves Zarco, Colón-Zarco's mother (pp. 362, 398 and 452, Trees I, V and XV, 9).

Dom Rui was followed by Dom Pedro de Menezes, whose first wife was Margarida de Miranda (p. 365, Tree II, 25), who bore Beatriz de Menezes. Beatriz married Dom Fernando de Noronha, brother of the Archbishop of Lisbon, Dom Pedro de Noronha, who was the lover of Colón-Zarco's sister-in-law, Isabel Moniz Perestrelo.

Admiral Dom Pedro begot one bastard son, the above-mentioned Dom Duarte de Menezes, who was also Admiral of the Realm (see pp. 365, 398 and 453, Trees II, V, and XVI, 26).

According to Duarte Nunes de Leão,[14] Dom Duarte de Menezes was followed as Admiral of the Realm by Rui de Mello da Cunha, married to Brites Pereira, daughter of the Admiral Carlos Pessanha. This Rui de Mello was the son of Brites de Brito (or de Mello) and Dom Álvaro da Cunha. Brites de Brito, in turn, was the sister of Estêvão de Mello (see p. 453, Tree XVI, 70), married to Teresa de Andrade (or Novais), daughter of the Admiral Dom Rui Freire de Andrade (p. 362, Tree I, 17) and first cousin of Isabel de Mello (sister of Martim de Mello – pp. 366 and 453, Tree XVI, son of 27), first wife of Dom Duarte de Menezes.

The Admiral of the Realm after Rui de Mello was Dom Pedro Rodrigues de Castro (1450), cousin of Dom Álvaro de Castro (see p. 402, Tree VI, grandfather of 48), whose daughter, Isabel de Castro (sister of Dom Álvaro de Castro, married to Isabel da Cunha), was the second wife of Dom Duarte de Menezes (see pp. 402 and 453, Tree VI and XVI, 26).

Dom Pedro was followed in 1454 by Lanzarote Pessanha, son of the above-mentioned Rui de Mello. In 1457, the appointee was Dom Nuno Vaz de Castel-Branco (Castelo Branco), brother of Dom Martinho de Castel-Branco (see p. 445, Tree XII, 51) and Mécia de Noronha (granddaughter of João Gonçalves Zarco) and cousin of Simão de Castel-Branco, married to Maria de Noronha, great-granddaughter of João Gonçalves Zarco, thus kinsmen of Colón-Zarco.

Dom Nuno was succeeded, in 1484, by Pedro de Albuquerque, who some genealogists think was the grandson of his namesake who was nicknamed 'Olive Oil'. The Admiral was the father of Francisco Coutinho de Albuquerque (see p. 445, Tree XII, 54), married to Beatriz da Câmara, bastard daughter of Rui da Câmara (first Captain of the Island of St Miguel of Azores, already mentioned) and aunt of Colón-Zarco. Other genealogists say that Pedro de Albuquerque was the uncle of Gonçalo de Albuquerque (married to Leonor de Menezes), father of Constança de Castro, wife of Dom Fernando de Noronha. This Dom Fernando was the bastard son of the Archbishop of Lisbon. Dom Pedro de Noronha and Isabel Moniz Perestrelo (see pp. 398 and 362, Tree V, descent of 3 and Tree I, 9).

It is thus proved that Salvador Zarco/'Cristóbal Colón' was not lying when he said that he was not the first Admiral in his family. Besides his grandfather João Gonçalves Zarco, all eight Admirals from the reign of King Dennis to that of John II were either his close or distant relatives.

There were other Admirals during this period, although the title had a different designation in Portuguese (*Capitão-mór do Mar*). In the reign of King John I, this post was held by Dom Afonso Furtado de Mendoça, who married Constança Nogueira (daughter of Afonso Nogueira, Mayor of Lisbon, and Joana Vaz de Almada) and begot Violante Nogueira the Mother Superior of the Convent of Santos where Filipa Moniz Perestrelo was boarded before her marriage. Afonso Furtado de Mendoça's mother-in-law was the sister of Dom João Vaz de Almada, Count of Avranches (see p. 411, Tree IX, father-in-law of B), and the aunt of Dom Alvaro Vaz de Almada likewise Count of Avranches, who was the Admiral of the Realm during the reign of King Duarte.

Dom Álvaro's first wife was Isabel da Cunha, daughter of the above-mentioned Dom Álvaro da Cunha, and his second wife was Catarina de Castro, daughter of Dom Fernando de Castro, Governor of the Household of Prince Henry 'the Navigator' and the first Count of Monsanto. His son Dom Fernando de Castro married Constança de Noronha, aunt of the above-mentioned Archbishop of Lisbon (see p. 398, Tree V, descents of 3 and 7). That makes ten of Salvador Fernandes Zarco's kinsmen who were Admirals before him. Since he must have known only the closest of them, he was probably unaware that he was the eleventh Admiral of the family.

After Colón-Zarco had been raised to the rank of Admiral of the Indies by the Catholic Monarchs, his cousin by marriage, Dom Francisco de Almada, was appointed Admiral of the Fleet in order to command an expedition to occupy the Bahamas, which had recently been discovered by Colón-Zarco. This expedition was nothing but a bluff to force the Catholic Monarchs to sign the Treaty of Tordesillas.

The already-mentioned Admiral Dom Pedro de Castro commanded

the fleet that took Princess Leonor (sister of Prince Fernando, Duke of Beja and so Colón-Zarco's aunt) to Germany in 1451 to marry Emperor Frederick III (see p. 444, Tree XI). The wedding ceremony was performed by Pope Nicholas V. The retinue included the sister of the Archbishop of Lisbon, Dom Pedro de Noronha (see pp. 362 and 398, Tree I and V, 9); Dom Lopo de Almeida, first Count of Abrantes, father of Dom João de Almeida, who married Inês de Noronha, daughter of Colón-Zarco's sister-in-law Isabel Moniz Perestrelo; and also Dom João da Sylveira, first Baron of Alvito (p. 365, Tree III, 40; see pp. 363 and 368) who, as Ambassador to the Count of the Holy Roman Empire, had arranged the details of the marriage between the Emperor Frederick and Salvador Fernandes Zarco's aunt.

Notes and References

1. See Barreto (1970), pp. 79–85.
2. Boar-hunting with hounds or mastiffs, see Soveral (1959), pp. 94–5.
3. Fowling with falcons or eagle owls.
4. 1391–1438; he wrote a book of *The Art of Horseriding* (1842).
5. Lima (1734), pp. 344–438.
6. Descended from the Aguiar and Sousa families.
7. Brandão (1683), Book 14, Chap. 15, p. 4; Soares (1731), Book 4, p. 419.
8. Morais (1688), p. 226.
9. Formerly a prosperous place; King John II convened the Cortes there in 1482.
10. Lima, 'Dom Luiz Caetano de (1734), pp. 321, 338, 523.
11. Bastard son of Count Henry of Burgundy and half-brother of King Alphonse I, first King of Portugal.
12. 'Captain' emphasised by the author.
13. Brandão (1683), Pt VI.
14. Leão, Duarte Nunes de (1600).

29 Stopovers at the Azores, Portugal and Arzila

After his caravel had run aground on 24 December 1492, Colón-Zarco set sail for home from the 'Gulf of Arrows' in the Antilles on 16 January 1493. He sailed on the 'Niña' and Martín Alonso Pinzón commanded the 'Pinta'. He set an extraordinary course which Morison[1] classified as a crass mistake, as he pointed his prow northward in the direction of the Arctic Circle. The crews were longing to get back to their families and receive the rewards promised by their Sovereigns, so his detour seemed doubly inexplicable.

When the defenders of the Portuguese theses mention that 'Colón', intending to see the Portuguese King, docked at Lisbon on his return from the Antilles instead of making straight for his departure point of Palos, or even Barcelona, where he would have found the Catholic Monarchs and have been able to transmit his great feat to them, the Catalan and Genoese supporters argue that Lisbon was the nearest anchorage to the Antilles. But, sailing from there, more precisely from Haiti, on parallel 20 degrees North, the shortest course for a navigator in the service of Castile would be east–northeast to the Canaries. As he now knew the location of the American archipelago, he could have made a straight run to Las Palmas, on parallel 28 degrees North, and taken on fresh supplies in this Spanish territory.

The Genoese and Catalans will reply that 'Colón' wished to return to Castile by the shortest route without touching the Canaries. But they forget the fact that Columbus mentioned in his log that they were running short of food, which, as it happened, occurred half-way home. Columbus's nearest landfall on the Spanish mainland was on parallel 36 degrees North. Yet he headed even further north to the Azores, which is on parallel 37 degrees North.

Las Casas relates that on 3 February Columbus realised that the Polar star was very high[2] as his ship was on the latitude of Cape St Vincent in Portugal. But Columbus held his course, putting forward the unconvincing argument that he could not change it owing to the roll of the ship. Las Casas adds that Columbus came to the conclusion that they 'had not yet reached the proximity of the Azores'. The word 'proximity' instead of 'latitude' indicates the intention of sailing close to the islands. Columbus also explained that he had come to that conclusion owing to the fact that the sky was overcast and that it was raining heavily. How could he have known the zone of the Azores so well?

From 4–7 February the two vessels ran into a tremendous storm and on 9 February Columbus wrote in his log that he had seen plants (i.e., like those of the Sargasso Sea). The storm abated. Pinzón's 'Pinta' was already further south, running eastward. But Columbus persisted in sailing north-east until sighting land. 'Unbelievable' cry the critics of a non-Portuguese Columbus; but he wrote in his log:

> my navigation and calculations worked out correctly . . . I was sure that I was near the islands of the Azores and that was one of them . . . I had pretended to have sailed farther in order to delude the pilots and the sailors that know how to read maps, because I want to maintain the control of the route to the Indies . . . because none of them knew exactly where they were, so that none of them could be certain if they were on course for the Indies.

Las Casas relates that during the storm Columbus had decided that the whole crew should draw lots to see who would go barefoot, if they were saved, on a pilgrimage of thanksgiving to Our Lady of Guadalupe. The draw was made with dried peas, one for each member of the crew in a sack and one of them marked with a cross.

Columbus was the first to draw and immediately drew out the pea with the cross. Another draw was made to see who would go to 'Santa Maria de Souto' (according to Columbus himself in his *Diario*), but the Genoese Fornari, in his translation of Fernando Colón's *Historia*, distorted this to 'Our Lady of Loreto', who was more venerated in Italy; and this time it was a sailor by the name of Pedro da Villa who drew the pea with the cross. A third draw was made, this time to see who would go to Santa Clara. Once more it was Columbus who drew the pea with the cross.

In fact, as he was the first one to draw, he could well have performed a piece of elementary conjuring and hidden the pea with the cross between two fingers. If this were so, why did he wish to be the pilgrim? Because there was a Convent of Our Lady of Guadalupe or another of Santa Clara in Spain that he wanted to visit? Or because there were hermitages of Our Lady of Guadalupe on both the islands of Terceira and Graciosa, to where the storm had driven them? Or because a hermitage dedicated to St Clare had been built on the island of São Miguel?

Did Columbus wish to foresee that one of the pilgrimages that had fallen to him could be made in the Azores as if it were predestined? Would this give his crew the impression that the Virgin Mary had anticipated his act of devotion? Columbus constantly demonstrated his devotion to the Virgin, perhaps because he was aware of his Jewish ancestry and feared that people would become suspicious – as they later did. He was unable to calculate his exact position during the storm and after it he did not know where he would drop anchor. But he surely knew from previous voyages that São Miguel and Terceira were about 60 miles apart and thought that he would be able to reach one of them.

THE 'NIÑA' AT SANTA MARIA

On the following day Columbus felt unwell, and shut himself in his cabin because 'I felt weak in the legs through constantly being cold and wet and through lack of food'. They finally sighted the island of Santa Maria, about 40 miles from São Miguel, on 15 February and the crew thought they were off Cape Roca (the Sintra Rock in Portugal), but Columbus told them it was one of the islands of the Azores Archipelago. He sent four men ashore in the longboat and they soon returned with three Portuguese who offered them chickens and fresh bread. The newcomers informed Columbus that the Captain of the island, João Castanheira (whose name was changed to Giovani Castagneda in the Italian translation)[3] greeted him 'saying that he knew him very well'[4] but had not come to welcome him personally because he lived too far away. He said he would appear next day.

When the next day dawned, Columbus, having been informed that there was a hermitage dedicated to Our Lady of the Angels, sent most of his crew ashore in the only longboat they possessed. They went barefoot and dressed only in their shirts. They did not return. At eleven o'clock, with only some of his crew aboard, Columbus decided to raise anchor and lost both his anchors in the process. He rounded a cape and, as though he knew exactly where he was, stopped in front of the hermitage. They saw a large group of armed knights and shortly after a Portuguese longboat full of armed men came alongside the 'Niña'.

Columbus manifested his indignation at the behaviour of the Azoreans and declared that the King of Portugal would punish their hostility, adding[5] that he was the Admiral of the Ocean Seas and Viceroy of India, carrying letters of introduction from the Catholic Monarchs to all the Princes and Lords of the world. He said that even without the mariners that were held on land he had enough crew to take the ship back to Spain. He told the locals that they would pay heavily for their audacity. To which the captain replied: 'I know neither King nor Queen of Castile . . . You are going to find out who the King of Portugal is and you must surrender'.[6]

Columbus swore that he would not leave without taking 100 Portuguese prisoners to Castile and that he would return and lay waste to the land. He set sail for the island of São Miguel on 20 February, but another storm forced him to turn back, mainly because he was worried by the loss of the anchors. He wanted to try and recover them, as he only had three experienced sailors, the rest of the crew being made up of 'Indians' untrained in seamanship. He reached Santa Maria on the afternoon of 21 February and this time Captain João Castanheira received Columbus and all his men, showing them friendship and offering them the food they so sorely lacked. When the 'pilgrim' mariners returned to the 'Niña', however, they claimed that they had heard their captors saying that the trap had been laid on the orders of the King of Portugal, who also planned to arrest Columbus.

This is highly unlikely, as John II could not have guessed that Columbus would take it into his head to make a detour to the Azores instead of heading straight for Castile. There would not have been time to warn the King, let alone for someone to return with such an order. If the King had given the order, Castanheira did not carry it out. Besides this, how can the change of attitude of the Captain of the island, disobeying royal instructions, be explained? The strangest thing of all about this Azorean episode is that it was not mentioned by one single Portuguese chronicler.

On the other hand, none of Columbus's contemporaries could have made up his stay on the island of Santa Maria in such minute detail. But even so, Portuguese chronicles relegate the event to oblivion.

Although the episode could not have been invented, it does contain some false details. Columbus was extremely careless and related some very dubious occurences:

1. Captain João Castanheira sent three sailors in his name as he lived some distance away from the port, a fact that has been proved. According to Gaspar Frutuoso,[7] João Castanheira, a 'gentleman', lived in the interior of the island, on the lands he had been granted.
2. The crew that was left on board the 'Niña' was extremely reduced and inexperienced and the ship was in poor shape to return to Spain.
3. Columbus saw a lot of armed knights on the beach near the hermitage and other soldiers in a longboat.

If this were true, how could he have threatened to take 100 Portuguese hostages and then return to 'lay waste' to Santa Maria and so cause a war between Portugal and Castile? How can João Castanheira's change of attitude, plus his seeming disobedience of royal orders, between 19 and 22 February, be explained? He was never punished and was later made a judge. The Portuguese had received instructions to arrest all foreign ships caught on the Guinea coast. Did Castanheira think that the 'Niña' was one of those vessels?

When they had sailed for the Antilles, three Portuguese caravels had shadowed Columbus's fleet in the area of the Canaries and the Spaniards thought at the time that the Portuguese would try to arrest them. But their fears were unfounded. Morison[8] concluded that King John II had ordered the arrest of Columbus because he was fed up with the fact that Castilian ships used to sail in the wake of the Portuguese and had more than once attacked them. But he was wrong. Portuguese ships had been boarded in 1474, during the reign of Alphonse V, but these attacks had ceased in 1480 following a 'Royal Charter' and the signing of a Peace Treaty that was confirmed in Toledo. Columbus wrote in his log that João Castanheira had said he knew him 'very well', and this led me to find out something about this Captain of the Island of Santa Maria.

ALBERGARIA, CASTANHEIRA AND RUI DA CÂMARA

According to Frutuoso, who confirms the narration of Colón-Zarco and his son Don Fernando, João Castanheira was a squire of Dom Manuel, the Duke of Beja and the future King. He was the interim Captain of Santa Maria in the absence of João Soares de Albergaria (see p. 346), who had gone to Portugal in June 1492 to marry his second wife, Branca de Sousa Falcão of Beja. Albergaria, Captain of Santa Maria and São Miguel, had been in Madeira in 1470–4 and had been received by the family of the second Captain of Madeira, João Gonçalves da Câmara, Colón-Zarco's uncle. Frutuoso wrote:

> João Soares de Albergaria went to Madeira . . . with his sick wife, spending so much money in trying to cure her that he found it necessary to sell one of the said islands and . . . Rui Gonçalves da Câmara bought the Island of São Miguel, which was more barren [with a smaller population] than Santa Maria, for two thousand cruzeiros in cash and 60 tons of sugar, which must have been very valuable at that time to have been the price of an island of such size.

Rui Gonçalves da Câmara was longing to have a Captaincy like his brother. As a second son, Madeira could not be his. But this opportunity arose to buy São Miguel and he was given authorisation to do so by Princess Beatriz, widow of Prince Fernando, Duke of Beja, and stepmother of Colón-Zarco – who, we must remember, was Rui Gonçalves da Câmara's nephew. João Soares de Albergaria was widowed and he remained on Santa Maria, while Rui Gonçalves da Câmara settled on São Miguel.

It is natural, therefore, that Colón wanted to drop anchor at São Miguel and not Santa Maria. It is likely that he wished to ask his uncle to inform the King that he had fulfilled his instructions to 'rediscover' the Antilles. But the storm had blown him 70 km off course and driven him to another island, the captain of which, who knew Columbus very well, was absent. According to Morison, the departure of the 'Niña' was delayed in the hope that the 'Pinta' and the 'Santa Maria' would appear and the whole 'India' fleet could be arrested. But Morison was a twentieth-century American and did not think in terms of fifteenth-century means of communication – a fault that is frequent among Columbian scholars. Kings of that time did not send out telegraphic circulars to all their outlying districts and Columbus's case was not a subject of a royal press conference.

Castanheira may have wanted to find out the reasons for Columbus's presence in the islands and held him up in order to seek advice from Rui Gonçalves da Câmara. Although Columbus makes no reference in his log to the fact that he had sent a message to Castanheira through one of the Portuguese he had received on board the 'Niña', the interim Captain of Santa Maria had plenty of time to send a caravel to São Miguel and ask the Admiral's uncle for advice. The Azoreans were experienced seamen and used to the waters of the archipelago. Columbus himself had tried to reach São Miguel and had been forced to turn back only because his sails were in shreds, he had lost his anchors and his crew were green. Why did he undertake an expedition in such unfavourable circumstances? And how could he have been so sure that he would not be arrested by the Captain on reaching the well-defended island?

COLUMBUS VISITS HIS FAMILY

I assume that Rui Gonçalves da Câmara's reply had put Castanheira at ease and had explained the situation to him. Columbus, however, was unable to speak to his uncle personally, so decided to attend on his Lord, King John II. He thus sailed for Lisbon instead of Palos. Let us read of the arrival of the 'Niña':

1. While Columbus was sailing east on 3 March 1493, a strong wind blew up, but 'he managed to sail 60 leagues' before his sails were torn. He sighted Cape Sintra (Cape Roca) the following day and decided to shelter in the Tagus.

This is the version that Columbus wrote in his log-book, and which his son Fernando and Las Casas copied. On arrival in the Tagus estuary, he sent the longboat to Belém with a letter requesting King John II's permission to anchor off Restelo, which he did at daybreak on 5 March near the 'great nau of the King of Portugal, which was also anchored in the Tagus and is the most well-armed with artillery and weapons that I have ever seen'. Then 'her mate, who was called Bartolomeu Dias of Lisbon', came alongside in a small bark armed like a caravel. He invited Columbus to board the bark, so that he could go and explain to the Ministers the reason for his coming; but Columbus declined, saying that he was the Admiral of the Ocean Sea and Viceroy of India and carried letters from the Sovereigns of Castile for all the Princes of the world. Dias returned to the nau and informed the captain, Álvaro Damán, who personally went to the 'Niña', giving Columbus 'a great welcome' with drums, trumpets and flutes and offering him everything he needed.

2. It can be concluded from this text that the King had authorised Columbus to anchor off Belém (the name that gave rise to a port in Verágua, Panama) on the previous day. It is here that appear the first discrepancies that have always confused Columbian historians.

   (a) Why did Columbus refer in such vague terms to Bartolomeu Dias, with whom he had certainly sailed for Mina with a fleet of eight ships loaded with material to build the Castle of St George (a name he had given to an island in the Antilles) in December 1481 and may also have voyaged with him to Genoa in June 1478? It is also presumed that Columbus helped Dias map out the route that would take the latter around the Cape of Good Hope, after receiving the famous missive that began 'my very good friend in Seville' from King John II. It is also strange that he should have added the navigator's birthplace 'of Lisbon' to his name.

   (b) It happens that while the Bartolomeu Dias of Cape of Good Hope fame was sailing in the southern hemisphere, a navigator of the same name was commanding the nau 'Figa' in operations against the Biscayans.[9] He was the same man that, in 1490 and 1495, was the mate of the 'São Cristóvão' described by Garcia de Resende:[10] 'The King had a thousand-ton nau built, the most well-armed and the biggest and the best that had ever been seen, with such strong beams and thick planking that it could withstand the impact of cannonballs, and carried so many huge calibre bombards and other artillery that it was spoken about almost everywhere'.

   The mate of the 'São Cristóvão', therefore, was not the Bartolomeu Dias who was a friend of Colón-Zarco.

(c) The captain of the nau Álvaro Damán (Damião), may have been Álvaro Damião de Aguiar, the son of Álvaro Esteves, the first to reach the Equator and who gave a bay the name of Santa Maria das Neves (St Mary of the Snows), similar to the name that Columbus gave to another bay in the Antilles – Nossa Senhora de las Nieves (Our Lady of the Snows). According to Morais, Álvaro Damião 'went to India' (see p. 493, Tree XVIII).

Maybe because he found the name 'Damán', written by Columbus's scribe on board, strange, Fernando Colón surprisingly altered it to 'Dacuña' in his *Historia*. Fernando must have ignored the existence of Álvaro Damião de Aguiar and known only Dom Álvaro da Cunha. The latter, however, was not a sailor, but his son was Admiral Rui de Mello da Cunha, cousin of Dom Duarte de Menezes and of Columbus, as I demonstrated on p. 385.

Álvaro Damião, about whom I shall speak again later, was already considered to be good enough to be appointed as captain of a nau in 1493. On the other hand, an Admiral remained on board only when invested with the command of a fleet and this was not the case when Columbus entered the Tagus.

3. On 6 and 7 March many gentlemen of the court and the people of Lisbon went to the 'Niña' to see the Indians. What did Columbus do during those forty-eight hours? He related that it was only on 8 March that the King sent Dom Martinho de Noronha with a letter inviting the Admiral to visit him and indicating that he had given orders for the ships to be supplied with everything they needed. Columbus went to spend the night at Sacavém. But at whose house?

We know that Diogo Dias, brother of Bartolomeu Dias (of the Cape of Good Hope), who would later be the scribe on Vasco da Gama's flagship on his first voyage to India[11] and captain of Cabral's supply ship when he rediscovered Brazil, was the 'royal treasurer of Sacavém' at that time. Did Columbus pay a visit of friendship and courtesy?

4. At daybreak on 9 March, Dom Martinho took Columbus to Azambuja, 'nine leagues from Lisbon'. From there he went to the country palace of Vale do Paraíso for his meeting with the King, who received him 'with much honour and asked him to be seated', a distinction that was granted only to certain nobility at that time. After their discussion, the King asked Dom João de Menezes, Prior of Crato, 'the highest personage present', to take Columbus and lodge him at his house. But the Prior lived at Santarém, which lies about 25 km from Azambuja.

5. More discrepancies appear. Although Rui de Pina[12] is guilty of

frequent historical imprecisions as he based a lot of his writings on 'I heard' (vaguely mentioning the rounding of the Cape of Good Hope, for instance, and omitting the name of Bartolomeu Dias), he deserves to be quoted here as he says that King John II was supposedly highly displeased when he saw that the Indians Columbus had brought with him were not black like the Africans. Pina only wrote what the King told him to write and John II would have pretended to be upset. Fernando Colón, however, says that 'the King seemed pleased to hear the details of his victory, even though he thought that, by the terms of the treaty signed with the Sovereigns of Castile, the new lands belonged to him. Colón replied that he knew nothing of what had been agreed and that he had only been ordered not to go to Portugal's Mina in Guinea, an order he had fulfilled to the letter, to which the King in turn replied that everything was alright and was sure that everything would be arranged for the good of everyone'. It was a farce that had been rehearsed, as Columbus had repeatedly referred to the Treaty of Toledo and had been in Lisbon with the King only five years before (i.e., eight years after the signing of the Treaty).

6. The chronology that Columbus attributes to these events is highly suspect. He puts back his meeting with the King and then unnecessarily prolongs his stay at Vale do Paraíso, certainly in an effort to cover up his journey to Santarém. It is interesting to note that Columbus called three channels to the north of Cuba Santarén, St Nicholas and Our Lady of the Assumption and that the only two parishes in the town of Santarém at that time were those of S. Nicolau and Nossa Senhora de Assumpção.

Columbus wrote that he had heard mass with the King on the morning of 10 March; but it was only on 11 March that he had lunched with the Sovereign, who 'showed him great affection' on saying his farewells. It is difficult to believe that the King of Portugal would show such friendship to a Genoese wool-carder, in an age when there was an accentuated distinction between the classes.

7. The crew of the 'Niña', anxious to rejoin their families and receive the praises of their Sovereigns and people, spent all this time in the Tagus, but Columbus was not in the slightest hurry. That afternoon, still accompanied by Dom Martinho, he made his way to Vila Franca de Xira and then to the Monastery of St Anthony, in order 'to kiss the hands of the Queen, and the Duke and the Marquis were with her' and they welcomed him 'with great honour'. And that night, instead of returning to his ship, he decided to sleep at Alhandra, on the outskirts of Lisbon. Where had he been between the dates he thought convenient to mention?

According to his own words, King John II gave him money for the crew of the 'Niña' and he reached the ship, which was victualled and ready to sail, on 12 March. But he raised anchor only two days later. They docked at Saltés on the afternoon of 15 March and Seville received them in glory.

8. I have already proved that the Bartolomeu Dias of the 'São Cristóvão' was not the same one that rounded the Cape of Good Hope and that Dom Álvaro da Cunha could not have been the captain of the nau. The only thing left to do now is identify the other people that welcomed Columbus in Portugal (see p. 398, Tree V).

   (a) Dom Martinho de Noronha was the son of Beatriz (sister of Dom Duarte de Menezes) and Dom Fernando de Noronha, brother of the Archbishop of Lisbon, Dom Pedro de Noronha, who in turn was the father of the children of Isabel Moniz Perestrelo, Columbus's sister-in-law. The Prior of Crato, Dom João de Menezes, was the son of Dom Duarte de Menezes and a relative of Dom Álvaro da Cunha.

   (b) We know that Queen Leonor, born in Beja, who Columbus went to visit in order 'to kiss her hand', was his half-sister. The Duke could only have been the Duke of Beja, the future King Manuel I, who was also his half-brother. And the Marquis was certainly the Marquis of Vila Real, Dom Pedro de Menezes (who fought alongside King Alphonse V at the Battle of Toro), brother of Dom Martinho and Dom João. What a welcome these noblemen would have given to a Genoese wool-carder!

9. If it was not invented, the storm off Lisbon was exactly what Columbus wanted to justify his change of course, forcing him to sail (as he wrote in detail) with only one sail – the foresail, for which Columbus always used the Portuguese name, *'papa figo'*, and never the Spanish. Now duly informed, King John II could set about drafting the terms of the future Treaty of Tordesillas. And Colón-Zarco took the opportunity to see his family and friends.

THE LETTER TO SANTÁNGEL

It is essential to analyse the letter that Columbus sent to the New Christian Luis de Santángel, the Catholic Monarchs' Secretary for Finances, who did his utmost to try and arrange money to finance the fleet for the Indies.

(a) The letter was sent from the Canary Islands and was dated 15 February 1493. But Columbus did not touch on that archipelago on the return trip of his first voyage. He sailed straight for the Azores, where he dropped anchor on that very day, and then on to Lisbon.

398 CHECK AND MATE

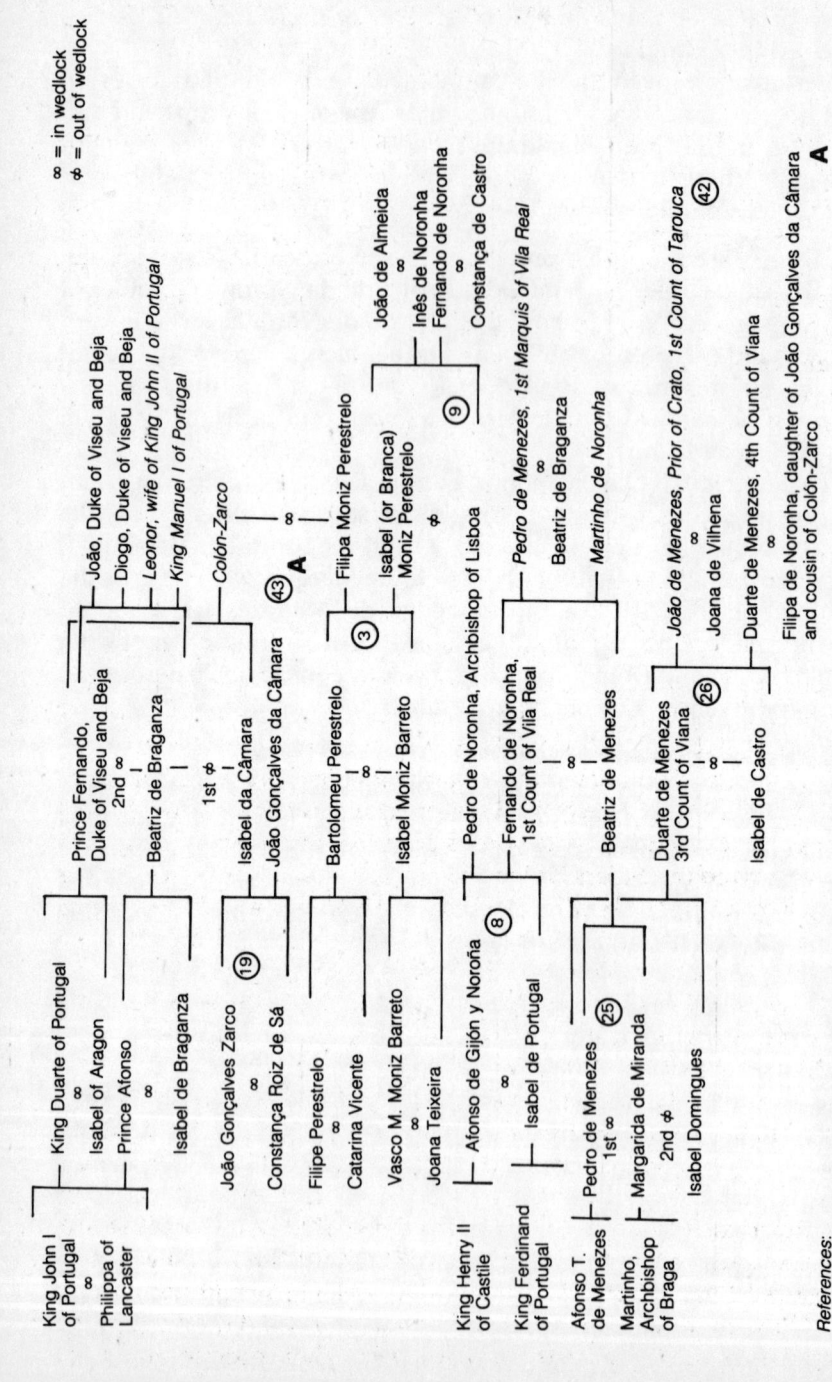

∞ = in wedlock
⚭ = out of wedlock

[Genealogical chart, rotated sideways on page, showing the following relationships:]

King John I of Portugal ∞ Philippa of Lancaster
 ├─ King Duarte of Portugal ∞ Isabel of Aragon
 │   ├─ Prince Fernando, Duke of Viseu and Beja ∞ Beatriz de Braganza
 │   │   ├─ João, Duke of Viseu and Beja
 │   │   ├─ Diogo, Duke of Viseu and Beja
 │   │   └─ Leonor, wife of King John II of Portugal ∞ King Manuel I of Portugal
 │   └─ Prince Afonso ∞ Isabel de Braganza
 ├─ João Gonçalves Zarco ∞ Constança Roiz de Sá
 │   ├─ Isabel da Câmara ⚭ Colón-Zarco (43)
 │   └─ João Gonçalves da Câmara (19) — A
 ├─ Filipe Perestrelo ∞ Catarina Vicente
 │   ├─ Bartolomeu Perestrelo ∞ Isabel Moniz Barreto
 │   │   └─ Filipa Moniz Perestrelo
 │   └─ Vasco M. Moniz Barreto ∞ Joana Teixeira
 │   └─ Isabel (or Branca) Moniz Perestrelo ⚭
King Henry II of Castile
 └─ Afonso de Gijón y Noroña (8)
     └─ Pedro de Noronha, Archbishop of Lisboa
         └─ Fernando de Noronha, 1st Count of Vila Real ∞
             ├─ João de Almeida
             ├─ Inês de Noronha
             └─ Fernando de Noronha ∞ Constança de Castro (9)
King Ferdinand of Portugal ∞ Isabel de Portugal
 └─ Afonso T. de Menezes
     ├─ Martinho, Archbishop of Braga
     └─ Pedro de Menezes (25)
         1st ∞ Margarida de Miranda
         2nd ⚭ Isabel Domingues
         ├─ Pedro de Menezes, 1st Marquis of Vila Real ∞ Beatriz de Braganza
         │   └─ Martinho de Noronha
         └─ Beatriz de Menezes ∞ Duarte de Menezes, 3rd Count of Viana (26)
             └─ Isabel de Castro ∞ Duarte de Menezes
                 ├─ João de Menezes, Prior of Crato, 1st Count of Tarouca (42)
                 ├─ Joana de Vilhena
                 └─ Duarte de Menezes, 4th Count of Viana
                     └─ Filipa de Noronha, daughter of João Gonçalves da Câmara and cousin of Colón-Zarco — A

References:
3, 7, 8 and 9 (Tree I); 25 and 26 (Tree II); 42 and 43 (Tree IV and VI).

▼ Who received Colón-Zarco in Portugal?

(b) The letter is accompanied by another sheet, dated 14 March, where Columbus relates that he had got caught in a terrible storm while in the 'Sea of Castile, but I ran for this port of Lisbon, where I am today, and our escape was a miracle'.

There are two discrepancies here: the letter is dated from the Canaries and not the Azores, and if Columbus made a run for Lisbon he could not have been in the 'Sea of Castile', which is to the north of the River Minho. He was obviously, and intentionally, in Portuguese waters before the storm sprang up, just as he had intentionally gone to the Azores. They are not errors on the part of the scribe, as the letter is clearly in Columbus's hand. It is obviously a trick on Columbus's part so that Santángel would think that he had gone to Lisbon by chance and not to report to King John II.

With Colón-Zarco's siglum deciphered and his true identity revealed, the fog that enshrouded the mysteries of Columbus has lifted and it is hardly worthwhile mentioning something new. But not completely.

THE SIEGE OF ARZILA

Colón-Zarco set sail from Cadiz for his fourth and last voyage of discovery to the Western Atlantic on 9 May 1502. He headed for the Canaries as usual and dropped anchor at Santa Catarina on 11 May. He then made a decision that has always puzzled historians. According to his son Don Fernando, Columbus commanded four ships with a complement of 140 men and that 'on the second day, we sailed to Arzila to the aid of the Portuguese who were said to be under heavy siege. But when we arrived the Moors had raised the siege, so the Admiral sent his Military Governor, his brother Bartolomeu, and me, as captains of the ships, ashore to visit the Commander of Arzila, who had been wounded. He thanked the Admiral for the visit and for the offers he had made. In return, he sent some of his knights to transmit his greetings to him, some of whom were kinsmen of the Admiral's deceased wife, Filipa Moniz' (see map of fourth voyage, p. 76, Fig. 6.4).

Some historians have assumed that Colón-Zarco's expedition was due to his Christian spirit. But without the authorisation of the Catholic Monarchs, how could he have dared to risk the lives of his crews and the success of the voyage through a shortage of men to keep his ships seaworthy and to fight the 'Indians' who had killed many of his sailors on the previous voyage? Would Colón-Zarco have been willing to help the besieged Portuguese? He knew that the Moors would not give up easily. How many dead or wounded – and in those days most of the wounded died due to lack of drugs, especially in hot countries that were swarming with infectious insects – would be left on the battlefield or in the Canaries to be treated? How would the Catholic Monarchs react on receiving such news?

As the besieged had driven the enemy off, Columbus gave them provisions – the only thing he had on board that was of any use to them at that moment – and in return the Commander of the garrison sent his thanks through 'kinsmen of Filipa Moniz, his wife' who had died eighteen years earlier. I should like to go further into the mystery of Colón-Zarco's decision to go and fight the Moors on land without royal consent. Did he do it simply because of his crusading spirit, or had someone ordered him to go on his rescue mission?

It is very difficult to imagine the son of a Genoese wool-carder being so worried about the situation of a Portuguese garrison, however they were 'hemmed in'. His mission was to get to the Antilles and find gold, precious stones and pearls, but he was willing to sacrifice his men in a fight that had nothing to do with them. Colón-Zarco being Portuguese, it could be easily understood if he had acted on the orders of King John II, whom 'he served throughout the whole 14 years' of this king's reign. But John II had died seven years before. We know that John II's successor, Manuel I, was Colón-Zarco's half-brother. Yet following the Treaty of Tordesillas and the discovery of the seaway to India in 1498, Colón-Zarco's secret mission had come to an end. He was now in the service of the Catholic Monarchs on his own account, trying to get his share of the riches coming from the Western Atlantic. It was neither his task nor was it convenient for him to fight in Africa. And already rich and powerful, King Manuel I (known as 'The Fortunate') had no need to ask Colón-Zarco to sacrifice his already reduced crews.

I think it can be concluded, therefore, that Colón-Zarco acted on his own initiative, and not only because he was imbibed with the crusading spirit. The siege of Arzila had finished when he reached the city, but he could have heard about the beginning of it while still in Cadiz or Huelva.

If the Commander of the Garrison were Dom Henrique de Menezes (great-grandson of Don Pedro Ponce, Lord of Marchena), it could be accepted that the rescue operation was carried out in memory of Dom Henriques's father, Dom Pedro, a cousin of Columbus and of one or two of the Admiral's friar friends of La Rábida. But Dom Henrique had been dead for twenty-three years. I would therefore have to resort once more to genealogy. Who was the Commander of Arzila in 1502? Who were the kinsmen of Filipa Moniz Perestrelo that he sent to thank Colón-Zarco?

I found out that the Commander at that time could have been Dom João de Menezes (cousin of the deceased Dom Henrique), son of Dom Duarte de Menezes and Isabel de Castro. It was Dom João who received Colón-Zarco at his palace at Santarém after the audience with King John II at Vale do Paraíso. He was no more the Prior of Crato at that time and he had asked for dispensation in order to marry Joana de Vilhena and had been appointed Commander of Arzila in 1493. But he assumed command of the garrison only in 1501.

Although Colón-Zarco was only his distant cousin, this was reason enough to pay him his respects as a host. But Dom João had left Arzila in April of the same year that he had been appointed in order to command a fleet of twenty ships that first headed for Oran and then went to relieve Venice from a siege by the Turks. Who had replaced him?

The new Commander of the garrison was Dom Duarte de Menezes (his grandfather's namesake), son of Dom João (who had been raised to the title of Count of Tarouca in reward for his deeds against the Sultan of Turkey in the Mediterranean), married to Filipa de Noronha, daughter of João Gonçalves da Câmara and, therefore, first cousin of Colón-Zarco.

Who were the officers that Dom Duarte sent to greet Colón-Zarco? Being a port, Arzila had three frontiers, so I went in search of the Frontier Commanders. They were the Count of Borba, the Viscount of Vila Nova de Cerveira and Dom Fernando de Castro.

1. The Count of Borba, Dom Vasco de Menezes Coutinho, had also been given the title of the first Count of Redondo in 1500, in recognition of his deeds in Africa. He was married to Catarina da Sylva and was the father of Margarida Coutinho, who would become the wife of the Commander-in-Chief of the Lancers, João Mascarenhas. Dom Vasco was the son of the Marshal of the Realm, Dom Fernando Coutinho and Maria de Noronha, Colón-Zarco's aunt.

    This is the second reason for Colón-Zarco's intervention. Maria de Noronha was the daughter of João Gonçalves da Câmara and Columbus's aunt. The count of Borba and Redondo, therefore, was his first cousin and the grandson of Dom Vasco Fernandes Coutinho, first Count of Marialva, in whose house at Matosinhos João Gonçalves Zarco had spent his infancy and may even have been born.

2. The Viscount of Vila Nova de Cerveira was Dom Francisco de Lima, son of Dom Leonel de Lima and Filipa da Cunha. He had been married to Isabel de Noronha, daughter of the second Count of Abrantes, Dom João de Almeida and Inês de Noronha. Inês, in turn, was the daughter of the Archbishop of Lisbon, Dom Pedro de Noronha and Isabel Moniz Perestrelo, which made her a sister-in-law of Colón-Zarco.

    Besides this, Dom Francisco de Lima had married Isabel de Mello in digamy. She was the daughter of Martim Afonso de Mello and sister of Maria de Mello, who was married to Diogo Moniz. Both these men were uncles of Colón-Zarco's late wife Filipa.

    In his brief allusion to Arzila, Don Fernando Colón was certainly telling the truth when he mentioned the presence of kinsmen of his father's wife. He only referred to Filipa, of course, as his father's name had to remain incognito.

3. The third captain was Dom Fernando de Castro, son of Dom Rodrigo de Castro, second Count of Monsanto, and Maria Coutinho, daughter of the above-mentioned Marshal of the Realm, Dom Fernando Coutinho and Maria de Noronha, Colón's aunt. Dom Rodrigo had also been Commander of Arzila, the Governor of Covilhã and had fought in the Battle of Toro.

And so another Columbian mystery is explained. Columbus sheered off his course for the Antilles and was willing to risk his own and his men's lives in a skirmish against the Moors in order to go to the aid of his family – and the handful of Portuguese that were under siege (this page, Table VI). He acted like the Portuguese he was – the son of a Master of the Order of Christ.

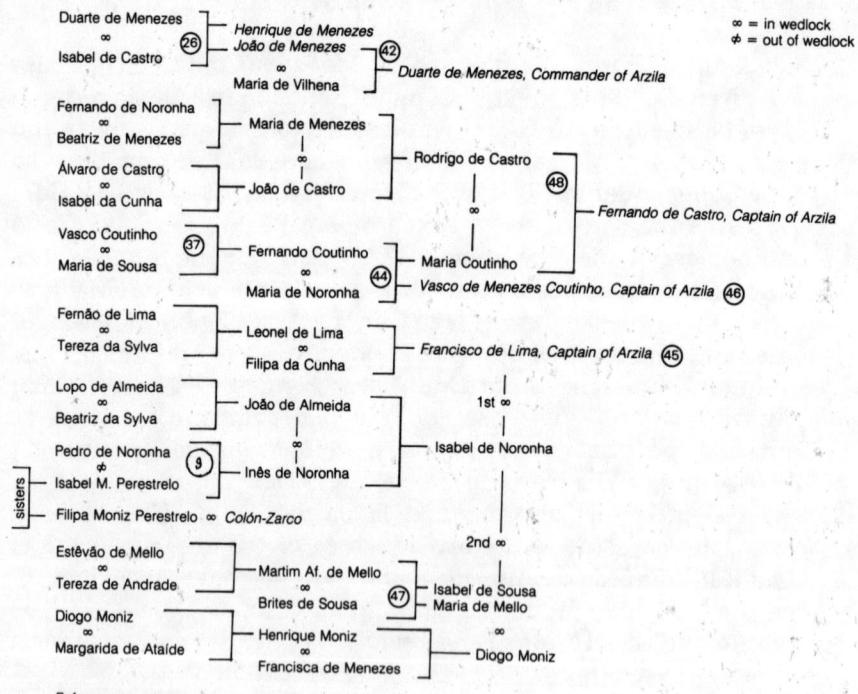

References:
9 (Tree I); 26 (Tree II); 37 (Tree III); 42 (Tree IV); 44 – Marshal of the Realm; 45 – Viscount of Vila Nova da Cerveira; 46 – 2nd Count of Abrantes; 47 – 8th Lord of Mello; 48 – Count of Borba.

VI Arzila

Notes and References

1. Morison (1945), Chap. XXII.
2. Las Casas (1875–6), Vol. I, Chap. LXVIII.
3. The original disappeared and he is called 'Juan Castañeda' in Las Casas version.
4. Colón, Don Hernando (1985), Chap. XXXVIII.
5. Las Casas (1875–6), Vol. I, Chap. LXVII, pp. 451–4.
6. Colón, Don Hernando (1985), Chap. XXXVIII, p. 229.
7. Frutuoso (1873), Book III, Chap. XIII.
8. Morison (1945), Chap. XXIII, p. 184, Maria (1972).
9. Campos (1987).
10. Resende (1545; 1798).
11. Barros (1932; 1945–6); Gois (1556); Caminla (1947).
12. Pina (1790).

Part VII
On the Trail of the Last Incognitos

30 *The Protectors of Columbus*

GENEALOGICAL RELATIONSHIPS

Although I am aware that the reader may find the science of genealogy unduly complex, I cannot help but resort to family trees in order to disperse the fog that enshrouds the life of Colón-Zarco. The elaboration of the genealogical trees synthesises genealogical transcriptions that are even more tedious, and makes it easier to see which families were a part of Colón-Zarco's life. Such family relations show also that Columbus did not create a complex cabalistic siglum merely to show that he was son of Prince Fernando and maternal grandson of the discoverer of Madeira.

Step by step, I have shown that the enigmas that have been considered to be impenetrable for 500 years have now been coherently explained with the identity that Colón-Zarco himself gave to his siglum. If the reader did not have these genealogical data at hand, he would lack the material that is the whole basis of my thesis.

Any historian, even one with only rudimentary notions of sociology, knows that fifteenth-century class barriers were almost insuperable;[1] such theses as that of Jacques Heers (p. 89), that attempt to show how the 'wool-carder Colombo' won the favour of Isabel Moniz so that he could woo her daughter Filipa are inexplicable, and suggest that the 'Monizes had little influence' at that time. The genealogical trees in reality show the high positions held by the people in question and the important marriages that took place, information that totally undermines Jacques Heers's suggestion.

Besides the Mellos, Monizes and Teixeiras, it is essential to refer to the Henriques, who came from (as is shown on pp. 409 and 410, by Trees VII and VIII) King Henry of Castile (bastard son of King Alphonse XI and Leonor Nuñez de Guzmán), as did the Noronhas (Trees I, 8 and 9; Tree II, son-in-law of 26). This Guzmán/Henriquez family split into two branches in the fifteenth century – the Portuguese branch in Beja and

Serpa and the Castilian one in Córdoba and Huelva. King Henry II's brother, Juan de la Cerda, also connected with the Henriquez, gave origin to the Portuguese branch of the La Cerdas (or Lacerdas).

I shall present only summarised parts of these genealogical relationships in this English edition and only when necessary. Tree IX shows the connection between all the people not directly connected with the thesis. I shall begin by analysing the genealogy of the descendants of Don Juan de La Cerda, brother of King Henry II (known as 'The Bastard') of Castile. He married Princess Maria Afonso, illegitimate daughter of King Dennis, half-sister of King Alphonse IV of Portugal and Princess Catarina, who married King Ferdinand of Castile; Princess Maria Afonso was also the half-sister of Prince Afonso of Portugal, who married Juana Ponce (daughter of Count Pedro Ponce, Lord of Marchena), who gave birth to Álvaro de Gusmão (Guzmán in Spanish).

These marriages were the beginning of the ties between the Houses of Portugal (Avis), Guzmán (Córdoba and Huelva), La Cerda (Seville) and Ponce (Marchena). And it was exactly to these places that Colón-Zarco made his way when he was preparing the 'discovery of the Antilles'. Before setting sail, the fleet was provisioned at Saltés (Palos) and on heading west should have passed to the south of the Azores and had no need to sheer south to the Canaries. It was not mere chance that led him to take that route. Tree VII (p. 409) shows the links among these families, although there are certain cases that are not included. One of these is that of Count Pedro Ponce, Lord of Marchena and father of Maria, who first married Fernando de Ledesma, bastard son of King Alphonse XI of Castile, and later married Dom Pedro Lobo (already mentioned above), Lord of Alvito, in digamy.

A great-grand-daughter of the Count, Isabel de Castro, was (as I have already mentioned above) the second wife of Dom Duarte Menezes, who was a relative of Colón-Zarco. This explains the relationship between the Friars Marchena and Colón-Zarco and the protection they afforded him. But Columbus had other patrons, among them the Dukes of Medinaceli and of Medina Sidónia, the Bishop of Palência (later Archbishop of Seville) and the Marquis and Marchioness of Moya.

How could a poor adventurer, peddling a doubtful project (which the scholars of Castile took seven years to accept), have deserved the benefaction of those Dukes who knew next to nothing about geography and nautical cosmology? Not only did they help him, but opened the doors of their palaces to him as well.

The constant marriages between the Portuguese and Spanish families that lived in the two countries' border regions and the struggles between political factions in which the lords frequently changed sides according to their interests and not those of their Sovereigns linked many families and resulted in domains changing hands. The branches of the Hen-

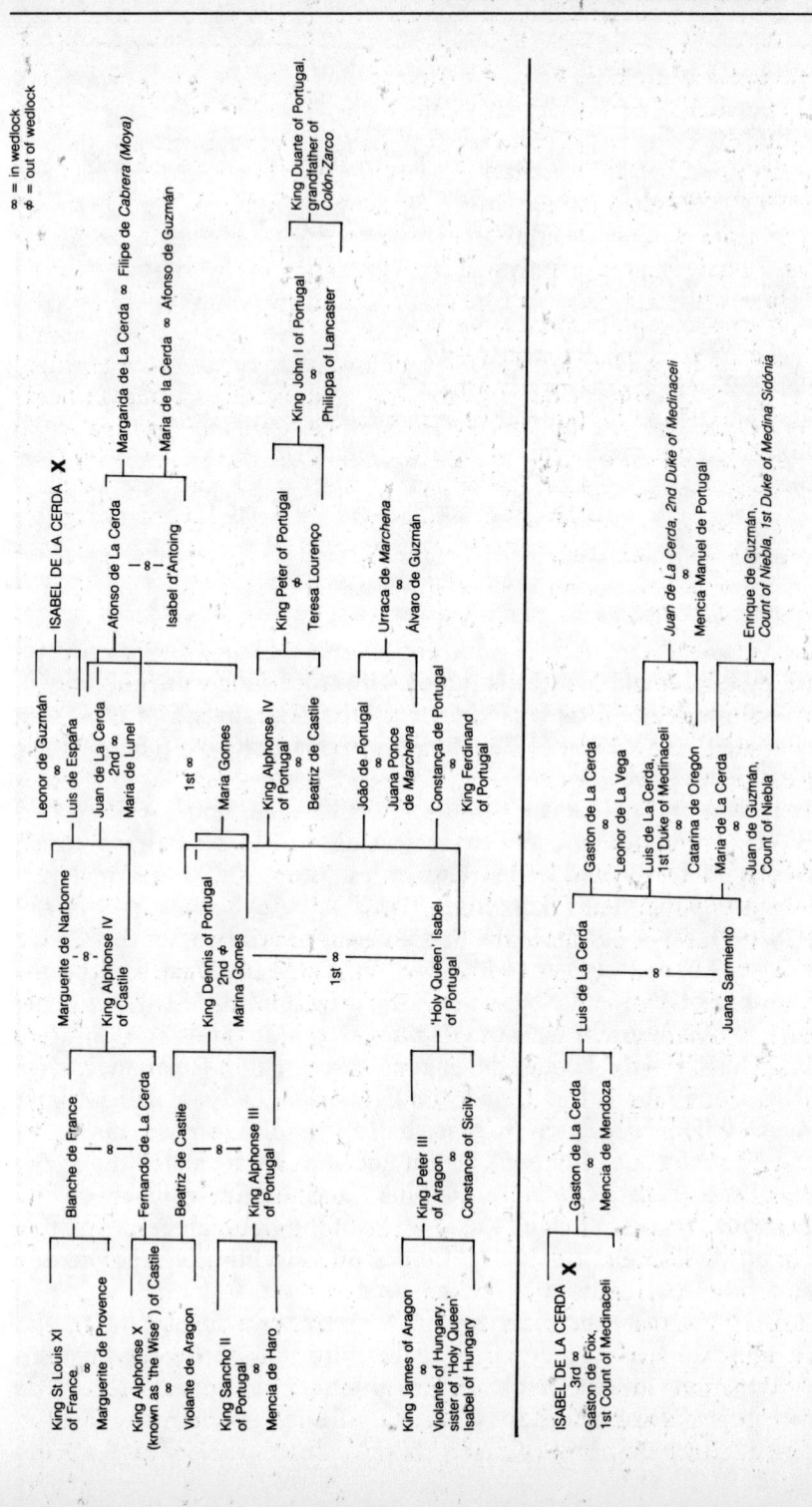

THE PROTECTORS OF COLUMBUS 409

∞ = in wedlock
φ = out of wedlock

King St Louis XI of France
∞
Marguerite de Provence
└─ Blanche de France
   ∞
   Fernando de La Cerda
   └─ Marguerite de Narbonne
      - ∞ -
      King Alphonse IV of Castile
      └─ Leonor de Guzmán
         ∞
         Luis de España
      └─ ISABEL DE LA CERDA ✗
         2nd ∞
         Juan de La Cerda
         ∞
         Maria de Lunel
         └─ Margarida de La Cerda ∞ Filipe de Cabrera (Moya)
         └─ Maria de la Cerda ∞ Afonso de Guzmán
      └─ Afonso de La Cerda
         φ
         Isabel d'Antoing

King Alphonse X (known as "the Wise") of Castile
∞
Violante de Aragon
└─ Beatriz de Castile
   ∞
   King Alphonse III of Portugal
   └─ King Dennis of Portugal
      2nd φ
      Marina Gomes
      1st ∞
      Maria Gomes
      └─ King Alphonse IV of Portugal
         ∞
         Beatriz de Castile
         └─ King Peter of Portugal
            φ
            Teresa Lourenço
            └─ King John I of Portugal
               ∞
               Philippa of Lancaster
               └─ King Duarte of Portugal, grandfather of Colón-Zarco
         └─ João de Portugal
            ∞
            Juana Ponce de Marchena
            └─ Urraca de Marchena
               ∞
               Álvaro de Guzmán

King Sancho II of Portugal
∞
Mencia de Haro

King James of Aragon
∞
Violante of Hungary, sister of 'Holy Queen' Isabel of Hungary
└─ King Peter III of Aragon
   ∞
   Constance of Sicily
   └─ 'Holy Queen' Isabel of Portugal
      1st ∞
      King Ferdinand of Portugal

ISABEL DE LA CERDA ✗
3rd ∞
Gaston de Foix,
1st Count of Medinaceli
└─ Gaston de La Cerda
   ∞
   Mencia de Mendoza
   └─ Luis de La Cerda
      ∞
      Juana Sarmiento
      └─ Gaston de La Cerda
         ∞
         Leonor de La Vega
         └─ Luis de La Cerda, 1st Duke of Medinaceli
            ∞
            Catarina de Oregón
            └─ Juan de La Cerda, 2nd Duke of Medinaceli
               ∞
               Mencia Manuel de Portugal
         └─ Maria de La Cerda
            ∞
            Juan de Guzmán, Count of Niebla
            └─ Enrique de Guzmán, Count of Niebla, 1st Duke of Medina Sidónia

VII Relation: Marchena/Cabrera (Moya)/Guzmán (Medina Sidónia)/La Cerda (Medinaceli)/Colón-Zarco

410   THE LAST INCOGNITOS

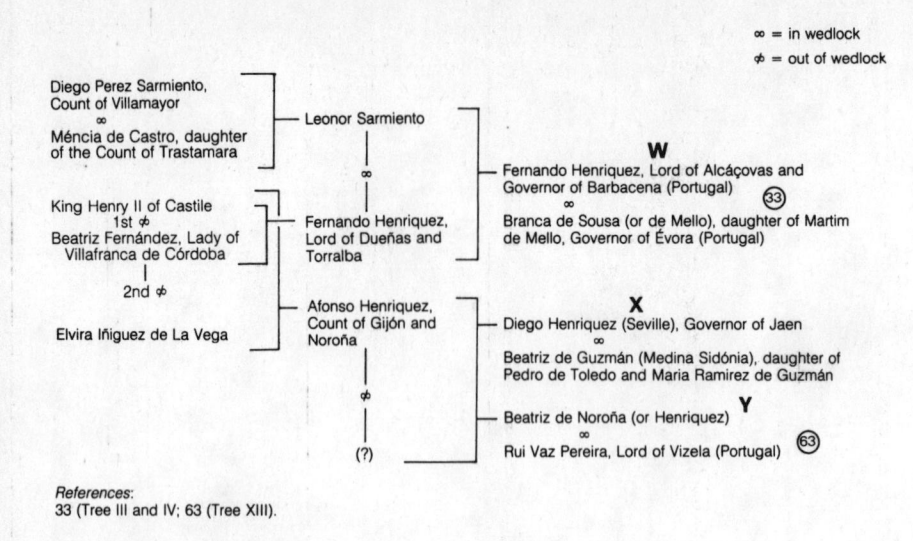

∞ = in wedlock
φ = out of wedlock

Diego Perez Sarmiento, Count of Villamayor
∞
Méncia de Castro, daughter of the Count of Trastamara
⎱ Leonor Sarmiento
∞
⎰ Fernando Henriquez, Lord of Dueñas and Torralba

King Henry II of Castile
1st φ
Beatriz Fernández, Lady of Villafranca de Córdoba

2nd φ

Elvira Iñiguez de La Vega

Afonso Henriquez, Count of Gijón and Noroña
φ
(?)

W
Fernando Henriquez, Lord of Alcáçovas and Governor of Barbacena (Portugal) ㉝
∞
Branca de Sousa (or de Mello), daughter of Martim de Mello, Governor of Évora (Portugal)

X
Diego Henriquez (Seville), Governor of Jaen
∞
Beatriz de Guzmán (Medina Sidónia), daughter of Pedro de Toledo and Maria Ramirez de Guzmán

Beatriz de Noroña (or Henriquez) **Y**
∞
Rui Vaz Pereira, Lord of Vizela (Portugal) ㊳

References:
33 (Tree III and IV; 63 (Tree XIII).

VIII Origin of Henriquez

riques (Enriquez in Spanish) family sprang from King Henry II (known as 'the Bastard') of Castile, either from the legitimate line of Queen Juana Manuel (La Cerda) or from the Monarch's affairs with Elvira de La Vega or Beatriz Fernandes.

Prince Fernando of Castile was the son of King Alphonse X (known as 'the Wise') and Queen Violante of Aragon. He was the paternal grandson of King Ferdinand (known as 'the Saint'), and maternal grandson of King James (known as 'the Conqueror') and son-in-law of Holy King Louis of France. He was given the nickname of 'La Cerda' ('the Bristle') as he was born with a tuft of hair on his chest. He was the grandfather of Luis de Espanha (son of King Alphonse IV), Count of Talmond and Clermont in France, Prince of the Canaries and Count of Niebla. His first wife, Leonor de Guzmán (daughter of Guzmán, known as 'The Good') bore him Isabel de La Cerda, Lady of Huelva and Gibraltar, who was given in marriage to Rodrigo of Asturias, Trastamara and Gijón at the age of ten and to Rui Ponce de Marchena at the age of twelve. King Peter of Castile wanted her hand, but only at forty-six did the lovely Isabel agree to wed Gastón (son of the Count of Foix), the first Count of Medinaceli. Luis de La Cerda, fifth Count and first Duke of Medinaceli (1497) married twice, his second wife, Catarina de Orejón (his former mistress) bearing him Juan de La Cerda, who was the second Duke and Colón-Zarco's patron. The third Count of Medinaceli's daughter, Maria de La Cerda, had in the meantime married Juan de Guzmán, first Duke of Medina Sidónia and Lord of Huelva.

Juan de Guzmán, who conquered Gibraltar from the Moors of Gra-

THE PROTECTORS OF COLUMBUS   411

∞ = in wedlock
φ = out of wedlock

Genealogical chart showing:

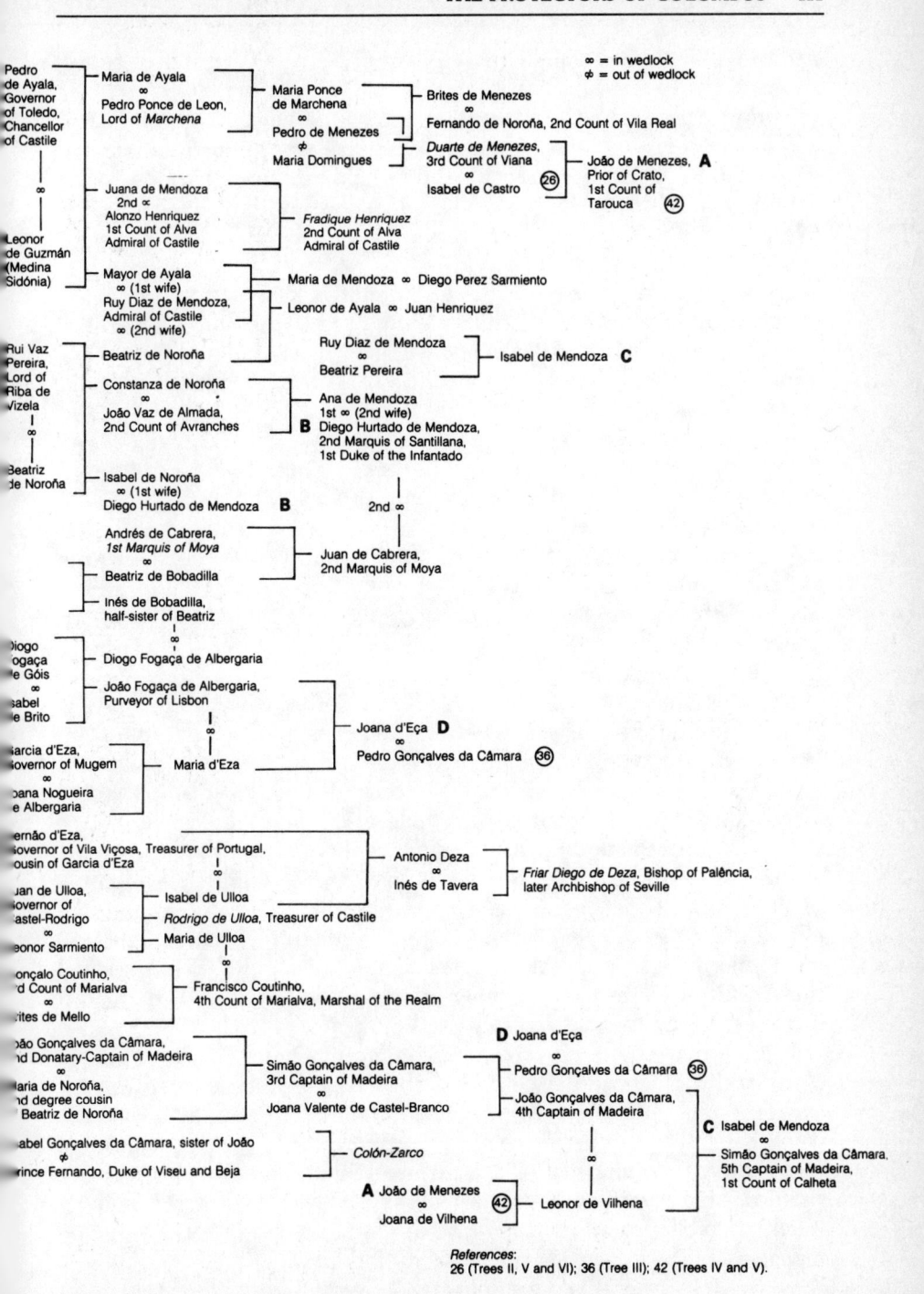

- Pedro de Ayala, Governor of Toledo, Chancellor of Castile ∞ Leonor de Guzmán (Medina Sidónia)
  - Maria de Ayala ∞ Pedro Ponce de Leon, Lord of *Marchena*
    - Maria Ponce de Marchena ∞ Pedro de Menezes φ Maria Domingues
      - Brites de Menezes ∞ Fernando de Noroña, 2nd Count of Vila Real
      - *Duarte de Menezes*, 3rd Count of Viana ∞ Isabel de Castro (26)
        - João de Menezes, **A** Prior of Crato, 1st Count of Tarouca (42)
  - Juana de Mendoza 2nd ∞ Alonzo Henriquez, 1st Count of Alva, Admiral of Castile
    - *Fradique Henriquez* 2nd Count of Alva, Admiral of Castile
  - Mayor de Ayala ∞ (1st wife) Ruy Diaz de Mendoza, Admiral of Castile ∞ (2nd wife)
    - Maria de Mendoza ∞ Diego Perez Sarmiento
    - Leonor de Ayala ∞ Juan Henriquez
    - Ruy Diaz de Mendoza ∞ Beatriz Pereira
      - Isabel de Mendoza **C**

- Rui Vaz Pereira, Lord of Riba de Vizela ∞ Beatriz de Noroña
  - Beatriz de Noroña
  - Constanza de Noroña ∞ João Vaz de Almada, 2nd Count of Avranches
    - Ana de Mendoza 1st ∞ (2nd wife) **B** Diego Hurtado de Mendoza, 2nd Marquis of Santillana, 1st Duke of the Infantado
  - Isabel de Noroña ∞ (1st wife) Diego Hurtado de Mendoza **B**
    - 2nd ∞
      - Juan de Cabrera, 2nd Marquis of Moya
  - Andrés de Cabrera, 1st Marquis of Moya ∞ Beatriz de Bobadilla
  - Inés de Bobadilla, half-sister of Beatriz

- Diogo Fogaça de Góis ∞ Isabel de Brito
  - Diogo Fogaça de Albergaria
  - João Fogaça de Albergaria, Purveyor of Lisbon
    - Joana d'Eça **D** ∞ Pedro Gonçalves da Câmara (36)

- Garcia d'Eza, Governor of Mugem ∞ Joana Nogueira de Albergaria
  - Maria d'Eza
- Fernão d'Eza, Governor of Vila Viçosa, Treasurer of Portugal, cousin of Garcia d'Eza
- Juan de Ulloa, Governor of Castel-Rodrigo ∞ Leonor Sarmiento
  - Isabel de Ulloa
  - *Rodrigo de Ulloa*, Treasurer of Castile
  - Maria de Ulloa
    - Antonio Deza ∞ Inés de Tavera
      - *Friar Diego de Deza*, Bishop of Palência, later Archbishop of Seville

- Gonçalo Coutinho, 3rd Count of Marialva ∞ Brites de Mello
  - Francisco Coutinho, 4th Count of Marialva, Marshal of the Realm

- João Gonçalves da Câmara, 2nd Donatary-Captain of Madeira ∞ Maria de Noroña, 2nd degree cousin of Beatriz de Noroña
  - Simão Gonçalves da Câmara, 3rd Captain of Madeira ∞ Joana Valente de Castel-Branco
    - *Colón-Zarco*
    - **A** João de Menezes ∞ Joana de Vilhena (42)
    - **D** Joana d'Eça ∞ Pedro Gonçalves da Câmara (36)
    - João Gonçalves da Câmara, 4th Captain of Madeira ∞ Leonor de Vilhena
      - **C** Isabel de Mendoza ∞ Simão Gonçalves da Câmara, 5th Captain of Madeira, 1st Count of Calheta
- Isabel Gonçalves da Câmara, sister of João φ Prince Fernando, Duke of Viseu and Beja

References:
26 (Trees II, V and VI); 36 (Tree III); 42 (Trees IV and V).

IX Relation: Noroña/Câmara/Ulloa/Deza/Moya/Menezes

nada in 1455, had a daughter who married Diego Henriquez, son of Afonso Henriquez, Count of Gijón.[2]

Don Manuel de La Cerda, brother of Luís de La Cerda, took Beatriz Fernandez, Lady of Vila Franca de Córdoba, as his mistress and she gave birth to Afonso Henriquez, who in turn married Branca de Sousa, daughter of Martim Afonso de Mello (mentioned above), Governor of Évora, in Portugal. These two lovers were also the parents of Henrique Henriques, who was in the service of Kings Alphonse V and John II of Portugal and who married Filipa de Noronha, João Gonçalves Zarco's daughter and Colón-Zarco's aunt. Another son of the couple was Diogo Henriques, who married Beatriz de Guzmán, bastard daughter of Henrique de Guzmán, second Count of Niebla and Lord of Huelva, in Seville and who was, therefore, first cousin to Colón-Zarco.

As can be seen from the genealogical trees, the Royal House of Portugal was closely related to the Marchenas, Guzmáns and La Cerdas. Stress must be laid on the case of Juan de La Cerda, second Duke of Medinaceli (mentioned above), who married Mência Manuel de Portugal, daughter of the second marriage of Dom João Manuel, Governor of Santarém and former Lord of the Bed-Chamber to the young Prince Manuel, the future King of Portugal. His mother, Isabel, was the daughter of Dom Afonso de Menezes and Maria of Aragon, a relative of the Catholic Monarchs.

The first Duke of Medinaceli, Luís de La Cerda, who had relatives at Serpa, near Cuba, and was closely related to the Mellos and the Henriques, wrote a letter to Cardinal Pedro de Mendoza, in which he states he had promised 'the Admiral three or four well-equipped ships and he did not want any more than that'. He also wrote and spoke to the Catholic Monarchs in an effort to convince them of the feasibility of the 'project of the Indies'. The theory was scientifically admissible and only practice proved that there was no passage to the Pacific. Through the ignorance of its members, however, the 'Examining Board' of the Monarchs rejected the proposal for years. Only when Colón-Zarco reached the Antilles did the same scholars rejoice, convinced that Spain had discovered the Orient. Even in this, they were proved wrong.

Colón-Zarco could not accept the generous offer of the Duke of Medinaceli, in whose palace he lived for two years. Not wanting the expedition to be financed by a private individual, exactly as he did not want it financed by the Kings of England or France, he persuaded the Duke to consider the cost of such an expedition and that its financing should be the responsibility of the Crown. The New Christian Santángel (mentioned above) also put his case to the Sovereigns. But, it must be repeated, it was only after Colón-Zarco had visited Portugal in 1488 to consult King John II did he manage to persuade Queen Isabella of Castile to cede him three caravels. Colón-Zarco was at last in a position to be able to carry out the mission with which the Portuguese King had entrusted him – that only the Spanish Sovereigns would be responsible for the 'discovery' that would alter the division of the world.

In the synthesised genealogical Tree VII (p. 409) of the Medinacelis, it can be seen that Margarida de La Cerda, daughter of Afonso, married Filipe of Castile, Lord of Cabrera, as was the first Marquis of Moya, who was married to Beatriz de Mendoza y Bobadilla.

The names in these genealogical trees are shown in their Spanish or Portuguese spelling: Henriquez, Guzmán, Deza, Hurtado and Mendoza in Spanish correspond to Henriques, Gusmão, d'Eza or d'Eça, Furtado and Mendoça or Mendonça in Portuguese, just as Marchena corresponds to Marchem, Marcham or Marchão, Aragón to Aragão, Alarcón to Alarcão, Harana, Arana or Araña to Aranha and Noroña to Noronha.

Referring to the Marquis and Marchioness of Moya, Salvador de Madariaga said (p. 195): 'Nobody could be closer to the Queen [Isabella] than they were'. He added that Cabrera was a 'convert' and mentioned others that held high posts in the Court, like the Secretary for Finances, Luis de Santángel, and Friar Fernando de Talavera, later Archbishop, who followed the two Marchena 'cousins' as the Queen's confessor.

ALL COLUMBUS'S PROTECTORS WERE HIS RELATIVES

Believing in Madariaga's research and knowing that the designation 'convert' had stuck for several generations, I looked for a family relationship between the Câmaras, who were still considered to be 'New Christians', and the Marquis and Marchioness of Moya. From the genealogy of the Henriquez (see Trees IV, p. 367, VIII, p. 410; and XIII, p. 445), I found that:

1. – Diego Henriquez (X) and his sister Beatriz de Noroña (Y) were first cousins of Fernando Henriquez (W – 33), as they were all grandchildren of King Henry II of Castile.
2. – Juan Henriquez (Z) was the grandson of Diego (X) and brother of Carlos de Guzmán (60), Commander of the Order of Christ, who lived in Beja, Afonso de Guzmán (61), Mayor of Badajoz, and Fernando de Sotomayor, who also lived in Badajoz and left bastard children there.
3. – As Juan Henriquez (2) was Maria de Noronha's father, he was Colón-Zarco's uncle, which meant that Columbus was a relative of the Guzmáns and, therefore, of the Duke of Medina Sidónia.
4. – João de Menezes (Trees IV, V and IX – 42A), Prior of Crato and son of Duarte de Menezes (Tree V – 26), was the father of Leonor de Vilhena, first cousin of Colón-Zarco.

5. – Joana d'Eza (Tree IX – D, p. 411) was Colón-Zarco's first cousin and the grand-daughter of Catarina Coutinho, daughter of Dom Gonçalo Coutinho, in whose house João Gonçalves Zarco lived in his youth.
6. – Joana's mother, Maria d'Eza, was the sister-in-law of Diogo Fogaça, who was married to Inês de Bobadilla, sister of Beatriz, Marchioness of Moya. As Diogo was an uncle of Juan de Cabrera, second Marquis of Moya, the latter was therefore a cousin of Colón-Zarco, as João Fogaça, Governor of Lisbon, brother of Diogo and husband of Maria d'Eza, was the father-in-law of Pedro Gonçalves da Câmara, Colón-Zarco's first cousin.

It can now be understood why the Moyas offered their protection to the 'discoverer of America'.

I was unable to find any relationship between the Marchioness of Moya, Beatriz de Bobadilla, and her namesake who was first married to Hernán Peraza de Ayala, Lord of the Canaries, and then to Alfonso de Lugo. Beatriz's daughter, Inéz de Peraza managed to gain possession (1454) of the Canarian island of Gomera for her son Gullém de Peraza, later first Count of Gomera (1459).

Fernando Colón (Chapter LXXXVIII of the *Historia*) relates that the ships for his father's fourth voyage had been completely provisioned. If Colón-Zarco delivered his provision to the Portuguese garrison at Arzila and then took on fresh supplies at Gomera, it may have been Beatriz de Bobadilla's grandson who supplied Colón-Zarco with whatever he needed. I would like Spanish genealogists to enlighten me on my conjecture, seeing that I have no access to official documents.

At the time of the *Capitulaciones* (see p. 184) Columbus shocked everyone by inexplicably requesting the same privileges as the Admiral of Castile, Alonso Henriquez, first Count of Alva de Liste, instead of mentioning the name of the contemporary incumbent, Fradique Henriquez, Alonso's son. It happened that Alonso Henriquez had supported the cause of Princess Juana and King Alphonse V of Portugal against Princess Isabella, the future Queen of Castile, in the struggle for the succession to the Castilian throne. He was the great-grandson of King Henry II and a relative of the Henriquez shown in Tree VIII. He followed his brother-in-law, Diego Hurtado de Mendoza, in the post of Admiral of Castile.

Fontán Gonzalez[3] and other Spanish researchers say that Alonso Henriquez had been appointed Admiral of Portugal by King Alphonse V, but his name does not appear on the list of Portuguese Admirals compiled by Luis Caetano de Lima. This is because the Portuguese King lost the Battle of Toro due to his receiving erroneous information that led him to abandon the field after the Prince, the future John II, had already conquered the castle. His ambition to be King of Portugal and

Castile in ruins, Alphonse V found himself abandoned by the defenders of Princess Juana's cause.

If, therefore, Columbus were the said Cristóforo from Genoa, he would never have made the mistake of invoking the name of the already deceased Alonso Henriquez instead of his son, who was King Ferdinand's grandfather and still occupied the post of Admiral. The adventurer 'Colombo' would surely have been duly informed so as not to confuse the incumbent Admiral with his predecessor, whom he could never have known.

Colón-Zarco's mistake can be explained through the family relationship that is shown in Tree IX (p. 411).

Admiral Alonso Henriquez, Fradique's father, was married to Juana de Mendoza, daughter of Pedro Gonzalvez de Mendoza and Constanza de Ayala and sister of Pedro de Mendoza, who in turn married Isabel de Zuñiga. This last-named couple's daughter, Isabel de Mendoza, was married to Simão Gonçalves da Câmara, Colón-Zarco's uncle. Fradique Henriquez, therefore, was Colón-Zarco's cousin.

Colón-Zarco was thus quite right when he said 'I am not the first admiral in my family'. Tree IX also shows the relationship between the Admirals of Castile and the Moyas. The whole of Colón's stay in Spain was spent among the nobility – and among his family.

The only person left to speak about now is Colón's last patron, the Bishop of Palência, later Archbishop of Seville.

As already mentioned, the name Deza corresponded to d'Eza or d'Eça in Portugal and the same person can be found in both Spanish and Portuguese genealogies under the respective orthography.

Fernando d'Eza was the grandson of Prince João, son of King Peter I and Inês de Castro, and Maria Telles, sister-in-law of King Ferdinand I of Portugal. His grand-daughter Maria d'Eza married João Fogaça and gave birth to Joana d'Eza, Colón-Zarco's cousin (p. 411, Tree IX, 36).

The same Fernando d'Eza, who was famous for having been wed to six different women in Galicia and Portugal simultaneously without any of them protesting, descended from the Andalusian branch of the Deza family, which was linked to the families of Henriquez, Noroña, Manrique, Mendoza and Alarcón. From the Galician branch, which started using the 'de' before its name, came Colón-Zarco's patron, Friar Diego de Deza.

Fernán (the same as Fernando in Spanish) Peres and his brother Alonso Goméz Deza served King Alphonse V of Portugal and fought for him in the Battle of Toro against Isabella of Castile. The Portuguese Sovereign appointed Alonso Deza as his Admiral. Alonso Henriquez, however, as Alonzo Deza never actually took over the post in Portugal.

A son of Alonso Deza, Fernão d'Eza, became the Royal Treasurer. When Princess Beatriz, widow of Prince Fernando and Colón-Zarco's stepmother, went to Castile in 1479 to negotiate the Peace of Alcántara with Queen Isabella, Fernando d'Eza was in her retinue. He married Isabel de Ulloa, who gave him eleven children, one of whom was Antonio de Deza (or de Ulloa). He, in turn, married Inés de Tavera and begot Friar Diego de Deza, who studied at the University of Salamanca and later became Professor of Theology. He was the Vicar of Seville in 1485 and, on being introduced to the Catholic Monarchs by his uncle Luis de Ulloa, he was appointed as tutor to Prince Juan. He became Vicar-General of Castile in 1499 and three years later was raised to the position of Bishop of Palência. He died in 1524 at about the age of eighty.

Luis de Ulloa was the brother of Juan de Ulloa, Lord of Castelo-Rodrigo in Castile. His daughter, Maria, married the Portuguese Francisco Coutinho, Count of Marialva, son of Gonçalo Coutinho (mentioned above). Maria was a sister of the Isabel who married Fernão d'Eza. Colón-Zarco, therefore, was a relative of Luis de Ulloa.

When 'Cristóvam Colom' arrived in Spain in 1484, he went to see Luis de Ulloa in Salamanca. In the following year, Friar Diego de Deza defended his plan for the discovery of the Orient before the 'Examining Board', the scholars of which, according to Las Casas, 'knew little about astronomy and mathematics'.

Friar Diego lodged Colón-Zarco at the Monastery of San Estéban and paid his travelling expenses. He was also part of the Committee that was formed to draw up the *Capitulaciones* that granted privileges to Colón-Zarco. It is thought that Colón-Zarco's son Diogo became Prince Juan's page due to his influence. But this was not absolutely necessary, as Colón-Zarco was a relative of the Catholic Monarchs through his father Prince Fernando.

The spontaneous friendship shown to Colón-Zarco by the Friars Marchena, the Dukes of Medinaceli and Medina-Sidónia, the Marquis and Marchioness of Moya and Friar Diego de Deza would have been out of the question if he had been the Genoese wool-carder 'that went from door to door selling coloured prints'. They were all related to him through the Câmaras and they all knew that he was the grandson of King Duarte and the discoverer of Madeira, besides being an experienced mariner of the Order of Christ. This is why they trusted in his scientific knowledge and the nebulous plan to reach the Orient by sailing west. But they maintained the necessary secrecy of his identity and supported him only because they wanted, with the invocation of the Templar spirit, to spread Christianity among pagan people and, at the same time, bring greater glory and wealth to Spain. And maybe, as in the case of the Moyas and Santángel, to find a place of refuge for the New Christians that were being mercilessly persecuted by the Inquisition. They must certainly have judged him to be an enemy of King John II, who had executed his half-brother Diogo, and never suspected that he might have been a secret agent in the service of this same Portuguese sovereign.

Madariaga emphasises the ease with which Colón established friendly relations with wealthy men and high officials of the Court immediately he arrived in Spain. And he quotes Gabriel Sánchez, the Lord Chancellor, who like Moya and Santángel was a New Christian and came from an eminent Jewish family of Aragon. It is assumed that this Gabriel Sánchez was an uncle of Rodrigo Sánchez of Segovia, who sailed on the 'Santa Maria' as a comptroller under the orders of the Court, and was a cousin of the physician-surgeon who served on the same ship. The other doctor on board, Chanca, who with Alonso Hojeda was in the service of the Medinaceli and took part on the second voyage, was also a New Christian.

Several members of the Sánchez family were burned at the stake by the Holy Inquisition, as were some of the Santángels, including one Luis, who planned an assassination attempt on the Inquisitor Arbués. The same fate was reserved for one Garcia Chanca, who was probably a descendant or relative of the surgeon.

A Jew called Luis de Torres sailed as an interpreter as he knew some Arabic. Besides the Jews, several Portuguese were among the crews, the pilot of the 'Niña' being Sancho Roiz da Gama, probably the son of one of the Gamas of the Algarve and a kinsman of the Admiral Vasco da Gama who discovered the seaway to India.

Notes and References

1. See Stefano (1967).
2. Luis de Haro, *Nobiliário genealogico de reys e titulos* (1830); Pedro de Medina, *Chronica de los duques de Medina Sidónia* (1861).
3. Gonzalez (1985).

31 *Adulterers and Adulterators*

The ever-increasing death rate among the Portuguese nobility in the wars against Castile,[1] civil wars like the Battles of Alfarrobeira (1449) and Toro (1476) and in the Crusade in North Africa, was made even greater by frequent shipwrecks during the maritime explorations and the Portuguese presence in Africa, where these knights built fortresses, explored rivers and hinterlands and lost their lives at the hands of the enemy or through diseases that were then incurable. In this stormy dawn of the Empire in the fifteenth century a paradox arose – although there were many more women than men in the kingdom, 'rich' widows, members of distinguished families, were passionately wooed and disputed over because they had accumulated inheritances, titles and lands and some, even of mature age, got tired of receiving claimants, sometimes much younger than themselves. If parents did not cloister their daughters inside walls of a convent, they ran the risk of losing them. The genealogists, however, devoted their efforts to marriages and descendants. I must do the same thing.

Guiomar de Mello, for instance, had five husbands: Estêvão Soares de Mello;[2] the Chief Justice Vaz de Castro; the Castilian nobleman Juan Lorenzo de Sarria, who settled in Beja; Martim Coelho da Sylva; and Diogo de Azambuja (the navigator who built the Fortress of Mina), six years younger than her.[3] On the other hand, a sister of hers, Maria de Mello (Lady-in-Waiting to the Duchess of Braganza, Constança) married Duarte de Abreu de Noronha of Estremoz, who contracted matrimony seven times in all, three of his wives being widows and heiresses through the death of their husbands, parents or brothers. The prolonged absence of their husbands at war or at sea also led many women to marital infidelity.

The poetry of the time of King Dennis, founder of the Order of Christ and one of the most notable poets and performers of the 'Songs of Love' and 'Songs of Friendship', which also flourished in Toulouse at the time of Counts Raymond V and VI (see p. 9), was no longer in fashion.

Despite a royal decree that banned duelling, the number of deaths through jealousy rose.

Fernão Lopes[4] narrates an event that occurred to a daughter of Dom Álvaro Pires de Castro (bastard of Pedro de Castro 'of War'), Count of Arraiolos, first Constable of the Realm, Lord of Cadaval and Governor of Lisbon, who had the 'wheel of knives' of St Catherine as the motto on his coat-of-arms and who married the daughter of Pedro Ponce, Lord of Marchena in Córdoba:

> Brites de Castro, for the love of whom King John I had his valet Fernando Afonso of Santarém burnt at the stake. Witnessing this execution, she asked the King the reason for it, to which he replied that he did not wish any greater revenge on her who was of such high class and had become the mistress of Fernando Afonso. She was so ashamed that she went to live with her mother in Castile. And she was already the widow of Dom Pedro Nunes de Lara, Count of Mayorca [Canaries].

It is said that Guiomar de Castro (Atouguia) was the mistress of King Henry IV of Castile (known as 'the Impotent', whose nickname seems to be a contradiction of the rumour). This monarch married Princess Joana, sister of Prince Fernando, Duke of Beja, and aunt of Salvador Fernandes Zarco, in 1455.

While he was still Prince of the Asturias, Henry rejected his first wife, Blanca of Navarra, in 1453, after thirteen years of marriage, on the grounds that she was sterile. But in the course of a scandalous court process, she accused him of sexual impotency, claiming that it was impossible for her to 'conceive without coitus', and the Bishop of Segovia annulled the marriage.[5]

QUEEN JOANA OF CASTILE AND HER DAUGHTER 'THE BELTRANEJA'

King Henry IV was married to Joana for six years before she got tired of his company and became the mistress of Don Beltrán de La Cueva, whom the cuckolded King raised to the rank of the Marquis of Ledesma, Duke of Albuquerque and the Grand-Master of the Order of St James of Castile. Tiring of her lover, Joana, who gave birth to a daughter in 1462, fled to Portugal, riding on Luis Furtado de Mendoça's horse. It was rumoured that the Queen also lay with this gallant knight before finally settling down with Dom Pedro of Castile, who gave her two more children.

Henry IV died in 1470, five years after a great part of the Castilian nobility had revolted and deposed him, placing a regent on the throne.

Princess Joana's daughter – Juana, second in the succession to the throne – was considered to be spurious, which led to Isabella 'the Catholic', already married to Prince Fernando, heir to the crown of Aragon, being acclaimed Queen in Segovia.

Princess Juana (who was nicknamed the 'Beltraneja' because she was thought to be the daughter of Beltrán de La Cueva) was betrothed to King Afonso V of Portugal. In 1475, a part of the Castilian nobility rose up again, led by the Archbishop of Toledo and the Marquis of Villena, and proclaimed Juana as Queen of Castile. This forced King Afonso V to defend the interests of his betrothed, as they were also his, and this led up to the disastrous Battle of Toro.

The Justice of the Realm at that time was regulated by the 'Afonsine Laws' of King Afonso V, the compilation of which had been started in the reign of John I and continued during that of King Duarte, a very important part being played by Prince Pedro. In the section referring to adultery, which was very widespread at the time, a deceived husband had the right to take his wife's life – only feminine infidelity was subject to punishment.

As I have already mentioned, Margarida de Brito, daughter of Joana Coutinho and Estêvão de Brito, was killed by her husband Dom Pedro de Sousa (p. 366, Tree III, 29) of Beja, who also took the life of the servant involved. He later married Joana de Mello, daughter of Afonso de Aguiar (brother-in-law of Isabel da Câmara, Colón-Zarco's mother), who was 'some years' older than he.

Yet even with such laws in force, a wife-killer was left with little choice and Margarida was already Dom Pedro's third wife. Despite all his struggles on the battlefield and in his family life, Dom Pedro lived to be over ninety.

Heitor de Oliveira, Lord of the Entailment of Oliveira, beheaded his wife, Violante de Miranda, daughter of Isabel de Brito (bastard daughter of João Nunes de Carvalhal, who was the preceptor of Prince Fernando, Duke of Beja, and his sister, the future Queen Leonor, in their childhood). The two murdered cousins, Margarida de Brito and Violante de Miranda were great-grand-daughters of Dom Martinho de Brito, Bishop of Évora.

Dom Diogo Coutinho, brother of Dom Vasco Coutinho, first Count of Marialva (p. 366, Tree III, 37), returned from Otranto and learned, through a half-burnt letter handed to him by a slave, that his wife had become the mistress of his brother, Don Francisco Coutinho (married to Beatriz de Menezes, grand-daughter of Dom Duarte de Menezes – see p. 365, Tree II, 26 and 27). He promptly killed her. He later remarried, this time Francisca de Guzmán, bastard daughter of Don Henrique de Guzmán, Duke of Medina Sidónia (see p. 410, Tree VIII).

Dom Francisco de Castro (known as 'The Hobby'),[6] who belonged to the family of the Lords of Vila Ruiva, killed his wife, Beatriz Pereira, daughter of João Mendes de Vasconcelos of Serpa. She had borne him three children before he strangled her 'without reason', thus leaving him free to marry Mécia de Sylveira, daughter of Nuno da Sylveira of Alvito.

I have mentioned these particular cases among the many that occurred during this period of warfare in Ceuta, Alcazar-Ceguer, Arzila and Tangier, when the warriors returned from the battlefront, either because they happened in the Duchy of Beja or because the people involved were in some way linked to Colón-Zarco.

BEATRIZ HENRIQUES AND NUNO PEREIRA OF SERPA

And I shall also mention the case of Nuno Pereira who, on returning from Tangier, was told that his first wife had given birth to 'the child of another'. To whom did he later get married?

Morais, when discussing Henriques mentions, although tersely: 'His first wife Beatriz Henriques' and refers to the fact that he was the Governor of Fronteira. It is Belchior de Andrade Leitão who speaks about his return from Tangier, giving us to understand that Beatriz gave birth during her husband's absence and that 'he slaughtered the bitch instead of the lamb'. But which lamb? I had to discover it in order to understand the metaphor. Had Nuno Pereira thought of sacrificing the fruit of his wife's affair? If the sentence referred to the lover, why call him lamb? Was that his surname ('Cordeiro')? I perused genealogical tree after genealogical tree without coming across the slightest correlation between a 'Cordeiro' and the people I had selected that were in some way linked to the Duchy of Beja. Beatriz Henriques had cousins in Andalusia. Was it a Castilian 'Cordero'? Faria (1961) claimed that the family was of Portuguese descent.

I focussed my attention on 'his first wife'. Who had Nuno Pereira married digamously after murdering his unfaithful spouse? And who was the offspring that was left motherless? As Beatriz was the sister of Henrique Henriques, Lord of Alcáçovas in 1474 and later 'Huntsman' of King Manuel I, her brother's marriage to Colón-Zarco's aunt, Filipa de Noronha, could well offer me a clue.

INTENTIONAL OMISSIONS

Just as the Portuguese chroniclers maintained complete silence in relation to Colón-Zarco's stay in the Azores, so did the genealogists conceal much information. Morais is a good example of this. He did not register the lineage of the Perestrelos and referred to them only through marriages with other families. He (under Câmara) omitted the existence of Isabel Moniz. Only Soares mentions the fact that she was the sister of Garcia Moniz. And Father António Cordeiro[7] is extremely vague in referring to Bartolomeu Perestrelo:

> This first captain survived his first wife and married Isabel Moniz in digamy . . . and the only issue from this second wife was Bartolomeu Perestrelo, the second of that name.

At the stroke of a pen, he obliterated the existence of the Perestrelo sisters, Violante, Filipa, Isabel and Catarina. Dom António Caetano de Sousa makes not a single mention of this Beatriz Henriques that Nuno Pereira murdered.

The only thing left for me to do was to find out if there was any bond of kinship between Nuno Pereira and any of the families that lived in the Alentejo and were contemporaries of Colón-Zarco. Some names which include the surname 'Pereira' are mentioned in the Portuguese edition. But I found no mention of Nuno Pereira, nor of whom he had married after killing his first wife.

Although he lived in Serpa and was linked to the Dias Pereiras, Nuno Pereira de La Cerda might appear in Castilian genealogies.

The only information I discovered was that both the La Cerdas in Spain and the Lacerdas in Portugal were descended from King Alphonse X (known as 'the Wise') of Castile and his wife Queen Violante of Aragon and, in Portugal, from one Don Juan de La Cerda, to whom King Dennis gave the hand of his bastard daughter Maria Gomes, conceived by Marinha Gomes (p. 409, see Tree VII). But at this time, that branch of the family was living in the north of the country and not in the Alentejo.

I then came across another clue. A nobleman by the name of Martin Gonzalvez de La Cerda fled from Córdoba at the time of King John I after committing a murder. Settled in Portugal, he married Violante Pereira, bastard daughter of the Prior of Crato, Dom Álvaro Gonçalves Pereira. This Violante Pereira had nothing in common with her namesake, daughter of Estêvão Soares, but my research had led me back to the Alentejo.

The big problem now was to place everyone in time. Our genealogical treatises supply scant chronological details, going no further than mentioning the King that occupied the throne at the time – and not always even then.

THE EXPEDITION TO TANGIER

It is known only that Nuno Pereira was the Governor of Fronteira and had lived in Serpa prior to this. Why had the King appointed him to such a position? He must surely have left a trail somewhere.

The last piece needed to complete the puzzle was the identity of the wife-killer Nuno Pereira. The chronicler and the genealogists I consulted did not throw much light on the subject but I now knew that it had been at Easter 1464 that he had returned from Tangier and found out that his wife had given birth to 'the child of another' and had 'killed the bitch instead of the lamb'.

The metaphor was explained – people of means used to slaughter a lamb at Easter, following an ancient ritual that is still practised in the north of Portugal today, especially in the area of Braga and the Province of Minho, although it has lost its religious character according to the Mosaic Law on the passover (Pesah, on the Nizan, first day of Spring) and is purely gastronomic. I also managed to establish two dates. Nuno Pereira had probably left Portugal before November 1463 and returned somewhere between 22 March and 25 April 1464.[8] The expedition to Tangier, however, had sailed from Lisbon to Lagos. In all, its organisation took five hard months.

I have a theory for Nuno Pereira's reason for killing his wife Beatriz. As she could not keep the child she had borne and attribute paternity to her husband, she gave it to someone else to bring up. When being denounced as an adultress, she had no defence.

GOVERNORS OF FRONTEIRA

Genealogies show only the family structures and not the appointments to governorships. But I managed to discover that the Governorship of Fronteira had been entrusted to Henrique Henriques de Miranda (see Figure 31.1).

I focused my attention on Andrade: 'he had come from Tangier'. And the greatest interest this person aroused in me was the fact that he had come from the La Cerdas of Castile, the family to which the Medinacelis, Colón-Zarco's protectors, belonged. If only I could find some link.

On 19 January 1464, the Count of Vila Real, Dom Pedro de Menezes (p. 365, see Tree II, 25), Donatary-Captain of Ceuta, departed from his stronghold to take part in the expedition against Tangier, which was led by his son, Dom Duarte de Menezes, Captain of Arzila and Count of Viana (pp. 365, 398, 411, 453 and see Trees II, V, IX and XVI), under the leadership of King Alphonse V (known as the 'the African'). This King had already attacked Alcazar-Ceguer in November 1463, and was now

424 THE LAST INCOGNITOS

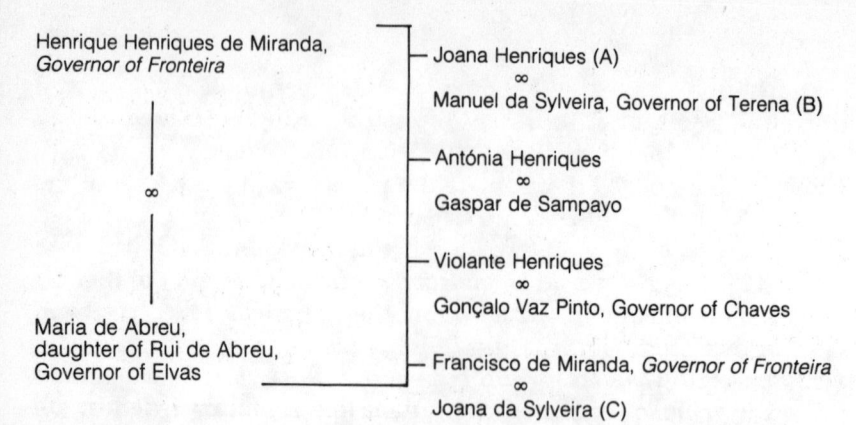

```
Henrique Henriques de Miranda,      ┬── Joana Henriques (A)
Governor of Fronteira               │       ∞
                                    │   Manuel da Sylveira, Governor of Terena (B)
         │                          │
         │                          ├── Antónia Henriques
         │                          │       ∞
         ∞                          │   Gaspar de Sampayo
         │                          │
         │                          ├── Violante Henriques
         │                          │       ∞
         │                          │   Gonçalo Vaz Pinto, Governor of Chaves
Maria de Abreu,                     │
daughter of Rui de Abreu,           └── Francisco de Miranda, Governor of Fronteira
Governor of Elvas                           ∞
                                        Joana da Sylveira (C)
```

References:
A – granddaughter of Fernando Henriquez (33 – Tree III) and grand-niece of Martim Afonso de Mello (27 – Trees II and III) and also of Isabel de Mello, wife of Duarte de Menezes; B and C – brothers and nephews of João da Sylveira (40 – Tree III).

31.1 Henrique Henriques de Miranda as Governor of Fronteira.

making an ill-prepared assault on Tangier. Among the knights on this expedition was one Nuno Pereira, but there is no mention of him being a 'La Cerda'. During the assault on Tangier, Dom Duarte de Menezes died while saving the King's life. Tangier fell to the Portuguese only in 1472. Alphonse V remained in Ceuta for many days, taking measures in relation to the Portuguese positions in North Africa. Accompanied by the Count of Vila Real, he returned to Lisbon in March 1464, 'in time for the Easter celebrations'.[9]

As I found no Nuno La Cerda, I continued the search through the brothers of Henrique Henriques de Miranda:

(a) Francisco de Miranda Henriques, Commander of the Order de St James at Elvas, married to Cecilia de Azambuja, daughter of Diogo de Azambuja – who had taken Vila de Alegrete from the Castilians and Safi from the Moors and had constructed the Fort of Mina on the Ivory Coast, where Colón-Zarco claimed to have been.
(b) Aires de Miranda, married to Briolanja Henriques, sister of the Beatriz Henriques murdered by Nuno Pereira.
(c) Gomes de Miranda, married to Isabel de Brito, illegitimate daughter of João Nunes do Carvalhal (Preceptor of Prince Fernando, Duke of Beja, and his sister, the future Queen Leonor, when they were children); they were the parents of Violante de Miranda, who was killed by her husband Heitor de Oliveira.

∞ = in wedlock

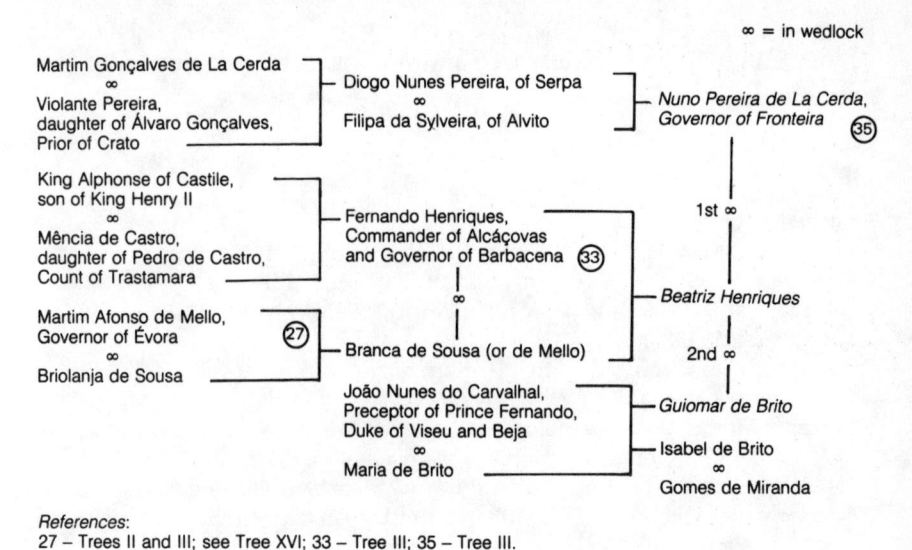

```
Martim Gonçalves de La Cerda ─┐
            ∞                  │── Diogo Nunes Pereira, of Serpa ─┐
Violante Pereira,              │              ∞                   │── Nuno Pereira de La Cerda,
daughter of Álvaro Gonçalves, ─┘   Filipa da Sylveira, of Alvito ─┘   Governor of Fronteira  ㉟
Prior of Crato

King Alphonse of Castile, ─┐
son of King Henry II       │                                           1st ∞
         ∞                 │── Fernando Henriques,        ─┐
Mência de Castro,          │    Commander of Alcáçovas    │
daughter of Pedro de Castro,─┘   and Governor of Barbacena ㉝│
Count of Trastamara                          ∞             │── Beatriz Henriques
                                             │             │
Martim Afonso de Mello, ─┐                   │             │
Governor of Évora       ㉗│── Branca de Sousa (or de Mello)┘    2nd ∞
         ∞               │
Briolanja de Sousa ──────┘
                             João Nunes do Carvalhal, ─┐
                             Preceptor of Prince Fernando,│── Guiomar de Brito
                             Duke of Viseu and Beja      │
                                       ∞                 │── Isabel de Brito
                             Maria de Brito ─────────────┘        ∞
                                                           Gomes de Miranda
```

References:
27 – Trees II and III; see Tree XVI; 33 – Tree III; 35 – Tree III.

X The two wives of Nuno Pereira

NUNO PEREIRA DE LA CERDA AND HIS TWO MARRIAGES

Encouraged by seeing a certain proximity between these Henriques de Miranda and Colón-Zarco, I continued looking for more close relations and found that a sister of Isabel de Brito (wife of Gomes de Miranda and sister-in-law of Beatriz Henriques), Guiomar de Brito, had married in digamy to Nuno Pereira de La Cerda, son of Diogo Nunes Pereira and Maria de Moura, whose genealogical tree is shown in Tree X.

I finally had proof that the Nuno Pereira whose ancestry the genealogists had concealed, as they had concealed the name of his second wife after he had killed his first, was the Nuno Pereira de La Cerda whose first marriage had also been omitted. Why the mystery? Why had his genealogy been split, making it seem as though there were two men and not one – and one that had fought at Alcazar-Ceguer, Tangier and Ceuta and then returned home to murder his wife.

The particular case of Nuno Pereira de La Cerda having killed Beatriz Henriques – and it was not said 'without reason' as was usual, if it were the case – was not a plausible motive to hide his identity, since many knights were outraged on returning from distant wars to discover that their wives had been unfaithful in their absence. But this particular case was certainly covered up. The sole link between the two 'Nunos Pereiras', in different volumes and under different headings (Henriques and La Cerdas), is that they were both Governors of Fronteira and as there were several, it is difficult to distinguish them in the genealogies, as they show only the family tree and not the posts people held.

Notes and References

1. Invasion of Galícia (1369); Invasion of Beira (1372); Siege of Lisbon (1373); Naval Battle of Saltés; Battles of Atoleiros and Valverde (1484); Battle of Aljubarrota (1385).
2. Must not be confused with his grandfather and great-great-grandfather, who were namesakes.
3. When Morais was compiling his genealogies in 1670, he referred to a grand-daughter of this Guiomar: 'a prisoner of the Inquisition of Évora (Alentejo) at the time of writing this book'. He was alluding to the wife of Jacome de Mello Perreia, Commander of the Cavalry in the Alentejo, who was also burnt at the stake. As already mentioned, most of the Pereiras were New Christians. The name of the lady who is presumed to have been Luisa Soares of Elvas is omitted.
4. Lopes (1921), Chaps 20, 35, 74.
5. Prescott (1943) Vol. I, Chap. 3, p. 97.
6. A bird of prey tamed and trained for hunting.
7. Cordeiro, António (1866), Vol. I, Book III, Chap. 3, p. 94.
8. Pina (1790), Chap. 153, p. 509; Leão (1600), Chap. 30, p. 231.
9. The Council of Nicae (325) determined that Easter would be celebrated 'on the 14th day of the moon (full) between the 22nd March and 25th April'.

32 *The Damsel of Córdoba*

Before setting sail on his fourth and last voyage in 1502, Colón-Zarco wrote to his son Dom Diogo: 'Don't forget Beatriz Henriquez;[1] for my sake, treat her as well as you would treat your own mother'. He recommends that she be granted an annual pension of 20 000 maravedis. Shortly before his death, in the terms of his will of 1506, Columbus addresses his son Diogo:

> I order you to take care of Beatriz Henriquez, mother of my son Don Fernando; give her an income so that she may live honourably and lack nothing. Take care of Beatriz Henriquez, who rests heavily on my conscience. I cannot mention the reason here.

Had Colón-Zarco really been in love with Beatriz Henriquez? He certainly seemed concerned about her at the end of his life. Don Fernando wrote nothing concerning his mother. All that we know about this 'damsel of Córdoba' from coeval writings is that Columbus left his two young sons with her while on his first voyage to the Antilles and, on his return, handed her the 10 000 maravedis he had earned.

Yet he did not marry her and wrote in the will: 'I cannot mention the reason here'.

Let us see what Morison wrote in *Christopher Columbus, Mariner*:

> The Haranas were a family long established in Córdoba and the vicinity as peasants, wine pressers and gardeners. Beatriz, daughter of a peasant named Pedro de Torquemada (a remote cousin of the Grand Inquisitor) and of Ana Nuñez de Harana, was born about 1465 in the hamlet of Santa Maria de Trasieras up in the hills northwest of Córdoba. Both her parents died when she was a child. With her elder brother, Pedro de Harana, who subsequently commanded a caravel on Columbus's Third Voyage, she went to live in Córdoba with her mother's first cousin Rodrigo Enriquez de Harana.

Rodrigo, though a wine presser by trade, was man of culture and intelligence who married above his station and lived beyond his means. He had a son, Diego de Harana, second cousin to Beatriz and subsequently Marshal of the Fleet on Columbus's First Voyage. These Haranas were friends and neighbours of Maestre Juan Sánchez, subsequently surgeon of the 'Santa Maria', and of a Genoese apothecary named Leonardo de Esbarraya, whose shop was near the Puerto del Hierro of Córdova. In those days apothecary shops were informal clubs for physicians, surgeons and amateur scientists. Columbus probably drifted into the *botica* because it was kept by a compatriot, and frequented it as the place where local scientists foregathered. He made friends with Diego de Harana, a member of this informal club, was invited to his father's house, and there met the young orphan who became his mistress.

[Documents prove . . . that Beatriz knew how to read and write, but nothing is known of her appearance, personality or character. We do not even know how long Columbus lived with her; perhaps not after his first voyage.

Madariaga (p. 190) criticised the excess of imagination of the authors that wrote about Beatriz but were not her contemporaries: 'romantic foolishness – Colón saving Beatriz's brother from a dangerous nocturnal quarrel – the charitable and convenient story of a secret marriage. They even gave the most cynical of reasons for this – Beatriz was 'unmarriageable'. Some made every effort to show that she was of *noble* birth'.[2] Madariaga wrote the word 'noble' in a different hand and immediately follows with the commentary: 'This she was not'.

He continued: 'Others went to the other extreme and created an easy maidservant of an inn'.

He then adds, in his own words:

As if Colón could love that! However, this point of view is based on a just consensus of the sexual morality of Spanish Christian women. A daughter that gave herself up out of wedlock in these strata of society, i.e. the lower nobility and the middle classes, would obviously lose her reputation, not only regarding her name and social standing, but in her own eyes as well. And so?

Beatriz Henriquez may have been Jewish. The sexual morality of the Jews was, of course, different from that of the Christians. It was not worse, although the Christians judged it to be so.

But Madariaga was wrong.

I have opted to write 'Henriquez' with the Portuguese spelling instead of Enriquez, as the name was spelt in Castile, in the same way that

Helena was written without the initial H. I am not suprised that Morison wrote Harana, even though Fernando Colón always spelt it Arana. La Torre did not exhibit any registration of Beatriz's baptism. Therefore there is no evidence that she was born in 1465. She could either have come into the world in that hamlet in that year or she could have been born a few months earlier, in 1464, somewhere else and then taken to Córdoba.

According to Morison and Madariaga, Beatriz's mother was surnamed Nuñes de Arana; the father, Torquemada. Why was she only called Henriquez? When Colón-Zarco wrote to his son Diogo in regard to Beatriz, he never called her Torquemada or Arana or Nuñez. The authors related: 'Her parents died when she was still a girl and with her elder brother Pedro – who would command a caravel on Colón-Zarco's third voyage to the Americas – was taken under the guardianship of her mother's cousin-german, Rodrigo Henriquez de Harana, established in Córdoba . . . His son Diego accompanied Colón on his first voyage as the master-of-arms of the fleet'.

They did not find any document about this matter; they were created only in books concerning Columbus.

Colón-Zarco's *Diário de a bordo* (logbook), which narrates the first voyage, mentions Diego de Arana only three times:

(a) 'I first sent a longboat ashore with Diego de Arana of Córdoba, master-at-arms of the fleet' (p. 168).
(b) (twice on the same p. 177) 'The Admiral placed a lot of trust in Diego de Arana [and two other officers] . . . He left them on the island of Hispaniola . . . and appointed Diego de Arana, born in Córdoba, and the other two as his lieutenants.

Beatriz's cousin, Diego de Arana, was never identified as Henriquez and neither was her brother Pedro. If they really were Henriquez, why did Columbus and his son Fernando never mention the fact? In his *Historia del Almirante* a book of 300 pages, Fernando Colón makes five brief references to Diego de Arana, in one of which he also refers to Pedro de Arana.

(a) 'Diego de Arana of Córdoba, master-at-arms of the fleet' (p. 132).
(b) 'he highly recommended Diego de Arana, son of Rodrigo de Arana of Córdoba, whom he had mentioned' (p. 135).
(c) 'to Navidade, where there were only Diego de Arana with ten men' (p. 172).
(d) 'so that in his absence they would not attack Arana and the 38 Christians that had stayed with him' (p. 176).
(e) 'Pedro de Arana, cousin of the Arana that died on the island [of] Hispaniola' (p. 327).

There are another three very brief references to Pedro de Arana. Don Fernando, therefore, never referred to them as Henriquez and even less did he allude to them as his cousins. On the other hand, he never hid the fact that Bartolomeau and Diogo Colón were his father's brothers. With one exception – in a document of the family, he admitted that he had been indicated as a cousin of Diego de Arana, son of Rodrigo de Arana and Constanza de Alarcón.

THE NOTARIAL DOCUMENTS

Only the following details can be obtained from the various notarial documents presented by La Torre:

1. Rodrigo Rodriguez de Arana was living in Córdoba in 1465 and owned a town house and country houses, with vines and olive groves and their respective presses, at Trasieras, Puerto Caballo and Valedeleche.
2. He married Constanza de Alarcón, daughter of Diego de Alarcón. Don Cristóbal de Mesa, a member of the 'Committee of Twenty-Four' was a witness to the wedding.

   Constanza's dowry amounted to 30 000 maravedis (which corresponded to a tenth of the cost of Colón's first voyage to the Antilles) and in 1467 she inherited from her mother, among other assets, a house in Córdoba that she sold for 50 000 maravedis.
3. In this same year, Rodrigo Rodriguez collected 3500 maravedis that he had loaned to the brothers Antón and Velasco Nuñez, let a house at Puerto Caballo for 300 per year and sold a house in Córdoba for 100 000 and properties at Puerto Caballo and Valedeleche. The sales realised a total of 198 000 maravedis.
4. His whereabouts were unknown between 1467 and 1473.
5. In 1473, his wife Constanza moved from the district of San Pedro to San Loryente and Rodrigo contracted a young lady as a companion for his son Diego.
6. Having been taken ill on his estate at Trasieras in 1477, he made a will in which he declared that he was the son of Juan Rodriguez de Arana and that he wished to be buried in the Church of San Pedro, as his parents had been.

   He ordered that his debts be paid and left his son Diego 'real estate and rents [from his properties] and shares [in banks]'.
7. In an unfinished will, in which the testator's name is not mentioned, an aunt of Rodrigo Rodriguez de Arana left him 10 000 maravedis and bequeathed 6000 to Rodrigo's sisters Catalina, Elvira and Marina Rodriguez and 2000 to a niece surnamed Mayor-Henriquez. She

made other bequests to a nephew, Juan Garcia de Saucedo, Chaplain-Dean of the Church of San Pedro, and to the Mother Superior of the Sisterhood of Jesus 'for her expenses and for works of charity'.

Up to here, all the Aranas also have the surname Rodriguez, although the unknown aunt of the *'lagareiro'* (either a man who owned a wine or olive press or was in charge of the presses on an estate) had another nephew with the surname of Henriquez. None of the documents suggest that Beatriz's 'brother' was a Henriquez. He is always known as Arana, as is his cousin Rodrigo Rodriguez; and he has only one cousin called Mayor-Henriquez. In those days, the name Mayor was given exclusively to the first-born daughter of a family. Everything indicates, therefore, that the Arana aunt had one niece surnamed Henriquez and one nephew and three nieces called Rodriguez.

8. It can de deduced that Rodrigo had married again in 1489, as in that year 'Lucia Nuñez, wife of Rodrigo Rodriguez de Arana' made a will in which she bequeathed all her assets to her sister Leonor Gutierrez, mother of Don Diego de Gôngora (of a noble family of Córdoba), left Rodrigo two houses and a country property at Trasieras and bequeathed 500 maravedis to the Monastery of Santa Inês.
9. While at Trasieras in the same year, Rodrigo fell ill and made a new will, which was witnessed by the same Don Cristóbal de Mesa and in which he repeats his wish to be buried in the Church of San Pedro of Córdoba. Apart from the clergy who had officiated in it, only certain distinguished families had the right to be buried in the interior of a church in those days.

Let us now look at what we know about Ana Nuñez de Arana, married to Pedro de Torquemada.

10. While living in the Santiago district of Córdoba in 1471, Ana Nuñez made a will, in which she said that she was the 'sister' of Mayor-Henriquez, to whom she bequeathed a fifth of her assets, leaving the rest to her children, Pedro de Arana and Beatriz Henriquez. She also expressed her wish to be buried in the Church of San Pedro, even though she was not living in that parish at the time, as her parents were buried there.
11. Some months later, an inventory was made of the goods she had left on the Trasieras property, mention being made of: 'an abandoned house, used furniture and carpets, a broken sword and some candlesticks'. It can thus be inferred that the sister Mayor-Henriquez was a spinster and lived with her old mother in the district of San Pedro.

This is everything that can be extracted from La Torre's documents – and precious little it is!

None of the documents tells us that Pedro de Torquemada was what La Torre calls a 'humble market-gardener'. We know only that his wife had a house in Córdoba and properties at Trasieras. But it is common knowledge that only gentlemen had the right to wear a sword. Torquemada could very well have kept his as a souvenir, as for many men their sword was an object of great intrinsic value and of honour and esteem. Men also felt a great sense of pride in using the swords of their fathers or grandfathers, although the blades were changed in keeping with the evolution in the 'art of combat'.

If the 'market-gardener' had found it, there would have been no reason for him in keeping it; he could easily have sold it. And how can it be explained that 'used carpets and candlesticks' were found in an abandoned country house? But it is easy to see that they were not the trimmings of a humble country house. It is known that they had lived on the Trasieras estate, which lay 15 km from Córdoba, and it is highly probable that he died there. But it was likewise with Rodrigo Rodriguez de Arana, who was not a market-gardener, but had noblemen as witnesses and the right to be buried in a church.

Starting from the assumption that Columbus was a Genoese woolcarder, La Torre made up his meeting with Pedro de Arana at the apothecary's in Córdoba, which he claimed, was frequented by Columbus's Genoese compatriots. And in 1933, ignoring the habits and customs of the fifteenth century, he also invented the legend of Beatriz Henriquez's miserable parents.

Regarding the 'dissolute life' of the *lagareiro* Rodrigo Rodriguez de Arana, to whom La Torre gave the name 'Henriquez', let us have a look at what contemporary history tells us.

The war of succession for the Spanish Crown lasted from 1464 until 1476. The Dukes of Medinaceli and Medina Sidónia and the Bishop of Seville, who defended the cause of King Henry IV's Portuguese mistress Guiomar de Castro, lined up against Princess Isabella, the future Catholic Queen, although Medina Sidónia changed sides following the death of Isabella's brother Afonso.

After the death of Prince Afonso, who had been recognised as heir by the rebellious nobles in detriment to Princess Juana, who was betrothed to King Alphonse V of Portugal, Princess Isabella was proclaimed as heiress to the throne by her supporters in 1468 and King Henry IV, who had been deposed three years earlier, recognised her as such.

Isabella married Ferdinand of Aragon and many of the nobility joined her ranks, including the Duke of Medinaceli.

The King died in 1474 and Isabella was proclaimed Queen. But the Castilian nobility, led by the Archbishop of Toledo and the Marquis of Villena, rose up against her the following year and proclaimed Princess Juana as Queen. This led up to the Battle of Toro, which took place in 1476.

The Duke of Medina Sidónia had meanwhile built a sumptuous palace, where he would lodge his relative 'Colón' in 1485. Some years before he had revolted against the Catholic Monarchs and wanted to set himself up as 'King of Seville', with the result that he was condemned to remain there in exile. But he played a valiant part in the War of Granada and redeemed himself in the eyes of the Sovereigns.

The nobility and landowners were responsible for the recruitment of troops in those days, which implied the spending of large sums of money. Only this can explain the sudden sale of Rodrigo Rodriguez de Arana's assets and his absence between 1467 and 1473. He had gone to fight in the War of Succession and had returned because he was either sick or wounded.

The *Relaciones Genealogicas* of Don Antonio Suarez de Alarcón which, although incomplete, supplied me with precious information, which I present here in a very synthesised form.

- *Torquemada* – The first titled Torquemada was Don Gonzalo de Torquemada, astrologer and knight of King Alphonse X (known as 'The Wise'), who gave him the town from which he took his name in 1258. He married Urraca Garcia and from him descended a long line of knights and illustrious clergymen and prelates. The Inquisitor of Spain, Thomas de Torquemada, was a contemporary of the so-called 'father' of Beatriz Henriquez, whom La Torre transformed into a local farmer, and a cousin of Constanza de Alarcón.

  To my surprise, I found that another Constanza de Alarcón, daughter of Martin Rodriguez de Alarcón and Maria Afonso Carrillo, had married Mosén Don Juan de Torquemada (Libro III, p. 252).

- *Alarcón* – In the fifteenth century, there were hundreds of Alarcóns in the Castilian nobility. Their genealogy begins with one Sibila Sforgia daughter of a knight of Ampurdam, whose second husband was King Peter IV of Aragon. She was the mother of King John I of Aragon and grandmother of Isabel of Aragon and Urgel (who married Prince Pedro, son of King John I of Portugal) and great-grandmother of Princess Isabel, who became the wife of King Alphonse V of Portugal, Colón-Zarco's uncle.

  She was also the great-grandmother of the Constanza de Alarcón that was married to Juan de Torquemada and Fernando de Alarcón, husband of Catalina de Cuña who, according to the genealogist 'had many children whose line I shall not describe because I have not got it organized; I have only seen the papers that mention Guiomar de Alarcón' (Libro III, Cap. IX, pp. 241–2).

  Regarding other ladies of the Alarcón family, the author likewise says: 'the documents referring to them have disappeared'. This was not a unique case, as since the thirteenth century the nobles of Aragon had left the kingdom for other Iberian courts at times of political instability. Dozens of the noble houses of Spain and Portugal descended from the Alarcóns.

Of Constanza de Alarcón's four brothers, I should like to mention Juan de Alarcón, who was the Standard-Bearer of Pope Martin V, whose original name was Otto Colonna. This Pope was the son of Dom Agapito Colonna, Archbishop of Lisbon, of whom I have repeatedly spoken and who I shall mention again (see p. 499).

Five hundred years after these happenings, and with the dissemination of family names, it is difficult for someone today to trace his or her origins. But it was much easier to do so in the fifteenth century, especially as marriage between cousins was common practice. This is why I relate the name of Martin Rodriguez de Alarcón (Constanza's father) to that of Rodrigo Rodriguez de Arana, husband of Constanza de Alarcón, Beatriz's Henriquez's 'aunt'.

The disappearance of the documents referring to the nephews of this Constanza de Alarcón married to Juan de Torquemada was discouraging, as among the names of the same branch of the family the name Diego (the name of Rodrigo Rodriguez de Arana's son) is found more than once. I can only hope that Spanish researchers come across a more complete genealogy. But I did find out that the Guiomar mentioned by the author married Don Pedro de Zafra, Mayor of Mondejar, Knight of the 'Committee of the Twenty-Four' that served the Catholic Monarchs; a cousin of hers, also called Martin Rodriguez de Alarcón, like Constanza's father, of Granada, Captain of the Guard of the Catholic Monarchs, had taken as his second wife Elvira Hurtado de Mendoza (daughter of Juan Hurtado de Mendoza and Inés Manrique), who was Lady of the Bedchamber to both Queen Maria and her niece Queen Leonor, who were successively married to King Manuel I of Portugal, Colón-Zarco's half-brother.

- *Arana* – This surname is frequently found related to the Guzmáns, Mendozas, Alarcóns and other noble houses of both Spain and Portugal, being Aranha in the latter country, from the Spanish Araña.

Juan de Urries y Ruiz de Arana, author of the book entitled *Enlaces de Reys de Portugal con Infantas de Aragon*[3] was the Marquis de Ayerbe. Unfortunately I was unable to consult the genealogy of this family. But in Morais's *Pedatura Luzitana* (vol. 3, no. 1, pp. 393–4), I found: 'Leonor Dias Pereira [of Viana], married to Afonso Rodrigues, mother of Maria Fernandes Pereira, wife of Rodrigo de Aranha'. It is a strange coincidence that can be explained only by the fact that the Spanish and Portuguese branches of the families were interlinked at that time.

On 17 August 1525, Fernando Colón, Beatriz Henriquez's son, had a notarial minute drafted[4] in which he makes an 'irrevocable bequest' to Pedro de Arana of 'some houses, a wine cellar, a press, a vat, some tuns and a market garden that he inherited from his mother'.

We can thus conclude that Beatriz Henriquez was not so poor as she had been made out to be. It must be noted that, according to Fernando Colón, his brother Diogo ceased to fulfil the wishes of their father's will and 'she did not receive the said ten thousand maravedis during the last three or four years of her life and I do not recall the case very well'.

Diogo Colón referred to his stepmother with a certain discomfort and, as he had forgotten to assure her her pension, determined 'that the amount of the debt be calculated and that it be paid to her heirs'.

When Diogo began to dishonour his payments, Beatriz Henriquez sold two houses to the Canon of Córdoba Cathedral, Juan Ruiz, for 52 000 maravedis, the contracts being signed on 6 November 1519. And she still possessed assets to bequeath to her son Fernando, who in turn left them to Pedro de Arana.

An interesting detail appears in the notarial document. The bequest made by Fernando in favour of Pedro received the approval of the Mayor of Córdoba, Juan de Torquemada. This is another coincidence, as the Mayor was the namesake of the already mentioned husband of Constanza de Alarcón, the children of whom the genealogists were unable to name. It cannot be the same person, however, as the latter 'was still alive in 1492', which means that he must have died not long after. But there is still a chance that he was the father of Pedro de Torquemada.

The proof that Beatriz Henriquez received a first-class education can be seen in the one she gave to her stepson who, when carrying out his duties as a page to Prince Juan, was praised for 'his excellent manners and culture' by, according to Las Casas, the Queen herself.

It must also be emphasised that Pedro de Arana, son of Diego de Arana and grandson of Rodrigo Rodriguez de Arana, referred to Beatriz Henriquez as a servant would refer to an employer or a gentleman to a lady: 'Don Hernando Colón, son of my ladyship Beatriz Henriquez'. Before mentioning the ample assets, he refers to Columbus's son as his cousin. It is thus assumed that this Pedro de Arana was the son of Diego de Arana who was killed by the local natives on the island of Hispaniola.

As it was not a document that was to be included in his chronicles, Don Fernando was discreet enough not to mention this fact. One wonders, however, if Pedro de Arana was not Beatriz Henriquez's cousin through his father or through his mother.

The last hypothesis to consider is that Pedro de Arana was merely the cousin of Diego de Arana, son of Pedro de Torquemada and Ana Nuñez de Arana, Diego being the foster-brother of Beatriz Henriquez. Such kinship would euphemistically allow him to call himself a 'third-degree cousin' of Don Fernando de Colón: 'His son Diego [de Arana] accompanied Columbus on his first voyage as master-at-arms of the fleet'. Until now, had nobody ever noticed how strange it was that the son of a Córdovan *lagareiro*, living over sixty miles from the nearest port (Malaga, before 1492, was still in the hands of the Muslims), could have learned the duties of seamanship and later should have had a post of such great responsibility on board a ship? He was the master-at-arms, which meant that he had to administer the expenses of the whole fleet and maintain discipline on board the ship during the voyage, under the direct orders of each captain.

And Pedro de Arana, the said brother of Beatriz Henriquez, had commanded one of the six ships that made up the fleet on the Admiral's third voyage. Why did Colón-Zarco choose such officers for such a risky undertaking if they were going to be seasick because they were not experienced seamen or were unable to deal with crews of Biscayans, Galicians and Portuguese that were never easy to control, as history has repeatedly shown? The Admiral trusted them, however, and had his reasons for doing so. We must not forget that to be a scribe in the fifteenth century, even in the Court, it was enough for a man to be able to read and write and have 'good manners' in order to arrange a post through family influence or favours. But to be commander or the master-at-arms of a ship demanded skills in seamanship and navigation and qualities of leadership. No Admiral would take the risk of appointing a layman to one of these positions just because he was a distant relation through a mistress that had borne him a son. This leads me to think that the two Arana cousins were in fact kinsmen of the Henriquez family (to which belonged the Admirals of Castile) and that their destiny was a life before the mast.

COUSINS AND LOVERS

I must speak again about the Governor of Fronteira, Nuno Pereira de La Cerda, the Portuguese kinsman of the Medinacelis of Córdoba. He was the grandson of Martim Gonçalvez de La Cerda, who had fled from that city after committing a crime; and he married Maria Gomes, bastard daughter of King Dennis of Portugal.

Nuno Pereira de La Cerda left Alcazar-Ceguer at the time of the first attempt to take Tangier (i.e., January 1461) or the second (November 1463). If it were the former, he was abroad for three years. If it were the latter, he was away from home for at least ten months, because of the long preparations that the attack required. On his return, he was told that his wife 'had given birth to the child of another'. Beatriz Henriques had cousins in Córdoba – the Henriquez de Guzmán, a branch of the Medina Sidónias. Studying the geneaological relationships (see Tree XIII, p. 447), my suspicions naturally fell on Don Carlos Henriquez de Guzmán, who had settled in Beja (Portugal). He married Cecília de Almada. He later changed his name to the Portuguese 'Gusmão' and was raised to the rank of Commander of Proença, of the Order of Christ, by King Alphonse V.

It is about twenty miles from Beja to Serpa, a half-day's ride. During a short rest, the 'child of another' may have been conceived. The child was certainly a girl. I am sure that the reader has already understood the aim of my sub-thesis – Beatriz Henriques's child was given the name of Beatriz Henriquez. The newly-born was taken to Córdoba, probably by way of Elvas and Badajoz, where a brother of Don Carlos de Guzmán, Afonso Henriquez de Guzmán, was the Magistrate of the city (he married Mência de Soto, sister of Don Fernando de Soto, who would be Military Governor of Florida), and another, Don Fernando Henriquez de Sotomayor, had also married and left family (see Tree XIII, p. 447, 33, 43, 60, 61 and 62).

It was essential that Beatriz hide the 'fruit of her sin' before her husband's return. She must have pleaded for the help of her first cousins, sons of her uncle Don Garcia Henriquez, who had five legitimate children and one legalised bastard, João de Noronha, who lived in Beja before going to Madeira, where he married Inês de Abreu, daughter of João Fernandes de Andrade of the Arco (see pp. 241 and 268). An aunt of these six Henriquez, Maria de Noroña, was married to João Gonçalves da Câmara (nicknamed 'the Pearmain' because he was redfaced and blond), second Donatary-Captain of Madeira, brother of Isabel da Câmara and uncle of Colón-Zarco. This means that Beatriz Henriques was a cousin of Columbus, as was her daughter, the 'damsel of Córdoba'.

Where could a wet-nurse for Beatriz Henriquez be found? She was entrusted to the Torquemadas, who owned an isolated farm and also had a small child – Pedro de Arana. Ana Nuñes suckled both of them. And the girl was baptised as their daughter.

Joaquin Torres Ascencio[5] came to the conclusion that 'Beatriz knew how to read and write, which was unusual among girls of her class or origins'. At last someone had placed her in her time. The poor could attain some degree of culture only through the church in the Middle Ages. It is obvious that Don Fernando Colón's mother was no 'rustic country-woman'. She had received an education above her station. Her cousin Arana refers to himself respectfully as a 'servant' of her ladyship. The orphan girl that was 'received into the house of Rodrigo de Arana in pity' was considered to be a 'lady' by his grandson and not a servant who should be grateful. This is where Madariaga made his mistake, in thinking that she could not have had noble blood[6]. As Beatriz became

Columbus's mistress, he merely tried to prove that she was Jewish, since the 'converts' were laxer in their morals. But noble blood? Never! What a pity it is that Madariaga did not decipher Colón-Zarco's siglum, since his book, in many respects, is one of the most rational that has been written regarding Columbus.

Notes and References

1. Morison (1945b).
2. Cerro (1933); Manzano (1964); *Datos nuevos referentes a Beatriz Enriquez de Arana y los Aranas de Cordoba* (Madrid, 1900–2).
3. Madrid (1899).
4. Del Cerro (1964), p. 237.
5. Ascencio (1933).
6. Beatriz Henriques, Beatriz Henriquez's mother, was great-granddaughter of King Alphonse of Castile (see p. 425, Tree X).

33 Distortions Corrected

The distortion of certain parts of the *Historia del Almirante* perpetrated by the Genoese Baliano de Fornari are either intentional or show profound ignorance. Don Fernando, for instance, would never have made the mistake of saying that 'Moguer' (near Huelva, which he knew so well, Palos, the River Tinto and the Monastery of La Rábida) was in Portugal (p. 74). Don Fernando himself had been there, many times.

THE NAME OF DON FERNANDO'S GRANDFATHER

Another distortion of Fornari must be pointed out here (*Historia del Almirante*, pp. 60–1). Don Fernando could never have written the following text:

> it happened that a lady of noble blood, Filipa Moniz, Mother Superior of the Convent of All Saints, where the Admiral used to hear mass, became so friendly with him that they got married. As her father, Pedro Moniz Perestrelo, had already died, his mother-in-law gave him all her husband's writings and charts.

I shall not bother to consider the disobedience of the navigator of the Order of Christ in relation to the rule of secrecy, leaving 'his writings and charts' so that his wife could give them to any foreigner that happened to drop in. But I find it strange that Don Fernando did not know the name of his father's father-in-law, who was a famous navigator in his own right, the co-discoverer of the Archipelago of Madeira and the first Donatary-Captain of Porto Santo. His name was Bartolomeu, not 'Pedro'. There was not any Pedro in the Moniz/Perestrelo family.
And Don Fernando certainly knew that his brother's grandfather could never use the name of 'Moniz,' which belonged to his wife.

Besides this, how could he commit such an indiscretion in saying that Dom Diego was the son of a former nun who had broken her vows in order to wed an adventurer? Don Fernando, moreover, was a friar and it would not have suited him to spread the story that a Mother Superior had been seduced by a poor but slick-tongued merchant, even if it did make his father a great 'conquistador'. I shall prove that Don Fernando Colón knew all his family well, including his 'two Portuguese grandmothers'.

THE TWO PORTUGUESE GRANDMOTHERS OF DON FERNANDO

Genealogical research frequently leads one to unexpected historical discoveries. It must be pointed out that Dom Diego, like his father, was not surnamed 'Colón' but 'Colom'. The spelling of his name as Colón appeared in Castilian documents only after his naturalisation. As can be seen, the son was named according to the spelling in Portuguese documents. Even if his name, whether real or a pseudonym, were 'Colom', the secretaries began writing it as they heard it pronounced. It appears in some documents as 'Culão' or 'Culã', and the same happened in documents concerning his father and brother. The text of a petition by Fernando Colón presents his surname distorted twice: 'Culã' and also 'Cunha'.

In *Documento sobre a ída para Castela da Imperatriz Dona Isabel, filha d'El Rei Don Manuel* [I] *e a mulher do Imperador* [Carlos V][1] is the following passage of a letter that António de Azevedo Coutinho, Portuguese Ambassador to Castile, wrote to King John III on 14 April 1526:

> Sire . . . those persons about whom your Highness wished to know and about whom I have been informed by my secretary, are the same . . . [the paper is torn in this place] if they have not since died and I am only interested in two, *dõ Fernando de Culã* and the licentiate Manuel, as he is a contumacious villain [*vilã*], for whom accepting the Host would be the same thing as my taking a biscuit.
>
> Cunha has just spoken to me and said that he would like justice to be done, as his grandmothers had been Portuguese and, for Portugal, had lost when King Alphonse V came in [Castilian territory] but expected to receive indemnity from the Empress; and I assured him that he would. He will not lose his case, but win.

The Emperor Charles V (I of Spain) granted the illegitimate son of Columbus and the grandson of two Portuguese ladies an annual income of 500 gold 'pesos' 'to help you in your sustenance and with the library that you have set up in the city of Seville'. This was following the disgrace into which Columbus had fallen in Spain. This occurred so that, although Columbus had been discredited, someone would glorify him in the famous 'Columbian Library', so that Fernando would narrate his father's life – but not all of it.

In Azevedo Coutinho's letter, we can see that 'Culã' is the same as 'Culão' or Colão.[2] The sounds 'om', 'ão' and sometimes 'ã' were written indiscriminately in the fifteenth century. While the above-mentioned transformation must be admitted as a possibility, I am personally inclined to accept the opposite – the secretary of the Portuguese Ambassador to Castile (a Castilian?) tried to write down the name Colón in Portuguese, giving rise to the distorted form of 'Culão'. It must be noticed that the Portuguese word for villain is written *'vilã'* in the same text instead of the correct *'vilão'*.

The alterations of surnames through translations was quite frequent in those days. Las Casas translated the name of the navigator Fernão Dulmo as 'Hernan Dolinos' and in the Italian translation of the book that was lost it appears as 'Fernan Dolmos'.

To close this subject, I shall cite the case of Pedro Vasques, the Galician who, according to Portuguese documents, sailed with Diogo de Teive to Newfoundland in 1452, whose name in the *Historia* of Fernando Colón, after being translated into Italian and then into Castilian, and in Las Casas's book (partly copied), appears as 'Pietro Velasco', which led many authors to think that he was Italian owing to the 'Pietro', even though his surname was also Spanish. It must be stressed that the fact that Columbus called the captain of the nau that received him in the Tagus in 1493 Álvaro Damán and that Fernando Colón called him Álvaro da Cunha in his *Historia* is even stranger.

According to Saúl dos Santos Ferreira and Ferreira de Serpa, the surname of 'Cunha' is a mistake of the palaeographer who did not understand the word 'Culã' (a non-existent surname in correct Portuguese); it is obviously an unintentional distortion. I think there is another reason for the error, as the palaeographer was the expert historian Anselmo Braancamp Freire.

1. In a letter written to António Carneiro on 6 April 1524, Ambassador Azevedo Coutinho referred to one Barrentos (Hernando de Barrientos, Castilian) and to one Da Cunha (Cristóbal Vásquez de Acuña, also Castilian) and said: 'These two seem to be good men, but Manuel and the attorney they call Pisa look very bad and untrustworthy'.
2. Two years later, in the already mentioned letter of 14 April 1526, Coutinho referred once more to one Manuel, who must be the same one judging by the unfavourable terms in which he is mentioned. He also refers to a Cunha, but it cannot possibly be the same Cristóbal Vásquez de Acuña, as he had already said that he was interested in

only two people, Don Fernando de Culã and Manuel. We can therefore come to the conclusion that the scribe at the embassy (on writing the draft) wrongly read *Cuña* instead of *Culã* and wrote the Portuguese *Cunha*. Only the letter of 1524 mentions the Castilian 'de Acuña' and another one refers to *Culã* (Colón) only two years later, but the latter did not refer to any 'de Acuña'.
3. In this letter written to King John III, brother of the Empress Isabel, Coutinho informed the King about a petition made by Fernando Colón and not by a Castilian by the name of de Acuña. It is not known whether the latter had Portuguese grandmothers whose husbands had fought in the Battle of Toro; but even if he had, why would the ambassador inform the King that he had asked for an indemnity? King Alphonse V had lost the war fought in Castile and had not confiscated lands in his kingdom that had belonged to Portuguese that had fought on his side. On the other hand, the Catholic Monarchs had confiscated all the assets of the Castilians that had taken the side of Alphonse V. Besides this, there were no bonds of kinship between King John III and de Acuña that would lead the King to bother with the latter, even if the King could not bring his influence to bear on his sister.
4. Empress Isabel, one of the most beautiful ladies of her century, had married Emperor Charles V a month earlier. If the petition had been made by Fernando Colón, it may be thus deduced that Columbus's natural son was not so ignorant of his father's family as he made out in his *Historia*. We shall see if there is any reason for this interpretation.

If he was the bastard son of Beatriz Henriquez, who – according to most historians – was not Portuguese, we might conclude that these two grandmothers make Columbus at least half Portuguese. And as these Portuguese ladies had lost everything they owned at the time of the civil war, the Genoese wool-carders called Colombo could have had nothing to do with the 'discoverer of America'. How could the Portuguese Ambassador (who needed a confirmation from the Portuguese Royal Archives about the implication of Columbus's family in those battles) assure Fernando that he would be indemnified on the orders of the Empress? What connection could there be between Columbus and the sister of King John III?

'Two Portuguese grandmothers' could either mean one paternal and one maternal or a grandmother and a great-grandmother, or might even include a great-aunt. When I began my investigations, I still accepted the possibility that Beatriz Henriquez was a Córdovan of humble origin, so that I was quite content to be led to believe that Don Fernando was referring to the paternal branch of his family: 'he knew that his father was at least half Portuguese'. Yet I could not imagine how Isabel da Câmara could have been harmed by the civil war. I could not find any indication that her husband, Diogo Afonso de Aguiar, or any of the Zarco, Teixeira, Moniz or Perestrelo brothers, every one of them resident in the Archipelagos of the Azores or Madeira, had taken part in the campaign or possessed any assets in Spain that could cause them any material loss.

I then got entangled in a web of questions:

(a) Had Don Fernando dared to try to dupe an Empress by means of an unfounded petition?
(b) Had he dared to exploit the credulity and benevolence of the Portuguese Princess by presenting his petition at the precise moment that Charles V was wildly in love with his beautiful wife and had consummated his marriage six weeks before? Why had he not presented his petition earlier?
(c) Had the friar allowed himself to use the influence of Eros in an attempt to increase his wealth with the assets of imaginary grandmothers?

Faced with the improbability of such an attitude on the part of Don Fernando, I studied the petition more closely. And another doubt arose:

(d) Why had it been Don Fernando, second son and illegitimate, who had presented the petition, and not his brother Dom Diogo, Columbus's first-born and legitimate son?

Since they came from different mothers, maybe it would be worthwhile looking into the illegitimate maternal branch. As it had already been shown that Don Fernando's mother was Beatriz Henriquez, daughter of the Portuguese Beatriz Henriques, I thought it might be a good idea to delve into the Henriquez genealogy (see Tree XI, p. 444).

I had demonstrated only that Don Carlos Henriquez must logically have been the father of the 'damsel of Córdoba', but I had not proved it. A sister of his, Maria de Noronha, had married João Gonçalves da Câmara, brother of Isabel da Câmara and was, therefore, great-aunt of Don Fernando (see Tree XII, p. 445). But this kinship led us nowhere, especially as the second Donatary-Captain of Madeira had not been involved in the civil war.

The improbability of the Battle of Alfarrobeira, where Prince Peter had been killed by his brother's army and when King Alphonse V was still a boy, led me to investigate the Battle of Toro and I noticed two things in the genealogies:

1. Don Fernando Henriquez (Tree III and IV, 33; XIII and, 33), bastard son of Don Afonso of Castile and grandson of King Henry II, had

444 THE LAST INCOGNITOS

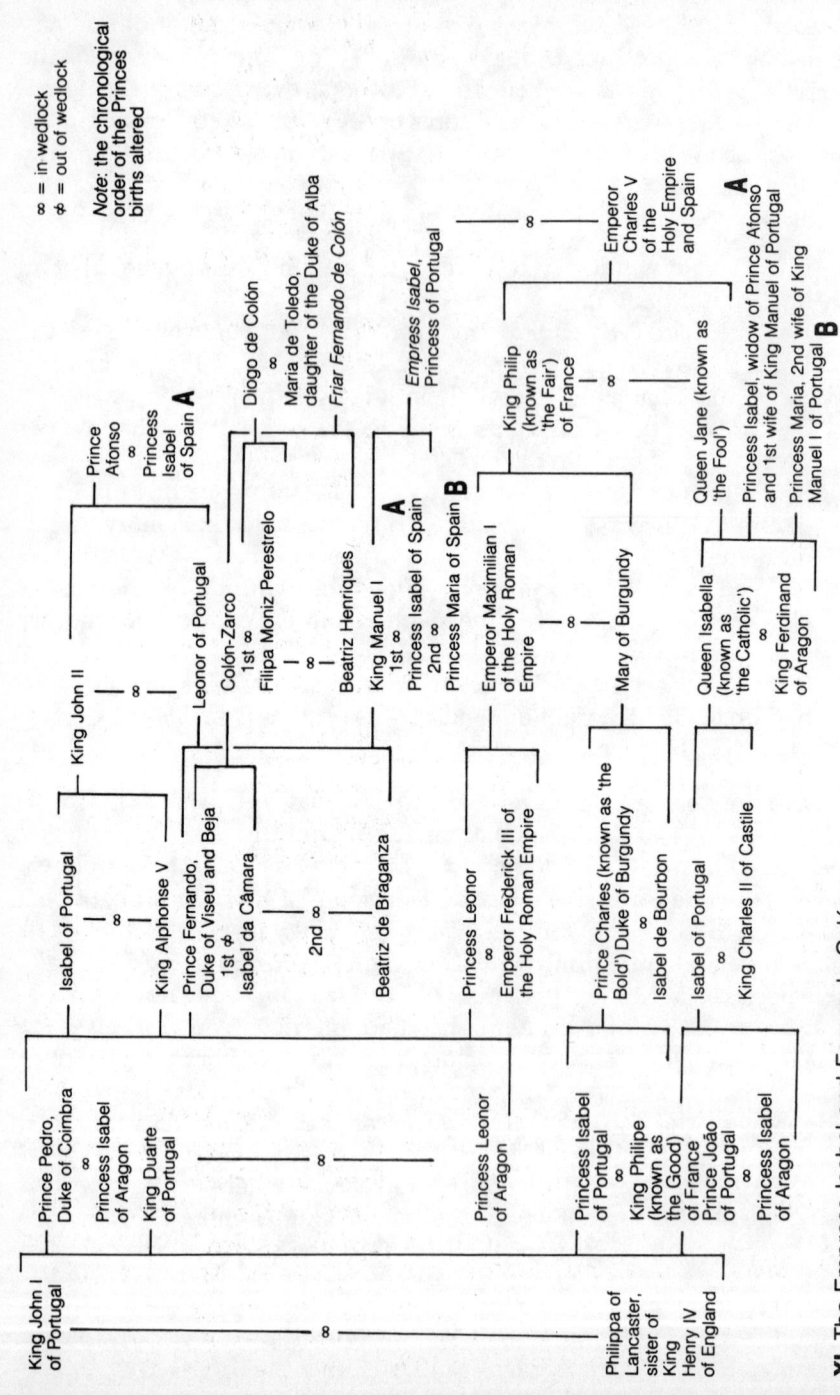

∞ = in wedlock
∞̸ = out of wedlock

Note: the chronological order of the Princes births altered

[Genealogical chart: "XI The Empress Isabel and Fernando Colón"]

Key figures shown in the chart:

- King John I of Portugal ∞ Philippa of Lancaster, sister of King Henry IV of England
- Prince Pedro, Duke of Coimbra ∞ Princess Isabel of Aragon
- King Duarte of Portugal
- Isabel of Portugal ∞ King John II
- King Alphonse V
- Leonor of Portugal
- Prince Fernando, Duke of Viseu and Beja; 1st ∞̸ Isabel da Câmara; 2nd ∞
- Beatriz de Braganza
- Prince Afonso ∞ Princess Isabel of Spain **A**
- Colón-Zarco 1st ∞ Filipa Moniz Perestrelo
- Diogo de Colón ∞ Maria de Toledo, daughter of the Duke of Alba
- Beatriz Henriques
- *Friar Fernando de Colón*
- King Manuel I; 1st ∞ Princess Isabel of Spain **A**; 2nd ∞ Princess Maria of Spain **B**
- *Empress Isabel, Princess of Portugal*
- Princess Leonor ∞ Emperor Frederick III of the Holy Roman Empire
- Emperor Maximilian I of the Holy Roman Empire ∞ Mary of Burgundy
- Prince Charles (known as 'the Bold') Duke of Burgundy ∞ Isabel de Bourbon
- King Philip (known as 'the Fair') of France ∞ Queen Jane (known as 'the Fool')
- Queen Isabella (known as 'the Catholic') ∞ King Ferdinand of Aragon
- Princess Isabel of Portugal
- King Philipe (known as the Good) of France
- Prince João of Portugal ∞ Princess Isabel of Aragon
- Isabel of Portugal ∞ King Charles II of Castile
- Emperor Charles V of the Holy Empire and Spain **A**
- Princess Isabel, widow of Prince Afonso and 1st wife of King Manuel of Portugal
- Princess Maria, 2nd wife of King Manuel I of Portugal **B**
- Princess Leonor of Aragon

DISTORTIONS CORRECTED 445

```
João Gonçalves ┬ João Gonçalves ┬ Simão Gonçalves ┬ João Gonçalves da Câmara (49) ∞ Leonor de Vilhena
Zarco          │ da Câmara (43) │ da Câmara       │ João Roiz de Noronha       ∞ Isabel de Abreu de Andrade
Discoverer of  │                │ 1st ∞           ├ Manuel de Noronha (50)
Madeira        │                │ Joana Valente   │ Luís Gonçalves de Ataíde ∞ Violante da Sylveira
and its first  │                │ 2nd ∞           │
Donatary-Captain│               │ Isabel da Sylva │
               │                ├ Pedro Gonçalves ┬ António Gonçalves da Câmara
               │                │ da Câmara (36)  │ 1st ∞ Maria de Castro
               │                │ Joana d'Eza     │ 2nd ∞ Isabel de Abreu de Andrade
               │                ├ Manuel de Noronha ┬ Maria de Noronha ∞ Simão de Castel-Branco
               │                │ 2nd ∞             │ Ana de Sousa ∞ Pedro Afonso de Aguiar
               │                │ Maria de Sousa    │
               │                ├ Mécia de Noronha ∞ Martinho de Castel-Branco (51)
               │                ├ Filipa de Noronha ∞ (1st wife) Henrique Henriques (52)
               │ Maria          └ Maria de Noronha ∞ Fernando Coutinho
               │ de Noroña
               ├ Rui Gonçalves ┬ João Roiz ┬ Ruz Gonçalves da Câmara ∞ Filipa Coutinho
               │ da Câmara     │ da Câmara │
               │ 1st ∞         │ ∞         ├ Garcia de Mello (53)
               │ Maria de      │ Inês de Mello
               │ Bettencourt   └ Beatriz da Câmara ∞ Francisco Coutinho de Albuquerque (54)
               │ 2nd ⚭
               │ Elvira Gonçalves
               ├ Garcia Roiz ┬ João Roiz da Câmara ∞ Violante Teixeira (55)
               │ da Câmara   ├ Aldonça da Câmara ∞ Bartolomeu Perestrelo (56)
               │ ∞           └ Constança da Câmara ∞ Diogo Cabral
               │ Violante
               │ de Freitas
               ├ Beatriz Gonçalves
               │ da Câmara
               │ ∞            └ Joana Cabral ∞ Brito Pestana
               │ Diogo Cabral
               ├ Isabel Gonçalves ┬ Colón Zarco ┬ Diogo Colom
∞              │ da Câmara        │ 1st ∞       │
               │ 1st ⚭            │ Filipa Moniz Perestrelo
               │ Prince Fernando  │ 2nd ⚭       └ Fernando Colón
               │                  │ Beatriz Henriquez
               │ ∞                ├ Rui Dias de Aguiar ∞ (his cousin) Leonor Homem de Sousa
               │                  ├ Pedro Afonso de Aguiar
               │                  ├ Diogo Afonso de Aguiar ∞ Isabel de Castel-Branco
               │ Diogo Afonso     └ Inês da Câmara (57) ∞ Lopo Vaz de Camões
               │ de Aguiar
Constança      ├ Elena Gonçalves da Câmara (58) ∞ Martim Mendes de Vasconcelos
Roiz de Sá     └ Constança Gonçalves da Câmara ┬ Leonor Homem de Sousa
                 ∞                              │ 1st ∞ Duarte Pestana
                 Garcia Homem de Sousa          └ 2nd ∞ Rui Dias de Aguiar
```

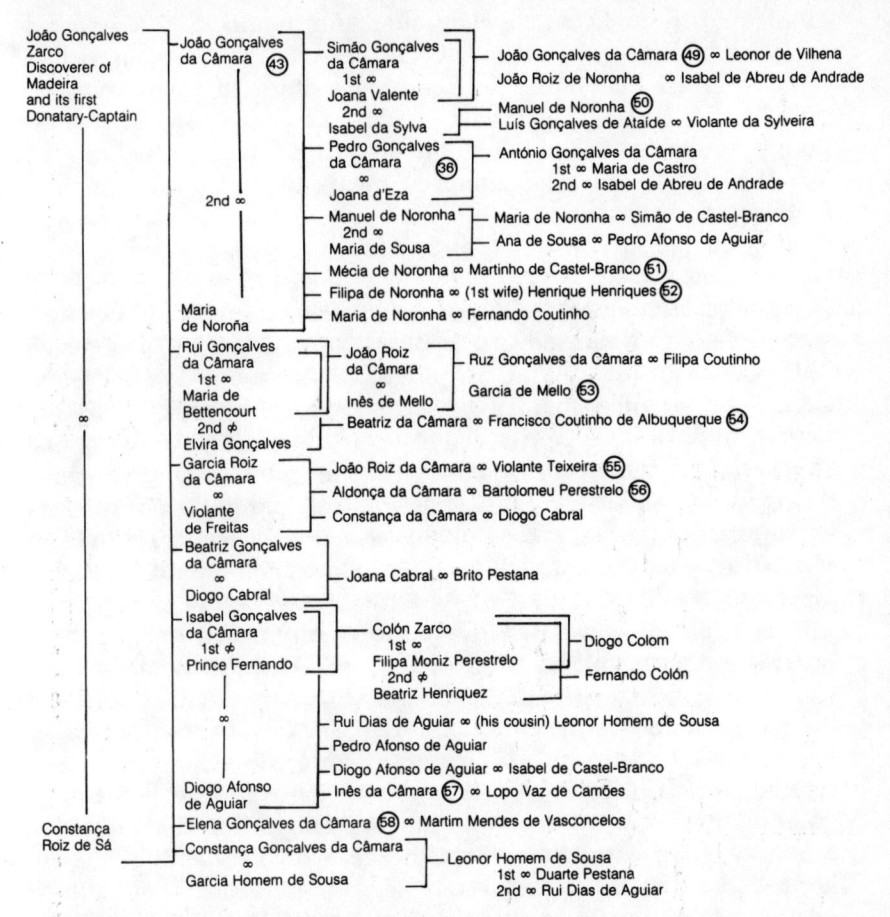

References:
36 (Tree III) and 43 (Tree IV); 49 – 4th Captain of Madeira; 50 – Bishop of Lamego and Chamberlain of Pope Clement VIII; 51 – Count of Vila Nova; 52 – 1st Lord of Alcáçovas, Governor of Barbacena and Ambassador of Portugal in Castile; 53 – who died with his wife and daughters in a shipwreck between Madeira and Lisbon; 54 – Marshal of the Realm like his ancestors; 55 – see Tree XIV; 56 – 2nd Captain of Porto Santo; 57 – kinswoman by marriage of Luis de Camões (22 – Tree I); 58 – Colón-Zarco's aunt, who connected the Câmara and the Vasconcelos families; Prince Henry (known as 'the Navigator') ordered him and others to marry local damsels and chose Zarco's daughter, Elena, for him.

XII Câmara

received half of the town of Dueñas from his sister, Leonor, Lady of Dueñas and Torralba, in 1406, when he married Leonor Sarmiento, daughter of Don Diego Perez Sarmiento de Villamayor, Lord of Salinias, Eneijo and La Bastida on the death of his father-in-law. On becoming a widower, he went to Portugal and when he took Branca de Mello (or de Sousa), daughter of Dom Martim Afonso de Mello, Governor of Évora, as his second wife, King Alphonse V gave him the town of Alcáçovas and the lands of Barbacena, which received a charter as a town only in 1519. The three eldest offspring of the couple and the only males, Henrique, Afonso and João, supported the King of Portugal in the Battle of Toro so the Castilian victors confiscated all their lands in Spanish territory. If there were no closely-related heirs, everything would go to the Crown. By omission, no clause excluded adulterine offspring once they had proved that they had not been legitimised owing to the murder of their mother (and there was a case of an adulteress who was widowed, legitimised her spurious son and then remarried – Inês Soares of Lamego).

2. Don Carlos Henriquez de Guzmán (Tree XIII, 60) moved to Portugal around 1460 and married Cecília de Almada, daughter of Artur de Brito, Governor of Beja, and the pious Constança de Almada. 'Dom' Carlos (he adopted the Portuguese surname 'Gusmão') was the eldest son of Don Diego Henriquez de Guzmán, Commander of Los Santos of the Order of St James (Castile), and Maria Vargas y Sotomayor, Lady of Torre de Laños. He was said to have possessed an impulsive character and would have serious differences with Dom Vasco de Menezes Coutinho, Count of Borba and Redondo (Tree III, grandson of 37).

Dom Carlos, however, was a favourite of Alphonse V and fought at his side in the calamitous Battle of Toro in 1476, in the wing of Dom Garcia de Menezes (see p. 432), and the King gave him the Command of Vila de Proença of the Order of Christ. The Castilians, meanwhile, had confiscated the lands of Torre de Laños; his wife Cecília considered herself to be highly ill-done by. She was Don Fernando's grandmother only by marriage, but her religious and moral education, together with her family's lineage, would certainly not allow her to forget that Columbus's son had the right to his grandfather's inheritance, and even more so as he was the innocent fruit of an extremely unfortunate ancestry. It was this that may have led Don Fernando to profess, even after proving himself to be a good cosmographer, geographer and navigator.

It is therefore demonstrated that Fernando Colón did not try to dupe Empress Isabel of Spain and Austria, who was his first cousin, as she was the daughter of King Manuel I, half brother of Colón-Zarco. It can also be concluded that he not only knew of the identity of his pseudo-Cordovan mother and the tragedy of his adulterous grandmother, but the true identity of his grandfather as well, since he presented a petition for losses caused by the civil war to his 'Portuguese grandmothers', from whom he would seemingly be the heir to some lands in Spain. He did not identify them in the document as he was sure the Empress would know who they were.

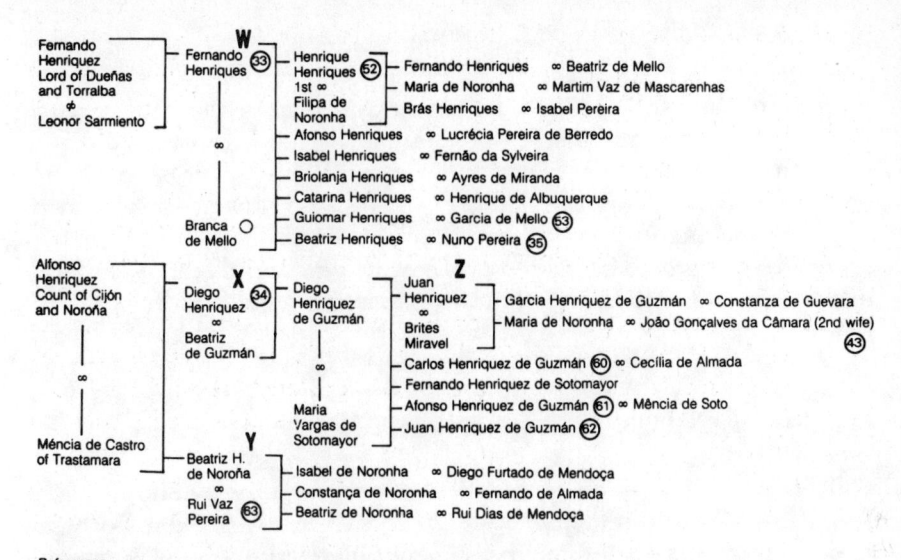

```
Fernando                Fernando  W    Henrique
Henriquez               Henriques 53   Henriques 52 ─ Fernando Henriques    ∞ Beatriz de Mello
Lord of Dueñas                         1st ∞         ─ Maria de Noronha     ∞ Martim Vaz de Mascarenhas
and Torralba                           Filipa de     ─ Brás Henriques       ∞ Isabel Pereira
   ф                                   Noronha
Leonor Sarmiento                       ─ Afonso Henriques   ∞ Lucrécia Pereira de Berredo
                           ∞           ─ Isabel Henriques   ∞ Fernão da Sylveira
                                       ─ Briolanja Henriques ∞ Ayres de Miranda
                                       ─ Catarina Henriques ∞ Henrique de Albuquerque
                        Branca  O      ─ Guiomar Henriques  ∞ Garcia de Mello 53
                        de Mello       ─ Beatriz Henriques  ∞ Nuno Pereira 35

Alfonso                          X
Henriquez
Count of Cijón          Diego  34    Diego       Juan
and Noroña              Henriquez    Henriquez   Henriques  ─ Garcia Henriquez de Guzmán   ∞ Constanza de Guevara
                          ∞          de Guzmán              ∞
                        Beatriz                  Brites     ─ Maria de Noronha   ∞ João Gonçalves da Câmara (2nd wife)
                        de Guzmán               Miravel                                                          43
                                                            ─ Carlos Henriquez de Guzmán 60 ∞ Cecília de Almada
      ∞                            ∞                        ─ Fernando Henriquez de Sotomayor
                                                 Maria      ─ Afonso Henriquez de Guzmán 61 ∞ Mência de Soto
Mência de Castro                                 Vargas de  ─ Juan Henriquez de Guzmán 62
of Trastamara           Beatriz H.               Sotomayor
                  Y     de Noroña    ─ Isabel de Noronha    ∞ Diego Furtado de Mendoça
                          ∞          ─ Constança de Noronha ∞ Fernando de Almada
                        Rui Vaz  63  ─ Beatriz de Noronha   ∞ Rui Dias de Mendoça
                        Pereira
```

References:
33 and 35 (Tree III); 34, 43 (Tree IV); 52 and 53 (Tree XII); 59 – daughter of Martim Afonso de Mello (27 – Tree II and III); 60 – Commander of Proença (Order of Christ), Constança's son-in-law (31 – Tree III); 61 – Administrator of Badajoz, on the Spanish frontier to Elvas (Northeast of Beja); 62 – who lived in Cordoba where he begot several bastard children; 63 – Lord of Riba de Vizela (Portugal).

XIII Henriques

As already said, the Battle of Toro ended the pretensions of Princess Juana and her betrothed, King Alphonse V of Portugal, and placed Queen Isabella 'the Catholic', married to Ferdinand de Aragon since 1469, on the throne of Castile. As far as these Sovereigns were concerned, their defeated enemies had been rightly deprived of their possessions in Spain. But the marriage of Princess Isabel and Charles V made it possible to request indemnities for the lands confiscated.

If 'Cristóforo Colombo' were Don Fernando's progenitor, how could the latter have presented such a petition to the Empress? He would surely have been obliged to furnish irrefutable evidence of his ancestry. What hope did he have of receiving indemnities from a Sovereign who would consider the loss of lands for helping a foreign king as a just punishment? But Empress Isabel was not the cousin of a Genoese wool-carder or a Catalan pirate – and these had no Portuguese grandmothers.

'I cannot mention the reason here' Colón-Zarco wrote, showing how much he repented not having married his son's mother. It was not because she was 'unmarriageable' – as Madariaga mentioned but denied – or because she was Jewish and the Inquisition could come knocking at her door. It was solely because she was the daughter of her namesake –

the one that was killed because of her adultery. Because she was the daughter of a Henriques, the cousin of the Medina Sidónias. Because the assassin of her mother was a La Cerda, a cousin of the Medinacelis. Because they were all kinsmen of Columbus. And because he did not want anyone to be able to find out his real identity. It was better to continue with the disguise that had been arranged for him – an assumed Genoese and the lover of a 'damsel of Córdoba'. Everything was obscure. No one would be able to probe or delve into his secrets. Columbus had been branded with another name; he would never again be himself.

MOLIARTE, COLÓN-ZARCO'S BROTHER-IN-LAW

Only one piece was left to identify – Miguel Moliarte. Shortly after João Esmeraldo had settled in Madeira, this German arrived in Portugal (around 1483) and made straight for the Archipelago. As far as is known, he was neither a farmer nor a sailor. The ease with which he received authorisation from King John II to go to the island is strange, as is the manner in which he, as a foreigner, became friendly with the widow of the Donatary-Captain of Porto Santo, Bartolomeu Perestrelo, and married her daughter Violante (or Brigolada), when there were innumerable Portuguese in the two Atlantic Archipelagos who would have given anything to find a lady of good stock whom they could have wedded.

It is a repeat of the 'Cristóforo Colombo' story. Were the Perestrelo sisters cross-eyed, crippled or hunchbacked? The third one, who was described as extremely beautiful, was 'given away' without even a ring put on her finger and her noble Archbishop lover surely had a very wide choice given the excess of females at the time, the exact opposite to the situation on the islands.

Let us take a look at two cases:

1. After being widowed, Manuel de Noronha (son of João Gonçalves da Câmara 'the Pearmain', second Donatary-Captain of Madeira and first cousin of Colón-Zarco) went to Beja to look for a new consort. He found Maria de Noronha, daughter of Dom João Henriques, of Cordovan ancestry.
2. João Roiz da Câmara, son of Rui Gonçalves da Câmara, second Captain of the Island of São Miguel and also first cousin of Colón-Zarco, also went to Beja to marry Inês de Mello, daughter of Rui Dias of Serpa.

These names are already known to the reader, as is the fact that Isabel da Câmara lived in the same area as the mistress of the future Duke of

Beja and gave birth to the future Admiral before settling down in Madeira with Diogo Afonso de Aguiar.

It is certain that the two men may have chosen to go to Beja knowing that they would find ladies of good stock who, at the same time, were known to their aunt Isabel. But the real reason for their journeys was the shortage of suitable consorts on the islands. Violante Moniz Perestrelo had all the requisites for her neighbours to ask her hand in marriage and had no need to marry a newly-arrived foreigner.

When the German arrived in Portugal he introduced himself as Moliarte, a name that is certainly not German, but in Spain he used the one of 'Moeller' or 'Müller'. This use of two surnames brings to mind the fact that Martin Behaim had studied with Camile Johan Moeller or Müller of Monterregio, the cosmographer and alchemist also known as 'Molyart' (see p. 119). Of course they could not be the same man, but, perhaps they were father and son or kinsman. I think this is an admissible possibility. And I remember that, just as the surname of Perestrelo came from 'Palastelli' (the 'mantle of the stars'), 'Moliarte' meant 'the lead of the art' (*moli* + *artis*) – the base with which alchemists wished to carry out the '*chrysopoeia*' and the '*argyopoeia*'. Just as the surname 'Esmeraldo' was redolent with esoteric symbolism (see p. 342).

Perestrelo, Moliarte and Esmeraldo certainly belonged to that international congregation whose 'brothers' permuted cabalistic messages of a lyric-hermetic style like *The Sybil's Prophecy* that Valentim Fernandes sent to Konrad Peutinger and which helped me to decipher Colón-Zarco's siglum five centuries later. Otherwise, how can the sudden link of an unknown German with a notable family of Porto Santo be explained? Had someone requested that he come to Portugal – not to one of the cities, but to the islands – in order to entrust him with the mission that was under the aegis of the Order of Christ? Would it have been the King himself who had contracted him after vetting his ability? Would that link have been one of the conditions of the contract so as not to raise suspicions that he was an agent of Portugal?

What is certain is that Moliarte settled in Huelva with his wife, Colón-Zarco's sister-in-law, and set himself up in business – around 1484 as far as is known. At the end of 1484, after being widowed, Colón-Zarco was already in Castile with his son Diogo and made for Córdoba. He then went to Huelva and La Rábida in search of his Marchena cousins. But the two friars, Antonio and Juan Peres (the custodian of the monastery) were indoor men, versed in astrology, theology and possibly politics. They could supply the brains, but not the legwork. But this was being done by Colón-Zarco's brother-in-law Moliarte. He studied the land and its inhabitants, assessing their qualities, listening to opinions, selecting the most able for when they received royal permission to sail across the Atlantic.

Moliarte was the spearhead of an advance party to study the lie of the land, an action that would be crucial in the voyages to the west. Everything was superbly planned. Information was obtained about shipowners, captains, pilots, experienced seamen, chandlers. Friendly relations that could be useful to Colón-Zarco were established. Contacts were made with Jews and Christians, noblemen and plebians, clergymen and adventurers, physician-surgeons. As a merchant, Moliarte could, without arousing suspicion, have access to the docks of Palos, cross the river Guadiana to the Algarve, through Ayamonte to Santo António de Arenilha[3] or further up the river in front of the Templar castle of Castro Marim or travel by road to Beja through Rosal de La Frontera (see Figure 33.1).

Having accompanied Colón-Zarco on his second expedition to the Antilles, however, Moliarte fell out with him and his name appears on a petition signed by a number of people, in which they complained that Colón-Zarco demanded too much of his crew when they were on land. He must have been ill at the time, as he died just after returning to Huelva. His widow, Violante, remarried in 1495 another foreigner, Francisco de Bardi, son of Giácomo de Bardi of Florence.

The 'Genoists' believed the entry of this 'bourgeois merchant' into Violante Moniz Perestrelo's life would prove the Italian birthright of his brother-in-law Columbus. Unfortunately for them, however, Giácomo Bardi had settled in Oporto and requested authorisation from King John II to use his family's arms, which bore a unicorn on the shield and on the crest[4]. There is no documental proof that he was a merchant in Portugal or that he was anything except a gentleman with the right to use a coat-of-arms. Francisco's younger brothers served in India in the reign of King Manuel I. Being merchants in Spain, both on the part of Moliarte and Bardi, was a ploy to cover up their mission as agents of King John II of Portugal. Both lost money in the venture, for which they were certainly not prepared. This can be concluded from the fact that Moliarte owed Columbus money, as can be seen in the personal note he wrote in Spanish (and not in Italian) (see p. 89), and from the fact that Bardi left his wife in misery when he died, which led Columbus to determine in his will that his son Diogo should pay her a life pension.

Trees XVI–XVII show the Teixeira, Moniz and Mello family relationships, and a complete tree of Colón-Zarco's descent.

DISTORTIONS CORRECTED 451

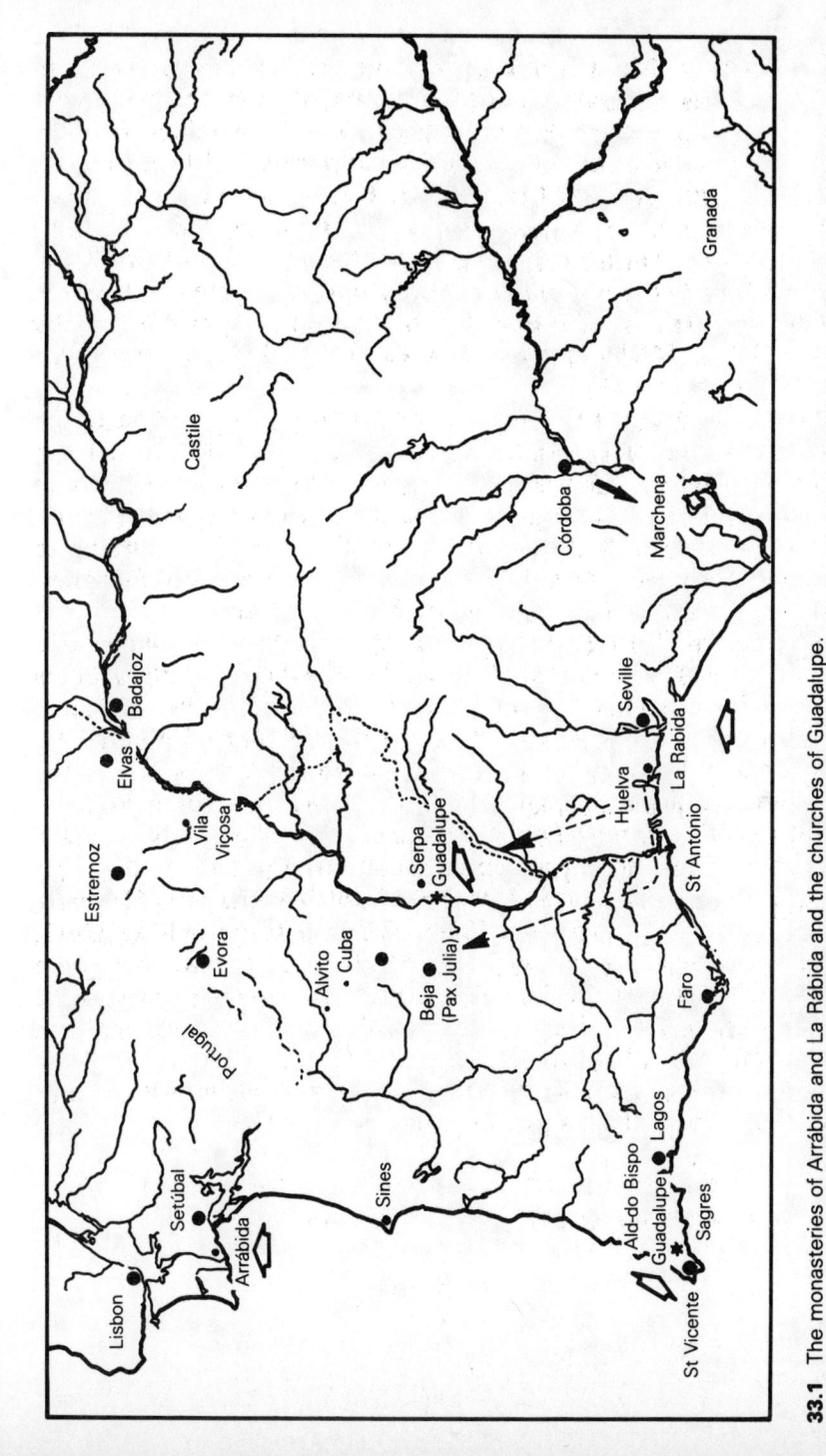

33.1 The monasteries of Arrábida and La Rábida and the churches of Guadalupe. A part of the Kingdoms of Portugal, Castile and Granada prior to 1492, showing the short distance between Beja and Huelva, Marchena to the south of Córdoba, the two Franciscan monasteries of Arrábida (Setúbal) and La Rábida (Huelva) and the two hermitages of Our Lady of Guadalupe near Beja and at Sagres, where Prince Henry and Prince Fernando respectively gathered the mariners of the Order of Christ.

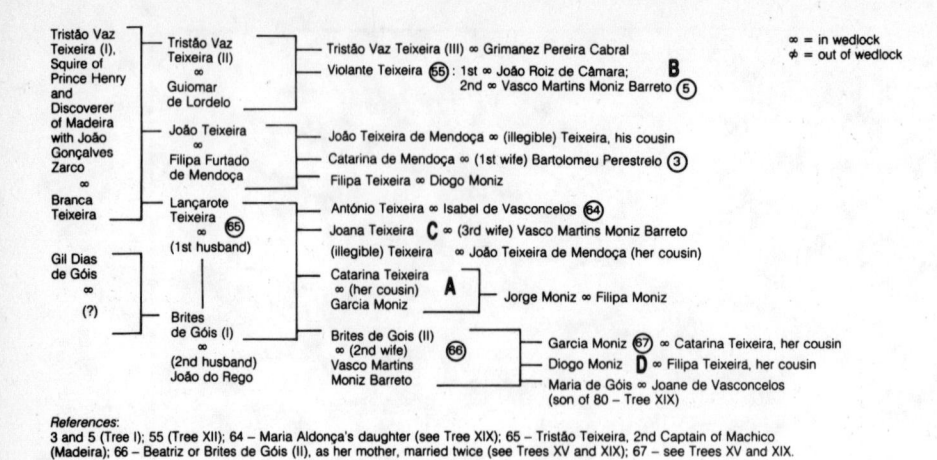

452 THE LAST INCOGNITOS

```
Tristão Vaz          Tristão Vaz          Tristão Vaz Teixeira (III) ∞ Grimanez Pereira Cabral           ∞ = in wedlock
Teixeira (I),        Teixeira (II)        Violante Teixeira 55 : 1st ∞ João Roiz de Câmara;              ⌀ = out of wedlock
Squire of            ∞                                        2nd ∞ Vasco Martins Moniz Barreto  5  B
Prince Henry         Guiomar
and                  de Lordelo
Discoverer
of Madeira           João Teixeira         João Teixeira de Mendoça ∞ (illegible) Teixeira, his cousin
with João            ∞
Gonçalves            Filipa Furtado        Catarina de Mendoça ∞ (1st wife) Bartolomeu Perestrelo  3
Zarco                de Mendoça
∞                                          Filipa Teixeira ∞ Diogo Moniz
Branca               Lançarote
Teixeira             Teixeira  65          António Teixeira ∞ Isabel de Vasconcelos  64
                     (1st husband)        Joana Teixeira  C  ∞ (3rd wife) Vasco Martins Moniz Barreto
Gil Dias                                   (illegible) Teixeira   ∞ João Teixeira de Mendoça (her cousin)
de Góis
∞                                          Catarina Teixeira
(?)                                        ∞ (her cousin)    A         Jorge Moniz ∞ Filipa Moniz
                     Brites                Garcia Moniz
                     de Góis (I)
                     ∞                     Brites de Gois (II)          Garcia Moniz 67 ∞ Catarina Teixeira, her cousin
                     (2nd husband)         ∞ (2nd wife)        66       Diogo Moniz  D  ∞ Filipa Teixeira, her cousin
                     João do Rego          Vasco Martins                Maria de Góis ∞ Joane de Vasconcelos
                                           Moniz Barreto                (son of 80 – Tree XIX)
```

References:
3 and 5 (Tree I); 55 (Tree XII); 64 – Maria Aldonça's daughter (see Tree XIX); 65 – Tristão Teixeira, 2nd Captain of Machico (Madeira); 66 – Beatriz or Brites de Góis (II), as her mother, married twice (see Trees XV and XIX); 67 – see Trees XV and XIX.

XIV Teixeira

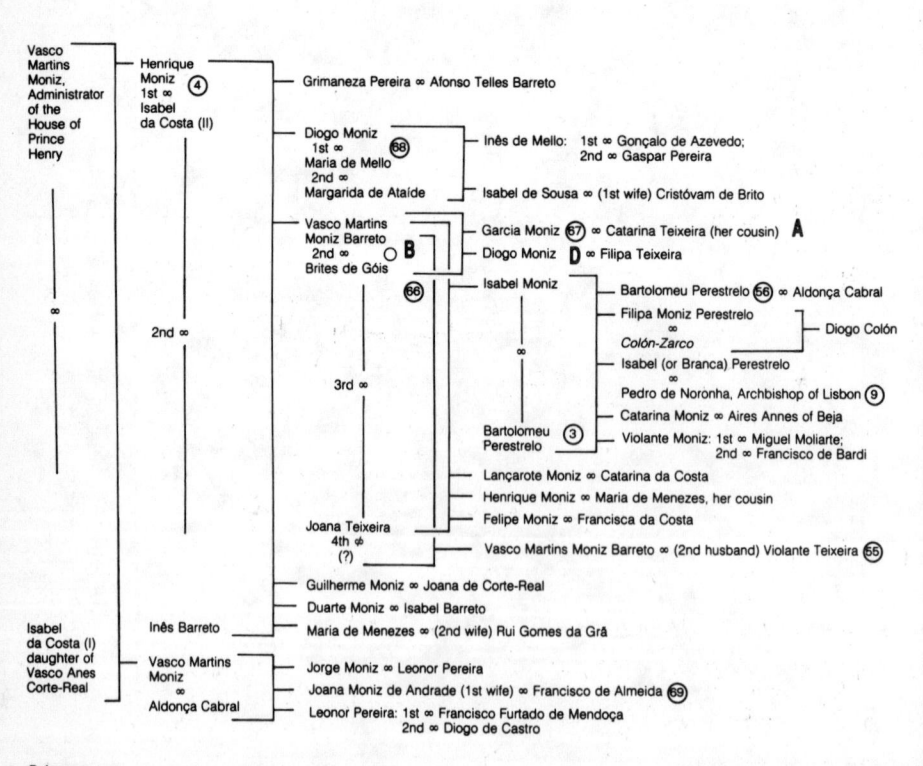

```
Vasco              Henrique
Martins            Moniz
Moniz,             1st ∞   4             Grimaneza Pereira ∞ Afonso Telles Barreto
Administrator      Isabel
of the             da Costa (II)
House of                                 Diogo Moniz         Inês de Mello:  1st ∞ Gonçalo de Azevedo;
Prince                                   1st ∞        68                     2nd ∞ Gaspar Pereira
Henry                                    Maria de Mello
                                         2nd ∞               Isabel de Sousa ∞ (1st wife) Cristóvam de Brito
                                         Margarida de Ataíde
                                         Vasco Martins       Garcia Moniz 67 ∞ Catarina Teixeira (her cousin)  A
                                         Moniz Barreto   B   Diogo Moniz  D  ∞ Filipa Teixeira
                                         2nd ∞  O
                                         Brites de Góis      Isabel Moniz
                                           66                       Bartolomeu Perestrelo 56 ∞ Aldonça Cabral
                    2nd ∞                                           Filipa Moniz Perestrelo
∞                                                                   ∞                          Diogo Colón
                                         3rd ∞                      Colón-Zarco
                                                                    Isabel (or Branca) Perestrelo
                                                                    ∞
                                                                    Pedro de Noronha, Archbishop of Lisbon  9
                                                                    Catarina Moniz ∞ Aires Annes of Beja
                                         Bartolomeu   3             Violante Moniz: 1st ∞ Miguel Moliarte;
                                         Perestrelo                                 2nd ∞ Francisco de Bardi
                                                              Lançarote Moniz ∞ Catarina da Costa
                                                              Henrique Moniz ∞ Maria de Menezes, her cousin
                                                              Felipe Moniz ∞ Francisca da Costa
                                         Joana Teixeira
                                         4th ⌀                Vasco Martins Moniz Barreto ∞ (2nd husband) Violante Teixeira  55
                                         (?)
                                         Guilherme Moniz ∞ Joana de Corte-Real
                   Inês Barreto          Duarte Moniz ∞ Isabel Barreto
Isabel                                   Maria de Menezes ∞ (2nd wife) Rui Gomes da Grã
da Costa (I)
daughter of        Vasco Martins         Jorge Moniz ∞ Leonor Pereira
Vasco Anes         Moniz
Corte-Real         ∞                     Joana Moniz de Andrade (1st wife) ∞ Francisco de Almeida  69
                   Aldonça Cabral        Leonor Pereira: 1st ∞ Francisco Furtado de Mendoça
                                                        2nd ∞ Diogo de Castro
```

References:
3 and 9 (Tree I); 55 (Trees XII and XIV); 56 (Tree XII); 66 and 67 (Tree XIV; see Tree XIX); 68 – Filipa Moniz Perestrelo's grand-uncle, who was married to Martim de Mello's daughter (47 – Tree VI); 69 – 1st Viceroy of India.

XV Moniz

DISTORTIONS CORRECTED

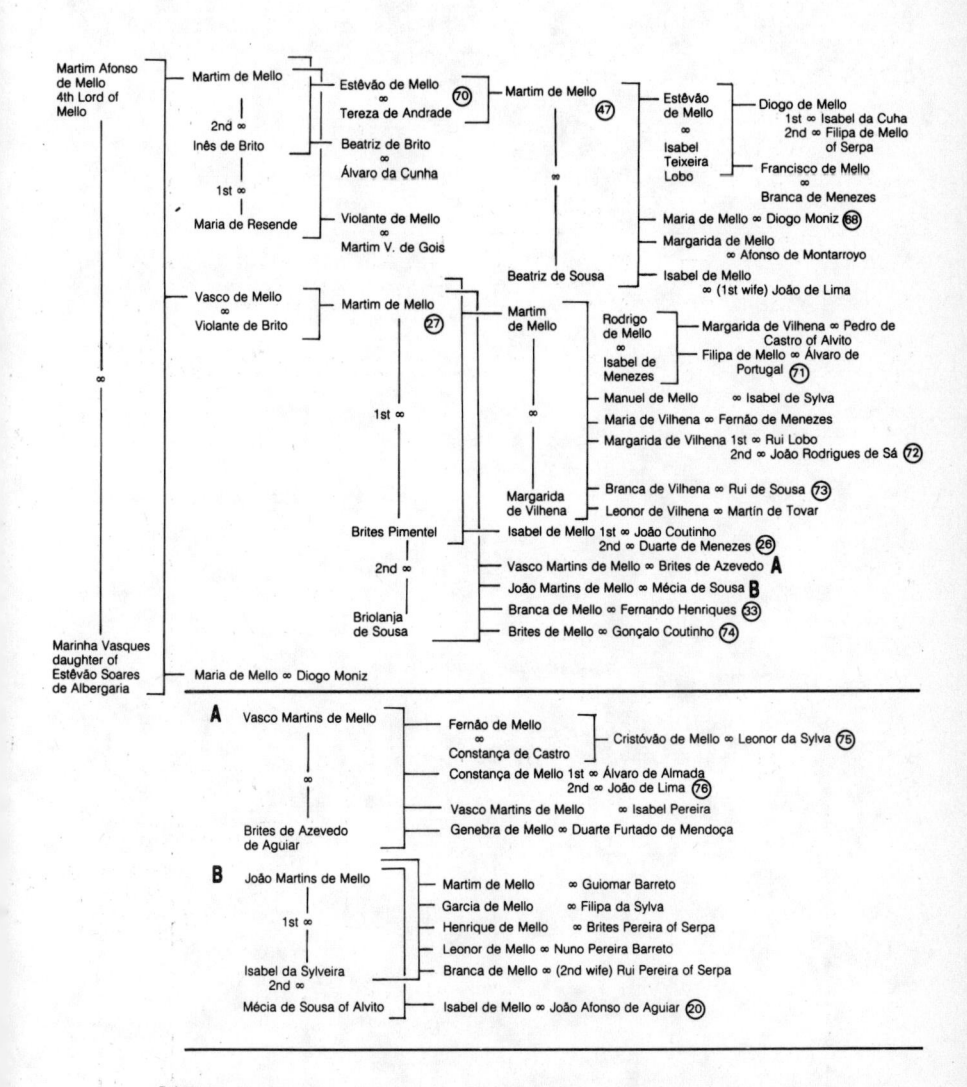

Genealogical tree of the Mello family:

- **Martim Afonso de Mello**, 4th Lord of Mello
  ∞ **Marinha Vasques**, daughter of Estêvão Soares de Albergaria
  - **Martim de Mello**
    - 2nd ∞ **Inês de Brito**
      - **Estêvão de Mello** ∞ **Tereza de Andrade**
      - **Beatriz de Brito** ∞ **Álvaro da Cunha**
      - **Violante de Mello** ∞ **Martim V. de Gois**
    - 1st ∞ **Maria de Resende**
  - **Vasco de Mello** ∞ **Violante de Brito**
    - **Martim de Mello**
      - 1st ∞ ...
        - **Martim de Mello** (27)
          - ∞ **Beatriz de Sousa**
            - **Martim de Mello** (47)
              - **Estêvão de Mello** ∞ **Isabel Teixeira Lobo**
                - **Diogo de Mello**
                  - 1st ∞ **Isabel da Cuha**
                  - 2nd ∞ **Filipa de Mello of Serpa**
                - **Francisco de Mello** ∞ **Branca de Menezes**
              - **Maria de Mello** ∞ **Diogo Moniz** (68)
              - **Margarida de Mello** ∞ **Afonso de Montarroyo**
              - **Isabel de Mello** ∞ (1st wife) **João de Lima**
            - **Martim de Mello** ∞ ...
              - **Rodrigo de Mello** ∞ **Isabel de Menezes** (71)
                - **Margarida de Vilhena** ∞ **Pedro de Castro of Alvito**
                - **Filipa de Mello** ∞ **Álvaro de Portugal**
              - **Manuel de Mello** ∞ **Isabel de Sylva**
              - **Maria de Vilhena** ∞ **Fernão de Menezes**
              - **Margarida de Vilhena**
                - 1st ∞ **Rui Lobo**
                - 2nd ∞ **João Rodrigues de Sá** (72)
            - **Margarida de Vilhena**
              - **Branca de Vilhena** ∞ **Rui de Sousa** (73)
              - **Leonor de Vilhena** ∞ **Martín de Tovar**
            - **Brites Pimentel**
              - **Isabel de Mello** 1st ∞ **João Coutinho**; 2nd ∞ **Duarte de Menezes** (26)
              - **Vasco Martins de Mello** ∞ **Brites de Azevedo** **A**
              - **João Martins de Mello** ∞ **Mécia de Sousa** **B**
            - 2nd ∞ **Briolanja de Sousa**
              - **Branca de Mello** ∞ **Fernando Henriques** (53)
              - **Brites de Mello** ∞ **Gonçalo Coutinho** (74)
  - **Maria de Mello** ∞ **Diogo Moniz**

A — **Vasco Martins de Mello** ∞ **Brites de Azevedo de Aguiar**
- **Fernão de Mello** ∞ **Constança de Castro**
  - **Cristóvão de Mello** ∞ **Leonor da Sylva** (75)
- **Constança de Mello**
  - 1st ∞ **Álvaro de Almada**
  - 2nd ∞ **João de Lima** (76)
- **Vasco Martins de Mello** ∞ **Isabel Pereira**
- **Genebra de Mello** ∞ **Duarte Furtado de Mendoça**

B — **João Martins de Mello**
- 1st ∞ **Isabel da Sylveira**
- 2nd ∞ **Mécia de Sousa of Alvito**
  - **Martim de Mello** ∞ **Guiomar Barreto**
  - **Garcia de Mello** ∞ **Filipa da Sylva**
  - **Henrique de Mello** ∞ **Brites Pereira of Serpa**
  - **Leonor de Mello** ∞ **Nuno Pereira Barreto**
  - **Branca de Mello** ∞ (2nd wife) **Rui Pereira of Serpa**
  - **Isabel de Mello** ∞ **João Afonso de Aguiar** (20)

References:
20 (Tree I); 26 and 27 (Tree II); 33 (Tree III); 47 (Tree VI); 68 (Tree XV); 70 – brother-in-law of Martim de Góis, so establishing the connection with the Góis family (see Tree XIX); 71 – brother of Fernando, 3rd Duke of Bragança; he was father of the 2nd Count of Tentúgal (Tree II) and also an ascendant of Jorge de Portugal, Count of Gelves, who married Isabel de Colón, being progenitors of seven consecutive generations of Colón-Zarco's heirs; 72 – the poet who wrongly believed he was a descendant of Cecília Colonna, Colón-Zarco's great-grandmother; 73 – descendant of Martim de Mello (27), father-in-law of Rui de Sousa (28 – Tree III) and Colón-Zarco's great-grandfather; 74 – Vasco Coutinho's great-grandson (37 – Tree III); 75 – who after married Henrique Henriques (52 – Tree XII); 76 – grand-uncle of the Captain of Arzila, Francisco de Lima (45 – Tree VI).

XVI Mello

454   THE LAST INCOGNITOS

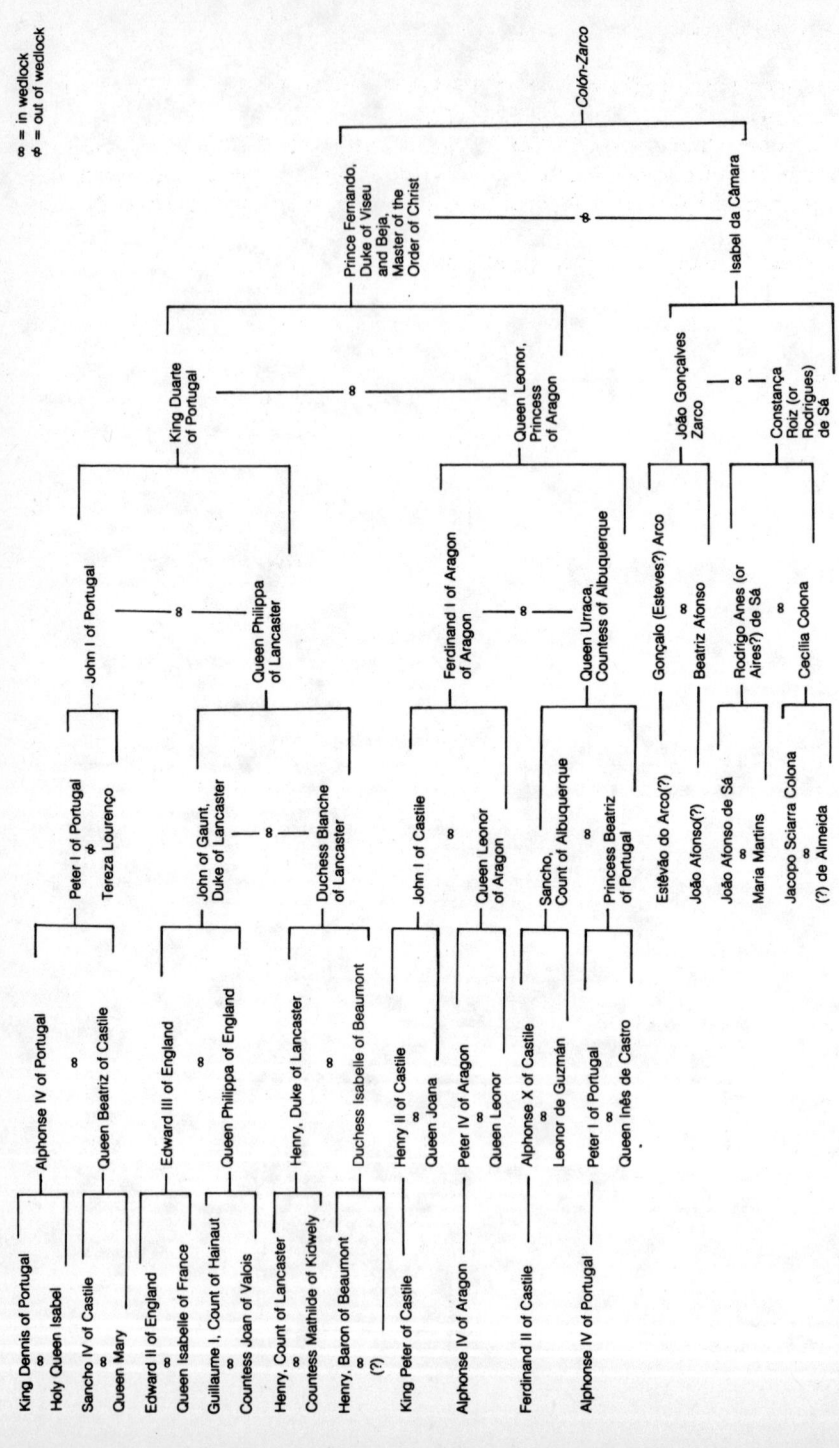

⚭ = in wedlock
⚮ = out of wedlock

XVII Colón-Zarco's ancestry

Notes and References

1. *Boletim, Classe de Letras*, Lisbon Academy of Sciences, 13, pp. 637–8.
2. Dornelas (1925), vol. XX, pp. 407–22.
3. It became known as Vila Real de Santo António in 1774, but in 1433 King Duarte had given Prince Henry (known as 'the Navigator') a tithe of the catch of the fishing village of Monte Gordo, near Santo António de Arenilha.
4. Faria, António Macedo de (1961), p. 83.

34   The Heraldic Secret

>   The mole cannot have the same view of the world as the eagle.
>
>   (Seneca)

Heraldry, which first appeared in ancient Egypt, became especially important with the Western and Eastern Crusades. The coats-of-arms of the Knights of the Temple that are reproduced in this chapter were strongly influenced by Hebrew symbology.

The knight Granier de Cassagnac[1] said: 'The coat-of-arms is the most expensive, richest and most difficult of all languages: it is a disciplined and magnificent language, with its own syntax, grammar and orthography'. By this he meant that a coat-of-arms involves not only its symbols but also its layout, its suitability, the nature of its crest and, above all, its colours, all of which have a special meaning.

According to Henry Corbin[2], 'The symbols have the advantage in that they are inexhaustible and that their interpretation cannot be contained. They cannot be denied or discussed like logic'.

Throughout this book, I have referred to the meaning of several symbols such as the wheel, the leather strap (baldric or belt), the fishing net (very often misinterpreted as a shrimp net), the set-square, the pair of compasses, the chain (of iron links) and the rope (of a ship), etc. All these symbols have a special meaning in heraldry and are sometimes represented with certain modifications of the original figure. The set-square, for example, which is usually of a simple 'L' shape, gave rise to the heraldic chevron. The compasses form a slightly convexed prong. The net was turned into a chequered pattern, which is clearly visible in the image of the navigators of the Order of Christ in the polyptych of Nuno Gonçalves (see Plate 11, panel 2).

When heraldic symbols were attributed to an initiate of an Order of Knighthood in the Middle Ages, it represented the 'mirror' of the virtues that he wished to achieve or would try to preserve. I have already referred to the mermaid (pp. 314–16), which was sometimes represented by the dolphin (as in the case of the Távora's crest) or the seal (as in the case of Câmara/Zarco). The mermaid was usually looking into a mirror, which, according to Templar alchemists, reflected enlightened and creative intelligence (Chokhman – Wisdom). At the same time, it depicted prudence, being the lunar symbol of truth.

The phoenix appeared in the coat-of-arms of the Teixeiras, the family of the navigator Tristão Vaz Teixeira (p. 452, Tree XIV).

The above-mentioned 'mirror' was referred to by the Jesuit priest Father António Vieira[3].

> The stars[4] are mirrors of the Sun; the rivers are mirrors of the trees . . . With all his stoic severity, Seneca said that mirrors existed so that the young man who had been born with good looks could see his gentility and would not become ugly with his vices; and he who had been born ugly would be able to make up for and correct that defect with the beauty of his virtues.
>
> This same doctrine had been expounded by Plato and Socrates, who placed mirrors in their schools so that their disciples could see themselves and so copy the virtues they taught (see Figure 34.1)

When appraising his coat-of-arms, the knight Templar was obliged to see it as the mirrors of virtues. And if his shield had come to him through inheritance, it was his task to maintain its honour by remaining free of vices.

HERALDIC SYMBOLOGY

One of the favourite heraldic symbols of the Templars was the baldric or belt, the bar or band on a shield.[5] Besides the specimens reproduced in this chapter, the heraldic baldric was also adopted by the Ataídes, Azambujas and other families. (Figure 34.3, p. 462, shows the arms of the Andrade, Perestrello, Aguiar, Mello, Esmeraldo, Mendoça, Cão and Sá families.)

In the concept of Templar alchemy, the net[6] was related to Jesus, with the miracle of the fishermen's catch and the comparison of the '*Salvador*' (Saviour) to the 'fisher of men'. His successor St Peter was the first one to wear the pontifical ring with a fish engraved on it, which has since been worn by all the Bishops of Rome. The first symbols of identification used by the Christians were the fish and the net, the cross being adopted later. Christ made his Apostles 'fishers of men', and 'they prepared their own nets'.[7]

The eagle was the bird of light, the image of the Sun, the altitude and

458 THE LAST INCOGNITOS

34.1 The mirror of the soul.
'The mirror where vanity is displayed reflects the hedonism of the Devil'.

(Fifteenth-century German woodcut)

the amplitude of the atmosphere. It could also represent the King, meaning the highest sovereign power (although the spiritual power of the Church was on a higher plane), as well as an Order of Knighthood, signifying self-denying and transcendental heroism. This symbol, therefore, was adopted by many Sovereigns and knights, especially 'huntsmen'. The eagle was sometimes depicted by two wings alone – or at times only one – with or without a clawed foot. The coats-of-arms of the Bobadillas, Grãs , Vilhenas, Abreus and others bore a clawless eagle.

The lion, which appeared on the coats-of-arms of the Bettencourts, Lusignans, Castel-Brancos, Serpas, La Cerdas and others, also symbolised the power of the Sovereign, the brilliance of the Sun, the glitter of gold, the discerning strength of the Word and Justice. It was also used by initiates in four different forms – with or without a crown, in a threatening attitude (rampant) or a passive attitude (passant).

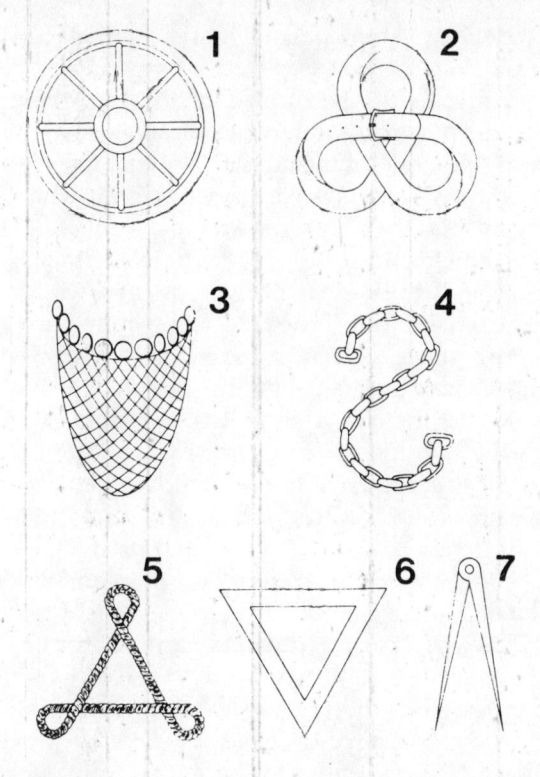

34.2 Heraldic symbology.

1. *Wheel*, which could also be cogged with knives.
2. *Buckled strap or belt*; 'talim' (baldric); heraldic band or bar.
3. *Fisherman's net*, sometimes with a curved frame; heraldic chequer.
4. *Chain*, the links of which appear separately in heraldry and symbolise a serpent.
5. *Rope*, or ship's cable.
6. *Set-square*, which may be 'L'-shaped or inverted in the form of a chevron.
7. *Compass* of the temple builders, with a range of 26 degrees.

The crowned lion, like the one on Valentim Fernandes's emblem (see Figure 19.2), represented the Hebraic 'Lion of Judah' (to which the alchemic cabalists gave a tail forked into two serpents) symbolising the strength that Christ instilled into believers. The lion rampant represented ardent, victorious vigour with a certain sense of independence. The passant depicted more humility, virility attenuated by subjection to an overlord. The Vargas, for instance, had a passant lion bearing undulate stripes (crossed by five waves of the ocean) on their coat-of-arms. As they were not mariners, this symbol of the sea corresponds to *Sephiroth* III (Understanding) subject to a superior 'Mother'.[8]

Columns were a symbol of strength. They were adopted for the coats-of-arms of the Colonnas, such as the Archbishop of Lisbon, Dom Agapito Colonna (see p. 499), father of Pope Urban V, Otto Colonna in secular life, and Dom António Esteves de Aguiar, Bishop of Ceuta (p. 493, see Tree XVIII). They depicted energy and stability, the link between Heaven and Earth, the connection between the power and the control of action, as power without control leads to anarchy, and control without power is useless. Not included in the examples reproduced, the coat-of-arms of the navigator Nicolau Coelho, with whom Colón-Zarco probably navigated in the Western Atlantic, contained two columns surmounted by shields bearing the Portuguese 'quinas' (five escutcheons), which could mean that he was of royal blood and had descended from a Colonna. But these columns merely represent the 'padrões' that the Portuguese erected in the lands they discovered.

The iron chain[9] was adopted, among others, by the Zuñigas (the Portuguese branch of the family spelt its name Zuniga).

The tower[10] (the Portuguese 'côba', see Figure 24.1) appeared on the coats-of-arms of the Figueiredos and Bobadillas, as well as the Zarco/Câmara and Serpa families, the coat-of-arms of the last-named bearing a salient lion.

The wheel[11] represented the sphere and its planiform image of the cosmos, and could be cogged with twelve knives laid out like the signs of the zodiac. This was a part of the coat of arms of the Castros of Penha Verde.

The rope[12] appeared in the arms of the Coimbras. The shield of the founder of the family, Dom Diogo Dias (1429), was made up of an artichoke in flower[13] surmounted to the left by a five-pointed star and encircled by a rope. His descendant, Dom Friar Henrique de Coimbra, Bishop of Ceuta (1427), eliminated those esoteric symbols and replaced them with a lamb.

The arm (member – colon) or half-arm, usually enclosed in a bracer, signified courage against the enemies of Christ – just as the artichoke symbolised the courage of St George against the dragon. Families that adopted the arm included the (Bartholomeu) Dias, Corrêas, Lousadas, Sanches, and the New Christians Salazar of Castela-Vieja and Mendes Zacuto.

The five-petalled rose[14] was widely used in Portuguese, as it was in British heraldry. I shall refer to this later.

The cauldron[15] appeared in the arms of the Guzmáns (Gusmões), Laras, Freires de Andrade and others.

The serpent was in the coat-of-arms of the Camões, the Andrades and others. A multi-headed serpent was sometimes used as a crest.

The castle was very widely used and among the families that had it in their coat-of-arms I shall mention the ones that have been referred to in this book – the Lobos, Lencastres, Menezes (of Marialva), Sousas (of Córdoba), Berredos, Albuquerques and others.

The leather strap,[16] which was a belt or baldric, appeared in the arms of the families related to the Corrêas ('*correia*' in modern Portuguese, 'strap' in English).

Five-, six- and eight-pointed stars appeared in many arms, including those of the Coutinhos.

The ermine, insignia of the peerage (black spots on white fur) indicate the older nobility and are to be found in the arms of the Barretos, Guzmáns, Castanhedas, and the Teives (of Diogo de Teive the navigator). It is thought that this type of insignia came from a foreign royal house. Some arms also included tear-drops.

I am not going to try and unravel the complexities of the seven heraldic colours[17] which are related to multiple symbolic concepts such as the stars, zodiacal signs, metals, precious stones, the months of the year, physical characteristics, human temperament and morals, or of the correlations of several colours in the same coat-of-arms, even when not related to heraldic symbology.[18] The conjugation of the various symbols of each colour and of the various colours of the arms, as well as their inter-connection with the heraldic figures of the crest and the shield and the divisions of the shield (see Figure 34.4), made up the personal identity of the knight to whom they were granted. Such symbolism must not be confused with the heraldry adopted by Masonry in the eighteenth century, which was based on medieval symbology.

A branch of a family that had already been raised to the peerage sometimes received a new coat-of-arms, but retained the original symbology. One of the rare exceptions was the case of the Andrades do Arco (Figure 34.3a), who did away with the serpents and leather strap of the original arms and replaced them with a centaur, the zodiacal symbol of Sagittarius, loosing an arrow from a bow (arch). The two serpents biting the band face each other. The one facing left is Good (the Saviour) and the one facing right represents Evil, the confrontation of esoteric dualism. Besides serpents, other symbols like lions, dragons, wolves (or dolphins, representing mermaids) may be at each end of the band. The links of a chain may be opened or closed, and in the latter case they may be depicted as serpents. In both cases they have the same symbolic meaning of dualism. They are sometimes linked to the net, as in the case of the Sampayos, with open links (Saviour) and chequered (Mercy).

Eagles can appear with bezants (small gold circles) and the number of bezants distinguishes one family from another, as can the tincture – this was the case of the Mellos (Figure 34.3d) and the Almadas. There are coats-of-arms related to the surname of the family, like Cyrne (swan), Gato (cat), Lobo (wolf), Pato (duck), Peixoto (little fish), etc.

The surname Cão (dog) – which the navigator Diogo Cão's descendants conveniently lost – did not come from the animal, but from the Hebrew *Qoum* (a person who is acquired as a slave), which gave rise to the Biblical name Cain. It is pronounced 'can' in French. Although the

34.3 Coats-of-arms of families genealogically linked to Colón-Zarco.

a *Andrade* – *crest*: serpent; *shield* per band with serpents at each end and *field* with two cauldrons. The Andrades descended from the Western Crusader Knights, the Counts of Trava and Trastamara, Lords of Andrade, Deuse, Ferrol and Villalba in Galicia. Rui Freire de Andrade was Master of the Order of Christ. Another branch went to Madeira and bought the lands of the Arco, symbol of Sagittarius.

b *Perestrelo* – *crest*: lion rampant; *shield* per bend with three stars and *field* of two ternaries of ten-petalled roses. The Perestrelos descended from an Italian astrologer, Pallastella.

c *Aguiar* – *crest* and *field* with identical eagle. The Aguiars descended from Dom Mendo Peres de Aguiar, who served King Alphonse I of Portugal and married Maior de Portocarreiro. The Aguilars of Córdoba were a branch of this family.

d *Mello* – *crest*: eagle with bezants; *field* with Templar double cross and bezants. The Mello family went back further than King Alphonse I. Their name came from the bird '*melro*' (blackbird), which was called '*mello*' in the old Luso-Galician language.

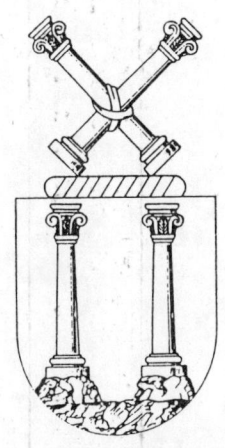

e *Esmeraldo* – *crest*: lion rampant; quartered *shield* per three bands; sinister chief per embattled and dexter base per chequered. The Esmeraldos descended from Flemish knights. João Esmeraldo settled on Madeira; it is thought he was an alchemist.

f *Furtado de Mendoça* – *crest*: wing and clawed foot of an eagle; *shield* per saltire, middle chief and base per bars and flanks per inverted 'S' representing serpents. The Furtado de Mendoças descended from Western Crusader Knights.

g *Cão* – *crest*: two crossed columns; *field* with two columns. The Cãos descended from Gonçalo Cão, who served King John I of Portugal. His son, the Discoverer Diogo Cão, served Prince Henry and King John II. The genealogies omit other ancestors and it is thought that he was a descendant of the Archbishop of Lisbon, Dom Agapito Colonna, and a Jewess.

h *Sá* – *crest*: chequered bull's head; squared *field*. The oldest known ancestor of this family is Rodrigo Annes (or Aires) de Sá, who served King Dennis of Portugal. One of his descendants was Constança Roiz de Sá (or de Almeida), who was the wife of João Gonçalves Zarco and grandmother of Colón-Zarco.

QUARTERED **CHEVRON** **BASE POINT**

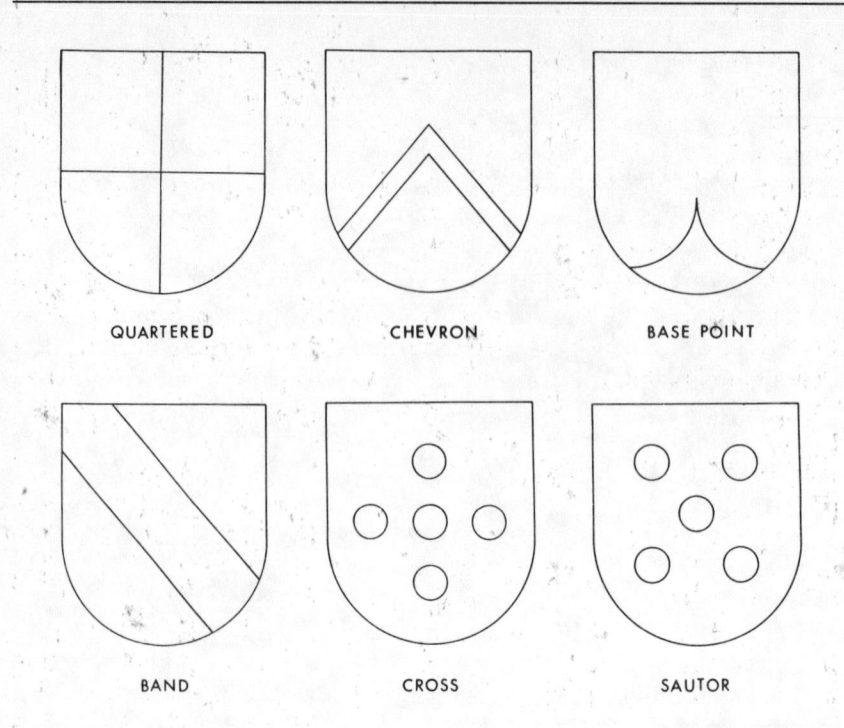

BAND **CROSS** **SAUTOR**

34.4 Divisions of the heraldic shield.

name Caim exists in Portuguese and Spanish, the phoneme *qoum* was also spelt 'cam', which degenerated into 'cão' in Portuguese because of the similarity to the name of the animal. The name has nothing to do with *cã* (white hair) that could have been a nickname/surname, but which was used only in the plural (*cãs*).

The Cãos' coat-of-arms shows two columns – exactly the symbol of the Colonnas, Italian princes and prelates (see Figure 34.3g). Was a descendant or kinsman of the above-mentioned Bishop of Lisboa, Dom Agapito Colonna, father of Pope Martin V, but with Jewish blood in his veins through his maternal line? The genealogies mentioned the descendants of many prelates, but the navigator Diogo Cão appeared from nowhere and then mysteriously disappeared to nowhere, even though he possessed a coat-of-arms. Was he a kinsman of the Castilian knight by the name of 'Cam', who served the Royal House of Castile and later fought at Arzila and Tangier as a 'mercenary' under the orders of King Alphonse V of Portugal?

Bartolomeu Dias (a patronymic transformed into a surname) also had a coat-of-arms, but left no descendants in the genealogies. There were many Old Christians called Dias before him, but it is not known whether he was 'Old' or 'New'.

Even occupying high positions in the Realm or close to the King, often hereditary, some Portuguese New Christians did not use a surname. If they were scholars, their one and only name was preceded by the designation of 'Master', which an artist also had the right to use. In order to be more easily identified, New Christians adopted the name of the place where they lived, such as the gates of the city, etc. They later adopted the names of towns and villages or trees and plants. In some cases they used a patronymic as a surname, following the custom of the Old Christians of the twelfth and thirteenth centuries – Esteves, Eanes (or Anes), Gonçalves, Dias, etc. And many, as a precaution, took the name Santos (Saints).

COLÓN-ZARCO'S COAT-OF-ARMS

The *Provisión*, a diploma of 20 May 1493, which is in the Colonino Museum of Seville and is reproduced in vol. II, chap. 20 of Navarrete's work, reads as follows:

> a castle and a lion that we grant you for your arms: the dexter chief will be vert a castle or; the sinister chief will be argent a lion rampant vert; the dexter base will be wavy azure of the sea a islands or; the sinister base will be *the arms you have already had*, which will be known as your arms and the arms of your sons and descendants for ever.

Besides having three new quarters added to his family coat-of-arms by 'Your Lords, the King and Queen', Colón-Zarco was also appointed Grand Admiral of the Ocean Sea, Viceroy and Governor of the Indies and Terra Firma, Captain-General of the Sea, Knight of the 'Golden Spur', a Grandee of Spain and *'Hidalgo Coberto'*, which meant that he had the privilege of wearing his hat in the presence of the Sovereigns when they were equally attired and after greeting them.

The oldest reproduction of Colón-Zarco's coat-of-arms that is known was transcribed from the sixteenth-century *Cartulários* in the book written by Henry Harrisse (see Figure 34.5a). This nineteenth-century biography said that the coat-of-arms had been designed in the presence of Colón-Zarco himself in 1502. But there is no evidence for the fact. If Colón-Zarco really did accept the alterations made by a herald, then he was an accomplice to a deception, as a coat-of-arms must fully conserve the symbology attributed by the Monarch that authorised it.

The differences between the 1502 coat-of-arms and the textual specifications of the *Provisión* are easy to see, through the following elements that were added:

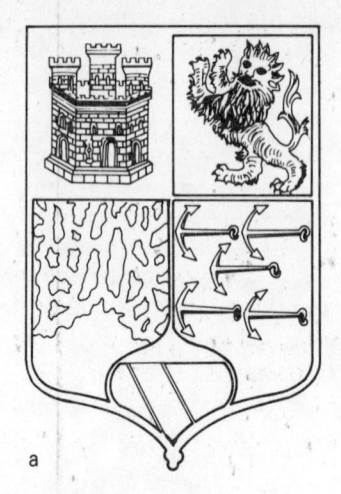

34.5 Colón-Zarco's coat-of-arms.

    a According to Harrisse (1502): the date is presumably a forgery.
    b According to Oviedo (1535).

> An amazing innovation took place with Colón-Zarco's coat-of-arms, more precisely with the base point, with a chief of gules and a counter-chief or per bend azure. Did Colón-Zarco, in 1502, really suggest the addition to the arms that the Catholic Monarchs had granted him? On observing the addition, it can be seen that it has the unusual shape of a Middle Eastern, Semitic, architectural cupola, the shape of which led Harrisse to observe that it had 'never been seen at any time or anywhere': it is the esoteric 'chamber' ('*camara*'), and the cupola that was the Templar seal. The band is the Templar baldric or belt (*talim*). The chief gules of the point symbolises courage and loyalty. The ground or of the counter-chief is the '*chrysopoeia*' achieved. Colón-Zarco had achieved his mission to Portugal and the desire of the wealth of a new world. The lion rampant, depicting 'virility, independence', is then replaced by the lion passant, showing the obedience to which he was subject. It is almost impossible to believe that Colón-Zarco authorised these additions to his arms.

(Ulisses 2000, *supplement, 33/34, Rome, 1986*)

1. The castle is presented per fenestra azure.
2. There is a field per bend inserted in the middle base. The configuration of this middle base deserved Harrisse's commentary: 'never seen at any time or anywhere! Only one shield had included a chevron of this type in Portugal before the fifteenth century – that of Manicongo, King of the Congo, christianised by King John II[19] and who was murdered by Pedro Vaz Bizagudo (see p. 60).
3. The waves of the sea are represented on a shield by wavy blue lines. If we follow the description of the *Capitulaciones*, the dexter base of Colón-Zarco's shield should be of wavy azure islands. The shield, according to Oviedo, merely shows azure islands without waves.[20] In 1493, the date of the *Capitulaciones* Colón-Zarco had discovered five islands. The arms presented by Harrisse in 1502 show many

islands plus a strip of terra-firma. This is historically correct, but there is no document that proves that the Spanish Sovereigns authorised this updating.

All these speculations are sterile, since the copies of the *Cartulários* are highly suspect (see p. 223).

Another interpretation of Columbus's coat-of-arms appeared in 1531, this time in Gonzalo Fernandez de Oviedo's work (see Figure 34.5b). On this shield, the lion is no longer rampant, but passant, and is crowned.

Santos Ferreira observed that neither in Spain nor in Portugal was any nobleman granted a coat-of-arms that included symbols found in the royal arms, unless they had a different tincture. He suggested that the castle was the symbol of the Kingdom of Castile and the lion that of the Kingdom of León, gules and or being respectively replaced by vert and argent, and he was correct.

The crowned lion, not very common in Portuguese and Spanish heraldry, probably had its origin in the esoteric 'Lion of Judah', which appeared in the arms of the House of Burgundy and Provence, of the nobility of Bavaria, the Rhineland, Flanders and Bohemia, particularly Moravia (present-day Czechoslovakia). The printer Valentim Fernandes used it. In Portugal, it appeared only on the arms of the New Christians Girãos (Girón in Aragon), Hagens (Germany), Kampeners (Antwerp) and the Ponce de Leons (Marchena) prior to the sixteenth century. It also appeared in the arms of the Lusignans (Cyprus and Jerusalem, kinsmen of Columbus's wife, Filipa), but only in a counter-quarter.

If Columbus had seen the design of the new arms in 1502, how could he have agreed to the insertion of a strip of terra firma in the quarter containing the islands when, in fact, he would discover it only two years later? The reference 'your arms that you have already had' in relation to the sinister base quarter indicates that Colón-Zarco had produced evidence of his noble origin and corroborates the thesis of Lobo d'Ávila and Santos Ferreira. 'Bezants' are *angoroths* in Hebrew, while little anchors are *ancoroths*. The bezants or *'quinas'* of the royal Portuguese arms, therefore, were transformed into little anchors. Columbus had adopted the arms of his father, Prince Fernando, and replaced the roundels argent with little anchors or. He was thus able to conceal his name Zarco and his Jewish origin.

Nobody else used an anchor in their arms, the origin of which was attributed to an anonymous Crusader, according to António Faria (1961), (p. 273).

Figure 34.6 shows two crests of the Henriques family (Figure 34.6a and 34.6b) the coat-of-arms granted to Columbus by the Catholic Monarchies (Figure 34.6c), and the coat-of-arms fourth quarter already possessed by Columbus's family (Figure 34.6d).

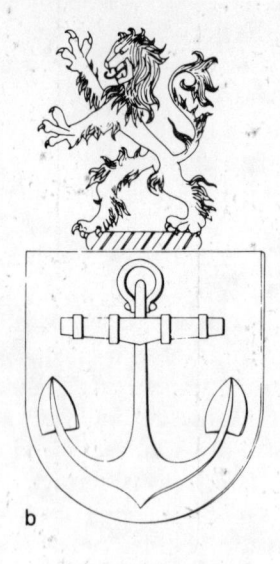

34.6 The two coats-of-arms of the Henriques, Colón-Zarco's and Columbus's family's coats-of-arms.

- a *Henriques I – crest*: lion rampant; *field* per chevron with castle, flanked by two lions rampant. There were two branches of the Henriques, one of Cordoba (Spain) and another of Beja (see p. 000). This shield was used by both the Portuguese and the Spanish branches.
- b *Henriques II – crest*: lion rampant; *field* with anchor (the original anchor was identical to that of the ancient Phoenicians, also used in Portugal). This shield is a variation of that of the branch used in the fifteenth century and disappeared from Portugal after Colón-Zarco had adopted anchors in sautor for his arms.
- c The coat-of-arms granted to Columbus by the Catholic Monarchs; Columbus put anchors instead of roundels on his arms as he could not use the same symbols as the Portuguese royal family.
- d The coat-of-arms with the bezants used by Columbus's family.

Notes and References

1. Quoted by Sorval (1891).
2. Corbin (1978).
3. Vieira, Fr António (1854–8), *Sermão do Demónio Mudo, pregado no Convento de Odivelas, Religiosas do Patriarca São Bernado, no anno de 1651.*
4. Speaking to nuns of little scientific knowledge, Vieira used the word 'stars' for 'planets'.
5. Vieira, Fr António (1854–8), Vol. 5, p. 452, designates the strap as a *taly*, from the Hebrew *thali*, meaning a wide strap. Cardoso, Jorge *Agiólogo Lusitano*, Vol. III, p. 55 (1742), however, spells it *talim*. It corresponds to the French *bauldrier* (today *baudrier*, which became *boldrié* in Portuguese) and *talabarte* in Italian (see reference on p. 460).
6. See Sephiroth IV, *Chesed*, 'Mercy' (pp. 263–518).
7. Vieira, Fr António (1854–8), *Sermão da Sexagésima; Facim vos fieri piscatoris hominum;* (St Matthew, IV, 21) *reficientes retia sua.*
8. See p. 258 and *Binah*, pp. 263–71 and 517.
9. See *Chesed*, 'Mercy', pp. 263–518.
10. See *Chochmah*, 'Wisdom', pp. 263–517.
11. See *Chesed*, 'Mercy', pp. 263–518.
12. See *Chesed*, 'Mercy', pp. 263–518.
13. See *Geburah*, 'Courage, strength, austerity, instilled fear', p. 488; and also *Netzach*, pp. 263–7 and 269.
14. See p. 267.
15. See *Chesed*, 'Mercy', pp. 263–518.
16. See *Netzach*, 'Victory', firmness, concealed intelligence, pp. 263–7 and 269.
17. In heraldry, the colours are termed as follows: blue = *azure*; red = *gules*; black = *sable*; green = *vert*; purple = *purpura*; gold or yellow = *or*; silver or white = *argent*.
18. Viel (1972); Gassicour (1907); Chassant (1978); Genouillac (1877); Alleau (1976); Portal (1974); Renesse (1894–1905); Bouton (1873).
19. Two more appeared in the reign of King John III – the Lucenas and the Cantos (from a Flemish family Kant).
20. The Marinhos and the Távoras were the only Portuguese families to have the waves of the sea represented in their arms prior to the sixteenth century.

35 A Question of Physiognomy

If we review the history of Genoa from the time of the fragmentation of northern Italy into autonomous States and Boccanegra's dictatorship in 1259 – two centuries before the birth of 'Cristóforo Colombo' – we shall see that the Republic reeled from constant internal upheavals and internecine wars against Venice and Milan. Meanwhile, Genoese colonists settled on Chios between 1304 and 1346 and Lesbos from 1354 to 1662, and many of the colonists returned to Genoa with their descendants at the end of the occupation. On the other hand, the Ligurian people always maintained cordial relations with the merchants of the Levant. Many rich Moors took the beautiful Genoese brunettes as their wives, while considerable numbers of the Republic's plebeian classes married slaves and Levantine servants. In Genoa, there were no blond Venetian 'Desdemones' among them. The former colonists of the Aegean Sea would surely have been attracted by the sensual beauty of the Heleno–Ottoman islanders. With this portion of Greek and Turkish blood running in their veins, the Genoese could be rightly proud of their fiery, dark beauty.

In 1458, the Doge invited the French to occupy the Republic. Were there blond men in the occupying forces? It is highly likely. In sufficient numbers to leave Northern genes among the populace? This is highly unlikely, as much as they seduced or raped the local beauties. According to his biographers, 'Cristóforo Colombo' was born around 1451. This means that the French arrived too late to leave the future Admiral in Susana Fontanarossa's womb and if the French military presence did upset the Doménico Colombo household, it could have affected the pigmentation only of the little Bianchinetta – that is, if she was born seven years after her brother Cristóforo.

The Italian Peninsula indubitably produced the greatest of the Renaissance painters, since the Flemings started to dispute their mastery only in the middle of the sixteenth century. It is interesting to note that the Italian masters always adorned their saints, angels and princes with blond hair. But they did this with the conviction that these golden tones instilled their characters with a greater finesse. I have not been able to discover – scientifically – whether angels are blond, but there is one thing I am almost sure of – Jesus Christ was not, even though the Italian artists used this fantasy in their pictures. They even gave Him blue Nordic eyes, though they could never have doubted that He was of pure Jewish blood. As far as their princes were concerned, they really did have lighter-coloured hair – a result of the many matrimonial connections with royal and princely houses of Central and Northern Europe. This fact contributed to the conception that flourished among Italian artists that the virtues of gentility and finesse were a question of class and not race.

Yet these same painters (except in rare cases and confined to Florence) depicted the Italian plebeian classes, which certainly included wool-carders and tavern-keepers, as dark-skinned and with very curly black hair. Only courtiers were depicted as blond.

From the sixteenth century, Italy became the pasture of foreign armies – Germans, Englishmen, Frenchmen and Flemings pitched their camps over the whole country and, certainly lavishing their pay on things other than drink, left their mark. The Napoleonic period gave rise to enormous miscegenation but even after 400 years of constant cross-breeding, statistics show that a majority of the Ligurian people are still dark-skinned, despite the fact that they live in the north of the Italian Peninsula.

THE BLONDISH RED-HAIRED ADMIRAL

When 'Cristóforo Colombo' came into the world in the fifteenth century, these foreign invaders had not yet debased the pigmentation of the Genoese, who certainly had black or very dark hair. Let us admit that 'Cristóforo Colombo', a wool-carder's son, had been born blond, through some unusual genetic mutation, and possessed the same aristocratic physiognomy and gentility and golden hair idealised by three Gaddis, two Lorenzettis, Daddi, Orcagna, Fra Angelico, Uccello, Pisanello, Masaccio and Fra Lippi – almost all of them Florentine painters[1] – who depicted the Italian plebeians as dark-skinned and black-haired. If Doménico's son really was fair-skinned and blond, he could well have served as a model for the artists who wished to paint a prince. But his chances of succeeding in that profession would have been about 5 per cent.

Let us now look at the case of Colón-Zarco, admitting that he was telling the truth when he conceived the siglum with which he can be identified. He would have had little chance of being blond on his mother's side, although a kinsman of his grandmother Constança, the 'Sá of the Galleys', had a red beard and a cousin, 'the Pearmain', was blond and red-faced. On his father's side, however, blonds and redheads abounded, as can be seen in Colón-Zarco's genealogical tree (see Tree XVII, p. 454).

Portuguese princes were born blond and gradually got darker as they grew older, until their hair became grey. I shall consider only the ancestors of Prince Fernando, Duke of Beja and Colón-Zarco's father, beginning with King Dennis, married to the Holy Queen Isabel of Aragon, who was blond. The Castilian Inês de Castro, wife of King Peter I of Portugal was also blond.[2] She was descended from Edward II of England and Queen Isabel of France.

It is not known whether they were blond, but their daughter-in-law, Queen Philippa of Haumont (Valois), was. One of them must have been red-haired as their grandson, John of Gaunt, Duke of Lancaster, was a blondish red-head and his daughter, Philippa, wife of King John I of Portugal and grandmother of Prince Fernando, Duke of Beja, was blond.

In the polyptych of Nuno Gonçalves (see Plate 11), we can see streaks of the blondish-red hair of King Alphonse V and Prince Fernando, Duke of Beja (Colón-Zarco's uncle and father), despite the fact that their hair was partially repainted brown by the restorer. It can also been seen that Prince Fernando (known as 'The Saint'), the Martyr of Fez and the main figure of the polyptych (in which he appears twice), great-uncle of Colón-Zarco, was a blondish red-head, as was the young Prince João, future King John II. According to the chroniclers of the time, the King was similar in colour to his cousin Salvador Fernandes Zarco/'Cristóbal Colón'.

Las Casas, who knew the 'Discoverer of America' so well, described him as follows:

> He was taller than the average, with a long and authoritative face; his nose was aquiline and he had greenish eyes; his skin was white and turned red when burned; his beard and hair were red when he was young, although they became white with his arduous life . . . and he had a venerable appearance, a person of great class and authority and worthy of total reverence.

Don Fernando Colón portrayed his father:

> His hair was blond when he was young but it turned white at the age of thirty.

Although Gomara[3] did not meet Columbus personally – he was born six years after Columbus's death – he spoke with people who had lived close to him and said:

The Admiral was tall and well-built; his face was long, red, freckled, tired and serious and he worked extremely hard.

As can be seen, both Don Fernando and the chronicler emphasise that the Admiral aged prematurely and that owing to his hard life he was white-haired at the age of thirty.

Columbus was described as white-skinned, with a face that reddened easily. As 'rubio' means blondish red-head in Spanish, Las Casas agreed with Don Fernando when he said that his father was 'blond' when young.

Columbus would then have been an anomaly if he is considered to have come from an Italian city on the Mediterranean, and from the plebeian class. He had a venerable appearance, a person of high class and authority, worthy of total reverence. Las Casas added that the Admiral was sober and moderate in eating, drinking and dressing and did not curse or swear when angered, in spite of living amongst crude mariners and he emphasised: 'His rudest words were merely "Go with God!" . . . "Does it not seem so?" . . . "What about that then?"' What an extraordinary education the wool-carder Doménico and the tavern-keeper António had given their son and nephew. What beautiful manners Susana Fontanarossa had taught her little blond son that played with the other children of his own age in the poor streets of Genoa. He must have been told to keep his ears closed so as not to hear the language of the customers in his uncle's tavern and his eyes open in order to see the evils of drinking.

The chronicler Oviedo also knew Columbus, although not as well as Las Casas, and confirmed the descriptions of Las Casas and Don Fernando. Nineteenth-century historians – speaking as though 'Cristóforo Colombo' were their contemporary – admitted he was blond; what they could not explain was his temperament and his fine manners. We can therefore allow Salvador Fernandes Zarco a 95 per cent chance of being a blondish red-head, complete with the freckles that are so common amongst the British.

As far as his physique and expression are concerned, he can be easily identified as a descendant of the House of Avis. His aquiline nose, his greenish eyes, his white skin that burned easily in the sun, his hair – this is certainly not a portrait of a plebeian from a sun-drenched port on the Italian Riviera; see the portrait of Maximilian, The Holy Roman Emperor in Figure 35.1.

PORTRAITS OF COLUMBUS

All portraits of Columbus are posthumous and the only one considered to be 'authentic' is to be found in the Madrid National Library. It was

35.1 Maximilian, The Holy Roman Emperor.
Transmission of hereditary facial characteristics in this branch of the House of Avis (the second Portuguese dynasty) was remarkable: a wide, high *nasion* (bridge of the nose) that is a continuation of the forehead; a hooked nose with wide nostrils; a curvilinear *septum* (partition between the nostrils) concave where it meets the upper lip; the lips are thick, the upper one having the shape of a 'Cupid's bow' and the lower being protuberant; the *gonion* (angle of the maxilla below the ear) is round rather than angular. Compare with Plate 15.

reproduced in *La Ilustración Española y Americana*, in the valuable monographic study of Don Juan Pérez de Guzmán entitled *'Los Retratos de Colón'* ('Portraits of Columbus').[4] There are another two portraits in which the depiction may be said to look like Columbus but in which the painters have narrowed his lips and made his hair blonder. One is a reproduction in the Giovanna Gallery in Como, which is a clearly altered copy of the portrait in the Madrid National Library. The other, which gives him a Nordic (but not 'Aryan' appearance, owing to the shape of his nose) is to be found in the Madrid Naval Museum.[5]

There is another Italian portrait, painted by Sebastiano del Piombo in the seventeenth century. The artist showed absolute disregard for previous iconography and portrayed an acceptable Genoese woolcarder, complete with protruding, blood-shot eyes.

THE PANELS OF THE HOLY PRINCE

If 'Cristóforo' has been the most internationally discussed topic during the last 500 years, its counterpart in Portugal in the last fifty has been the identification of the figures that are portrayed in the panels of Nuno Gonçalves (see Plate 11). Many intellectuals have tried to unravel the mystery and I cannot but help broach the subject. The thesis of José António Saraiva, expounded in the *The Panels of the Holy Prince*, is the only coherent one as far as I am concerned, but following the philosophical studies of Afonso Botelho and António Quadros, there is a little left to say.

The central figure of the panel that is wrongly called 'of the Archbishop', which António Quadros has intelligently entitled 'The Mission of the Orders of Christ and Avis'[6], is repeated in the panel that is alongside. José António Saraiva and António Quadros maintain – and I agree – that it represents Prince Fernando (known as 'The Saint'), the Martyr of Fez. He is flanked by two knights who are wearing breastplates and are armed with pikes, as well as by a youth in a studded doublet and two kneeling knights, one holding a pike and the other a sword.

The knights wearing breastplates are the Braganza brothers, the one on the left being Dom Fernando II, Duke of Guimarães (son of Dom Fernando I, Duke of Braganza), who commanded an expedition to North Africa in 1471. To the right is his brother Dom João, Marquis of Montemor-o-Novo. They both have the anthropomorphic characteristics of the Braganzas that can be seen in all the portraits of Dom Nuno Álvares Pereira – a long, straight nose, giving the illusion of a receding chin.

The other three knights (with studded doublets and dark collars) have clearly aquiline noses with a wide frontal bone, narrow bridge and downturned mouth, although this is not so accentuated in the young man's face. The two kneeling figures in the foreground have a rope between them, the centre of the 'pillar of Mercy' and the Templar symbol of the Discoveries that would become an important element of the Manueline style, together with the Cross of the Order of Christ. The rope is rolled up, which symbolises the fact that the ships have weighed anchor and left for the high seas. The kneeling knight on the left is Prince Fernando, second Duke of Viseu and first Duke of Beja, Master of the Order of Christ, heir and successor of his uncle, Prince Henry the Navigator, fifth Constable of the Realm – and the father of Salvador Fernandes Zarco/'Cristóbal Colón'. Behind him is his younger son Prince Diogo, who would later be the third Duke of Beja. As he was the son of Princess Beatriz and grandson of Dom Fernando, Duke of Braganza, his nose is slightly straighter. Facing the Duke of Beja is his older son Dom João, who would succeed his father to his Dukedoms and the Mastership of the Order of Christ.

Compare the physiognomy of Colón-Zarco with that of the three Dukes of Beja and the Holy Princess Joana (who was both his niece and cousin), who is to been seen to the left of the central figure in the panel to the left.

I also agree with António Quadros when he gives the name of 'The Panel of the Brotherhood of the Holy Ghost' to the part that is usually entitled 'The Panel of the Fishermen'. This title perfectly fits the prostrate Franciscan monk and the navigators of the Order of Christ behind him, particularly devoted to the Holy Ghost and wrapped in the net (see p. 459, Fig. 34.2, no. 3 and Plate XI, Panel 2) that gave rise to the heraldic chequer design (see p. 463, Fig. 34.3-h).

The central figure, The Holy Prince, is holding a baton, which António Quadros claims is that of the Order of Avis. The baton of command symbolises magic power: one gives orders in obedience only to a superior Being and in consideration of the choice between Good and Evil. It is linked to secrecy, which is the virtue of silence, good example, discretion, the passion for Truth, faith maintained in the face of the worst provocations, abstinence and fasting, self-denial being the most important abstinence of all. The only known Saint that is represented with such a baton is St John the Evangelist. The Ancient Greeks depicted the scholar-physician Aesculapius with a baton and a serpent (a symbol adopted by doctors), which signify the power of command and life respectively.

It is the Gospel of St John that the Holy Prince is holding. And the baton is the badge of the soldiers of the Holy Grail, the *non-ceptrum* symbol of the mystic royalty of Percival, who embodies the 'clear valley' (i.e. Clairvaux). Percival is St Bernard, who is personified in the polyptych in the figure of the Holy Prince. This is the Saint who preached the Crusade that the Portuguese carried out by land and by sea.

THE JEW AND HIS EMBLEMATIC 'MARGARITA'

The Jew in the polyptych (see Figure 35.2) is wearing the type of hat that was usually worn by economists, physicians, lawyers, printers, etc. of the Court. Although one fifteenth-century engraving depicts King John II with a Jew wearing a turban, it is common knowledge that the Portuguese Jews preferred to wear Christian-type clothes (see p. 24–7), some of them dressing with ostentatious luxury. The fact that he is clean-shaven is explained by Moritz Steinscheider;[7] even when holding offices of Hebrew Theocracy, the Jewish cabalists of the Peninsula shaved, despite the fact that it was contrary to the Law of Moses. The one depicted in the panel was obviously not a traditionalist. He may have been the Chief Rabbi, the successor of Dom Judah Cohen and Master Mosém. Is it Zacuto?

On his breast, the Jew has an emblem that represents the 'Margarita' (marigold), the symbol of the Hebraic decenary. In the original painting,

35.2 The Jew and his 'Margarita'. The Jew in the polyptych before its restoration. The sephirothic ten-pointed star was distinctive of Peninsular Jews of high social standing and replaced the more common 'Shield of David'; in the Iberian Peninsula it was called a 'Margarita' (marigold), and was yellow. The book the Jew is holding has imaginary characters, but they are thought to represent the *Pentateuch*, one of the first works printed in Portugal.

which was altered either by the need to conceal the fact that there was a Jew in the Panels of the Holy Prince or by the imperfection of the restorer, this 'ten pointed star' should have been yellow in order to indicate the religion of the personage, just as other Jews who were not of the nobility had been obliged to wear a 'Star of David' so that they could be more easily identified. In this particular case, we are not even dealing with a New Christian, despite his hat and clean-shaven face, as he is holding a sacred Hebrew book.

It must not be forgotten that Colón-Zarco gave the name of 'Margarita' to an island in Venezuela (New Sparta). He did not call it the 'Island of "Margaritas"', which would have meant the Island of Pearls (see maps on pp. 77, Fig. 6.5). He chose to give it an esoteric, cabalistic name – the wild ten-petalled marigold. One of the Balearic Islands, where a colony of Jews had settled, is called Margarita, although it is thought that the name was given owing to its geographical shape.

Yet it was not only due to the presence of this Jew among the members of the royal family that led to the polyptych of Nuno Gonçalves being removed from Lisbon Cathedral and hidden in a cellar of the Monastery of St Vincent-without-the-Walls. Another reason justified its removal from public view: two of the figures in the panels are Dom Fernando, Duke of Braganza, and Dom Diogo, Duke of Viseu and Beja (respectively cousin and half-brother of Colón-Zarco), who would certainly haunt King John II, their cousin and brother-in-law respectively, and keep the decapitation of the former and the stabbing of the latter, at the hands of the King himself, alive in the memory of the people.

The coercive conversion of the Jews would have been more tragic if it had not been for the Cult of the Holy Ghost. In the light of Daniel's prophecy, they did not see in Jesus Christ the Messiah, but an Angel sent to Earth to see that the Law was fulfilled, but not altered. The Cult of the Holy Ghost was crucial in persuading them to join the converted New Christians, some of whom were already members of the Order of Christ. It was not by chance that this cult was predominant in all the areas of the country where the New Christians settled and it assumed particular relevance in the Azores. In Portugal at the time, the cult of the Holy Ghost resembled the veneration of the symbology of the Grail, especially among knights. In turn, this was allied to the cult of the Magi (who are misrepresented as the 'three Wise Kings' at Christmas, with the aim of achieving an identical ecumenical ideal). And when Colón-Zarco reached the mouth of a river on the Day of Kings in 1502, he called it Belém, the name of the beach on the River Tagus from where the caravels and naus of the Portuguese Discoveries departed.

ST BERNARD'S MISSION

The guiding principle of the Polyptych of the Holy Prince is the mission of St Bernard – *Discover* and *Evangelise*. The painting appeals for tolerance and invokes the Holy Ghost through the Gospel of St John, consoling and forgiving the folly of Alfarrobeira, where the Princes of the House of Avis settled their differences on the battlefield and where one of the most illustrious of those Princes, Pedro, lost his life.

In his sermon *'De gratia et libero arbitrio'* (1127), St Bernard expounded the relationship between Goodwill and Freedom.

In his *'De diligendo Deo'* he preached the excellence of Love, even carnal love – the lowest in the scale of values – the principle of which he accepted because it was natural. But the highest degree of Love is mystical union, a divine process that leads the soul to perfect understanding. In the scale of Love, the overseas feats of the Portuguese, although debased by the ill-will of some men, go from soldiers and poets like Luís de Camões to Saints like João de Brito. Nuno Gonçalves painted a rope at the feet of the two Masters of the Order of Christ, father and son. The rope is rolled up so that it seems to be a work of manual magic in Templar 'Manueline', a decorative style that flourished in Portugal at the time of King Manuel I; it was inspired in the Discoveries and is redolent with maritime symbology such as cable mouldings, corals, etc.; two of its main elements are the armillary sphere (the emblem of Kings John II and Manuel I) and the Cross of the Order of Christ . If prayer and hymns demand discipline and guidance, sculpture obliged man to acquire a specific 'manual magic'. A simple monk was not enough – the artistic monk became necessary. Most of the essential elements of Gothic architecture came from St Bernard, as did Manueline, although from roots that went further back in time. The rope rolled up at the feet of the Holy Prince is Voyage, Discovery. António Quadros emphasises the existence of ten magical words in the poetical language of Portuguese: 'Sea', 'Nau', 'Voyage', 'Discovery', 'Search', 'Orient', 'Empire', 'Love', 'Yearning', 'Occult'. All these key words of Portuguese thought are to be found in the polyptych of Nuno Gonçalves, under the protection of the ethereal strength of the Holy Ghost. Salvador Fernandes Zarco/'Cristóbal Colón' was imbued with the spirit of those words.

Notes and References

1. A period of Italian painting that lasted from 1300 to 1469.
2. After the troops of Napoleon had violated the tomb of Inês de Castro in the

Monastery of Alcobaça in a search for jewels, they left her mortal remains scattered over the floor. The Queen's skull still contained some blond hairs.
3. Francisco López de Gómara (or Gómora) (1512–72), author of *Hispania Victrix. Primera y segunda parte de la historia general de las índias con todo el descubrimiento, y cosas notables que han acrescido desde que se ganaron hasta el año de 1551. Con la conquista de México, y la nueva Espanã*, Vol. XXII (Madrid, 1885).
4. Madrid (12 October 1982), pp. 250ff.
5. 'Los Retratos de Colón'.
6. Quadros (1986–7), Vol. II, double print between pp. 176 and 177.
7. See Steinschneider (1893), medieval Jewish translations which include manuscripts of many European cities, with a list of 100 astrologers and including the *Almanaque Perpetuo* of Abraão Zacuto, printed in Leiria, Portugal, in 1496; it contains a remarkable introduction which refers to tradition and customs.
8. See Quadros (1983).

36 The 'Brothers' and 'Cousins' Gambit

At the beginning of this book, I referred to the fact that the Crusaders and their Muslim foes fraternised during periods of truce in Jerusalem. I also compared King John II's political manoeuvres to a game of chess, so that on the 'board' of the world he would 'capture' Brazil and the Oriental trade. My careful movement of kings, queens, bishops, knights, rooks and pawns has been a game of chess, in which the move of every piece or argument has been an attempt to solve sequent problems of the mysterious chessboard that was the life of Salvador Zarco/'Cristóbal Colón'.

I shall now invite the reader to share in the deciphering of some enigmas.

THE 'PRE-DISCOVERERS' OF THE ANTILLES

1. There is a document (a facsimile of which is on p. 121) in the Hebrew Theological Seminary Museum in New York that deserves the closest attention. It is the calendar and tables of the declinations of the Sun by Abraão Ben Zacuto, astronomer of King John II of Portugal, which were handed to Colón-Zarco nine days before his trip to the Antilles in the service of the Castilian Sovereigns. These solar and lunar tables not only made the Admiral's astronomic calculations easier, but made them more precise as well, to the point that they saved the lives of the Admiral and his men on his fourth voyage (see p. 92). As the local Indians were aware that some of the Spanish colonists had risen against the Admiral and had marooned him and his crew on Jamaica, they decided to attack him. But Zacuto's Tables told Columbus that there would be a lunar eclipse within three nights. At dusk on the third day, he summoned the local chiefs and, through an interpreter, warned them that he had the power to make the Moon disappear. The chiefs were unmoved,

but when the phenomenon occured they were terrified and begged Columbus to ask his Christian God to save them from his heavenly anger.[1]

These Tables were printed by Samuel d'Ortas (see p. 121) in a workshop in the suburbs of Leiria (Portugal) and published on 25 July 1492, exactly nine days before Colón-Zarco set sail from Palos (3 August 1492). Why should the King of Portugal be generous enough to help a 'foreign adventurer' discover India for his rivals, the Monarchs of Castile?

2. Dom Fernando Colón said:

> On arriving at the island that the Admiral baptised with [his] name of San Salvador, the local people surrounded the ship . . . some swimming, some in canoes, carrying parrots, skeins of spun cotton, spears and other sundry objects, exchanging them for glass beads, rattles and other worthless trinkets . . . each skein weighed about 25 pounds and was extremely well spun . . .; they happily exchanged anything they had for a piece of broken glass, to the point that one of them handed over 16 skeins of cotton for 3 Portuguese 'brancas'.[2] From San Salvador, the Admiral sailed on to another island that he called Concepción and there they saw a man that . . . carried a string of green glass beads and two strings of white glass beads in a little basket.

Dom Fernando's explanation was: 'it seemed that he had come from San Salvador to Concepción and then intended to go on to Fernandina in order to give the local population the news of the Christians'.[3]

Colón-Zarco himself confirmed the propensity of the local people to trade:

> They brought spears and skeins of cotton to trade, which they exchanged for pieces of broken glass or china. Some of them had pieces of gold in their noses, which they happily exchanged for rattles, glass beads or other worthless objects . . . and I saw them exchange 16 skeins of cotton for three Portuguese 'ceitis'.[4]

I have already mentioned the strange fact that, in order to trade with the peoples of Japan, China and India, of whose wealth and civilisation he had sung the praises, Colón-Zarco took coloured beads for barter and Portuguese coins to purchase things if need be. What was even stranger was the fact that the local Indians accepted the coins.

And now we have the picture of a spry Indian from an island that Columbus had not yet visited hopping from island to island with a basket of glass beads, which were not made in the Antilles, anticipating the European fleet and advertising the 'generous' trade offered by the white man.

3. Both Colón-Zarco and his son Dom Fernando mentioned the fact that the inhabitants of the Antilles 'have no iron of any type . . .; I call the sticks they carry darts or spears according to their size, although they are tipped not with iron, but with fish bones or teeth'.[5]

However, some sailors that had penetrated into the interior of the Island of Guadaloupe and visited the local people returned to the ships and related what they had seen. Dom Fernando narrated the story, without forgetting that his father was the *first* discoverer of that archipelago: 'the most marvellous thing they found was an iron casserole'. His explanation was: 'I think that as the pebbles and flints of that land have the colour of shiny iron, some silly person thought it looked like iron, although it could not have been as no iron of any type has been found among these people, and I do not know what the Admiral has to say'.[6]

4. The local natives had their ears and noses pierced with gold ornaments and wore bracelets of the same material on their arms. When Colón-Zarco arrived at Hispaniola, the King (*cacique*) of the island went to meet him with 200 of his almost naked men and said that 'the land where gold is to be found lies farther to the west . . .; after having eaten, one of his servants brought a belt that was the same as those of Castile, except that the buckle was made of gold, and gave it to me, together with two pieces of wrought gold'.[7]

5. Dom Fernando narrates that a large group of Indians armed with spears, surrounded the Admiral and his men, but an elder one suddenly appeared, shouting at the top of his voice and trying to hold back his colleagues; he walked towards the mariners and 'went up to one of the Christians who had a loaded crossbow and showed to his comrades; . . . he told them that the crossbow would kill all of them and from a great distance. He drew a sword from its sheath and said the same thing. On hearing this, they all fled, leaving the Indian shaking in fear'.

6. Referring to the island to the west where there was gold, Columbus thought: 'Another big island, which must be Cipangu, from what I can make out from the Indians I have on board, who call it Cóba [Cuba] and where there are many naus and many seamen.[8]

Had the Asians managed to sail from the Pacific to the Atlantic through a non-existent Panama Canal? As can be seen, besides the fact that the local natives bartered spears, spun yarn and gold in exchange for glass beads and Portuguese coins, they also had at least one casserole that seemed to be made of iron, they were aware of the deadly range of the fifteenth-century crossbow and the sharpness of the swords. European glass beads were already known on other islands; one of the naked natives had a belt with Castilian-like buckles, although it was made of gold, and big naus with large crews had been seen in Cuba.

As it is impossible to make casseroles with pebbles or flints, it must be admitted that they (or it) must have been made of iron, which is non-existent in those parts. How can these phenomena be explained? Who had been there before 'Cristóbal Colón? With casseroles, crossbows, buckles and Portuguese coins!

7. The Indians of the Antilles used big and small canoes carved out of tree trunks, the biggest one transporting about 40 paddlers. The Castilians called them *'piraguas'* and the Italians *'pirogas'*. Why was it that in his log Colón-Zarco always referred to them as *'almedías'*, a Luso-Arabic word used only by the Portuguese to designate the 'dongos' found on the River Zaire (or Congo)?

8. The *Capitulaciones* that were signed at Santa Fé by the Catholic Monarchs and 'Cristóbal Colón' on 17 April 1492, four months before his first voyage in their service, is undoubtedly a genuine document and a contract whose terms the future Admiral undertook to fulfil. A part of the introduction reads as follows: 'The things requested that Their Highnesses give and authorise Dom Xpoval de Colón, because he has already discovered lands in the Oceans and Seas and on this voyage, with the help of God, he will find them again'.

And further on, he conceded what he had already discovered to the Sovereigns: 'that Your Highnesses are the Lords of the said Ocean Seas'.

The Castilians were the lords only of the coast of the Mediterranean, Galicia, Biscay and the Canaries. The Treaty of Alcáçovas prevented them from going south of the Canaries. As they were not dealing with the South Atlantic or the coasts of Equatorial Africa, therefore, Columbus must have been referring to the Western Seas.

When Columbus returned from his first voyage – with only seven islands discovered – the Catholic Monarchs, who were in Barcelona, confirmed his privileges and powers and wrote, with the same precision as was usual: 'the said islands and terra firma of which you spoke and discovered, and the other islands and terra firma that will be found through your industry in future, in the said part of the Indies'.

Had Columbus lied to the Catholic Monarchs when he claimed to have discovered islands and terra firma in the west before his first voyage to the Antilles?

9. Columbus having chosen a cliff overlooking the sea and dominating the port of Santo Tomás as a suitable point to build a fortress, the sailors started digging and, to the amazement of everyone 'they found "nests" of clay and straw that, instead of eggs, contained three or four spherical stones, as big as oranges, that seemed to have been made for artillery, which surprised them very much[9] . . . as if they had been placed there just a few years before'.[10]

   They were not the enormous stones spheres that the Indians of the mainland worshipped and openly exposed as a representation of the God-Sun. These spheres were smaller and had been hidden in a place that was suitable for the construction of a fortress. They were stone cannon-balls. Had someone set up a gun emplacement there before Columbus's arrival and tried to preserve his ammunition in case of future need? All the hidden balls were exactly the same size and chiselled to perfection and it must have been a long and difficult task making them.

10. Columbus could hardly understand or be understood by the local natives. How could he, therefore, have told his men of the existence of the 'Amazons' (the name that the Portuguese gave to the river in Brazil), a people exclusively made up of women, on another island? And this island really did exist. It was called 'Matinimo' and had a matriarchal social system. The island would be found by the fleet not long afterwards, exactly where Columbus had predicted: 'this island must be no more than fifteen or twenty leagues to the southeast from our point of departure, but the Indians are not able to point out its exact direction'.

11. On the island of Hispaniola, they found children that were different from the rest of the population: 'very white, like the people in Spain; some of them are as white as us'.[12]

    What explanation could be found regarding the strange pigmentation of these children in the midst of a 'redskin' population?

    The only writer that has been honest enough to quote these phenomena is Luís Arranz Márquez, in his introduction to Columbus's *Diario de a bordo*.

    Did Columbus really want to hint that someone (or even he himself) had discovered America before him? But if this was the first time he had sailed in the service of the Catholic Monarchs, who had supplied the ships for such voyages? Could it have been the mythical St Brendan?

12. The *Diario* relates only the first of Columbus's voyages. How much information like this was suppressed by Fornari (see p. 183) in Fernando Colón's *Historia*, which relates the other three expeditions?

It is interesting to mention that this phenomenon also happened in Massachusetts, from where the Corte-Real brothers never returned (see pp. 62–3). In 1643, Roger Williams wrote the first dictionary of the language spoken by the Indians of the Wampanoag tribe that lived in the area where the Dighton Rock was found and explained that their name came from *wompi* (white) and *nanoag* (people). He said that although the Indians 'were burned by the sun, they had been born white'. Williams, who had gone to America with the Pilgrim Fathers, also remarked that the Wampanoags were the only Indians that got on well with the newcomers.[13]

There are also three copies of a letter written by Giovanni Verrazano, who had sailed up to Narragansett Bay in 1524, although he did not mention with whom. Although these three copies are not all exactly the same, they all manifest surprise that the Indians had received them very amicably and he described them thus: 'they are very white; some tend to be white, others tend to be red'.[14]

CHRONOLOGY OF COLUMBUS'S LIFE

I shall now try and put some facts of Colón-Zarco's life in chronological order, beginning with the descendants of his father:

Prince Fernando (successor of Prince Henry the Navigator), second Duke of Viseu, first Duke of Beja, Master of the Order of Christ and Constable of the Realm: he was born in 1433 and died in 1470; married his first cousin Beatriz, grand-daughter of King John I and Dom Afonso, second Count of Barcelos and first Duke of Braganza (son-in-law of the Constable Nuno Álvares Pereira, first Count of Barcelos) born in (?) and died in 1506; they had nine children:

1. *Dom João* (his father's successor), third Duke of Viseu, second Duke of Beja, Master of the Orders of Christ and St James and Constable of the Realm: his date of birth is unknown, but it is thought that he was twenty-three when he died in 1472; he must have come into the world in 1449, about two years after his parent's marriage.
2. *Dom Diogo* (his brother's successor), fourth Duke of Viseu, third Duke of Beja, Master of the Orders of Christ and St James and Constable of the Realm: born in 1457 (?), he was killed by his cousin and brother-in-law, King John II, in 1484.
3. *Leonor*: she was born in 1458 and died in 1525; she married the future King John II (born 1455, died 1496) at the age of thirteen; they had one son, Prince Afonso who was born in 1475 and killed in a riding accident in 1491.
4. *Isabel*: born in (?) and died in 1472; she became the second wife of

Dom Fernando, third Duke of Braganza (born in 1430 and beheaded in 1483); they had four children.
5. *Dom Duarte*, who died very young.
6., 7. and 8. Two boys and a girl, who died in childhood.
9. *Dom Manuel* (successor of his brother Dom Diogo and his cousin and brother-in-law King John II), fifth Duke of Viseu, fourth Duke of Beja, Master of the Orders of Christ and St James, Constable of the Realm and King of Portugal, Manuel I: he was born in 1469 and died in 1521; his first wife was Isabel, daughter of the Catholic Monarchs and widow of Prince Afonso (see 3 above), who was born in 1470 and died in 1498; they had one child. He took Maria, Isabel's sister, who was born in Córdova in 1482 and died in 1517, as his second wife; they had eleven children, one of whom was the future King John III. His third wife was Leonor, niece of his first two wives and eldest daughter of Queen Juana (known as 'the Insane') and King Philip (known as 'the Handsome') of Spain, whom he married in 1518. She was born in 1498 and died in 1559; they had two children.

When King Manuel asked for the hand of his second wife, the Catholic Monarchs demanded the expulsion of the Jews from Portugal as one of the conditions of the marriage contract.

When the massacre of the New Christians took place in Lisbon in 1506, King Manuel, who was in Setúbal at the time, immediately published a sentence against the representatives of the Lisbon City Council, decreeing the extinction of the 'House of the Twenty-four' (the City Administrators), the Guilds and the Administration of the Hospitals. Only two years later, bowing to the pleas of his wife, did the King grant his pardon and revoke his sentence.

Let us also take a look at the marriages of Dom Pedro de Sousa, first Count of Prado, whose dates of birth and death are unknown. It is known only that he was still alive in 1569 and was over ninety years of age (see p. 366, Tree III, 29), which allows us to calculate that he died in 1571 at about 92 years of age and was born around the year 1479.

He was the son of Rui de Sousa (first Lord of Sagres and Beringel, Royal Bailiff of King John and later the Steward of his Household, knight in the wars in Africa, Portuguese Ambassador in Castile and England) and his second wife, Branca de Vilhena. Dom Pedro was the second Lord of Sagres, Beringel and Prado, Governor of Beja and Alcácer do Sal, Captain of Azamor and Alcazar-Ceguer (North Africa).

He had three wives:

1st   *Mécia Henriques* (daughter of Fernão da Sylveira and Isabel Henriques, and grand-daughter of Fernando Henriques (pp. 366 and 447, see Trees III and XIII, 33), who died about 1499. It is thought that they had married the previous year, when Dom Pedro was nineteen years of age.

2nd *Margarida de Brito* (daughter of Estêvão de Brito and Joana Coutinho), who was murdered by her husband in the same year they were married (1500) 'without reason'.

3rd *Joana de Mello* (daughter of João Afonso de Aguiar and Isabel de Mello – see p. 367, Tree IV, 20), who was two or three years older than her husband and must have been about twenty-four when she married in 1501.

Let us, at the same time, look at the descendancy of *João Afonso de Aguiar*, married to Isabel de Mello and brother-in-law of Isabel da Câmara:

1. *Anonymous*, as his name was erased from the genealogies, but thought to be Bartolomeu de Aguiar/Colón. As he was five or six years younger than Salvador Fernandes Zarco (his pseudo-brother 'Cristóbal Colón'), he must have been born in 1454.
2. *Anonymous*, as his name was erased from the genealogies, but it is assumed that he was Diogo de Aguiar/'Colón'. Being five or six years younger than his brother, Bartolomeu, he must have been born in 1460.
3. *Joana de Mello*, married to Dom Pedro de Sousa, first Count of Prado. As she was two or three years older than her husband, she must have been born in 1476.
4. *Leonor de Mello*, born around 1477. Married Gomes de Figueiredo about 1496, at the age of nineteen.

The following list of dates and births can thus be elaborated:

Children of Prince Fernando, Duke of Viseu and Beja:	Year
Salvador Fernandes Zarco/Colón (bastard), born in Cuba or Vila Ruiva	1448
Dom João (legitimate), born in Beja	1449(?)
Dom Diogo (legitimate), born in Beja	1457
Children of João Afonso Aguiar, Purveyor of Évora:	
Bartolomeu de Aguiar/Colón, born in Vila Ruiva	1454
Diogo de Aguiar/Colón, born in Vila Ruiva	1460
Joana de Mello, born in Évora	1476
Leonor de Mello, born in Évora	1477

It can be seen that Salvador Fernandes Zarco/'Cristóbal Colón' came into the world probably a year before his half-brother Dom João. As

Prince Fernando married in 1447, it can be admitted that his wife Beatriz insisted that he break off his affair with Isabel, when the former was with child. It is not known for how long Isabel remained in Cuba or, more probably, in Vila Ruiva (*Terra rubra*), perhaps until her son could be weaned or until he was five years old, the time that João de Aguiar's first son Bartolomeu, was born. She may have left the boy in the care of her friends and neighbours or hosts (future brothers-in-law) and departed for Madeira, to where Diogo Afonso de Aguiar followed her and married her. On the other hand, they may have eloped.

Salvador Fernandes/Colón-Zarco, who began his naval apprenticeship at the age of fourteen, may have gone to Madeira some years later to improve his skill by practice in new waters, as was usual among the navigators.

This sub-thesis allows us to adjust all these dates, which is impossible with the Genoese and Catalan theses. It also adjusts all of Columbus's biographical phenomena, not through 'the good offices of the Holy Ghost', but because Columbus conceived a cabalistic siglum that was free of falsehood.

Using Colón-Zarco's autobiographical details as a base, plus some elementary arithmetic, it was not very hard to calculate his year of birth. It should be just as easy to do the same concerning Dom Fernando Colón's mother, Beatriz Henriquez, since it is known that he was born in 1488 and that she was about twenty-two at the time. Don José de La Torre, quoted by Morison, fixed the date as 1465. Beatriz, however, was 'about 22' when she met Colón-Zarco (i.e. in 1486). This means, therefore that she was born in 1464 and not in 1465.

As Colón-Zarco sailed with Bartolomeu Dias, it is possible to calculate the probable dates of the voyages from the following official documents:

1. A decree of 21 June 1478, in which the Crown Prince (the future John II), who since 1474 had been in charge of affairs concerning the Discoveries, awarded Bartolomeu Dias a fifth of all the plunder he took for payment in lieu of 12 000 reals that he had paid to ransom a slave of the Prince that was imprisoned in Genoa.
2. In December 1481, a fleet of eight ships sailed for Mina carrying material for the construction of St George's Castle. According to the chronicler João de Barros, one of the captains of the fleet was Bartolomeu Dias.

At the end of his letter to the Council of Castile, on his return from his third voyage in 1500, Colón-Zarco wrote: 'I came from afar to serve these Princes (the Catholic Monarchs) and left a wife and children that I have never seen again'. From this text, it can be concluded that Filipa Moniz Perestrelo had given birth to at least one other child besides Diogo and that he had died in the same year as their mother, possibly from the plague, after Colón-Zarco had left for Castile in 1484. It is possible that Colón-Zarco had departed alone, leaving his sister-in-law Violante, Moliarte's wife, to take Diogo to Huelva.

I shall now compare the Genoese chronology with that which I have tried to establish, so that fundamental discrepancies may be pointed out.

	'Colombo'	Zarco
Birth of Cristóbal Colón	1451	1448
Birth of Bartolomeu Colón	1456	1454
Birth of Diogo Colón	(?)	1460
Columbus's studies in Lisbon and at sea	–	1462–7
Columbus's first stay in Madeira	1477	1468 or 9
Voyages to Africa, the Azores and the Sargasso Sea	(?)	1469–73
Columbus's stay in Lisbon	(?)	1474–5
Voyage to Greenland	(?)	1476–7
Voyage to the Mediterranean	(?)	1478
Voyage to Madeira (sugar trade)	1478	–
Marriage to Filipa	1478	1478
Birth of Columbus's first son Diogo	1479	1479
Residence on Porto Santo (in the house of João Esmeraldo)	1479	1479/80
Voyage to Guinea and Mina	(?)	1481
Voyages to the West	(?)	1481/82
Trip to Madeira to fetch Filipa	1482	1482
Return to Lisbon with wife and son	1483	1482
Death of Columbus's wife Filipa	1484	1484
Proposal to King John II	1484	1484
Columbus's departure to Castile	1485	1484
Moliarte's departure for Huelva	(?)	1484
Residence in the house of the Medinacelis	1485/86	1484/86
Settled in Córdoba	1486	1486
Stay in Lisbon with King John II	1488	1488
Birth of Dom Fernando de Colón	1488	1488
Death of Cristobal Colón	1506	1506
Death of Bartolomeu Colón	1515	1515
Death of Beatriz Henriques	1525	1525
Death of Columbus's son Dom Diogo de Colón	1526	1526
Death of Columbus's son Dom Fernando de Colón	1539	1539

Columbus did not include his year of birth in his siglum as he obviously thought it unnecessary. But his autobiographical notes are more than enough for a historian to come to a conclusion. If it were not for the fact that it denounced a fraud, the difference of three years in Columbus's age would be insignificant. By ignoring the autobiographical notes that prove when Columbus was born, the 'Genoists' have clearly demonstrated that 1448 is not suitable for their theories. Despite the thousands of 'Colombos' with which Italy was swarming in the fifteenth century, they managed to find only two brothers by the name of Bartolomeu and Cristóforo (and created a Giacomo/Diego) – the wool-carders of Genoa – whose ages and knowledge did not even coincide with those of the navigators and cartographers that appeared in Spain. But it was those two – like Vespucci – that have become fossilised in History.

On delving into the Aguiars' genealogy, it can be seen that a cousin of Joana de Mello (daughter of the Purveyor of Évora, João Afonso de Aguiar, and wife of Dom Pedro de Sousa, first Count of Prado), called Damião de Aguiar, was related to the Vasconcelos through his marriage to Francisca de Mendoça e Vasconcelos. And as (see p. 493, Tree XVIII, 58) one Martin Mendes de Vasconcelos married Elena Gonçalves da Câmara, daughter of João Gonçalves Zarco, I became interested in his respective family ties.

According to the genealogist Count Dom Pedro, the Vasconcelos were a very old family that appeared in genealogies from the time of King Alphonse II (1211–23). In the fourteenth century, Joane Mendes de Vasconcelos married Aldora Afonso Barreto, daughter of Dom Vasco Afonso Alcoforado and Beatriz Barreto. An issue of this marriage, Aldonça de Vasconcelos, became the wife of Dom Afonso Telles de Menezes and gave birth to Queen Leonor Telles, wife of King Ferdinand.

A cousin of that Joane Mendes de Vasconcelos, Martim Mendes de Vasconcelos, Lord of Alvarenga, was the father of his namesake who Prince Henry the Navigator 'sent to the Island of Madeira to wed Elena Gonçalves da Câmara, Colón's aunt (see p. 445, Tree XII, 57).

As can be seen, the Prince had the power to order the marriage of whomsoever he wished, even in Madeira, and was not the slightest bit interested in whether the couple would be happy together. It is not surprising, therefore, that King John II decided that Filipa Moniz Perestrelo should leave Porto Santo in order to marry Salvador Fernandes Zarco (see p. 345), who had to be seen as a respectable citizen and not an out-and-out adventurer, in the same way that he sent his German agent and alchemist, Moliarte, to the same island to marry Filipa's sister Violante. They were pieces in his game of chess, and each one had to be moved according to the state of the game. But let us return to the Vasconcelos.

The marriage of Martim Mendes to Elena da Câmara started the Vasconcelos family 'of the Islands', which became scattered over the Archipelagos of Madeira and the Azores and related to the Monizes and the Roiz de Sás. The branch that were the 'Commanders of Seixo' intermarried with the Menezes, Pereiras and Castros, all families of Beja. The Vasconcelos, 'Lords of Esporão', became related to the Sylveiras of Alvito and the 'Squires of Vidigueira', to the Vaz Pintos, the Lords of Ferreiros and Tendais, to the Freire de Andrades of Évora and to the Britos of Elvas, thus continuing the union of families from the islands with those of Alentejo and producing some of the most outstanding families of the Portuguese nobility (see Tree XVIII, p. 493).

The names of these families have been repeatedly mentioned in this book and partial genealogical trees have been presented. It was while elaborating the relationship between the Aguiars and the Vasconcelos that I came across an interesting opportunity to identify one of the people that is referred to in Columbus's *Diario de a bordo* and was still enveloped in mystery – Álvaro Damán.

THE PROBABLE IDENTITY OF ÁLVARO DAMÁN

Diogo Afonso de Aguiar, who married Isabel da Câmara (Colón-Zarco's mother), has his first name stapled over an erasure in which the name 'Damião' can still be read (see note on p. 376, facsimile of Morais's *Pedatura* and Figure 28.1). As already mentioned, Isabel's surname was also erased and replaced by the name 'Câmara'. I should like to refer to two of this couple's children: Inês Dias da Câmara (married to Lopo Vaz de Camões, cousin of the poet Luis de Camões – see p. 362, Tree I, 22) and Rui Dias de Aguiar, half-sister and half-brother of Colón-Zarco.

The name Rui Dias de Aguiar aroused my interest owing to a possible relationship to Rui Dias de Gois. It happens that this second Rui Dias was the grandfather of the famous Portuguese writer Damião de Gois, who deserves a special mention. Born in Alenquer in 1502, he was appointed Lord of the Bed-chamber of King Manuel I (Colón-Zarco's half-brother) at the age of eleven. He journeyed around Europe in 1523 and fifteen years later married a Flemish noblewoman, Joan of Hargen, daughter of Nicholas of Tillimburg, Lord of that town. He settled in Louvain in Belgium, but in 1544 was called to Lisbon to take up the post of Head of the Archives of the Realm (Torre de Tombo) and to write the Chronicles of Kings Manuel I and John III. In his *Biographic Dictionary of Portuguese Musicians*[15] Ernesto Vieira includes the name of Damião de Góis as a composer and an organ player.

'BROTHERS AND COUSINS' 493

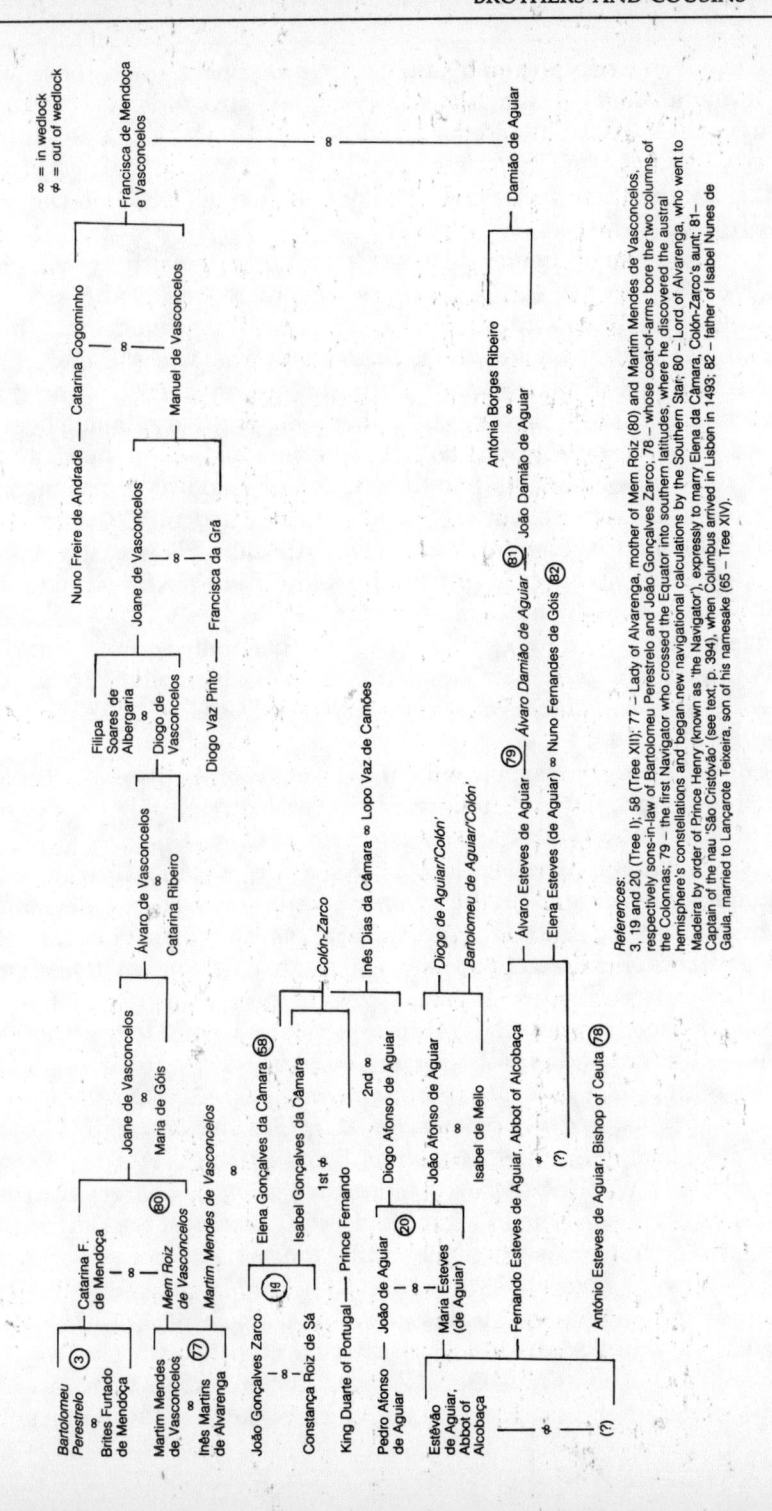

∞ = in wedlock
⌀ = out of wedlock

```
Bartolomeu
Perestrelo ─③
    ∞
Brites Furtado ── Catarina F.
de Mendoça        de Mendoça
                       ∞
Martim Mendes ── Joane de Vasconcelos
de Vasconcelos     Maria de Góis
                                            Nuno Freire de Andrade ── Catarina Cogominho
Mem Roiz ─⑧⓪                                              ∞
                                  Filipa                              Francisca de Mendoça
Inês Martins      Martim Mendes de Vasconcelos    Soares de      Joane de Vasconcelos    e Vasconcelos
de Alvarenga⑦⑦                                 Albergaria          ∞
                    ∞                              ∞              Manuel de Vasconcelos
                                                Diogo de
João Gonçalves Zarco  ─⑱  Elena Gonçalves da Câmara ─⑧⑤    Vasconcelos
                           Isabel Gonçalves da Câmara     Diogo Vaz Pinto ── Francisca da Grã
Constança Roiz de Sá           1st ⌀
                           ∞
King Duarte of Portugal ── Prince Fernando
                           2nd ∞
Pedro Afonso ── João de Aguiar ── Diogo Afonso de Aguiar
de Aguiar                                                                         Damião de Aguiar
              ⓶⓪         João Afonso de Aguiar                                    ∞
Estêvão         Maria Esteves    Isabel de Mello                          Antónia Borges Ribeiro
de Aguiar,      (de Aguiar)                                                       ∞
Abbot of           ∞   (?)                                         Álvaro Damião de Aguiar ─⑧①
Alcobaça                                                                       João Damião de Aguiar⑧②
              ∞                Fernando Esteves de Aguiar, Abbot of Alcobaça
         (?)
              António Esteves de Aguiar, Bishop of Ceuta ─⑦⑧      Álvaro Esteves (de Aguiar) ─⑦⑨
                                                                       ∞
                                                                   Colón-Zarco
                                                               Inês Dias da Câmara ∞ Lopo Vaz de Camões

                                                               Diogo de Aguiar/Colón
                                                               Bartolomeu de Aguiar/Colón
                                                               Elena Esteves (de Aguiar) ∞ Nuno Fernandes de Góis ⑧②
                                                               Álvaro de Vasconcelos
                                                               Catarina Ribeiro
```

References:
3, 19 and 20 (Tree I); 58 (Tree XII); 77 – Lady of Alvarenga, mother of Mem Roiz (80) and Martim Mendes de Vasconcelos, respectively sons-in-law of Bartolomeu Perestrelo and João Gonçalves Zarco; 78 – whose coat-of-arms bore the two columns of the Colonnas; 79 – the first Navigator who crossed the Equator into southern latitudes, where he discovered the austral hemisphere's constellations and began new navigational calculations by the Southern Star; 80 – Lord of Alvarenga, who went to Madeira by order of Prince Henry (known as 'the Navigator'), expressly to marry Elena da Câmara, Colón-Zarco's aunt; 81 – Captain of the nau 'São Cristóvão' (see text, p. 394), when Columbus arrived in Lisbon in 1493; 82 – father of Isabel Nunes de Gaula, married to Lançarote Teixeira, son of his namesake (65 – Tree XIV).

XVIII Relation: Aguiar/Vasconcelos/Câmara/Perestrelo

Damião was one of the many anthroponyms chosen by the Jews and the surname Góis in conjunction with the patronymic Dias was a toponym that was adopted by a certain Diogo Gonçalves, who, at the time of Count Henry, father of the first King of Portugal, married Maria Amião, daughter of Dom Amião, Lord of the town of Góis. Had Damião de Góis Jewish ancestors?

Góis was a great friend of Erasmus of Rotterdam (the celebrated author of *The Elegy to Madness*), considered to be one of the greatest thinkers of his time, with whom he had sung and played organ in a circle of friends in Fribourg, Switzerland. He was also a friend of St Thomas More. He was, however, accused of heresy and denounced as being interested in the doctrines of Luther. One of the arguments he put forward to prove his Catholic faith was that he had given an organ to 'The House of the Holy Ghost' so that the congregation could accompany their holy services with music. But he was arrested in 1571 and sentenced to life imprisonment by the Holy Inquisition. He was allowed to return home on 1 January 1574 and was found dead next morning next to a little fire with which he warmed himself. The 'Holy Office' claimed that death was caused by a stroke, but it was generally accepted that he had been poisoned. His inquisitorial process was described by Henrique Lopes de Mendoça in his *Biographical Study: Damião de Góis and the Inquisition*.

Let us return to the erasure-ridden genealogy of the Aguiars. I shall make three brief comments on Tree XVIII (p. 493).

1. It is not known whether Pedro Afonso de Aguiar, Governor of Montomor-o-Velho, was the brother or cousin of Dom Estêvão de Aguiar, Abbot of Alcobaça (see p. 375), as the genealogies of this clergyman are extremely vague. Although it is known that Dom Estêvão was the father of Maria Esteves de Aguiar (wife of João Afonso), there is no proof that he was the father of Dom Fernando Esteves de Aguiar, his successor at Alcobaça and heir to all his goods – though the fact that Dom Fernando was Dom Estêvão's heir and successor tells us quite a lot. This Dom Fernando was the Royal Almoner in his old age, as well as being a member of the Royal Council of King Alphonse V, Colón-Zarco's uncle. The genealogies are always very sparing in regard to the illicit affairs of the clergy and prelates, often leaving the names of the mother, or mothers, of their children in the void of obscurity.

2. It is not known who the mistress of Abbot Dom Fernando Esteves de Aguiar was, but Andrade tells us that he was the father of Álvaro Esteves (de Aguiar) and Elena Esteves de Aguiar. In turn, Álvaro Esteves (to whom I shall refer in more detail later on) was probably

the father of Álvaro Damião de Aguiar, grandfather of João de Aguiar and great-grandfather of Damião de Aguiar.[16]
3. Damião de Aguiar, maternal grandson of Gonçalo Ribeiro (a veteran who took part in the unsuccessful attack on Maroma in North Africa at a very advanced age), was Lord Counsellor of the Realm, Lord Chancellor, Governor of Cadaval and Commander of the Order of Christ. He married Francisca de Mendoça e Vasconcelos. His paternal grandfather was Álvaro Damião de Aguiar son of the Álvaro Esteves who reached the Equator in 1471, a feat that would oblige navigators to study new methods of orientation, as the Polar Star disappeared at this latitude and they had to calculate latitude through the meridian height of the Sun until they came into view of the Southern Cross.

Álvaro Damião de Aguiar, Caravel captain (p. 493, see Tree XVIII, 81), was probably the person that received Colón-Zarco on behalf of King John II when the former arrived in Lisbon on his return from his first trip to the Antilles. Columbus's scribe, or someone else, spelt his name Álvaro Damán, just as somebody had given the name of 'Castañeda' to Captain Castanheira of the island of Santa Maria in the Azores.

The higher classes were very few in number at that time and everyone knew everyone, especially if they were kinsmen, no matter how distant, if they were members of the Order of Christ or, above all, if they were navigators. The Bartolomeu Dias that welcomed Columbus when he sailed into the River Tagus on his return from the Antilles was certainly not the one that had rounded the Cape of Good Hope. But he was surely a relative of his famous namesake and would have been well acquainted with Columbus, this being the reason that the King sent Dias to greet him. Álvaro Damião was also a suitable choice, as he was not only a cousin of Diogo and Bartolomeu de Aguiar/Colón, but also of Salvador Fernandes Zarco/'Cristóbal Colón' himself. Another convergent point for my sub-thesis – Colón-Zarco spent his whole life among cousins.

Encouraged by this discovery, I decided to delve into the ancestry of Erasmus's condemned friend Damião de Góis, son of Isabel de Alemy and maternal grandson of Álvaro Gomes de Limi (or Alemy), son of Nicholas, Lord of Alemy in the County of Flanders, who came to Portugal in the service of Princess Isabel, Duchess of Burgundy, mother of Emperor Maximilian I of Austria, sister of Prince Fernando, Duke of Beja, and, therefore, Colón-Zarco's aunt.

1. The Teixeiras, as I explained above, were related to the Henriques, Câmaras and Monizes (see Tree XIX, p. 497). Tristão Vaz Teixeira, who 'rediscovered' Madeira with Zarco, had a son and a grandson

with his own name. So I present them by the Roman numbers I, II and III.
2. Tristão I (married to Branca) was the father of Lançarote Teixeira, who married Beatriz de Góis, great-grandmother of Filipa Moniz Perestrelo.
3. Antonio Teixeira, son of Beatriz de Góis and Lançarote, married Isabel de Vasconcelos, and they thus became great-aunt and great-uncle of Filipa (granddaughter of António's sister Joana).
4. Tristão II (Guiomar's husband) was the father of Violante Teixeira, who married in digamy Rodrigo da Câmara (see p. 452, Tree XIV, G), Colón-Zarco's uncle.
5. Elena Esteves de Aguiar (see p. 493, Tree XVIII-82), married to Nuno de Góis, was the daughter of Abbot Fernando Esteves de Aguiar, sister of Álvaro Esteves de Aguiar and great-aunt of Damião de Aguiar (see p. 493, Tree XVIII-79), who took Francisca de Vasconcelos as his wife.
6. Francisca was the cousin of Martim Nunes de Vasconcelos, husband of Elena Gonçalves da Câmara, Colón-Zarco's aunt (see p. 445, Tree XVIII-58).

   Seeing that the Góis were connected to the Teixeiras and the Vasconcelos, all of whom were related to Filipa or Colón-Zarco, I was encouraged to continue my research.
7. Filipa Moniz Perestrelo was a niece of Fernão Vaz Corte-Real (brother and uncle of the explorers of the Canadian coast and true discoverers of North America), because he married Judith de Góis, sister of Joana Teixeira (see p. 497, Tree XIX-83); she was also a relative of Damião de Góis, through their respective great-grandmothers, the sisters Catarina and Brites (Beatriz), daughters of Gil Dias de Góis.
8. Rui Dias de Aguiar (half-brother of Colón-Zarco) was a cousin of Lourenço Pantoja (son of Lopo de Góis), who lived in Alvito (see p. 497, Tree XIX), and was the father of Luis Pantoja de Góis, who married Antónia Mendes. This lady was the daughter of Fernão de Magalhães Teixeira, a relative of the Portugese navigator Fernão de Magalhães (Ferdinand Magellan) who discovered a passage from the Atlantic to the Pacific round the southern tip of the American continent while leading the expedition that made the first circumnavigation of the world. He did not live to complete the voyage, being killed in the Philippines.
9. If – as I presume, although I have no concrete documentary proof – Elena Esteves de Aguiar was the grand-daughter of Dom Esteves de Aguiar, Abbot of Alcobaça, then Damião de Góis was also a kinsman of Diogo and Bartolomeu de Aguiar/'Colón'.

Like two genealogists before me,[17] I have my doubts regarding the origin of Abbot Dom Estêvão de Aguiar, who was the father of Maria

'BROTHERS AND COUSINS' 497

∞ = in wedlock
φ = out of wedlock

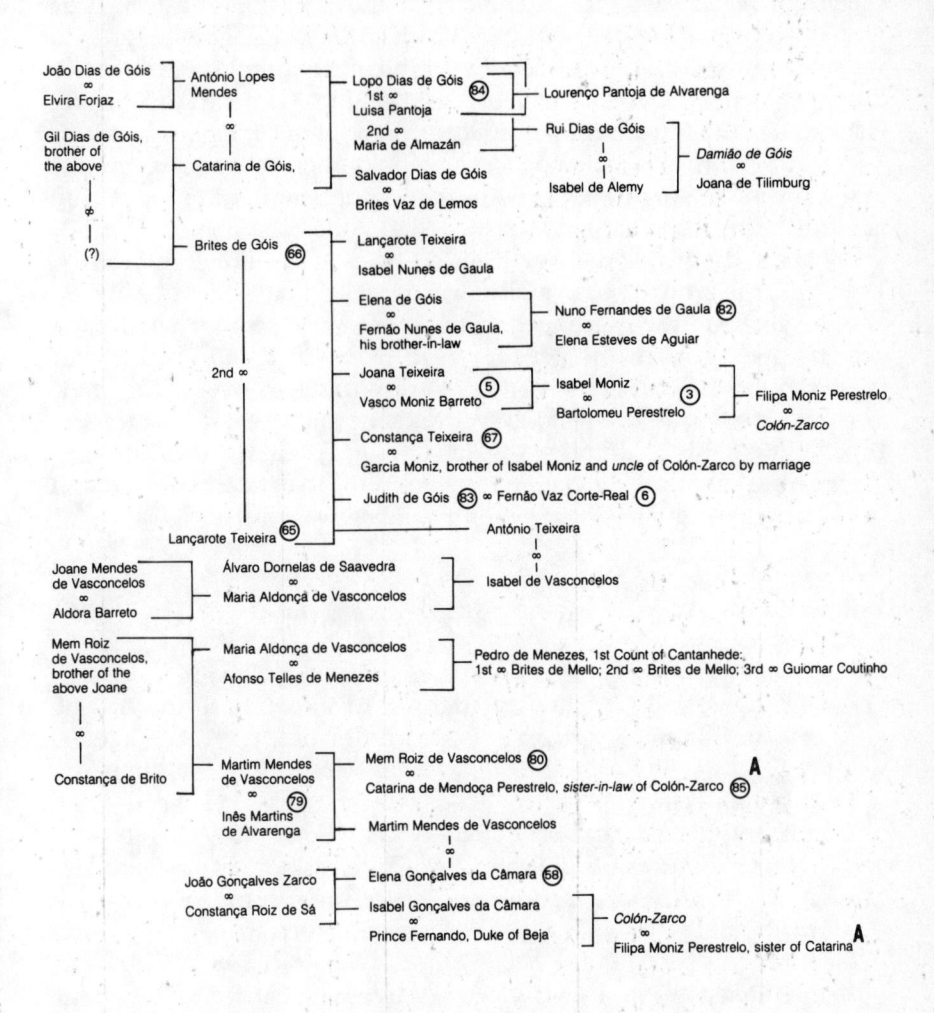

```
João Dias de Góis ──┐   ┌─ António Lopes ──┐   ┌─ Lopo Dias de Góis ─(84)─┐
        ∞           │   │   Mendes         │   │  1st ∞                    ├─ Lourenço Pantoja de Alvarenga
Elvira Forjaz ──────┤   │                  │   │  Luisa Pantoja           ─┘
                    ├───┤       ∞          ├───┤
Gil Dias de Góis,   │   │                  │   │  2nd ∞                   ─┐   ┌─ Rui Dias de Góis
brother of          │   │                  │   │  Maria de Almazán         ├──┤          ∞           ┌─ Damião de Góis
the above ──────────┘   └─ Catarina de Góis,│   │                          ─┘   │                    │        ∞
                                            │   └─ Salvador Dias de Góis           └─ Isabel de Alemy   └─ Joana de Tilimburg
   ∞                                        │        ∞
   │                                        │      Brites Vaz de Lemos
  (?)                                       │
                                            │   ┌─ Lançarote Teixeira
                        Brites de Góis (66)─┤   │        ∞
                                            │   │  Isabel Nunes de Gaula
                                            │   │
                                            │   ├─ Elena de Góis           ─┐   ┌─ Nuno Fernandes de Gaula (82)
                                            │   │                           ├──┤          ∞
                              2nd ∞         │   │  Fernão Nunes de Gaula,  ─┘   └─ Elena Esteves de Aguiar
                                            │   │  his brother-in-law
                                            │   │
                                            │   ├─ Joana Teixeira          ─┐   ┌─ Isabel Moniz            ┌─ Filipa Moniz Perestrelo
                                            │   │        ∞                  ├─(5)─┤       ∞         ─(3)──┤          ∞
                                            │   │  Vasco Moniz Barreto     ─┘   └─ Bartolomeu Perestrelo  └─ Colón-Zarco
                                            │   │
                                            │   ├─ Constança Teixeira (67)
                                            │   │        ∞
                                            │   │  Garcia Moniz, brother of Isabel Moniz and uncle of Colón-Zarco by marriage
                                            │   │
                                            │   └─ Judith de Góis (83) ∞ Fernão Vaz Corte-Real (6)
                                            │
                   Lançarote Teixeira (85)──┤       ┌─ António Teixeira
                                            │       │        ∞
Joane Mendes        ┌─ Álvaro Dornelas de Saavedra ─┤  Isabel de Vasconcelos
de Vasconcelos      │          ∞
     ∞         ─────┤   Maria Aldonça de Vasconcelos
Aldora Barreto      │
                    │
Mem Roiz            │   Maria Aldonça de Vasconcelos ─┐   ┌─ Pedro de Menezes, 1st Count of Cantanhede:
de Vasconcelos,     │            ∞                    ├──┤  1st ∞ Brites de Mello; 2nd ∞ Brites de Mello; 3rd ∞ Guiomar Coutinho
brother of the      │   Afonso Telles de Menezes     ─┘
above Joane  ───────┤
     ∞              │                                              A
     │              │   Martim Mendes ─┐   ┌─ Mem Roiz de Vasconcelos (80)
Constança de Brito  │   de Vasconcelos │   │          ∞
                    │        ∞         ├──┤  Catarina de Mendoça Perestrelo, sister-in-law of Colón-Zarco (85)
                    │   Inês Martins   │   │
                    │   de Alvarenga(79)    └─ Martim Mendes de Vasconcelos
                    │
                    │                    ┌─ Elena Gonçalves da Câmara (58)
João Gonçalves Zarco│                    │
        ∞           ├── Câmara ──────────┤  Isabel Gonçalves da Câmara  ┐   ┌─ Colón-Zarco
Constança Roiz de Sá                     │            ∞                 ├──┤          ∞                                A
                                         └─ Prince Fernando, Duke of Beja    └─ Filipa Moniz Perestrelo, sister of Catarina
```

References:
3 (Tree I and V); 5 and 6 (Tree I); 58 (Trees XII and XVIII); 65, 66 and 67 (Tree XIV); 82 (Tree XVIII); landlord in Alvito and Beja, later Purveyor of the Town of Oporto; 83 – Judith is a name clearly Jewish; her husband (6) was son of João Vaz and brother of Miguel and Gaspar Corte-Real, explorers of the Western seas (Newfoundland); 85 – daughter of Bartolomeu Perestrelo and his first wife, Brites (or Catarina) Furtado de Mendoça; she was only half-sister of Filipa, Columbus's wife.

XIX Relation: Góis/Teixeira/Moniz/Câmara

Esteves (married to her cousin João Afonso de Aguiar), Dom Fernando Esteves de Aguiar, his successor and heir, and Dom António Esteves de Aguiar, Bishop of Ceuta, who, as far as is known, left no family.

The two investigators wondered why the arms of the Abbot bore two columns, similar to the shield of Dom Agapito Colonna, Bishop of Lisbon, and the arms of the navigator Diogo Cão (see Figure 34.39). Is this another unusual coincidence or would the Abbot have by any chance been the son (or nephew) of the Bishop and brother (or cousin) of Otto Colonna (Pope Martin V)?

Abbots and Bishops did not grow on trees in the fifteenth century. They had their own class, which was closely linked to the nobility. Those Aguiars of the Alcobaça branch do not appear in the genealogies very frequently, even though they held high posts and were closely related to royalty and nobility. It is not really known where they came from or if there was any real motive for them to conceal their ancestry. Together with this assumption – which does not claim to prove anything – appears a 'shadow' of the chance that his descendants Diogo and Bartolomeu de Aguiar/'Colón' could have been Colonnas.

THE COLONNAS

The sixteenth-century poet João Rodrigues de Sá,[18] Governor of Gaia, Lord of Sever and Matosinhos, dedicated two stanzas to the coat-of-arms of João Gonçalves Zarco:

> Two wolves wish to scale a keep tower on a field the colour of an orchard which are the arms of a lineage very worthy of the honour.

> Câmara is his surname, well-known in Portugal and on the Island of Madeira. He was the first one ever to have received them.

As mentioned above, João Gonçalves Zarco was married to Constança Rodrigues (or Roiz) de Sá, daughter of Rodrigo Anes de Sá and Cecilia Colonna; and João Rodrigues de Sá was the great-great-great-grandson of the same Rodrigo Anes de Sá, Lord of Matosinhos.

It is known that João Gonçalves Zarco, probably the son of Gonçalo Esteves (of the Arco), had lived, or had been born, in the manor of Matosinhos, the residence of the Coutinhos, Lords of the Leomil Estate. Gonçalo Esteves married Brites, daughter of João Afonso de Sá, Comptroller of the Household of King John I and, despite his advanced age, he distinguished himself in the capture of Ceuta. While still a squire, João Gonçalves Zarco had fought alongside Prince Henry 'the Navigator' and Dom Vasco Fernandes Coutinho in the conquest of Ceuta.

João Gonçalves Zarco was undoubtedly of good stock. His father-in-law, Rodrigo Anes de Sá, was the grandson of the Berredas and the Lords of the 'Sá Estate' on the outskirts of Guimarães at the time of King Alphonse IV and Peter I. It is thought that João Afonso de Sá moved to Matosinhos at the time of King Ferdinand and later took the part of future King John I in the struggles against Leonor de Telles (King Ferdinand's widow) and Castile.

João Rodrigues de Sá wrote other verses, in which he claimed he was a descendant of the Colonnas. And Francisco Sá de Miranda wrote a poem to his cousin João Rodrigues de Sá e Menezes and refers to the beauty of a cousin of theirs, Vittoria Colonna[19], and claimed that they were all Colonnas.[20]

Some genealogists have maintained that Cecília Colonna was the daughter of one Jacopo Sciarra (Colonna), so I decided to see who were the Colonnas of the time. Giácome Colonna, Cardinal of Pope Nicholas III (1278), had six nephews and nieces, three of whom were sons of his brother Pietro Colonna:

1. Stephan (*Estêvão*) Colonna, who was killed together with his son in Rienzi when he tried to take Rome during the Papal struggles.
2. Giácome Colonna, Bishop of Lombez, who died in 1341.
3. Sciarra Colonna, who collaborated with King Philip (known as 'the Bean') against Pope Boniface VIII and died in 1329. He was the father of Giácome (Jacopo) Sciarra and grandfather of Agapito Colonna, Bishop of Ascoli and Brescia. Agapito was in Portugal in 1369–70 to arrange the peace between King Ferdinand and King Henry IV of Castile. He was Bishop of Lisbon from 1370 to 1380, but lived in the city only until 1377. As mentioned above, he had already fathered a son, Otto (or Eudes) Colonna, who was crowned as Pope Martin V in Constance in 1417. He also had another son, Pietro, in Italy.

Did he not leave Portuguese descendants? From where had the coat-of-arms with two columns which the navigator Diogo Cão and the Bishop of Ceuta, António Esteves, could exhibit come from? Were these descendants of a Bishop's brother? I can see no reason for the fact that his illegitimate descendants were not registered in the genealogy of Portuguese families, unless the mother – or mothers – of his sons was (or were) Jewesses. Affairs of this type were as common in the genealogy of the following century, when the Inquisition made its presence felt.

I have emphasised the name 'Estêvão ('Stephano' in Italian) as it was very common among the Colonnas. The Bishop of Ceuta was António Esteves de Aguiar. 'Esteves' is a patronymic, meaning 'son of Estêvão', and his presumed father, the Abbot of Alcobaça, was Estêvão de Aguiar.

We also know that the Bishop of Ceuta had a coat-of-arms with the Colonna's two columns. And were the brothers Bartolomeu and Diogo Aguiar descendants or relatives of Dom Agapito?

Other genealogists say that Cecília Colonna was the great-granddaughter of Sciarra Colonna, grandfather of Agapito. Although this is a possibility, the information above lends more weight to the fact that he was her uncle and Jacopo Sciarra's brother.

The only thing that is known is that Cecília Colonna married Rodrigo Anes de Sá, to whom King Ferdinand gave the Castle of Gaia, near Matosinhos, in 1367. It is said that he went to Rome in 1371 to present the King's compliments to Pope Gregory IX,[21] who had been elected the year before. It is thus that someone insinuated that it was there that Rodrigo de Sá had met and married Cecília. Everything points to the fact that this embassy is a pure figment of the imagination, as the Pope was still in Avignon at this time; there were still six years of Papal struggles before he would leave France and resettle the seat of the Church in Rome.

It can be concluded, therefore, that Rodrigo de Sá did not bring his bride from Rome and that he married her in Portugal, probably in 1367 or 1377. There is no doubt that Cecília's daughter, Constança Rodrigues (or Roiz) de Sá, wife of João Gonçalves Zarco, was a Colonna. This couple's daughter, Isabel Gonçalves da Câmara was a Colonna. Which means that Salvador Fernandes Zarco/'Cristóbal Colón' was also a Colonna. The duplication of surnames once again comes to the fore. According to some authors, Constança de Sá also used the surname 'Almeida', which means that Cecília may have been borne by a lady of that family.

It is curious that the name of the city where Otto Colonna, the defender of Portugal's interests, was crowned as Pope Martin V, is Constance, like the name of his great-niece or cousin Constança de Sá, João Gonçalves Zarco's wife. And new elements come to light! In deciphering Colón-Zarco's siglum, only one word seemed unsuitable, although admissible – S (C) IA.[22]

On identifying his mother, Salvador Fernandes Zarco wrote:

ISA S (C) IA   CAM

A Latin dictionary contains only common etymons and a very scarce number of historical illustrious Roman personages, such as Caesar, Cicero, etc. With the elements that I then possessed and using solely the initials of the ternary, I had deciphered:

ISAbella ScIA CAMara

Isabel known as Câmara

I naturally interpreted this form of identification as a consequence of the reluctance of João Gonçalves Zarco's descendants to use this New Christian surname. Yet I was not entirely satisfied. The sixteenth-century genealogists had deduced that the daughter of the discoverer of Madeira would use the name Gonçalves, but Colón-Zarco indicates another surname in his siglum which confirmed that he was a Colonna:

ISABELLA SCIARRA CAMARA

Isabel Sciarra Câmara

It is convenient to recall that in the first chapter of his *Historia* Fernando Colón clearly stated: 'My father shortened his name'; and added 'And those that know most about his origins say he came from Placenza, where there are some honoured people of his family buried in tombs that bear arms and epitaphs of the Colombos because, in fact, that was the surname of his forebears'. Colón-Zarco's cabalistic message confirms the hyperbolical information of his son. The expression 'those that know most' cancels out all the previous hypotheses and Fernando Colón, on referring to the tombs with arms and epitaphs of his '"forebears"', perceives the invalidity of those who invented the story that Colón was of '*humilibus parentis*' in Genoa.

The substitution of the surname 'Colonna' by 'Colombo' is evident, as it otherwise makes Fernando Colón seem to be an imbecile. A decayed fifteenth-century nobleman could merge into anonymity or slide into crime, but would never stoop to a humble workshop job.

Fernando Colón said about his father: 'it is obvious that he never exercised a mechanical or manual art', and added 'he spent all his childhood and youth in navigation and cosmography and his adulthood in discovering lands'.

The notable Colonnas, who had family branches in Placence, Florence and Rome, counted *condottieri*, prelates and admirals among their members. Being ruined and some of them being trained for war, their metamorphosis into mercenary soldiers, brigands or corsairs, as suggested by the Catalan thesis, is admissible. But chained to the fiction of the wool-carder, historians have limited themselves to reciting the arguments of the Genoese and have not explored the possibility that Columbus was one of the Colonnas. Italians can now be proud – and rightly so – that Columbus was descended from that noble Italic family. Columbus transformed Colonna ('column' in Latin) into Colom in Portuguese and Colón in Castilian. He chose for his siglum the Greek '*colon*' – 'member', 'ark', the grammatical 'colon' – which in Hebrew was *zarga*.

If Abbot Dom Estêvão de Aguiar was, as I presume, grandson or great-grandson of Dom Agapito, Bartolomeu and Diogo Aguiar/'Colón' were kinsmen of Salvador Fernandes Zarco/'Colón'. They were all Colonnas with their name 'shortened'.

'Stephen' in Portugese as we have seen is *Estêvão*, and also 'big rock-rose' (*esteva*), a plant whose wood was used for making arrows. The patronymic of Estêvão or Esteva is Esteves, which later, like other patronymics, became a surname.

It must be remembered that Portuguese Jews adopted the names of trees and plants in order to conceal their Hebrew names. An Esteves family divided into two branches, one of them moving to the north of the country to settle in the area of Bragança and the other moving south to Beja. These were the two regions that were particularly chosen by the Jews when they were forced to adopt Christianity and settle far from the big urban centres.

At the time of the wars between Portugal and Spain (see p. 466), the two branches of the Esteves adopted the surname Xara, without genealogists being able (being unwilling) to explain the reason. The '*esteva*' bush also became known as a '*xara*', which also meant arrow, without the dictionaries mentioning its etymological origin.

With their new surnames, some of the older Esteves from Alentejo (Beja) moved to the Spanish Estremadura, while some from the Trás-os-Montes (Bragança) moved to Catalonia. During the Spanish persecution of the 'Converts' in the fifteenth century, however, many of these Xaras returned to their places of origin. They had already been Christians for many generations.

Dom Juan Garcia de Xara, Commander of the Order of St James of Castile, thought it healthier to settle in Portugal only in 1521. He owned a manor at Vale de Freimixil and had a coat-of-arms made up of a band with serpents biting each end and containing two crowned lions passant, a rock-rose on the sinister field and a castle on the dexter base.

It is important to mention that Dom Juan de Xara was married to Isabel de Torquemada, sister of Friar Thomas de Torquemada, Grand-Inquisitor of Spain and implacable persecutor of the 'Converts', even though (according to Madariaga's investigation) he also came from Jewish stock, as I have already mentioned.

But the most important detail of all is the fact that the surname 'Xara' is a Portuguese corruption of 'Sciarra'. This fact allows me to prove my sub-thesis that the Esteves de Aguiar (p. 493, see Tree XVIII) were Colonnas.

It also allows me to presume that Jacopo (Jacob or Giácome) Sciarra, Cecília Colonna's father, had married a New Christian surnamed Almeida. This would explain why Constança (Cecília's daughter and João Gonçalves Zarco's wife) called herself both Roiz (Rodrigues) de Sá and Roiz de Almeida. It also explains why the genealogists took so much care to conceal her Jewish ancestry, exactly as they had done in relation to her husband, the discoverer of Madeira.

A study of Colón-Zarco's biography shows that the 'discoverer of America' could be of no other nationality than Portuguese. Everything points to the fact that the only lie he ever told was when he claimed that he was going in search of Cypangu. This stratagem was a part of his mission, and only after Vasco da Gama had reached India and Cabral had officially 'discovered' Brazil did he begin to look for terra firma – and to the South.

If 'Cristóbal Colón' were not Salvador Fernandes Zarco in the service of the King of Portugal and if the Portuguese had not already navigated the waters of the Antilles before him (or even with him), a much larger book than this one, dealing exclusively with the doubts that exist regarding Columbus's exploits, could be written – a succession of inexplicable facts and insoluble enigmas of the life of this invented, non-Portuguese 'Columbus'.

Colón-Zarco was not boasting when he claimed royal descent in his siglum. It is his descent from the royal line and from the Zarco–Câmara family that allows us to fit all the pieces of his puzzle together and thus make up the picture of who he really was. Obviously, in the same way that 'Cristóforo Colombo' from Genoa could have been born a blondish red-head due to a genetic anomaly, he could also have been a double of the princes of the Royal House of Avis. But a genetic anomaly would not have allowed him to acquire the education, culture and scientific knowledge that he possessed.

The critics have always been amazed by 'Colombo's' understanding of theology and that the *Profecias* had sprung from the course of his thoughts and his pen. Being Salvador Zarco, however, the son of Prince Fernando, and having received a painstaking education, this talent and propensity for religion begin to make sense. And even more so when we think that he was descended from the Holy Queen Isabel, the Holy Prince Fernando was his uncle, the Holy Princess Joana was his first cousin and Queen Leonor, who became the wife of King John II at the age of thirteen and who was, according to chroniclers, studious and merciful and could be compared to the Holy Queen Isabel, was his half-sister.

Nobody has managed to prove that Columbus studied at the University of Pavia, nor would it have been chronologically possible for him to have done so. He could have learned Theology, Latin, Greek and Law there, but in no way could he have acquired his knowledge of cosmography or the nautical cryptography of the Order of Christ in Italy of the time, since those sciences were still relatively unknown there. Only the Portuguese could have taught him that. It is a fact that there is no record of a student by the name of 'Cristóforo Colombo' having studied there at that or any other time.

He 'forgot his mother tongue' so quickly that he never expressed himself in Italian. And his poor Castilian had a Portuguese syntax and it was always replete with Portuguese expressions.

I am willing to accept that a defender of a thesis claiming a non-Portuguese Colombus may allege that 'his' 'Cristofóro Colombo' was a red-headed double of the Princes that are portrayed in the polyptych of Nuno Gonçalves and that he had studied letters, theology and sciences through correspondence. But I cannot accept that a non-Portuguese 'Colombo' made every effort to create a cabalistic siglum so complex that it was 500 years before someone deciphered it. He would also find it difficult to accept the fact that his mother – whether Genoese, Catalan or Mayorcan – had become the mistress of a Colonna so that her children could adopt that name. Everything, including concessions, has a limit.

GAMBIT

The thesis of this book is that Salvador Fernandes Zarco/'Cristóbal Colón' was a secret agent of King John II, with the mission of forcing the Catholic Monarchs to sign the Treaty of Tordesillas.

The siglum of the Temurah of the Cabala dealing with the Admiral of the Indies and the Ocean Sea says that he was the bastard son of Prince Fernando, grandson of the navigator João Gonçalves Zarco and a Colonna through his mother's line, and also that he was born in Cuba (near Vila Ruiva) in 1448.

As sub-theses I have presented the strong probabilities of his being the cousin of illustrious figures, as were his protectors in Spain, that his 'brothers' Diogo and Bartolomeu Colón were Aguiars and that Beatriz Henriques was his third-degree cousin.

I have also claimed that the Portuguese had been to the Antilles before Colón made his first trip in the service of the Catholic Monarchs and that they had already discovered Brazil by that time.

I should also like to have documentary evidence that Salvador Fernandes Zarco/'Cristóbal Colón' did not serve King John II in order to escape the latter's vengeance because his half-brother Dom Diogo, Duke of Viseu and Beja, had betrayed the Order of Christ and the Kingdom by making a pact with the Catholic Monarchs for an invasion of Portugal, which would have effectively ended Portuguese domination of the sea routes to the Orient. I should also like to prove that he was chosen for his mission for other reasons than the fact that he belonged to the Order of Christ, of which his great-uncle, Prince Henry 'the Navigator', and his father, Prince Fernando, had been the main driving force after King Dennis.

THE 'PROMISED LAND': AMERICA

Right up to the end of his life, Columbus stubbornly maintained that he had discovered the seaway to the Orient by the West, preferring to be considered as a lunatic in order to safeguard his descendants from the curse that could fall on them because of the way he had deceived the Catholic Monarchs. Yet he gave new and rich lands to Spain that, after the conquest of Granada, would build an empire in America and the Western Pacific.

The main reason, in my opinion, that impelled Salvador Zarco to continue his quest of discovering the Orient by the West was not the miserly ambition of coming into the possession of more and more land. After the signing of the Treaty of Tordesillas and the death of King John II, his mission as a secret agent had finished. He was ill, exhausted and debilitated by fevers, physically senile. But he did not retire in order to enjoy the privileges of a rich, elderly man. His persistence in the search for 'terra firma' came from a sense of duty that went beyond the dictates of blood, nobility or patriotism. His worship of the Holy Ghost led him to try and attain the greatest ecumenical objectives.

Had Colón-Zarco been inspired by St Bernard when he wrote the *Profecias*? In his sermon *'De Natividade'*, the founder of the Order of the Temple, predicted: *'Vide autem, ne forte ipsi sint et tres Magi venientes jam non ab Oriente sed etiam ab Occidente'*: Three wise Kings would come, not from the East but from the West, to spread the faith.

Only King John II and the Catholic Monarchs[23] could be identified as evangelists in Colón-Zarco's time.

Colón-Zarco frequently quoted the prophet Isaiah:[24] *'Quia ecce Jerusalem exultationem et populum ejus guadiam . . . Ecce ego creo coelos novos, et terram novam'*, thus predicting the 'discovery of new lands and new skies for the exultation of Jeruselum and the joy of the people'. He also quoted St John the Evangelist, highly venerated by the Cathars and the Templars and whose Gospel is to be seen in the central panels of Nuno Gonçalves's polyptych (Plate 11) in the hands of the Holy Prince: *'Et vidi coelum novum, et terram novam . . . Et vidi civitatem Jerusalem novam descendentem coelo'*: a new sky, new stars and planets in the firmament of the other hemisphere; a new land of the West; a new Jerusalem descending from Heaven.[25]

It was a true prophecy of the Discoveries and of the Evangelisation of the people found. But it may not have been purely a mission of evangelism for Columbus. In the face of the ever-increasing persecution of the Jews in the kingdoms of the three new 'Wise Kings' of the West, Colón-Zarco wished to fulfil a prophetic dream of finding a 'Promised Land' for those that could escape the fires of the Inquisition – America.

This mission that he took upon himself to carry out, in accordance with his name Salvador (Saviour), which was the name he gave to the first island he discovered, was outside the scope of his royal service of evangelisation and increasing material wealth. His aim was the salvation of the Jews. It was also a Christian and Templar mission.

Just as a chess player sometimes operates a gambit – the sacrifice of one or two pieces in order to reach check and mate – the time has come for me to execute my gambit of the sub-theses, which have been subjectively proved by non-documentary rationale. I sacrifice my 'brothers and cousins' willingly. My decipherment of Colón-Zarco's siglum was check. The identification of Salvador Fernandes as a Colonna who had refined his name was mate – final and irrefutable.

Notes and References

1. See the solar eclipse in Rider Haggard's *King Solomon's Mines*; many other writers have used this idea.
2. Portuguese coin of small value; see Colón, Don Fernando (1985), pp. 112, 114.
3. Colón, Don Fernando (1985), p. 116.
4. Portuguese coin of small value; see Colón, Cristóbal (1985), p. 92.
5. Colón, Don Fernando (1985), pp. 139, 322.
6. Colón, Don Fernando (1985), pp. 165–7.
7. Colón, Cristóbal (1985), pp. 128–30.
8. Colón, Cristóbal (1985), p. 105.
9. Colón, Don Fernando (1985), p. 179.
10. Las Casas (1875), vol. I, Chap. XCI.
11. Colón, Cristóbal (1985), pp. 192–3.
12. Colón, Cristóbal (1985), p. 92.
13. *A Key into the language of America* (London, 1643).
14. 'di colore bianchissimo; alcuni redano put in bianchezza, altri in color flavo' (Silva, Manuel Luciano da, 1971).
15. Vieira, Ernesto (1900).
16. In the seventeenth century, the descendants of Damião de Aguiar, Lord of Povolide, had to prove that they were from the line of Dom Fernando Esteves de Aguiar, Abbot of Alcobaça, through his son João de Aguiar and that this last-named was not the
17. Jose Augusto da Sylveira Mascarenhas and Sebastião Gabriel Côrrea d'Andrade, the author's great-great-grandfather and great-grandfather respectively whose '*Notas Genealógicas*' were very useful.
18. A notable poet and doctor of Law of Coimbra University. He was born in 1495, went to Italy in 1521 and returned in 1523.
19. Vittoria Colonna, Marchioness of Pescara and poetess (known as 'the

Divine') gave him hospitality in her palace. See Morpurgo (1889).
20. Luiz de Mello Vaz de Sampayo denies that João Rodrigues de Sá and Francisco Sá de Miranda were great-great-great-grandchildren of Rodrigo Anes de Sá (who was married four times: to Catarina Anes (or Aires), Mécia Pires, Cecília Colonna and Berenguela Annes. Sampayo claims that they were descendants of Catarina Anes); see Sampayo (1971), n. 319, pp. cxlii-cxliii.
21. See Felgueiras Gayo (1875), Chap. XXVI, p. 117; see also Fortunato de Almeida, *Hierarchia Catholica Medii Aevi*, I, pp. 224, 304, 511 (1910); Esperança (1666); Lopes (1895), Chap. LII, pp. 164ff, Baluze (1927), Vol. II, pp. 770–1.
22. See Triangle V, p. 291–318.
23. Vieira (1854–8) identified the 3 magi as King John II and the Catholic Monarchs.
24. Isaiah, LXV, v. 17.
25. St John, *Apocalypse*, XXI, 1, 2, 24.

Appendix 1 Portuguese Kings of the First Two Dynasties

Alphonsine Dynasty

Alphonse I (1126–85)
Sancho I (1185–1211)
Alphonse II (1211–23)
Sancho II (1223–45)
Alphonse III (1245–79)
Dennis (1279–1325)
Alphonse IV (1325–57)
Peter I (1357–67)
Ferdinand (1367–1383)

Dynasty of Avis

John I (1385–1433)
Duarte (1433–8)
Alphonse V (1438–81)
John II (1481–95)
Manuel I (1495–1521)
John III (1521–57)
Sebastian (1557–78)

Appendix 2   Portuguese Vocabulary Used by Columbus

Portuguese	Castilian	Translation
Abastada	Abastecida	Supplied
Aberto	Abierto	Open
Abra	Puerto, Ensenada	Harbour
Acertamiento	Por casualidad	By chance
Acordo	Acuerdo	Agreement
Asente	Asiente	Assured
Adovar	Reparar	To remedy
Algun	Alguno	Some
Alivianar	Aligerar	To lighten
Almarraxa	Vajia para esencia	Vessel or pot for essence
Ameaçaban	Amenazaban	menaced
Amortecer	Desfallecer	To fail
Amostrar	Muestrar	To show
Anchor, Anchura	Largura de nave	Ship's width
Agujeta, Agulleta	Herrete, Passacordone	Aglet
Argumento	Conjectura	Conjecture
Arrecife	Restinga	Reef
Arriscada	Arriesgada	Daring, risky
Arroás	Delfin, Golfino	Dolphin
Atá (até)	Hasta	Until
Balços (abertas de vento)	Rachas de viento	Gust of Wind
Vaan (vão)	Vayan	(to go)
Ben (bem)	Biene	Well
Bier (vier)	Vinier	(to come)
Bimbre (vimbre, vime)	Mimbre	Osier twig
Boa, Bon	Buena, Bueno	Good
Bonaço	Apacible	Pleasant

Portuguese	Castilian	Translation
Boy (boi)	Buey	Ox
Cãs (cães)	Perros	Dogs
Carantona	Carátula, Máscara	Grimace, mask
Cativo	Enfermo	Sick
Caxina (cassina)	Arbusto purgativo	Laxative plant
Cobro (pôr a)	Poner en lugar seguro	To put in a safe place
Conten	Contiene	To contain
Corpo	Cuerpo	Body
Contraminar	Dehacer intrigas	To intrigue
Constrengese	Constriñese	To constrain
Corredios (cabelos)	Lisos, Lacios	Lank (hairs)
Crimes	Crimenes	Crimes
Cuento (milhar)	Millar	One thousand
De la dar	Dándole	To give him
Debuxar	Dibujar	To draw
Descuidar	Descargar cuidados	To get rid of problems
Defender	Garantizar	To guarantee
Deisara	Dexara	Let
Dese, Dey	Diese, Di	Gave
Despois	Después	After
Dizer	Decir	To say
Doliente	Malsano	Sick
Duveda	Duda	Doubt
E os dese a el dito	Y los diese al dicho	And give them to the same man
Ebangelos	Evangelios	Gospel
Emprestado	Prestado	Lent
Enxerir (inserir)	Injertar	To insert
Escurana	Niebla marina	Fog
Esmorecer	Desfallecer, Desmayar	To faint
Espelunca	Antro, Caverna	Den
Espeto (também por sol a prumo)	Punta de hierro, Assador, Pincho	Skewer/sun at midday
Estante	Residente	Resident
Estebe	Estuve	Was
Facenda (fazenda)	Hacienda	Farm
Falar	Hablar	To speak
Feyto	Hecho	Deed/exploit
Fexes (feixes)	Haces	Shears
Figado	Higado	Liver

Portuguese	Castilian	Translation
For (se), Foy	Fuer (si), Fué	If you go, went
Faleceu	Falleció	Died
Forno	Horno	Oven
Fresco (de)	Recientemente	Recently
Fugir	Huir	To run away
Golpe de suerte	Feliz oportunidad	Lucky occasion
Imprimir	Empremar, Impresionar	To print
Jaz, Jazendo	Yaz, Yaciendo	Lies, lying (down)
Julgar	Juzgar	To judge
Letra	Carta	Letter
Lume	Lumbre	Fire
Móa (mó)	Muela	Millstone
Moeda	Moneda	Money/coin
Moller	Mujer	Woman
Multidume	Muchedumbre	Crowd
Nácaras (conchas de madre-pérola)	Ostras	Oyster – shells
Ningun	Nadie	Nobody
Oubiran	Oyeran	Heard
Pampano	Pez	Pitch
Papahigo (papafigo)	Vela de nave	A kind of a ship's sail
Pardela	Espécie de gaivota	A kind of a seagull
Pareceu	Compareció	Turned up
Pella (amálgama)	Metales fundidos	Amalgam
Péndula	Pluma	Plume
Perigo	Peligro	Danger
Peden, Pediran	Piden, Pidieran	Asked
Por ende	Además	Besides
Practica	Conversación	Conversation
Presente (oferta)	Regalo	Gift
Qualquer	Cualquiera	Any
Refrescar a la religna	Ceñir al viento quando flamean las velas	To let go the sail
Repuxar	Rexazar	To draw back
Ribaldo (velhaco)	Bellaco	Rascal
Sacando	Exceptuando	Excluding
Salva (fazer a)	Provar la comida	To test the food
Salvo (senão)	Sino	Otherwise
Salto (assalto)	Acometida	Assault

Portuguese	Castilian	Translation
Segundo	Según	According to
Sirga	Cabo de remorque desde de la tierra	Tow-rope
Surtir	Fondear la nave	To anchor
Teendo	Teniendo	Having
Temporar	Templar, Considerar	To consider
Topo (do mastro)	Extremo del mastil	Mast top
Treboada	Tronada	Thunderstorm
Trinquete, Triquete	Vela de proa y su palo	Foresail
Trombones	Trebúns, Truenos	Trombones
Virazon	Viento de la mar	Breeze
Volta	Rumbo de navegación	Ship's course
Voltejar	Dar bordadas a la nave	To whirl (sailing)
Voz (a uma)	Unánimemente	Unanimously
Zerta (certa)	Cierta	Certain

Appendix 3 Toponymic Evolution

PIZZIGANO (1367) West of Ireland	BIANCO (1436) South-west of Ireland	PARETO (1455) South-west of Ireland	Royal Charter of 3 November 1460	Royal Charter c. 1471
Mayotas scia Braçir (1) AZORES	Maitas AZORES	Maidas AZORES	AZORES Santa Maria São Miguel Terceira	AZORES Santa Maria São Miguel
Braçir (2) Braçir (3) SO-CALLED FORTUNATE ISLANDS (MADEIRA and CANARIES)	Brazyl MADEIRA E CANARIES	Brazil Jesus Cristo FORTUNATE ISLANDS OF ST BRENDAN (MADEIRA and CANARIES)	São Dinis	Pico
Caprinia Paloma Canária/Forte Ventura Lanceroto (Lanzarote) Inferno (Hell)	Caprara (Caprala) Colombi Bentyras (Fortunate) Corbomarino San Zorzi Conici Porto Santo Deserta	Caprala Collumbi Ventura (Fortunate) Corvi Marini San Zorzo Conici (Conigi) Porto Santo Desertas	São Luis São Tomás Graciosa Santa Iria São Jorge MADEIRA Porto Santo Desertas Madera	Faial Froli (Flores) Graciosa Santa Cruz (Corvo) (*) São Jorge Porto Santo Desertas Lenha (timber)

* Only after 1471.

Source: Admiral Gago Coutinho, *A Náutica dos Descobrimentos*, vol. I, p. 170 (Lisbon, 1951–2); Commander Fontoura da Costa, *Descobrimentos Portugueses no Atlântico e na Costa Africana do Bojador ao Cabo de Santa Catarina*, in *Congresso do Mundo Português – Coleccção Henriquina*, vol. iii (Lisbon, 1958) p. 257.

Appendix 4   The Hebrew Alphabet

א	'Aleph 1	י	Yôdh 10	ק	Qôph 100
ב	Bêth 2	כ	Kaph 20	ר	Rêš 200
ג	Ghîmel 3	ל	Lāmedh 30	ש	Sîn 300
ד	Däleth 4	מ	Mêm 40	ת	Tāw 400
ה	Hē 5	נ	Nûn 50	ך	Kaph final 500
ו	Wāw 6	ס	Sāmekh 60	ם	Mêm final 600
ז	Záyim 7	ע	Ayin 70	ן	Nûn final 700
ח	Hêth 8	פ	Phē 80	ף	Phe final 800
ט	Têth 9	צ	Çādhe 90	ץ	Çādhê final 900
				א	'Aleph final 1000

APPENDIX 4

The remaining numbers are formed by joining together the corresponding figures and bear the name of the phonema resulting from the equivalent consonants. The meaning of each letter/number is explained below.

א	ALEPH	*Mother*; the Unthinkable Unit; the immaterial origin; *Man* as a collective unit, the *Principle*, the birth (nowadays and, after the invention of the microscope, the ovule).
ב	BÊTH	Archtype of all dwellings; the *Mouth*; (in Hebrew) the *House*.
ג	GHÎMEL	Relation between movement and form; the mould for *Matter*; the *Throat*.
ד	DĀLETH	Everything that exists, physically, in nature; all abundant matter, all division; the *Reciprocity*.
ה	HĒ	*Life*, everything that Is; everything related to limpidity; clarity; the *Light*.
ו	WĀW	Everything that reunites, that joins the air; fertilisation; the *Eye*.
ז	ZÁYIN	Every object that has a destiny, a purpose; the *Arrow*; the fertilising principle that procreates (nowadays, the spermatozoid).
ח	HÊTH	The existence; the reservoir of either *Energy* or substance.
ט	TÊTH	The safe place; *Sanctuary*; purpose; the primitive archetype of the female.
י	YÔDH	That which indicates *Power*; what is used to show that power; the existence already revealed in time (contrary to *Aleph* which is timeless); the *Finger*.
כ	KAPH	The concept of taking or containing something; the cup of the *Hand*.
ל	LĀMEDH	Everything that rises, develops, spreads, that becomes simultaneously thicker; the *Arm*, the *Phallus*, the *Member*.
מ	MÊM	*Mother* (as in *Aleph*, material instead of spiritual); everything that is fecund, creator; the *Water*, the *Sea*, the *Earth*.
נ	NÛN	Any created being; the *Son*, what comes from the seed; the *Fruit*, the *Plant*.
ס	SĀMEKH	*Weapons* in a rotary movement, with a specific purpose; the *Staff*.
ע	'AYIN	Everything related to noise, sounds, emptiness; the *Wind*.

516   APPENDIX 4

פ	PĒ	The inner sound; what is in the centre; the *Word*; the *Verb*.
צ	ÇĀDHÊ	An objective; the *Universal Matter*.
ק	QÔPH	Everything which is useful to Man or which protects him; a cutting tool; the *Axe*.
ר	RÊŠ	What has a movement of its own; the whole *Universe*; the *Head*.
ש	SÎN	The enlivening spirit; the understanding strength that follows a creative purpose; the *Fire*.
ת	TĀN	The finality (as the last letter of the Hebraic alphabet); the concept of the perfection of the *Creation*; the symbol of *Man*.

Appendix 5 The Ten Sephiroth

In the Western Cabala, the nature of the emanation of each *sephirah* is explained as follows in a very truncated way:

I   KETHER   (*Crown*) – masculine essence; top of the pillar of Equilibrium; the secret of secrets; the inner and hidden light.

Symbols: the divine spark, the swastika.
Virtue: the fulfilment.
Vice: none.

II   CHOKMAH   (*Wisdom*) – masculine essence; the top of the pillar of Mercy; the enlightening intelligence.

Symbols: the sceptre of Power; the sphere (and its astrolabic representations: cosmographic sphere, geographic globe and its armillary representation); the phallus, the tower, the vaulted ceiling, the compass, the equilateral set-square.
Virtue: devotion.
Vice: none.

III   BINAH   (*Understanding*) – feminine essence; the top of the pillar of Severity; it is a double essence, either Aima (sparkling and fertile mother) or Ama (barren woman or even virgin; the greater mother; the big deep sea; the vision of pain).

Symbols: the sacramental chalice; the Grail; the external vestments that conceal the body.
Virtue: silence.
Vice: greed.

518   APPENDIX 5

IV   CHESED
       or
       GEDULAH   (*Mercy* or *Magnificence*) – masculine essence; centre of the pillar of Mercy; centre of the vision of love.

                           Symbols: the pyramid; the sphere (as in II); the Greek cross (like the one of the Orders of Avis and Christ); the staff; the cable (ship's rope); the fishing net (spread or in a concave frame, a symbol already used in palaeo-Christianity); scuttle for burning coal; cauldron.
                           Virtue: obedience.
                           Vice: fanaticism.

V   GEBURAH   (*Courage, Strength, Austerity*, what creates *Fear*) – masculine essence; centre of the pillar of Severity; justice.

                           Symbols: the pentagon; the five-petalled Rose (the petals may be represented doubled in sepales, thus becoming ten); the artichoke in blossom.
                           Virtue: energy; bravery.
                           Vice: cruelty.

VI   TIPAERETH   (*Beauty*) – masculine essence; centre of the pillar of Equilibrium; the Child; the sacrificed God; the vision of the harmony of things and the mystery of the Crucifixion.

                           Symbols: the Cross of Calvary; the Rose on the Cross; the flat-topped pyramid; the cube.
                           Virtue: Consecration of the 'Great Work'.
                           Vice: pride.

VII   NETZAH   (*Victory*) – feminine essence; the bottom of the pillar of Mercy; occult intelligence; firmness.

                           Symbols: the candelabrum of seven arms; the five-petalled Rose and the artichoke in blossom (as in V); the Pelican feeding its young; the leather strap or belt.
                           Virtue: the absence of selfishness.
                           Vice: impudence; lust.

VIII   HOD   (*Glory*) – hermaphroditic essence; the bottom of the pillar of Severity; perfect intelligence; vision of splendour.

Symbols: the numbers, the versicles.
Virtue: the truth.
Vice: the lie; dishonesty (by contradictory opposition).

IX   YESOD   (the *Foundation*, the *Roots*) – masculine essence; third level of the pillar of Equilibrium; vision of the mechanism of the World.

Symbols: the Moon; perfume; reproductive organs.
Virtue: independence.
Vice: negligence.

X   MALKUTH   (*The Kingdom*) – feminine essence; lowest level of the pillar of Equilibrium; brilliant and excessive intelligence; it is dependent on Binah; threshold of the shadow of death; threshold of the garden of Eden; the lesser mother.

Symbols: the superimposed double cube ('what is above, is below'); the double personality; the Greek cross.
Virtue: discernment.
Vice: greed (as in III); cupidity.

Appendix 6 Latin Words that Correspond to the Arrangement of Letters in the Triangles

TRIANGLE I

1	2	3	4	5	6	7	8	9	10	11
AMI	IMA	MAI	MIA	JAM	MAJ	AMO	OMA	AMU	MAU	MAIA

1 AMI — *Amicire* = to cover; *Amicitia* = friendship; *Amiculum* = robe, cloak; *Amita* = aunt; *Amitini* = cousins; *Amittere* = to lose.

2 IMA — *Imaginare* = to imagine, to dream, to represent; *Imago* = portrait, model, figure, representation; *Imaguncula* = medal, face.

3 MAI — Maia = *Maia*, formerly Amaia (name of a goddess and a place near Oporto in Portugal); *Maius* = May.

4 MIA — *Mia* (barbarism for *Mea*) = my or mine.

5 JAM — *Jam* = later, now; *Jampridem* = a long time ago.

6 MAJ — *Majestas* = majesty, greatness, sovereignty, gravity, nobility.

7 AMO — *Amoenare* = to cheer; *Amolior/Amovere* = to withdraw, to rebut; *Amor* = love, charity, affection.

8 OMA — *Omasum* = paunchy.

9 AMU — *Amuletum* = amulet, charm; *Amusius* = ruler; *Amusium* = level, plumb line, square.

10 MAU — *Maurus* = Moor; *Mauritania* = Mauritania; *Mausuleum* = mausuleum.

11 MAIA — (already explained in MAI).

TRIANGLE II

1	2	3	4	5	6	7	8	9	10	11	12
OLI	LUI	UL(L)I	JUL	IL(L)AU	AULI	LIV	VIL	VOL	VUL	OLO	IL(L)I

1 OLI — *Olisipo* = Lisbon; *Oli* (the same as Illis) = that place; *Olim* = formerly.

2 LUI   *Lui* = to pay a debt, to serve a sentence, to redeem.
3 UL(L)Y   *Ullyssiponensis* = born in Lisbon (Ullysipo).
4 JUL   *Julius* = July/Julius.
5 IL(L)AU   *Illaudatus* = not praised; *Illautus* = dirty.
6 AULI   *Aulicus* = courtier
7 LIV   *Livida* = livid, extremely pale; *Lividinans* = envious.
8 VIL   *Vilipendere* = to despise, vile, vileness.
9 VOL   *Vola* = palm of the hand; *Volare* = to fly; *Volere* = to want; *Volubilis* = inconstant; *Volutare/Volvere* = to stir; *Volutatio* = swaying of a vessel, restlessness, anxiety.
10 VUL   *Vulgare* = to disclose, to spread; *Vulgaris* = ordinary; *Vulnerare* = to wound, to offend; *Vulpes* = fox; *Vulpinus* = cunning, *Vulsus* = torn, ruined, destroyed; *Vultur* = vulture; *Vulgata* = Vulgate.
11 OLO   *Olor* = foul smell, swan.
12 IL(L)I   *Illic* = to or in that place; *Illicere* = to seduce; *Illigare* = to tie to, to entangle.

TRIANGLE III

	1	2	3	4	5	6	7	8
	MAX	CAM	SAM	SMA	MAS	AGM	GAM	MAG

1 MAX   *Maxima* = maxim.
2 CAM   *Camera/Camara* = chamber, vessel, vault, arched roof, place on Madeira Island.
             Cambire = to swap, to exchange, to fight, to leave, to set out. *Camilum* = basket of a bride's trousseau; *Caminus* = furnace, chimney, road, path (vulgar Latin); *Campus* = camp, field; *Camura* = jewel case; *Camus* = hanging rope, ruff.
3 SAM   *Samiata* = clean, polished; *Samuel* = Samuel.
4 MAS   *Mas* = male, masculine, manly; *Mascarpio* = shameless, dishonest.
5 SMA   *Smaragdus* = emerald.
6 AGM   *Agmen* = orderly crowd, squad, marching army.
7 GAM   *Gamma* = land mark.
8 MAG   *Magia* = magic; *Magis* = more; *Magister* = master, president; *Magisterius* = mastership, dignity; *Magnanimus* magnanimous, generous; *Magnarius* = merchant; *Magnificere* = to esteem; *Magnificus* = magnificent; *Magniloquus* = the one who says important things; *Magnus* = great; *Mago* = sea port of Minorca in the Balearic Islands; *Magda* and *Magdalena*.

TRIANGLE IV

	1	2	3	4	5	6	7	8
	ALO	AL(L)U	LAU	ALV	LAV	IL(L)A	ALI	AULA

1 ALO/AL(L)O *Alogia* = madness; *Alloquor* = to speak, to say, to make a speech, to comfort.
2 AL(L)U *Allucere* = to shine; *Allucitare* = to dream; *Alluctare* = to fight; *Alludere* = to joke, to hint; *Alluere* = to wash; *Alluvio* = flood.
3 LAU *Laudare* = to praise; *Laureatus* = anointed with laurel leaves, laureate; *Lausus* = weeping; *Laute* = delicate, splendid.
4 ALV *Alveus* = bed of a river, channel; *Alvus* = womb, excrement.
5 LAV *Lavare* = to wash; *Lavatio* = bath; *Lavernio* = thief.
6 IL(L)A *Illabefactus* = steady, firm; *Illabor* = to slip, to fall, to penetrate, to insinuate oneself; *Illaborare* = to work; *Illac* = over there, by someone's order; *Illacrymabilis* = cruel, callous; *Illaesibilis* = invulnerable; *Illaesus* = safe and sound, complete; *Illaqueatus* = entangled with, bound to, embarrassed with, tied to, *Illatebrare* = to hide; *Illaudibilis* = reproachful.
7 ALI *Ali* = in other way; *Alias* = another time, another place; *Alibis* = what can create; *Alienare* = to alienate, to separate; *Alio* = for another purpose; *Alius* = some one else; *Allicere* = to entice, to attract; *Allidere* = to come across; *Alligare* = to connect, to lure, to attach.
8 AULA *Aula* = palace, room, yard.

TRIANGLE V

1	2	3	4	5	6	7	8	9	10	11
COC	OCC	CIC	SOS	OSS	SUS	SIS	COX	COS	OSC	SOC

12	13	14	15	16	17	18	19	20	21	22
SCO	CUS	SUC	SCU	CIS	SIC	SCI	COG	GIG	AUSC	CAUS

1 COC (words unsuitable for the message)
2 OCC (abbreviations OC(C)A have already been considered in Triangle VI, no. 1); *Occedere* = to arrive, to meet; *Occeptere* = To start all over again; *Occidens* = West; *Occidere* = to kill, to harass, to wound, to beat; *Occidio* = slaughter, butchery;

Occidium = destruction, ruin; *Occillatio* = swinging; *Occillare* = to beat, to break; *Occipiere* = to begin, to start; *Occiput* = scruff; *Occisor* = murderer; *Occisus* = dead, lost, desperate; *Occlamare* = to scream; *Occludere* = to close; *Occubare* = to hide; *Occulcare* = to tread, to run over; *Occulere* = to conceal, to bury, to shut up, to keep secret; *Occumbere* = to die; *Occupare* = to occupy, to warn, to foresee, to seize; *Occurrere* = to come, to hinder, to resist; *Occursatio* = comings and goings of claimants to a noble, doing favours; *Occursator* = inappropriate claimant.

3 CIC   *Cicatricare* = to heal; *Cicura* = tame, appeased.
4 SOS   *Sos* (instead of *Suos*) = his, her, its, their; *Sospos* = safe; *Sospitare* = to be out of danger.
5 OSS   *Ossus* (instead of *Os*) = bone
6 SUS   *Sus* = up and down; *Suscipere* = to receive, to accept, to start, to set about to establish; *Suscire* = to send; *Suscitare* = to resurrect, to raise, to wake up, to stimulate, to restore; *Suspectare/Suspicere* = to suspect, to look up; *Suspendere* = to suspend; *Suspentio* = arched roof, vault; *Suspirare* = to sigh, to exhale, to breathe in, to evaporate, to wish; *Sustendere* = to spread mysteriously, to trap; *Sustendere/Sustingere* = to hold, to make steady, to defend, to feed, to comfort, to tolerate, to be patient, to resist, to delay; *Susurrare* = to blame, to whisper; *Sussurrans* = denouncer; Susana, Susan.
7 SIS   *Sistens* = steady; *Sistere* = to repress, to restrain.
8 COX   (words unsuitable for the message)
9 COS   *Cos* = rock, stone; *Cosmicus* = wordly; *Cosmus* = cosmos (Greek–Latin), perfume, coiffure; *Costae* = ship's ribs, ribs.
10 OSC   *Osca* = raw wool; *Oscedo* = bad habit; *Oscilare* = to oscilate, to swing; *Oscitare* = to be idle; *Osculare* = to kiss.
11 SOC   *Socer* = father-in-law; *Sociare* = to associate, to gather, to join; *Socius* = mate, partner; *Scordia* = madness, cowardice, laziness.
12 SCO   *Scobina* = file; *Scopulus* = target, rock, shallows; *Scopus* = target, purpose, goal; *Scordalus/Scorio* = impudent; *Scoria* = scum; *Scorpio* = scorpion; *Scortator* = licentious.
13 CUS   *Cusio* = to coin; *Cusiodare* = to sharpen; *Cuspis* = sword end, spear end, spit, sting; *Custodire* = to keep, to defend; *Custos* = gaoler, spy, perceptor.
14 SUC   *Succasus* = to fall; *Succedere* = to succeed, to submit; *Succendere* = to light, to inflame; *Succensere* = to grow exasperated; *Succentor* = agitator; *Successus* = to go to and fro, succession; *Succidia* = slaughter; *Succidere* = to saw to cut off close; *Succidus* = dirty; *Sucingere* = to tighten, to involve; *Succla-*

		mare = to scream, to hoot; *Succollare* = to carry on the back; *Succosus* = wealthy; *Succudere* = to forge; *Succumbere* = to fall, to die, to discourage; *Succurrere* = to help; *Succussare* to shake, to swing; *Succutere* = to shake, to affect.
15	SCU	*Sculna* = Judge; *Sculpere* = to carve; *Scurror* = to flatter; *Scutilus* = slender; *Scutum* = shield.
16	CIS	*Cis* = there, over here; *Cisium* = carriage; *Cispellere* = to put away; *Cista* = arch; *Cisterna* = reservoir; *Cistus* sargasso.
17	SIC	*Sic* = in such a way; *Sica* = dagger, sober; *Sicata* = sold out.
18	SCI	*Scia* = known as, said; *Sciens* = expert, skilful; *Scindere* = to tear, to crack, to violate; *Scintilla* = to spark; *Sciola* scholar; *Scistare* = to ask; *Scitum* = statute, arrangement; *Scita* = ordered, intelligent, beautiful.
19	COG	*Cogitare* = to think; *Cognatio* = relatives; *Cognito* = knowledge; *Cognobilis* = easy to know; *Cognomen* = surname; *Cognoscere* = to know; *Cogere* = to congregate.
20	GIG	*Giganteus* = giant; *Gignere* = to beget, to produce.
21	AUSC	(is considered in Triangle VI, no. 7 – AUS).
22	CAUS	(is considered in Triangle VI, no. 4 – CAU).

TRIANGLE VI

1	2	3	4	5	6	7	8	9	10
OC(C)A	COA	ACO	CAU	ACU	SUA	AUS	VAC	CAV	ACI

11	12	13	14	15	16	17	18	19
ISA	ASI	JAC	GAU	AUG	VAG	AGI	CAIA	GAIA

1	OC(C)A	*Occallescere* = to become callous, to become insensitive, to grieve; *Occasio* = Occasion; *Occasus* = fall, death, West, sunset.
2	COA	*Coaccedere* = To gather; *Coacervare* = to pile up; *Coascere* = to sour, to become violent; *Coactare* = to force, to compel, to coerce; *Coadunare* = to join, to amass; *Coaquare* = to be equal to; *Coastimare* = to assess together; *Coactanea/Coeva* = current, contemporary; *Coagimentare* = to put together, to tie, to reinforce; *Coagulum* = encouragement to love reciprocally; *Coalere/Coalescere* = to be brought up together; *Coarctare* to clasp, to hug, to take a short cut; *Coargere* = to scold, to persuade; *Coaspernare* = to despise.
3	ACO	*Acoeneta* = chambermaid, selfish; *Acolatus* = prodigal, licentious; *Aconae* = stones; *Aconoti* = easily; *Acontiae* = stars, bitterness.

4	CAU	*Caudex* = trunk, root, book, silly man; *Causa* = origin, reason, partiality, excuse, illness; *Causificare* = to excuse, to forgive; *Cauteriata* = fire-marked as punishment for a crime; *Cautes* = rock; *Cautio* = precaution, wisdom.
5	ACU	*Acuere* = to sharpen; *Aculus* = point, thorn; *Acumex* = sharpness, cunning; *Acus* = needle, the power to see distinctly, to foresee.
6	SUA	*Suadere* = to persuade, to advise, to warn; *Suaviari* = to kiss lovingly; *Suaviloquus* = to speak gently; *Suavitas* = gentleness, youthful beauty.
7	AUS	*Ausculari* = to kiss; *Ausculare* = to hear, to listen, to obey; *Ausonius* = Italian; *Auspiciabilis* = good omen; *Auspiciare* = to soothsay; *Austerus* = severe, cruel; *Ausum* = boldness, difficult task.
8	VAC	*Vacans/Vacuatus* = needy, empty, exhausted, deprived, unmarried woman, in the feminine; *Vacanter* = in vain; *Vacatio* = interruption; *Vacerrosus* = furious, desperate, mad; *Vacilans* = hesitant, some one who is about to fall; *Vacuus* = empty.
9	CAV	*Cavamen* = hole, pit; *Cavare* = to dig; *Cavefacere/Cavere* = to avoid, to defend against, to be cautious; *Caverna* = den; *Cavillare* = to mock.
10	AC(C)/ACI	*Accidens* = accident; *Accidentia* = event; *Accidere* = to cut, to mitigate, to ruin; *Accire* = to call; *Acingere* = to fasten, to prepare, to undertake; *Accipere* = to receive, to accept; *Accipiter* = goshawk, (plur.) = goshawks, Azores; *Acidula* = pin; *Acidus* = acid; *Acies* = blade edge, army, battle, device.
11	ISA	*Isabella* = Elisabeth; *Isagoa* = introduction, initiation.
12	ASI	*Asia* = Asia.
13	JAC	*Jacere* = to lie, to be buried, to be upset, to abase one-self, to be ill, to be abased or despised; *Jacere/Jecitare* = to throw, to fling, to spread the news; *Jactare* = to boast, to brag; *Jactura* = loss, damage, lure.
14	GAU	*Gaudium* = joy.
15	AUG	*Augere* = to add; *Augur* = to soothsay; *Augurium* = omen; *Augustatus* = sacred; *Augustus* = Augustus, holy, saint.
16	VAG	*Vagabundus* = wandering.
17	AGI	*Agilis* = light, swift; *Agitare* = to agitate, to disturb, to persecute, to treat.
18	CAIA	*Caia* = Portuguese town.
19	GAIA	*Gaia*, goddess, Portuguese city.

UPPER TRIANGLE

```
  1    2    3    4    5    6    7    8    9
 COP  SPO  POS  CUP  SPU  SUP  PUS  PUG  CIP

 10   11   12   13   14   15   16   17   18   19
 PIC  SPI  SIP  PIS  PIG  PAUX PAUC PAUS AUSP PAIX
```

1 COP — *Copia* = plenty, wealth, power, army, crowd; *Copulare* = to copulate, to join.

2 SPO — *Spodium* = scoria; *Spoliatere* = to despoil, to steal, to deprive, to kill, to discredit someone of well-known reputation; *Spondere* = to promise, to wed, to be doubtful; *Sponsor* = bridegroom, he who proposes marriage; *Sportula* = small gift.

3 POS — *Poscere* = to ask, to demand, to address, to accuse; *Positio* = position, posture, spot; *Positor* = builder, founder; *Possedere* = to possess; *Posse* = to have power, to have authority; *Post* = after; *Posteris* = descendants, heirs, posterity; *Posterior* = what follows, who succeeds; *Postfero* = who despises; *Prospactor* = heir, successor; *Postulare* = to ask, to beseech, to beg, to accuse.

4 CUP — *Cupere* = to wish, to covet, to love passionately, to love infamously; *Cupressifer* = who grows or brings cypresses, noble; *Cuprum* = copper; *Cupula* = small cistern, barrel.

5 SPU — *Spuma* = foam; *Spurcare* = to smear, to maculate, to be foul; *Spurius* = born out of wedlock; *Sputum* = spittle, saliva.

6 SUP — *Super* = what is above; superior; *Superaggere* = to amass, to accumulate; *Superambulare* = to trample under foot, to walk on; *Superare* = to win, to succeed; *Superbia* = haughtiness, pride, arrogance; *Supercilius* = stern, severe, high; *Superemicare* = to throw oneself on something or someone; *Superfulgere* = to glitter over; *Superstor* = survivor; *Supervacuus* = useless, unnecessary; *Suppetere* = to help, to lend a hand; *Suplicatio* = prayer; *Subpoeitere* = to repent a little; *Suppositio* = replacement of one thing by another; *Supprimere* = to suppress, to humiliate *Suprema* = the last moments of one's life, the last will, death; *Supremus* = the greatest in dignity or power.

7 PUS — *Pus* = outrage.

8 PUG — *Pugilatus* = boxing; *Pugillus* = handful; *Pugio* = dag-

		ger, unconvincing argument; *Pugna* = fight, combat, battle; *Pugnaciter* = stubbornly; *Pugnaculum* = tower, bastion; *Pugnax* = warrior, stubborn.
9	CIP	*Cipangus* = Cipango/Japan; *Cippus* = trunk of a tree; cypress.
10	PIC	*Picea,Piceaster* = pinetree, resinacious tree; *Pictura* = painting; *Picus* = griffin, fabulous creature, dragon.
11	SPI	*Spica* = spike; *Spicere* = to see, special; *Spiculum* = point of a spear; *Spinathorn/Spinifer* = which produces thorns, causes damage; *Spiritus* = breath, air, turning, wind, spirit, soul; *Spirare* = to breathe, to inspire, to want, to wish to show off; *Spissare* = to make thick, to condense, to obstruct.
12	SPI	*Siparma* = sail; *Sipare* = to spread.
13	PIS	*Piscatio* = fishing; *Piscina* = tank, lake, vat, tun; *Pisere* = to step on, to grind; *Pisone* = mortar, grail (name of a noble Italian family – Pizon); *Pistillum* = pestle; *Pistrilla* = grindstone.
14	PIG	*Pigeor* = to regret; *Piger* = lazy/careless; *Pigmentum* = colour, paint; *Pignerare* = to pile up; *Pignus* = pledge, token.
15	PAUX	*Paux* = little; *Pauxillatim* = little by little, slowly.
16	PAUC	*Paucies* = few times; *Pauciloquium* = laconic; *Paucitas* = rarity; *Paucis* = little, rare.
17	PAUS	*Pausa* = break; *Pausarius* = galley officer who beats the rhythm of the rowing.
18	AUSP	*Auspiciabilis* = of good omen; *Auspiciare* = to soothsay.
19	PAIX	(barbarism of PAX, instead of *pacis*, genitive case) = peace, of the peace, forgiveness, of the forgiveness (peace in ancient Provençal and in French).

LOWER TRIANGLE

```
   1   2   3   4   5   6   7   8   9   10  11
  SOB CUB BUC SUB GUB CIB BIC SIB BIS GIB BIG
```

1	SOB	*Soboles* = generation, family, ancestry, race, son; *Sobrietas* = sobriety moderation, temperate; *Sobrinus* = nephew, cousin, relative.
2	CUB	*Cuba* = bed; *Cubiculum* = chamber, room.
3	BUC	*Bucco* = self-praiser.
4	SUB	*Sub* = to be under; *Subacusatus* = accused, reproached; *Subactus* = softened, subjugated, vanquished, restrained,

violated; *Subagitare* = to stir, to move; *Subamare* = to love impatiently, bitterly; *Subaperior* = to divide, to share; *Subauscultare* = to overhear; *Subbini* = two things at the same time; *Subdivitus* = put in some one else's place; *Subdomare* = to dominate; *Subdolus* = deceiver, cunning, devious; *Subducere* = to steal, to rob; *Subductus* = out of danger; *Suber* = tree, cork oak; *Suberrere* = to wander; *Subito* = swift, extemporary; *Subjectio* = subjection, deceitful replacement of one thing by another; *Subjectus* = impostor; *Sublapsus* = one who escaped or ran away; *Sublatere* = to be hidden under something; *Subletare* = to mock, to scorn; *Sublimare* = to lift up; *Submerere* = not completely unworthy; *Submersus* = submerged, drowned; *Submovere* = to withdraw; *Subornatus* = procured dishonestly, bribed; *Subrepere* = to insinuate; *Subretio* = deceit, mistake, to enter furtively; *Subsignanus* = soldier who serves under a private banner; *Subsistere* = to withhold, to stop the enemy; *Subsonare* = to point secretly; *Subtenere* = to spread.

5	GUB	*Guberna* = helm; *Gubernare* = to steer; *Gubernatio* = ship's steering; *Gubernator* = pilot.
6	CIB	*Cibare* = to support, to keep, to nourish; *Cibicida* = rapacious.
7	BIC	*Bicorpus* = creature with two bodies.
8	SIB	*Sibus* = cunning, sharp.
9	BIS	*Bis* = twice; *Bisucilingus* = bilingual.
10	GIB	*Gibba* = hunchback.
11	BIG	*Biga* = two-horse cart; *Bignae* = twin sisters

UPPER HEMISPHERE

1	2	3	4
FER	FRE	EF(F)R	REF

1	FER	*Fera* = wild animal, ferocious, untamed; *Feralis* = sad, gloomy, dreary, threatening with death, cruel, destructive; *Ferax* = fertile, fecund, fruitful; *Ferens* = carrier, transporter, detainer; *Ferentaria* = swift, ready; *Ferire* = to wound; *Ferdinandus/Fernandus* = Ferdinand; *Ferotia* = haughtiness, valour, boldness, audacity; *Ferratus* = dressed in iron; *Ferrum* = iron, iron arm/weapon; *Fervefacere* = to cheer up, to enliven.
2	FRE	*Fremebundus* = furious, exasperated; *Frequentare* to attend to, to repeat the same concept, to join, to provoke; *Fre-*

quentia = frequence, competition/contest, crowd; *Fretum* = straits (sea); *Fretus* = support, confidence.

3 EF(F)R   *Effringere* = to break, to break in, to tear; *Effrons* = shameless.

4 REF   *Refacere* = to do again; *Refector* = renewer, repairer; *Refelere* = to answer back, to deny a crime, to disapprove; *Referire* = to wound again; *Referre* = to take again, to refer, to break the news; to withdraw, to resume, to reverse, to come back; *Referrere* = to stress, to matter, to be necessary.

LOWER HEMISPHERE

  1   2   3   4   5   6   7   8   9   10
ENS NES SEN ENC NEC CEN NEX ENG NEG GEN

1 ENS   *Ens* = being, thing, entity; *Ensifer* = sword carrier; *Ensis* = sword.

2 NES   *Nescire* = to ignore.

3 SEN   *Senaculum/Senatus* = senate; *Senarius* = made of six; *Senectus* = old age, weakness, austerity, melancholy, sadness; *Senescere* = to grow old; *Sensatus* = wise; *Sensus* = feeling, concept, reasoning; *Sententia* = sentence, condemnation, will; *Sentire* = to feel, to think, to know, to determine, to order, to suffer; *Sentis* = thorn; *Sentiscere* = to forebode, to warn.

4 ENC   *Encyclius* = circular, part of everything.

5 NEC   *Nec* = nor, no; *Necare* = to kill; *Necessitudo* = kinsman; *Necessare* = to compel, to force; *Nectere* = to tighten, to join, to brake.

6 CEN   *Cencere* = to judge; *Censio* = punishment, condemnation; *Censura* = censorship; *Centurio* = commander of 100 soldiers.

7 NEX   *Nex* = violent death, harm; *Nexare* = to interweave; *Nexus* = arrested due to debts.

8 ENG   *Engonatum* = portable sundial.

9 NEG   *Negletio* = despise, abandon; *Negligens* = negligent; *Negligo* = to despise; *Negocians* = merchant; *Negotium* = deal, despair, wariness.

10 GEN   *Genealogia* = Genealogy; *Gener* = son-in-law, brother-in-law; *Generalis* = general, universal; *Generator* = father; *Generare* = to breed; *Generosus* = generous; *Genesis* = birth; *Generalitas* = joy; *Genitrix* = mother; *Genium* = genius; *Gens* = people; *Gentilitas* = kinship; *Genua* = Genoa; *Genuus* = Genoese; *Genus* = generation, caste, family, origin.

Bibliography

Acosta, Isaak de, *As Conjecturas Sagradas* (Leyde, 1719).
Acqua, Carlo dell', *Cristóforo Colombo Studente all'Universitá di Pavia, etc* (Pavia, 1892).
Aelson, J., *Island Kirke fra dens grundaeggelse til Reformation* (Copenhagen, 1992).
Agnelli, Pietro, *Sulla questione 'La pia centinitá di Cristóforo Colombo'* (Studio Placenza, 1892).
Alba, Duquesa d', *Documentos escogidos del Archivo de la Casa de Alba* (Madrid, 1891).
Albornoz, Claudio Sanchez, *La España Musulmana según los Autores Islamitas y Christianos Medievales* (Buenos Aires, 1946).
Alleau, R., *Raymond de Lulle et l' Alchimie* (Paris, 1945).
Alleau, R., *Aspects de l'Alchimie Traditionelle* (Paris, 1953).
Alleau, R., *La Science des Symboles* (Paris, 1976).
Almagiá, Roberto, *Cristóforo Colombo* (Rome, 1918).
Almeida, Fortunato de, *História da Igreja em Portugal* (Coimbra, 1910).
Altolaguirre y Duval, see Duval.
Alveydre, Saint 'Yves d', *L'Archeometric suivie de la Théologie des patriarches* (Paris, 1950).
Amador de los Rios, see Los Rios.
Ameal, João, *História de Portugal* (Oporto, 1942).
Ameal, João, *Perspectivas da História* (Lisbon, 1960).
Amorim, Roa, *História de Tomar* (Tomar, 1965).
Angléria (or Anghiera), Pedro Mártir de, (trans.), *De Rebus Occeanicis et orbe novo; Fuentes historicas sobre Colón y América, etc.* (Buenos Aires, 1914).
Anselmo, António Joaquim, *Bibliografia das obras impressas em Portugal no século XVI* (Lisbon, 1926).
Anselmo, Artur, *Origens da Imprensa em Portugal* (Lisbon, 1981).
Aragão, Teixeira de, *Descrição Geral e Histórica das Moedas Cunhadas em Nome dos Reis, Regentes e Governadores de Portugal* (Lisbon, 1874).

Aragão, *Breve notícia sobre o Descobrimento*, etc., in *Memórias da Academia Real das Sciencias de Lisbon* (Lisbon, 1930).
Arana, Juan de Urries y Ruiz, *Enlaces de Reys de Portugal con Infantas de Aragon* (Madrid, 1899).
Arco, Ricardo del, *Fernando o Católico, artífice de la España Imperial* (Saragossa, 1939).
Areia, A. Vieira d', *O Processo dos Templários* (Oporto, 1947).
Armas, António Rumeu de, *El 'Portugués' Cristóbal Colón en Castille* (Madrid, 1982).
Arnades, Juan, *'Colón Catalán, según la leyenda,' Destino* (Barcelona, 1953).
Arnold, P. *Templiers et Rose-Croix* (Paris, 1955).
Arquivo Histórico da Madeira, *Ementa dos Livros da Vereação da Câmara do Funchal*, vol. III, no. 1 (Lisbon, 1933).
Arquivo Histórico da Marinha, *A Viagem de Diogo de Teive e Pedro Vasques de la Frontera*, vol. I (Lisbon, 1930).
Arquivo Nacional Do Tombo, *Alguns Documentos da Chancelaria de Dom Afonso V* (Lisbon, n.d.).
Arranz, Luis, *Historia* (Madrid, 1986).
Arribas, E., *Cristobal Colón, natural de Pontevedra* (Madrid, 1913).
Ascensio y Toledo, *see* Toledo.
Ascensio, Joaquin Torres, *Beatriz Enriquez de Arana y Cristobal Colón, Estudos y documentos* (Madrid, 1933).
Assereto, Hugo, *Nove proves de la Catalanitat de Colón. Les grandes falsetat de la tesi Genovesa* (Paris, n.d.).
Assereto, Hugo, *La data di nascita di Cristóforo Colombo* (Spezia, 1904).
Astengo, Archb. Andrea, *Delle memorie particolari e speciamente degli nomini illustri della Citá di Savona* (Savona, 1885).
Astruc, Canon, *La Conquête de la viconté de Carcassonne par Monfort* (Carcassonne, 1912).
Aulete, F.J. Caldas, *Dicionário Contemporâneo da Língua Portuguesa* (Lisbon, 1925).
Avezac, M. d', *Notice des découvertes faites au Moyen-Âge dans l'Océan Atlantique* (Paris, 1845).
Ávila, Arthur Lobo d', *Um Infante de Portugal* (with Saúl Santos Ferreira) (Lisbon, 1942).
Azevedo, João Lúcio, *História dos Cristãos-Novos Portugueses* (Lisbon, 1921).
Azevedo, João Lúcio, *Épocas de Portugal Económico, 'Apêndice'* (Lisbon, 1929).
Azevedo, Pedro Augusto de S. Bartolomeu, *O Apelido de Camões no século XV* (Lisbon, 1913).
Azurara (or Zurara), Gomes Eannes de (d. 1474), *Chronica da Guiné*, in Correia da Serra, *Inéditos da História de Portugal* (Lisbon, 1841).
Azurara (or Zurara), Gomes Eannes de, *Chronica dos feitos de D.*

Duarte de Menezes, in Correia da Serra, *Inéditos da História de Portugal* (Lisbon, 1841).
Bachmann, H., *Deutsche Reichsgeschichte im Zeitalter Friedrichs III und Maximilians I* (Leipzig, 1884/1894).
Bacon, Roger, *Le Miroir del'Alchimie* (Paris, 1960).
Bacon, Roger, *Lettre sur les prodiges de la nature et de l'art* (Paris, 1977).
Baer, Fritz, *Die Juden in Christlichen Spanien* (Berlin, 1936).
Baião, António Múrias, Manuel, and Cidade, Hernani, *História da Expansão do Mundo* (Lisbon, 1940).
Balaguer, Victor, *Castilla y Aragón en el descubrimiento de América* (Madrid, 1892).
Ballesteros y Beretta, *see* Beretta.
Baluze, E., *Vitae Paparum Avinionensium* (Paris, 1927).
Barante, J., *Histoire des Ducs de Bourgogne de la maison de Valois* (Brussels, 1835).
Barcia, Andrés, *Historiadores primitivos de la Índia* (Madrid, 1794).
Barreto, Mascarenhas, *Corrida – Breve História da Tauromaquia em Portugal* (Lisbon, 1970).
Barros, Henrique da Gama, *Hist. da Administração Pública em Portugal, séc. XII e–XV* (Lisbon, 1885).
Barros, João de, *Décadas da Ásia*, (Coimbra, 1932; Lisbon, 1945–6).
Basto, Martins, *Breve Resumo dos privilégios da nobreza* (Lisbon, 1854).
Basto, Rafael Eduardo de Azevedo, *Esmeraldo de Situ Orbis por Duarte Pacheco Pereira* (Lisbon, 1892).
Bauvois, E., 'La Chrétienté des Gardar au Moyen Âge', *Revue des questions historiques* (Paris, 1902).
Bayerri y Bertomeu, *see* Bertolomeu.
Beaujouan, Guy, 'La Science dans l'Occident Médiéval Chrétien', in *Historie Géneral des Sciences* (Paris, 1957).
Beaujouan, Guy, *Science Livresque et Art Nautique au XVème Siècle* (Paris, 1966).
Becker, Jerónimo, *La Patria de Colón. A propósito del libro del doctor don Rafael Calzada* (Madrid, 1921).
Bellefond, Villaut de, *see* Villaut de Bellefond.
Bellet, F. Brunet y, *Colón. Fue el verdadero descubridor de América. Donde nasció?* (Barcelona, 1892).
Beltrán y Rózpide, *see* Rózpide.
Bensaúde, Joaquim, *L'Astronomie Nautique au Portugal à l'Époque des Grandes Découvertes* (Lisbon, 1943a).
Bensaúde, Joaquim, *A Cruzada do Infante Dom Henrique* (Lisbon, 1943b).
Beretta, António Ballesteros y, *Cristóbal Colón y el descubrimiento de América* (Barcelona, 1945).
Beretta, António Ballesteros y, *Historia de España* (Barcelona, 1948).

Bernáldez (Bachelor), Andrés, *Historia de los Reys Católicos Don Fernando y Doña Isabel, etc.* (Seville, 1570).
Berthelet, René, *La Pensée de l'Asie et l'Astrobiologie* (Paris, 1938).
Bertomeu, Enrique Bayerri y, *Colón tal qual fué*, (Barcelona, 1961).
Bible, Vulgate; *A Bíblia Sagrada, segundo a Vulgata Latina*, António Pereira de Figueiredo (ed.) (Lisbon, 1842).
Biblioteca Publica de Ponta Delgada, *Tombo de Pedro Annes do Canto* (Ponta Delgada, Azores, n.d.).
Bidez, J. *Eros ou Platon et l'orient* (Brussels, 1945).
Blanc, Dr D., *Connaissance du Catharisme* (Paris, 1970).
Blanco, Andrés Gonzáles, 'Cristóbal Colón natural de la isla de Córcega', *Por esos Mundos*, 221 (Madrid, 1913).
Bloy, Léon, *Les Dernières Colonnes de l'Église* (Paris, 1930).
Bloy, Léon, *Oeuvres Complètes* (Paris, 1930).
Bocardus, Emory S. *Evolução do Pensamento Social* (Rio de Janeiro, 1960).
Bordeille, Viscount André de, *Dictionnaire de la Conversation et de la Lecture* (Paris, 1838).
Boron, Robert de, *Le Roman de l'Estoire dou Graal* (in ancient French) (Paris, 1927).
Borst, A., *Die Katharer* (Stuttgart, 1953).
Botelho, Afonso, *Estética e Enigmática dos Painéis* (Lisbon, 1957).
Bourdonne, Georges, *Les Templiers, histoire et tragedie* (Paris, 1972).
Bouton, R., *Traité de l'art héraldique ou science du Blason* (Paris, 1873).
Branco, Camilo Castelo, *see* Castelo Branco.
Brandão, Friar António, *Monarquia Lusitana, VII* (Lisbon, 1683).
Broomfield, Morton W., *Joachim of Flora – A Critical Survey of Canon Teachings, Sources, Biography and Influence* (London, 1957).
Bruner y Bellet, see Brunet y Bellet.
Brunetti, Almir dos Santos, *A lenda do Graal no contexto heterodoxo do Pensamento português* (Lisbon, 1974).
Brynidsen, Thorwald, *Era Colón Norte-americano?*', *Heraldo de Tortosa* (22 August 1934).
Burckhardt, Titus, *Alchimie* (Olten, 1960).
Buxtorf, J., *Lexicum hebraicum et chaldaicum* (Basel, 1955).
Cabert, P., *Actualité du Catharisme* (Toulouse, 1961).
Cabral, Fr Manuel de Pinho and Ramalho, J.H. *Magnum Lexicum Novissimum Latinum et Unitanum* (Paris, 1867).
Cadet De Gassicourt, F., with (Paulin, Baron de Roure), *see* Gassicourt.
Caldera y Solano, *see* Solano.
Calzada, Rafael, *El Rio Lerez* (Pontevedra, 1892).
Calzada, Valentin Mendez, 'La misteriosa firma de Cristóbal Colón, etc.', *La Razon* (Buenos Aires, 1927).
Caminha, Pero Vaz de, *Carta ao Rei Dom Manuel* (Lisbon, 1947).
Campos, Viriato, *Aclaração do Feito de Bartolomeu Dias, Português de Lei e Grande, Perito na Arte de Navegar* (Lisbon, 1987).

Cancellieri, Abbot Francesco, *Dissertazione epistolari bibliographische sopra Cristóforo Colombo* (Rome, 1809).
Canel, Eva, *Cristóbal Colón, natural de Pontevedra* (Madrid, 1913).
Canto, Ernesto do, 'Quem deu o nome ao Labrador?', *Arquivos dos Açores*, 4 (Ponta Delgada, 1894).
Cappas, Fr Ricardo, *Colón y los Españoles* (Madrid, 1915).
Carbia, Rómulo D., *La Patria de Cristóbal Colón* (Buenos Aires, 1923).
Cardoso, Antonio, *Viagem de Bartolomeu Dias em 1487/1488, vista por um marinheiro*, (Naval Academy of Lisbon 1990).
Cardoso, Antonio, 'Os Descobrimentos Marítimos, as Planta's e os Animais' (Naval Academy of Lisbon 1989).
Cardoso, Jorge, *Agiológio Lusitano* (Lisbon, 1742).
Caron, Michael and Houtin, Serge, *Les Alchimistes* (Paris, 1964).
Carreras y Valls, *see* Valls.
Cartier, Jacques, *Brief récit et succinte narrative de la navigation faicte et Ysles du Canada, etc.* (Paris, 1937).
Carvalho, Francisco Freire de, *Primeiro Ensaio sobre a História Literária de Portugal* (Lisbon, 1846).
Carvalho, Joaquim, *Miselânea, Obra Completa* (Lisbon, 1982).
Carvalho, Joaquim Barradas de, *O Tempo e os Grupos Sociais (um exemplo português da época dos Descobrimentos: Gomes Eanes Zurara e Valentim Fernandes'*, *Revista de História*, 15 (São Paulo, 1953).
Carvalho, Joaquim Barradas de, *As Fontes de Duarte Pacheco Pereira no 'Esmeraldo de Situ Orbis'* (Lisbon, 1982).
Carvalho, Manuel Mendes, *Theatro Genealógico* (Lisbon, false d.)
Casanova, Martin, *La vérité sur la patrie et l'origine de Christophe Colomb* (Bastia, 1881).
Casaril, *Rabi Siménon Ben Yochai et la Cabbale* (Paris, n.d.).
Castanheda, Fernão Lopes de (d. 1559), *Historia do Descobrimento & Conquista da Índia* (Coimbra, 1924).
Castelo-Branco, João, e Torres and Mesquita, Manuel de Castro Reveivade, *Resenha das Famílias Titulares do Reino de Portugal* (Lisbon, 1928).
Castelo-Branco, Jose Barbosa, *Costados de Famílias Illustres de Portugal, Algarve, Ilhas e Índias* (Lisbon, 1864).
Castiglioni, Carlo, *Cristóforo Colombo* (Brescia, 1948).
Castillo, Júlio de, *Lisboa Antiga: Conquista de Lisboa aos Mouros* (Lisbon, 1936).
Castro, A., 'Algunas observaciones acerca del concepto del honor en los siglos XVI y XVII', *Revista de Filologia Española* (Madrid, 1916).
Castro, Baptista de, *Colecção de Manuscritos da Biblioteca Pública de Évora* (Lisbon, 1935).
Catalá y Roca, *see* Roca.
Cesarini, A. Edouine, *Cristóbal Colón identifié Corse* (Nice, 1932).
Cerro, José de la Torre y del, *Beatriz Enriquez de Arana y Cristóbal Colón* (Madrid, 1933).

Chassant, A. (with Taussin, H.), *Dictionnaire des Devises Historiques et Héraldiques* (Geneva, 1978).
Chenu, M.D., *La théologie comme science au XIIIième siècle* (Paris, 1975).
Chevalier, Jean and Gherrboaut, Alain, *Dictionnaire des Symboles* (Paris, 1970).
Cicero, Marcus Tullios, 'De Legibus', in *Marci Tulii Ciceronis Opera*, etc. (Venice, 1772).
Cidade, Hernani (see Baião).
Claudien, C., *Recherches sur l'Histoire de l'Astronomie Ancienne* (Paris, 1885).
Coeltro, José Ramos, *A Mãe de Camões, Reservados da Biblioteca Nacional de Lisboa* (Lisbon, n.d.).
Cohen, Antoine, *Dante et le contenu iniciatique de la Vita Nuova* (Paris, 1958).
Colombo, Ezio, *Amerigo Vespucio e Vicente Jánez Pinzón alla scoperta dell'America* (Millan, 1876).
Colón, Cristóbal, *Diario de a bordo*, Luís Arranz Márquez (ed.) (Madrid, 1985).
Colón, Don Hernando (Fernando), *Historia del Almirante*, Luís Arranz Márquez (ed.) (Madrid, 1985).
Corbin, Henry, *Les Pèlerins de l'Orient et les vagabons de l'Occident; l'Orient des Pèlerins abrahamiques* (Paris, 1978).
Cordeiro, António, *História Insulana* (Azores, 1866).
Cordeiro, Luciano, *De la Part Prise par les Portugais dans la Découverte de l'Amérique* (Lisbon, 1876).
Cornish, F.W., *Chivalry* (London, 1911).
Corrêa, Gaspar, *Lendas da Índia, vol. III* (Lisbon, 1858).
Corrêa, Jácome, Marquis, of *Discussão Histórica das Medidas Geográficas no século XVI* (Lisbon, 1929).
Correia, Natália, 'Açores: o lugar do Espírito Santo', *Cultura Portuguesa*, 1 (1981).
Correnti, Cesar, *Cristóforo Colombo a Pavia: Note storico-critiche*, etc. (Pavia, 1882).
Cortesão, Armando Zuzarte, 'Subsídios para a História do Descobrimento da Guiné e Cabo Verde' *Boletim Geral das Colónias* (Lisbon, 1931).
Cortesão, Armando Zuzarte, *História da Cartografia portuguesa* (Coimbra, 1968–71).
Cortesão, Jaime, *História dos Descobrimentos Portugueses* (Lisbon, 1931).
Costa, Albino Neves da, *Pombal, Mentira Histórica* (Coimbra, 1968).
Costa, António Fontoura da, *Cartas das Ilhas de Cabo Verde, de Valentim Fernandes (1506)* (Lisbon, 1940).
Costa, Fr Bernardo da, *História Militar da Ordem de Nosso Senhor Jesus Cristo* (contrary to what it suggests it deals only with the subject of the extinction of the Order of the Temple) (Lisbon, 1771).

Coutinho, Gago, *Náutica dos Descobrimentos* (Lisbon, 1951–2).
Coutinho, Manuel de Azevedo, 'Carta transcrita, dirigida a D. João III, 1526' *Boletim da Academia das Ciências de Lisboa* (1946).
Craesbeck, Dr Francisco Xavier da Serra, *Espelho da Nobreza* (Coimbra, 1895).
Cuneo-Vidal, Rómulo, *Cristóbal Colón (Genovés) – Reconstitución Historica de los 'natales' del Descubridor de América, etc.* (Barcelona, 1924).
Cunha, Dom Rodrigo da, *Catálogo dos Bispos do Porto* (Oporto, 1623).
Cutts, E.C., *Scenes and Characters of the Middle Ages* (London, 1922).
David, Maurice, *Who was Columbus?* (New York, 1933).
Delabarre, Edmund Burke, *Recent History of the Dighton Rock* (New England, 1919).
Delabarre, Edmund Burke, 'The Earliest and Most Puzzling of New England Antiquities', *Bulletin of the New York Society for the Preservation of New England* (New York, 1923).
Del Cerro, José de la Torre y, *Cristóbal Colón. Siete años decisivos de su vida (1485/1492)* (Madrid, 1964).
Delpoux, P., *Les Sièges de Carcassonne et l'Inquisition* (Paris, 1951).
Delpoux, P., *Le Catharisme en Albigeois – Croisade et Inquisition aux XIII et XIVième Siècles* (Paris, 1954).
Deslandes, Venâncio, *Documentos para a História da Typographia Portuguesa nos Séculos XVI e XVII* (Lisbon, 1888).
Delumeau, Jean, *La Peur en Occident* (Paris, 1971).
Desmarquets, J.A., *Mémoires chronologiques pour servir à l'histoire de Dieppe* (Paris, 1785).
Dias, Malheiro, *História da Colonização Portuguesa – Introdução* (Oporto, 1932).
Diaz Del Castillo, Bernal, *Historia Verdadera de la Conquista de la Nueva – España, etc.*, (Madrid 1632; Paris, 1877–9).
Dillon, E. 'The Ordinances of Chivalry/Archeologia', (London, 1910).
Dinis, Fr Dias, *Monumenta Henriquina, vol. I* (Lisbon, 1931).
Dinis, Fr Dias, *As Crónicas Medievais Portuguesas – Adulterações de Rui de Pina* (Braga, 1952).
Dondaine, Fr P., *Liber de duobus principiis er rituel cathare* (Rome, Santa Sabina, 1939; Paris, 1972).
Dória, António Álvaro, *O Problema do Descobrimento da Madeira* (Guimarães, 1945).
Dornelas (or d'Ornelas) Afonso, 'Elementos para o estudo etimológico do apelido Cólon', *Boletim da Academia das Ciências, 20*.
Dornelas (or d'Ornelas) Afonso, *Memória sobre a residência de Cristóvao Colombo na Ilha da Madeira* (Lisbon, 1923).
Dozy, Raoul, *Histoire des Musulmans d'Espagne (711/1110)* (Leyden, 1932).
Duarte, El-Rei Dom, *Leal Conselheiro (1438)*, Imprensa Nacional/Casa da Moeda, (ed.) under the sponsorship of Fundação Calouste Gulbenkian

(Lisbon, 1982).
Duhem, Pierre, *Le Système du Monde* (Paris, 1940).
Dupuy, *Histoire de la Condénation des Templiers* (Paris, 1962).
Durban, Pierre, *Actualité du Catharisme* (Toulouse, 1968).
Duval, Angel de Altolaguirre y, *Cristóbal Colón y Pablo del Pozzo Toscanelli, etc.* (Madrid, 1903).
Duval, Angel de Altolaguirre y, *La Carta de Navigación atribuída a Cristóbal Colón – Estudio* (Madrid, 1925).
Edge, Captain, *Brief Discourses referring to the industries of Moscowitz merchants of London*, in Purchas, Pilgrimages, 13th vol. (London, 1625).
Elíade, Mircea, *Herreros y Alquimistas* (Madrid, 1959).
Ensch, Carmen, *L'Epopée albigeoise* (Toulouse, 1958).
Esperança, Manuel da, *História Seráfica* (Lisbon, 1666).
Eubel, C., *Hierarchia Catholica Medii Aevi* (Madrid, 1898 edn).
Evola, Julius, *El Misterio del Graal* (Barcelona, 1975).
Faria, Aida Neves, 'Análise Sócio–Económica das Comunidades Judaicas Portuguesas', *Boletim Faculdade de Letras* (Lisbon, 1963).
Faria, António Machado de, *Armorial Lusitano* (Lisbon, 1961).
Faria, Manuel Severim de, *Notícias de Portugal* (Lisbon, 1555).
Felner, Lima, *Colecção de Monumentos Inéditos para a História das Conquistas Portuguesas* (Lisbon, 1858).
Fernandes, Valentim, 'Manuscrito de Valentim Fernandes', *Academia Portuguesa de História* (Lisbon, 1940).
Ferreira, Dr Alexandre, *Suplemento histórico das memórias e notícias de célebre Ordem dos Templários para a História da Ordem de N.S. Jesu Christo* (Lisbon, 1735).
Ferreira, Joaquim, *Camões; dúvidas e acertos* (Oporto, n.d.).
Ferreira, Saúl (with Serpa, Santos Ferreira de) *Os Livros de Dom Tivisco e Confirmação Histórica* (Lisbon, 1938).
Ferreira, Saúl (with Ávila, Lobo d') *Um Infante de Portugal (Salvador Goncalves Zarco) Descobridor do Novo Mundo* (Lisbon, 1942).
Figuier, Louis, *L'Alchimie et les Alchimistes* (Paris, 1970).
Figueiredo, Fidelino de, 'A nacionalidade de Cristóvão Colombo', *Revista de História*, XV.
Fonseca, António Bélard da, *O Mistério dos Painéis* (Lisbon, 1963).
Fonseca, Quirino da, *Os Navios do Infante*, (Lisbon, 1958).
Fonseca, Quirino da, *A Caravela Portuguesa* (Lisbon, 1978).
Fornaison, Ernest, *Le Mystère Cathare* (Paris, 1964).
Fortune, Dion, *La Cabala Mistica* (Buenos Aires, 1965).
Foster, Michele, *Les Nombres – Le Système Sephirotique des Cabalistes* (Paris, 1983).
Fourgères, Etienne de, *Livres des Manières* (Paris, 1887).
Franco-Mendes, David, *Memórias, Anais e Documentos* (Lisbon, 1769).

Franco y Lopez, *see* Mora.
Frank, Adolph, *Le Kabbale ou la Philosophie Réligieuse des Hébreux* (Paris, 1967).
Freire, Francisco de Castro, *Memória Histórica da Faculdade de Matemática* (Coimbra, 1872).
Freire, João Mello, *Os Judeus em Portugal*, vol. II (Coimbra, 1928).
Freire, Manuel Braamcamp, *Crítica e História* (Lisbon, 1930).
Freitas, Lima de, *O Labirinto* (Lisbon, 1975).
Freitas, Friar Serafim de, 'De Justo Imperio Lusitaronum Asiatico', *Instituto Nacional de Investigação Científica*, (ed.) (Lisbon, 1983).
Frutuoso, Dr Gaspar, *Saudades da Terra, Livros II e III*, 16th-century manuscript (Funchal, 1873).
Funch-Brentano *Le Moyen Âge* (Paris, 1922).
Fuscanelli, R., *Le Mystère des Cathédrales en l'Interpretation Ésoterique des Symboles Herétiques Hermétiphes du Grand-Ouvre* (Paris, 1926).
Fuscanelli, R., *Les Demeures Philosophales* (Paris, 1930).
Gaibrós, Manuel Ballesteros y, *Cristóbal Colón* (Madrid, 1943).
Gaio, Manuel Felgueiras, *Nobiliário das Famílias de Portugal* (Lisbon, 1875).
Gallo, Enrique, *La Historia como Arte* (Buenos Aires, 1942).
Gallo, Enrique, *Historia de Cristóbal Colón – Análisis Crítica de las Fuentes Documentales y de los problemas Colombinos* (Buenos Aires, 1945).
Garcon, Maurice, *Le Symbolisme des Sabats* (Paris, 1971).
Gazenmüller, W., *L'Alchimie au Moyen Âge* (Paris, 1975).
Gassicourt, F. Cadet de (with Paulin, Baron du Roure), *L'Hermetisme dans l'Art Héraldique* (Paris, 1907).
Genouillac, H. Goudron de, *Grammaire Héraldique contenant la définition exacte de la Science des Armoiries* (Paris, 1877).
Ginsburg, Christian D., *Introduction to the Massoretico-Critical edition of the Hebrew Bible* (London, 1897).
Girão, Amorim, *Geografia de Portugal* (Coimbra, 1937).
Giustiniani, Agostino, *Psalterium hebraeum, graecum, arabicum et chaldeum* (Genoa, 1536).
Gobert, M.H., *Les Nombres Sacrés et l'Origine des Réligions* (Paris, 1962).
Godinho, Dr Vitorino Magalhães, *Documentos sobre a expansão portuguesa* (Lisbon, n.d.).
Gois, Damião de, *Chronica do Principe D. João que este foi destes feitos o Segundo do Nome*, etc. (Lisbon, 1905).
Golther, W., *Die Deutsche Dichtung im Mittelalter* (Stuttgart, 1912).
Gomara, Francisco López de, *Hispania Victrix, Primera y Segunda parte de la Historia general de las Indias con todo Descubrimiento y Cosas Notables*, (1551) vol. XXII (Madrid, 1859).
Gomes, Pinharanda, *História da Filosofia Portuguesa: A Filologia Hebraico – Portuguesa: A Patrologia Lusitana* (Oporto, 1981–3).

Gonzalez, Foután, *Don Cristóbal Colón, subdito de D. Isabel de Castilla y Galego de Nacíon* (Vigo, 1985).
Goudron de Genouillac, see Genouillac.
Grad, A.D., *Pour comprendre la Kabbale* (Paris, 1968).
Grad, A.D., *Le Livre des Principes Kabbalistiques* (Paris, 1974).
Grouvelle, Philipe de, *Mémoires Historiques sur les Templiers* (Paris, 1930).
Guimarães, J., Vieira, *A Ordem de Cristo* (Lisboa, 1901).
Guimarães, J., Vieira, *Marrocos e Três Mestres da Ordem de Cristo* (Lisbon, 1915).
Guiraud, J., *Le Concolamentum* (Paris, 1906).
Guiraud, J., *Histoire de l'Inquisition au Moyen-Âge* (Paris, 1935–8).
Gurvitch, Georges, *Definition du Concept de Classes Sociales* (Paris, 1962).
Guzman, Enrique de Guzmán y Gallo, see Gallo.
Habsburg, Archduke Otto von, *Portugal e África no Mundo de Hoje* (Lisbon, 1972).
Hagen, F.H. von der, *Ritterleben und Ritterdichtung* (Berlin, 1855).
Haeber, Konrad, *Typographia Ibérica del Siglo XV* (The Hague, 1903).
Hall, Mandy P., *An Encyclopedic Outline of Masonic, Hermetic, Qabalistic and Rosicrucian Symbolical Philosophy* (Los Angeles, 1977).
Hani, J., *Le Symbolisme du Temple Chrètien* (Paris, 1962).
Hanotaux, Gabriel, *Histoire des Colonies Françaises* (Paris, 1929).
Harrisse, Henry, *Notes on Columbus* (New York, 1864–6).
Harrisse, Henry, *Christophe Colomb, son origine, sa vie, les voyages* (Paris, 1884).
Harrisse, Henry, *Christophe Colon en Savone et les 'Memorie'* (Genoa, 1887).
Haven, M. *Arnauld de Villeneuve* (Paris, 1896).
Hearnshaw, F.J.C., 'A Cavalaria e o seu lugar na História', in *A Cavalaria Medieval*, see Prestage.
Heers, Jacques, *Christophe Colomb* (Paris, 1981).
Herculano, Alexandre, *História da Origem e Estabelecimento da Inquisição em Portugal*, Book III (Lisbon, 1844).
Hervás, José Perez, *La Cuna de Colón* (Lima, 1913).
Hogan, Manly P., *Great Symbol of Solomon* (New York, 1922).
Hooykaas, R., *Humanism and the Voyages of Discovery in 16th Century Portuguese Science and Letters* (Amsterdam, 1979).
Hovelace, Abel, *La Linguistique* (Paris, 1928).
Howard, Stephen, *The Knights Templar* (London, 1972).
Hoyos, Marquis of, *Colón y los Reyes Católicos*, conference in Madrid (1892).
Hruby, Kurt, *A Cabala e a Tradição Judaica* (Lisbon, 1979), see Tryon-Montalembert.
Humboldt, Alexander von, *Christophe Colom et la Découverte de l'Amérique* (Paris, 1833).
Hutin, Serge and Caron, Michel, *Histoire de l'Alchimie* (Vervier, 1971).
'Imprensa Nacional', *Alguns Documentos do Arquivo Nacional da Torre do*

Tombo, acerca das navegações e conquistas portuguesas (Lisbon, 1892).
Ingstad, Helge, 'Vinland ruins prove Vikings, found the New World', *The National Geographic Magazine* (November 1964).
Irving, Washington, *A History of the life and voyages of Christopher Columbus* (London, 1928–30).
Isaac, Jules, *Génese de l'Antisémitisme* (Paris, n.d.).
Izco, Wenceslao Aygnals de, *La Escuela del Pueblo, vol. 10* (Madrid, 1852).
Jacob, E.F., 'Os Começos da Cavalaria Medieval', in *A Cavalaria Medieval*, see Prestage.
James, G.R.R., *History of Chivalry* (London, 1830).
Jeffreys, John Gwin, *British Conchology, vol. III* (London, 1970).
Jonson, F., *Historia Eclesiastica Islandiae* (Copenhagen, 1772–8).
Jordão, Levy Maria, *Memórias Históricas sobre os Bispados de Ceuta e de Tânger* (Lisbon, 1858).
Klawe, Janina, *Communication to the Warsaw Faculty of Sciences* (Warsaw, 1989).
Napic, Dragomir, *Geografia* (Lisbon, 'Aster', n.d.).
Knight, Gareth, *A Practical Guide to Qabalistic Symbolism* (London, 1976).
Kreschmer, Konrad, *Die italienischen Portulane des Mittelalters; Ein Beitrag zur Geschichte der Kartographie und Nautik* (Berlin, 1909).
Laicus (pseudonym), *Cristoph Columbus, seine Leben und seine Entdekungen* (Einsiedeln, 1888).
Lapa, M. Rodrigues, *A Demanda do Santo Graal* (Lisbon, 1928).
Lapa, M. Rodrigues, *Das Origens da Poesia lírica em Portugal na Idade-Média* (Lisbon, 1929).
La Riega, Celso Garcia de, *Colón Español, su origen y patria* (Madrid, 1914).
La Torre, see Del Cerro.
La Tour-Landry, G. de., *The Book of the Knight* (English trans.) (London, 1868).
Las Casas, Friar Bartolomé de, *Opus Epistolarum Petri Martyris Anglerii Mediola* (1530).
Las Casas, Friar Bartolomé de, *História General de las Indias*, etc. (Madrid, 1875–6).
Larsen, Lofus, *The Discovery of North America Twenty Years Before Columbus* (New York, 1924).
Lavocar, M. *Procès des Frères er d l'ordre du Temple* (Paris, 1970).
Leão, Duarte Nunes de, *Chronica de D. Afonso V* (Lisbon, 1600).
Leão, Francisco da Cunha, *O que é o Ideal Português (O Ideal Português e o Homem)* (Lisbon, 1962).
Leitão, Belchior de Andrade, *Notas Genealógicas* (Lisbon, 1630).
Leite, Duarte, *História dos Descobrimentos* (Lisbon, 1950).
Leite, Duarte, *Viagens de Luís Cadamosto e de Pedro Sintra* (Lisbon, 1950).
Leite, Duarte, *Descobridores do Brasil* (Oporto, 1931).

Leon, Moisés de, *Zohar, Livro do Esplendor* (Toledo, 1304; Paris, 1732).
Leone, Metzner, *Pedro Álvares Cabral* (Lisbon, 1968).
Lery, Jean de, *Le Voyage au Brésil – 1556/1558* (Paris, 1563; 1927).
Lessa, Prof Doutor Almerindo, *Lições de Medicina Social e Saúde Pública (ISCSP)* (Lisbon, 1958).
Letrone, A., *Recherches Critiques, historiques et geographiques sur les fragments de Heron de Alexandrie* (Paris, 1851).
Lima, Dom Luís Caetano de, *Geografia Histórica de Todos os Estados da Europa* (Lisbon, 1734).
Lisboa, João de, *Livro da Marinharia* (Lisbon, 1540).
Lollis, Cesare de, *Scriti et Autografi, publicati con Prefazio e transcrizione diplomatica del Pro. Cesare Lollis*, in *Raccolta* (Rome, 1892–4).
Lopes, Fernão, *Chronica del-Rei D. João I* (Lisbon, 1931).
Lopes, Fernão, *Chronica de D. Fernando* (Lisbon, 1895).
Lopez, Luís Franco y, *see* Mora.
Lorgues, Count Roselly de, *Christophe Colomb, histoire de sa vie et les voyages, d'après documents authentiques tirés d'Espagne et d'Italic* (Paris, 1856).
Los Rios, José Amador de, *História Social Política y Religiosa de los Judios de España y Portugal*, vol. I, 'Xalxer ha Cabbalâh' (Madrid, 1876).
Lüderitz, Anne, *Die Liebestheorie der Provençalen bei den Minnesingern der Staufferzeit* (Berlin, 1904).
Lulle, Raymond de, *Le Codicile* (15th century) (Paris, 1953).
Machado, Carlos Roma, *A nacionalidade portuguesa e o nome de Cristóbal Colón* (Coimbra, 1934).
Machado, Dr Pedro José, *Dicionário Etimológico da Língua Portuguesa* (Lisbon, 1959).
Madariaga, Salvador de, *Christophe Colomb* (Paris, 1952).
Maia, Carlos Roma Machado de Faria, *see* Machado.
Maia, Francisco de Athayde M. de Faria de, *Capitães dos Donatários (1436/1766)* (Lisbon, 1972).
Major, Richard Henry, *Life of Prince Henry of Portugal* (London, 1868).
Manso, Viscount of Paiva, *História Eclesiástica Ultramarina*, Book I (África Septentrional, Bispado de Ceuta, Tânger, etc.) (Lisbon, 1872).
Manteuffel, T., *Naissance d'une Héresie, Les Adeptes de la pauvreté volontaire au Moyen-Âge* (Paris, 1970).
Manuel II (King of Portugal), *Livros Antigos Portugueses (1489–1600)* (London, 1929).
Manzano, J. Manzano y, *Cristobal Colón. Siete años decisivos de su vida* (Madrid, 1964).
Marchard, René, *De la Pierre Philosophale à l'Atome* (Paris, 1959).
Marcot, Ramón 'La verdadera Patria de Colón', *España Moderna* (Madrid, June 1919).
Marcot, Ramón, *Colón pontevedrés* (Havana, 1920).

Marín Luís Astrana, *Cristóbal Colón. Su Patria, sus restos y el Enigma del Descubrimiento de América* (Madrid, 1929).
Marques, Luis Arranz, *Historia del Almirante* (Madrid, 1984).
Martins, Felisberto and Figueiredo, José Nunes, *Gramática Grega* (Coimbra, 1952).
Martins, Oliveira, *A vida de Nun' Alvares* (Lisbon, 1917).
Mascarenhas Barreto, *see* Barreto.
Matoso, António, *História de Portugal* (Lisbon, 1939).
Mayer, K.B., *Classe y Sociedad* (Buenos Aires, 1969).
Medina, Manuel de Saralegui y, *'El Testamento de Colón'*, *Obras Póstumas*, IV (Madrid, 1928).
Mely, F. de, *'L'Alchimie parmi les Chinois et l'Alchimie Gréque'*, *Journal des Savants* (Paris, 1895).
Mendez, Calzada, *see* Calzada.
Mendonça, Henrique Lopes de, *Estudos sobre os navios Portugueses dos séculos XV e XVI* (Lisbon, 1898).
Mendonça, Henrique Lopes de, *Estudo Biográfico: Damião de Goes e a Inquisição; Estudos sobre os Navios Portugueses dos Séculos XV e XVI* (Lisbon, 1898).
Mendonça, José Godinho de, *Regras de Equitação* (Coimbra, 1879).
Mendoza, Diego Hurtado de, *Historia de la Guerra Contra los Moros* (Madrid, 1554).
Menendez Pelayo, *see* Pelayo.
Menezes, Dom Fernando (Count of Ericeira), *História de Tânger* (Lisbon, 1732).
Menezes, Manuel de, *Revisão ao problema da descoberta e povoamento dos Açores*, Angra do Heroismo (Azores, 1949).
Merêa, Paulo, *'De Portucale (civitas) ao Portugal de D. Henrique'*, quoted in Serrão, Joel (ed.), *Dicionário da História de Portugal* (Lisbon, 1933).
Mesquita, Manuel de Castro Pereira de, *Resenha das Famílias Titulares do Reino de Portugal* (Lisbon, 1922).
Michelet, V.E., *Le Secret de la Chevalerie* (Paris, 1930).
Migne, P.L.L., *'L'Eglise et la Synagogue'*, *Patrologie* (Paris, 1868).
Monge, Alf, *Norse medieval cryptography in runic carving* (London, 1970).
Moore, Henry, *The Kabbala* (London, 1979).
Moore, Thomas, see Morus.
Mora, Luis Franco y, Baron of, *'Cristóbal Colón español, como nacido en território pertenenciente al Reino de Aragón'*, *Boletin de la Real Academia de la Historia de Madrid* (October) (Madrid, 1886).
Morais (or Moraes), Cristóvão Alão de, *Pedatura Luzitana* (Lisbon, 1688).
Morison, Samuel Eliot, *Portuguese Voyages to America in Fifteenth Century* (Boston, 1940).
Morison, Samuel Eliot, *The Admiral of the Ocean Sea – A life of Christopher Columbus* (New York, 1962).

Morison, Samuel Eliot, *Christopher Columbus Mariner* (New York, 1974).
Morpurgo, Alessandro, *Vittoria Colonna* (Trieste, 1889).
Morus, São Tomás, *Utopia* (London, 1518; Portuguese edn, Lisbon, 1954).
Mota, A. Teixeira da, '*A Arte de Navegar no Atlântico e no Índico*', Anais do Club Militar-Naval (1957 vol.).
Mota, A. Teixiera da, *A Evolução da Ciência Náutica, durante os séculos XV e XVI, na cartografia portuguesa da época* (Lisbon, 1961).
Muratori, *Rerum Italicarum Scriptores* (Manuscript of Biblioteca di Genova, vol. 22, col. 301/2) (Milan, 1733).
Murias, Manuel, *see* Baião.
Navarrete, Martin Fernandez, *Colección de los Viages y descubrimiento que hicieron por mar los Españoles* (Madrid, 1825).
Navarrete, Martin Fernandez, *Documentos de Colón y de las primeras poblaciones; viages de Cristóbal Colón, com una carta* (Madrid, 1915).
Niel, Fernand, *Albigeois et Cathares* (Paris, 1955).
Novonha, A. Henriques de, *Nobiliaria da Ilha da Madeira* (Funchal, Madeira, 1950).
Oliva, Fernán Peres de, *Las obras del Maestro Fernán Peres de Oliva, rector que fué de Salamanca y Catedrático de Teologia de Ella, etc.* (Cordoba, 1635).
Oliveira, Eduardo Freire de, *Elementos para a História do Município de Lisboa*, Book I of *Provimento de Saúde* (Lisbon, 1815).
Oliveira, José Augusto de, *Conquista de Lisboa aos Mouros (1147)* (Lisbon, 1936).
Olmet, Fernando Antón del, *Historia del Renascimiento, vol. III* (Barcelona, 1916).
Ornelas, Afonso d', *see* Dornelas.
Orta, Garcia de, *Colóquios dos Simples e Drogas Medicinais da Índia (Sec. XVI)* (1963 edn).
Oucide, Laureano de, *Cristóbal Colón, su origen y su patria. Carta a Juan Salari* (Buenos Aires, 1910).
Ouy, A., *La Philosophie Secrète des Alchimistes* (Paris, 1972).
Oviedo, Gonçalo Fernandes de, *História General y Natural de las Indias* (Seville, 1547; Spanish edn, Madrid, 1749).
Padron, *Historia de la América* (Madrid, 1962).
Pakula, Marvin H., *Heraldry and Armour in the Middle Ages* (London, 1972).
Papus (pseudonym of, Dr Gerard d'Encausse), *Cabbale – Traditional Secrète de l'Occident* (Paris, 1978).
Paragallo, Próspero, *Disquisizione Colombine, vol. I* (Rome, 1893).
Paragallo, Próspero, *La Favula de Alonso Sanchez, Precursore e Maestro di Cristóforo Colombo* (Lisbon, 1896).

Paragallo, Próspero, *Cristóforo Colombo e la sua Famiglia* (Lisbon, 1889).
Paragallo, Próspero, *Cenni Intorno alla Colonia italiana in Portugallo nei secoli XIV, XV e XVI* (Genoa, 1907).
Pardo, Constantino de Horta, *La verdadera Cuña de Cristóbal Colón* (New York, 1912).
Pastor, A.R., 'A Cavalaria e as Ordens Militares de Espanha', in *A Cavalaria Medieval*, see Prestage.
Paulin, Baron of Rourede (with Gassicourt, F. Cadet de), *L'Hermetisme dans l'Art Héraldique* (Paris, 1907).
Pedroso, Armado Alvarez, *Nueva revisión de algunos de los que fueron Problemas Colombinos* (Trujillo, 1946).
Pelayo, Marcelino Menéndez, *Historia da los Heterodoxos Españoles* (Madrid, 1958).
Pene, G., *La Conquête du Languedoc* (Nice, 1957).
Perctlaine, H., *Histoire de la guerre contre les Albigeois* (Paris, 1833).
Pereira, Duarte Pacheco, *Esmeraldo de Situ Orbis* (Lisbon, 1975).
Pereira, Gabriel, 'Fragmentos relativos à História e Geografia da Península Ibérica', *Boletim da Sociedade de Geografia*, 5 (Lisbon, 1900).
Pereire, Dr Esteves, *O Livro de Marco Polo* (Lisbon, 1922).
Peres, Damião, 'Tratado de Tordesilhas', in *Congresso do Mundo Português*, vol. III, 'Descobrimentos Marítimos' (Lisbon, 1940).
Peres, Damião, *História dos Descobrimentos Portugueses* (Oporto, 1943).
Perrot, A.M., *Collection Historique des Ordres de Chevalerie* (Paris, 1820).
Pescia, Giuseppe, *E Genova o Terrarossa di Maconesi il luogo di nascita di Cristoforo Colombo?* (Dissertation, Chiavari, 1891–2).
Pestana Júnior, *D. Cristóval Colón ou Syman Palha, na História e na Caballa* (Lisbon, 1928).
Peygnot, Claude, *Le Nombre – Le Langage de Dieu* (Paris, 1987).
Pichel, Gina, *História Universal de Arte* (Milan) (Brazilian trans., S. Paulo, 1967).
Pidal, Ramón Menéndez, 'La Lengua de Cristóbal Colón', *Hispania* (Madrid, 1943).
Piel, Joseph, *Nomes dos Santos na Toponímia Portuguesa* (Lisbon, 1942).
Pimenta, Alfredo, *Elementos da História de Portugal* (Lisbon, 1934).
Pimenta, Alfredo, *Novos Estudos Filosóficos e Críticos* (Lisbon, 1935).
Pimental, Manuel, *A Arte de Navegar* (Lisbon, 1762).
Pina, Rui de (d. 1521), *Chronica d'El-Rei Dom Afonso V* (Lisbon, 1790).
Pinilla, Tomás Rodriguez, *Colón en España, Estudo Histórico-Crítico sobre la vida y trechos del Descubridor del Nuevo Mundo* (Madrid, 1884).
Pinto, José Loureiro, 'Enigmas da História de Coimbra, as Judiarias e problemas que suscitaram', *Arquivo Coimbrão*, vol. XII (1954).
Pisano, Master Matheus, 'Livro da Guerra de Seuta, vol. I', in *Inéditos da História Portuguesa* (Lisbon, 1970).

Pius II, Pope, *Historia Rerum Gestarum* (Rome, 1475).
Plancy, Colinde, *Dictionnaire Internal* (Paris, 1970).
Poisson, *Cinc Traités d'Alchimie* (Paris, n.d.).
Portalk, F., *Les Couleurs Symboliques* (Paris, 1974).
Porto da Cruz, Viscount of; see Visconde de Porto da Cruz.
Prescott, William, *Ferdinand and Isabella* (Philadelphia, 1743).
Prescott, William Hickling (trans.), *Historia del reinado de los Reyes Catolicos, etc.* (Madrid, 1855).
Prestage, Edgar, 'A Cavalaria em Portugal', in *A Cavalaria Medieval* (Oporto, n.d.).
Purificação, Fr António da, *Crónica da Antiquíssima Província de Portugal* (Lisbon, 1642).
Quadros, António, *Poesia e Filosofia do Mito Sebastianista* (Lisbon, 1983).
Quadros, António, *Portugal, Razão e Mistèrio/O Projecto Áureo ou o Império do Espírito Santo* (Lisbon, 1986–7).
Quintela, Inácio da Costa, 'Monarchia Lusitana', vol. 1112 of *Anais da Marinha Portuguesa* (Lisbon, 1830).
Rafn, Karl Christian, *Americas Oplagelse I del Tiende Aarhundred Efter de Nordische olds Kriffer* (*Mémoire sur la découverte de l'Amérique au Xème siècle*) (Copenhagen, 1841).
Rahn, Otto, *La Croisade contre le Graal* (Paris, 1933).
Randles, W.G.L., *De la Terre Plate au Globe Terrestre* (Paris, 1980).
Read, John, *Prelude to Chemistry* (London, 1936).
Reed, A.W., 'A Cavalaria e a ideia de gentil-homem', in *A Cavalaria Medieval*, see Prestage.
'Regimento da Guerra e dos Cavaleiros', *Ordenações Afonsinas*, vol. V (Coimbra, 1786).
Rego, Fr António da Silva, 'O Plano Henriquino das Índias', *Atlântida*, 4(3) (Angra do Heroismo, Azores, 1958).
Rego, Fr António da Silva, *O Ultramar Português no século XVIII* (Lisbon, 1967).
Remedios, Mendes dos, *Os Judeus em Portugal; Memórias para a História e Teoria das Cortes Gerais*, vol. IX (Coimbra, 1911).
Remón, Frei Alonso, *Historia General de la Ordem de Nuestra Señora de la Merces*, vol. II, Chap. 6 (Madrid, 1530).
Renesse, Théophile de, *Dictionnaire des Figures Héraldiques* (Brussels, 1895–1905).
Resende, Garcia de, *Chronica dos Valorosos e Insignes Feitos del Rey Dom Joham II* (Lisbon, 1545; 1798).
Resie, Count of, *Histoire et Traité des Sciences Occultes* (Paris, 1857).
Reuchlin, Jean, *De Verbo Mirifico*, mentioned in *De Arte Cabbalistica* (Paris, 1615).
Rey Sancho, see Sancho.
Ribeiro, J. Pinto, *Extracto de a Memória sobre a Humana Tolerância de Judeus*

e Mouros em Portugal; Dissertações Chronológicas e Críticas sobre a História de Portugal (Madrid, 1633).

Ribeiro, J. Pinto, *Discurso si es util e justo de desterrar de los Reinos de Portugal a los Christianos-Nuevos, convencidos do Judaísmo por le Tribunal del Santo Ofício y reconciliados por ele con sus familias* (Lisbon, 1645).

Ribeiro, Patrocínio, *A Nacionalidade Portuguesa de Cristovám Colombo* (Lisbon, 1927).

Ribeiro, Patrocínio, *O Carácter Misterioso de Colombo e o problema da sua nacionalidade* (Coimbra, 1917).

Risch, Fredrich, *Veröffentlichungen des Forschung institut für Vergleich ende Religionsgeschichte an der Universität*, vol. II, Chap. 11 (Leipzig, 1930).

Riega, Celso García de la, *La Gallega, Nave Capitana de Colón, etc.* (Pontevedra, 1897).

Riega, Celso García de la, 'Colón español y judio', *Boletim de la Soc. Geográfica de Madrid*, 40 (Madrid, 1899).

Riega, Celso García de la, *Colón español, su origen y su patria* (Madrid, 1914).

Rita, Friar A. de Santa Rita, see Santa Rita.

Roca, Pedro Catalá y 'Sobre los italianismos observados en la "Carta de Colón a Santángel"', *Studi Colombiani*, vol. II (Genoa, 1951).

Rocca, Giuseppe A., *Cristóforo Colombo y la sua patria* (Savona, 1892).

Roman, Hieronymo, *Pragmáticos e Leys Hechas y Recopiladas, etc.* (Medina del Campo, 1549) p. 3.

Roman, Hieronymo, *Historia y Témas Historicos de los Religiosos* (about the deeds of the Noble Generation and the Jews in the Conquests and Discoveries) (Madrid, 1595).

Rosenroth, Christian Knorr, *Kabbala Denudata (17th century)* (Rome, n.d.).

Rossi, Giuseppe Carlo, *Navegações de Luís Cadamosto* (Lisbao, 1954).

Rostrenen, Fr Gregoire de, *Dictionnaire de la langue Heltique* (Paris, 1732).

Róspide, Ricardo Beltrán y, *Cristóbal Colón Cristoforo Colombo* (Madrid, 1918).

Róspide, Ricardo Beltrán y, *Viajes y descubrimiento en la Edad Media* (Madrid, 1919).

Róspide, Ricardo Beltrán y, *Cristóbal Colón Genovés?* (Madrid, 1925).

Róspide, Ricardo Beltrán y, *Historia de España* (Barcelona, 1948).

Ruas, Henrique Barrilaro, *A Liberdade e o Rei* (Lisbon, 1972).

Ruas, Henrique Barrilaro, 'Cinco Modos de Patriotismo', *Rumo*, II (1962).

Ruas, Henrique Barrilaro, 'O Conde Dom Henrique', in Serrão, Joel (ed.), *Dicionário da História de Portugal* (Lisbon, 1967).

Rubio, Jeronimo 'João Pinto Delgado y situación de los Judios en Portugal en los siglos XVI y XVII, in *Miscelanea de Estudios Arabes y Hebraicos* (Granada, 1957).

Ruibal, Maria Julieta, *'Quem foi o pintor Nuno Gonçalves'?* in *Revista de Belas Arte*, 2nd set, no. 221/22 (Lisbon, 1966).
Saint-Martin, (Louis-Claude), *L'Homme de Désir* (Paris, 1979).
Salari, Juan, *La Cuña de Cristóbal Colón* (New York, 1912).
Salas, Carlos, *Pedro Mártir de Angleria. Estudo biográfico-bibliográfico* (Cordoba, Argentina, 1917).
Salinério, Giulio, *Anotaciones sobre Cornelio Tacito*, in *Raccolta* (Milan, 1864–6) (also mentioned by Harrisse, 1884 and Staglieno, 1881).
Sampayo, Luís de Melo Vaz de, *Subsídios para uma Biografia de Pedro Álvares Cabral* (Coimbra, 1971).
Sáncho, Luciano Rey, *España, patria infalible de Cristóbal Colón. Refutación de las pruebas Genovesas del escritor don Luis Astrana Marin* (Coruña, 1941).
Sanguinetti, Angelo, *Se Cristóforo Colombo abbia studiado in Pavia* (Genoa, 1880).
Sanguinetti, Angelo, *Ancora di Cristóforo Colombo studente all'Universitá di Pavia* (Pavia, 1882).
Sanguinetti, Angelo, *Delle sigle usate da C. Colombo nella sua firma*, Giornale Linguistico, 10th year, vol. V.
Sansonetti, Paul George, *Graal et Alchimie* (Paris, 1972).
Santarém, Viscount of, *Quadro Elementar; Alguns documentos para a História e Theoria das Cortes Gerais em Portugal* (Lisbon, 1928).
Santarém, Viscount of, *História e Theoria das Cortes Gerais*, vol. II (Lisbon, 1975).
Santarém, Viscount of, *Recherches historiques, critiques et bibliographes sur Americ-Vespuce et les voyages* (Paris, 1956).
Santa Rita, Fr A. de, *Dos Transdurienses na Índia* (Goa, 1835).
Santos, Domingos Gomes dos, *Do Valor Histórico de Rui de Pina* (Lisbon, 1932).
Sanz, Manuel Serrano y, *Origen de la Dominación Española en América* (Madrid, 1918).
Sanz, Manuel Serrano y, *'La naturaleza de Colón no es gallega'*, *Revista de Archivos, Bibliotecas y Museus* (March–April) (Madrid, 1919).
Saraiva, António José, *História da Cultura em Portugal*, vol. II (Lisbon, 1953).
Saraiva, António José, *A Inquisição em Portugal* (Lisbon, 1956).
Saralegui y Medina, see Medina.
Saunier, Marc, *La Legende des Symboles* (Paris, n.d.).
Scarlatti, Lita, *'Nuno Gonçalves, Cavaleiro da Casa d'El-Rei e seu Pintor*, in *Colóquio* (January-February 1970).
Schneider, Wolfgang, *Lexicon* (Weinleim, 1962).
Schwartz, Samuel, *A Tomadade de Lisboa* (Lisbon, 1953).
Segurado, Jorge, *Paineis de S. Vicente e Infante Santo* (Lisbon, 1984).
Sergio, António, *Ensaios*, vol. I (Coimbra, 1949).

Serpa, Santos Ferreira de (with Ferreira, Saúl) *Os Livros de Dom Tivisco e Confirmação Histórica*, (Lisbon, 1938).
Serrano, Adrián Sánchez, 'Donde está el error de Colón extremeño?', *El Diario Español* (13 October 1928) (Buenos Aires).
Serrano, Manuel *see* Sanz.
Serrão, Joel, *Dicionário da História de Portugal*, – 'Náutica' (Lisbon, 1975).
Servier, Jean, *L'Homme et l' Invisible* (Paris, 1961).
Shaeffer, H., *Historia de Portugal*, Soulange Bodin (trans.) (Paris, 1892).
Shears, F. S., 'A Cavalaria de França', in *A Cavalaria Medieval, see* Prestage.
Shulten, Adolf, *Viriato* (Oporto, 1940).
Shulten, Adolf, *Fragmentos Relativos à História e Geografia da Peninsula Ibérica* (Oporto, 1949).
Silva, António de Morais, *Dicionário da Língua Portuguesa* (Lisbon, 1891).
Silva, Dr Manuel Luciano da, *Portuguese Pilgrims and the Dighton Rock* (Bristol, USA, 1971).
Silva, Dr Manuel Luciano da, *The True Antilles: Newfoundland and Nova Scotia* (Bristol, USA, 1972).
Simões, Manuel Breda, *O Simbolismo da Tripla Coroação e os 'Impérios' do Espírito Santo* (Azores, 1984).
Soares, Dom António, *Relações Genealógicas* (Lisbon, 1731).
Solano, Valentín Calderera y, *Informe sobre los retratos de Cristóbal Colón, su traje y escudo de armas* (Madrid, 1832).
Sorokin, Pitrim A., *Sociedad, Cultura y Personalidad – Su Estrutura y su Dinamica General* (Madrid, 1960).
Sorval, Gerard de, *Le Langage Secret du Blason* (Paris, 1891).
Sousa, Cordeiro de, *A Inscrição da Pedra de Dighton* (Lisbon, 1938).
Sousa, Dom Antonio Caetano de, *História Genealógica da Casa Real Portuguesa* (Lisbon, 1739).
Sousa, Dom Antonio Faria de, *História Genealógica da Casa Real Portuguesa* (Lisbon, 1785).
Sousa, Dom Antonio Faria de, *Europa* (Lisbon, 1651).
Sousa, João de, *Vestígios da Língua Arábica em Portugal* (Lisbon, 1830).
Sousa, Friar Luís de, *História de S. Domingos* (Lisbon, 1623).
Soveral, Carlos Eduardo de, *A Nostalgia de Hesíodo – La Varende: Das Pujanças e Estesias Tradicionais* (Lisbon, 1959).
Soveral, Carlos Eduardo de, *Jasão e Medeia* (Lisbon, 1959).
Staglieno, Merchiese Marcello, *Il Borgo di S. Stefano ai tempi di Colombo e le case di Domenico Colombo* (Genoa, 1881).
Stebbing, H., *History of Chivalry and the Crusades* (London, 1926).
Stefano, Luciano de, *La Sociedad Estamental de la Baja Idad Media Española a la Luz de la Literatura de la Epoca* (Caracas, Venezuela, 1967).
Steinschneider, Moritz, *Die Hebraeischen Übersetzungen des Mittelalters und die Juden als Dolmetscher* (Berlin, 1893).
Stillman, J.M., *The Story of Alchemy and Early Chemistry* (New York, 1960).

Streicher, Fr Fritz, *'La Patria de Colón', Investigación y Progresso* (Madrid, 1929).
Streicher, Fr Fritz, *'Die Kolumbus-Autographe', Spanische Forschungen*, vol. I (Munich, 1928).
Tass, Agency, referred to in *Diário Español* (Tarragona, 5 June 1949).
Taussin, H. (with Chassant, A., *see* Chassant.
Tavares, Maria Ferro, *Os Judeus em Portugal no Século XV* (Lisbon, 1982).
Taviani, Paolo Emilio, *Christophe Colomb – genèse de la grande découverte* (Paris, 1980).
Taylor, H.O., *The Medieval Mind, Ideal and Actual Society*, IV (London, 1911).
Teran, Manuel de, *Imago Mundi – Geografia Universal* (Madrid, 1964).
Teste, Gino, *Dizionario di Alchimia* (Rome, 1950).
Thorndike, L., *A History of Magic, vol. II* (New York, 1934).
Toledo, Don Diego Ruíz de, *Códice diplomático-americano de Cristóbal Colón, etc.* (Havana, 1839).
Toledo, José Maria Ascensio y, *Cristóbal Colón, su vida, sus viages, sus descubrimientos, etc.* (Barcelona, 1891).
Torres, João Carlos Castelo Branco e, *Resenha das Famílias Titulares do Reino de Portugal* (Lisbon, 1928).
Touchaix, Charles, *Les secrets de la Kabbale* (Paris, 1978).
Tourniac, J., *De la Chevalerie au Secret du Temple* (Paris, 1975).
Tresmontant, C. *La Métaphysique du Christianisme et la Crise du XIII ième Siècle* (Paris, 1969).
Trevisano, Angelo, *Libretto di tutte le Navigatione di Re de Spagna* (Venice, 1504).
Tryon-Montalembert (with Hruby Kurt) *A Cabala e a Tradição Judaica* (trans.) (Lisbon, 1979).
Ughelli, Abbot Fernando, *Italia Sacra Sive de Episcopis Italiae et Insularum Adjatentitum* (Venice, 1717-22).
Uhajón, Francisco R. de, *La Patria de Colón según los documentos de las Ordenes Militares* (Madrid, 1892).
Ulloa, Luís, *Colomb, Catalan. La vraie genèse de la découverte de l'Amérique* (Paris, 1928).
Ulloa, Luís, *Noves proves de la catalanitat de Colom. Les grandi falsetats de la tesi genovesa* (Paris; n.d.).
Vacander, *Les Origines de L'Héresie albigeoise* (Paris, 1894).
Valles, Edgar, *'Especiarias do Oriente no Brasil e na África Portuguesa'*, *Garcia da Orta* 6(4) (Lisbon, 1964).
Valls, Ricardo Carreras y, *La Descoberta d'America – Ferrer, Cabot y Colom* (Reus, 1928).
Valls, Ricardo Carreras y, *El Català Xpo FERENS Colom de Terra Rubra, descobridor d'América* (Barcelona, 1930).

Van Praag, J.A., *Restos de los Idiomas Hispano–lusitanos entre los Sefardies de Amesterdam* (Madrid, 1931).
Varela, Consuelo, *Cristóbal Colón, Textos y Documentos Completos* (Madrid, 1984).
Vasconcelos, Frazão de, '*Diogo Gomes, Caravelista do Infante e o descobridor da Ilha de Sant'Iago de Cabo Verde*', Boletim Geral do Ultramar, 320 (April), (Lisbon, 1956).
Vasconcelos, Manuel y, *Vida y Virtudes del Rey D. Joam II* (Madrid, 1636).
Vasconcelos, Simão, *Chronica da Companhia de Jesus do Estado do Brasil* (Lisbon, 1663).
Vassel, J., '*Le Symbolisme des couleurs héraldiques*', Études Traditionelles (Paris, 1950).
Veer, de *Henry the Navigator* (Danzig, 1865).
Ventura, Augusto Gersão, *A Máquina do Mundo* (Oporto, 1944).
Verrazzano, Giovanni, *Letter* (1524), *see* Silva, Manuel Luciano da (1971).
Vidal, Rómulo Cúneo, see Cúneo-Vidal.
Vieira, Fr António, '*Sermões (Século XVII)*', in *Obras Completas*, 27 vols, (Lisbon, 1854–8).
Vieira, Fr António, *Sermões* (S. Paulo, Brazil, 1963).
Vieira, Ernesto, *Diccionário Biográfico de Músicos Portugueses* (Lisbon, 1900).
Viel, Robert, *Les Origines Symboliques du Blason* (Paris, 1972).
Vignaud, Henri, *Études Critiques sur la vie de Christophe Colomb avant sa découverte, etc.* (Paris, 1905).
Vignaud, Henri, *La Lettre et la Carta de Toscanelli sur la Route des Indes, etc.* (Paris, 1901).
Vignaud, Henri, *Americ Vespuce, etc.* (Paris, 1917).
Vilhena, D. Tomás de, *História da Ordem da Santa Cavallaria em Portugal* (Coimbra, 1920).
Villaut de Bellefond, P., *Relations des Côtes d'Afrique Appellées Guiné* (Paris, 1699).
Villemarque, Hersat de, *La Légende Celtique, etc.* (Paris, 1864).
Villeneuve, Armand de, *Le Rosaire des Philosphes* (Paris, 1972).
Villeneuve, Roland, *Sabats et Sortilèges – Érotologie de Satan* (Paris, 1970).
Visconde de Porto da Cruz, '*Janela que pertenceu ao Conselheiro Aires d'Ornelas*', in *Revista Portuguesa de Arqueologia* (Lisbon, 1936).
Viterbo, Sousa, *Elucidário* (Lisbon, 1865).
Viterbo, Sousa, *O Movimento Tipográfico em Portugal no século XVI* (Coimbra, 1924).
Volpe, D., *Movimento religioso e sette erecticali nella società medievale italiana (secoli XI–XIV)* (Florence, 1912).
Warustel, André, *Les Nombres et Leurs Mystères* (Paris, 1946).

Wassermann, Jacob, *Cristóbal Colón, el Quijote del Océano* (Spanish trans.) (Madrid, 1930).
Wechessler, E., *Das Kulturproblem des Minnesangs* (Halle, 1909).
Wiesenthal, Simon, *La Voile de l'Espoir* (Paris, 1972).
Williams, Roger, *The Key of the Language of America* (London, 1643).
Wiznitzer, Arnold, *Jews in Colonial Brazil* (Columbia, USA, 1960) (Brazilian trans., São Paulo, Brazil, 1966).
Witte, Canon M. de, 'Les Bulles Pontificales et l'Expansion Portugaise au XVième Siècle', *Revue d'Histoire Ecclésiastique*, XLVIII, 1953 (Louvain, Belgium, 1958).
Woodhouse, F.C., *Military Religious Orders* (London, 1879).
Zagallo, Manuel C. de Almeida Cayolla de Azevedo, *Cristóvão Colombo e a Ilha da Madeira – A Case de João Esmeraldo* (Lisbon, 1945).
Zaz, Enrique, *Galicia, patria de Colón* (Havana, 1923).
Zeromsky, Stefan, *Wiatr od Morza (Warsaw, 1922)*; (English trans., London, 1934).
Zúquete, Afonso Martins, *Nobreza de Portugal e Brasil* (Lisbon, 1960).

Index

Note: CC stands for Christopher Columbus. Relationships in brackets after names refer to him unless otherwise specified, and do not imply that the relationship is an established fact, but merely that it has been suggested.

Abraham, Don (astronomer) 253
Abraham, Rabbi of Beja 28
Abravanel, Isahac 28
Abreu, Inês de 437
Abyssinia 68, 227
Acuña, Cristóbal Vásquez de 441–2
Afonso, Diogo 57
Afonso, Prince of Portugal (son of John II) *tree* 443, 371
Africa 50–1, 55, 67, 71, 125
Aguiar family *tree* 367, *tree* 493, 374–82
   coat-of-arms *ill* 462
Aguiar, Álvaro Esteves de *tree* 493, 119, 395, 494–5
Aguiar, António Esteves de, Bishop of Ceuta *tree* 493, 498–9
Aguiar, Damião de *tree* 493, 491
Aguiar, Diogo Afonso de (step-father)
   tree 362, *tree* 366, *tree* 367,
   tree 445, *tree* 493, 138, 349, 443,
   449, 492
   marries Isabel da Câmara 352, 376–7, 382
Aguiar, Elena Esteves de *tree* 493, *tree* 497, 494, 496
Aguiar, Estêvão de, Abbot of Alcobaça *tree* 493, 375, 494, 499, 501–2
Aguiar, Fernando Esteves de, Abbot of Acobaça *tree* 493, 494, 496, 498
Aguiar, Gonçalo Anes de 380
Aguiar, João Afonso de *tree* 362, *tree* 366, *tree* 367, *tree* 453, *tree* 493, 349
Aguiar, Jorge de, Admiral 375, 379
Aguiar, Maria Esteves de 494

Aguiar, Pedro Afonso de *tree* 362, *tree* 445, *tree* 493, 375, 380, 494
Aguiar, Rui Dias de (half-brother) *tree* 366, *tree* 445, 492, 496
Alarcón family 433–4
Alarcón, Constanza de 430, 433–5
Alarcón, Constanza Rodriguez de 433
Alarcón, Diego de 430
Alarcón, Fernando de 433
Alarcón, Guiomar de 433–4
Alarcón, Juan de 434
Alarcón, Martin Rodriguez de 433–4
Alarcón, Martin Rodriguez de (II) 434
Albergaria, Diogo Fogaça de *tree* 411, 414
Albergaria, João Soares de 381, 392–3
Albumasar (astrologer) 253
Albuquerque, Afonso de 247
Albuquerque, Constança de Castro e 354
Albuquerque, Francisco Coutinho de *tree* 445, 385
Albuquerque, Gonçalo de *tree* 362, 386
Albuquerque, João Afonso de 361
Albuquerque, Pedro de 166, 361
Albuquerque, Pedro de, Admiral (grandson of above) 336
Alcácer do Sal 22, 37, 371
Alcáçovas, Treaty of 66, 68, 230, 484
Alcántara, Peace of 416
Alcazar-Ceguer 34, 38, 363–4, 423
alchemy *ill* 255, 25, 257–60, 353
Alcoforado, Vasco Afonso 491
Alemy, Isabel de *tree* 497, 495
Alexander II, Pope 20
Alexander III, Pope 5, 21, 42
Alexander VI, Pope 70, 72, 180, 222

553

554 INDEX

Alfragan (astronomer) 253
Algarve 356–7
Aljustel 37
All Saints' Bay 159
All Saints' Hospital (Lisbon) 30–1, 205
Almada, Aires de 72
Almada, Cecilia de *tree* 366, 437, 446
Almada, Constança Vaz de *tree* 366, 466
Almada, Francisco de 70, 386
Almada, João Vaz de 386
Almeida, Diogo de, Prior of Crato 345
Almeida, Estevão de 345
Almeida, Lopo de *tree* 402, 387
Almeida, Roiz de *see* Sá, Roiz de
Alpha, Cape *map* 341, 332
Alphonse VI, King of Castile 9
Alphonse XI, King of Castile 48, 171, 408
Alphonse I, King of Portugal *table* 34, 4, 9, 21–2, 36–7, 47
Alphonse II, King of Portugal *table* 34, 13, 37, 117
Alphonse III, King of Portugal *tree* 409, 23, 37, 46
Alphonse IV, King of Portugal *tree* 409, *tree* 454, 24, 47–9, 171, 408
Alphonse V, King of Portugal *tree* 344, *tree* 444, 62, 87, 314, 316, 355–7, 371, 381, 445–6, 472
  and battle of Toro 414–15, 442, 447
  and maritime expansion 117, 134–6, 153, 164
  at Tangier 363, 423–4
'Alphonsine Ordinances' 43, 420
Alpoim, João 27
Alvito (Portugal) 361, 363–4, 370
Amazon, discovery of 160
'Amazons' 485
America
  name of 160
  pre-Columbian discoveries of 55, 120, 131, 143–4, 149–61, 486
Amião, Maria 494
Amores, Juan de 56
Amoury, Armand 6
Andrade family 241
  coat-of-arms *ill* 462, 268
Andrade, João Fernandes de 130, 437
Andrade, Rui Freire de, Admiral *tree* 362, 385
Andrade, Teresa de *tree* 362, 385
Anghiera (Angleria), Pietro Martír de 82–3, 160, 179–80, 181, 184–5, 396
Anjou, René Duke of 102–3, 140
Annes, Lourenço 371
Annes of Beja, Aires *tree* 452, 348, 384
'Antilia' (island) *map* 154, *map* 172, *map* 174, 120, 152–3, 163–4, 166–7, 172–3, 321

Antille (island) 129
Antilles (West Indies) *map* 73, *map* 75, *map* 77, *map* 341 (*see also individual islands*)
  CC in 12, 124, 128, 141, 222–3, 388
  pre-Columbian knowledge of 58–61, 72, 124, 130, 153, 481–6
António, Master 28
Aragon, as birthplace of CC 98
Arana family 428–30, 434–6 (*see also* Harana)
Arana, Constanza de 430
Arana, Diego de 428–30, 434–6
Arana, Pedro de 217, 427, 429–30, 431, 435–8
Arana, Rodrigo de 430, 432–5, 437
Arbués (Inquisitor) 417
Arco, Fernão Domingos do 241
Arco, Gonçalo Esteves do *tree* 362, 498
Arco da Calheta (Madeira) 130, 241
Aredo, Diogo de *quoted* 17
Árias, João 125
Arles, Council of 6
armillary sphere *def* 237–8, 118, 269
arm (in heraldry) 460
Arzaquel (inventor of astrolabe) 253
Arzila (North Africa) *map* 76, 399–402
Ascension Island 58
astrolabe 118–99, 253
astrology 253–4, 254–5, 256
Asturias, Rodrigo 410
Ataíde, Constança de 367, 369
Ataíde, Martinho de 369
Ataíde, Pedro de 367, 369
Aval, Constance de 415
Aveiro, João de 67
Avila, Don Luis de 220
Ayala y Guzmán, Maria de 369
Ayala y Toledo, Teresa Maria de 220
Azambuja, Cecilia de 424
Azambuja, Diogo de 227, 418, 424
Azambuja (Portugal) 69, 124
Azcoitia, Pedro 208
Azevedo, Brites de *tree* 453, 378
Azores *map* 59, *map* 76, on *maps* 171, 153, 162 (*see also* Brazil (Azores); Flores; Santa Cruz; Santa Maria; Terceira)
  CC in 124, 389–90
  discovery of 56–7
  on maps 125, 139, 169
  population 29, 138–9
Azurara 363

Baga Sea 153, 155–6
Bahamas 70, 386
Balearic Isles 4
Baptiste, João 139

INDEX 555

Barbados *map* 77, 61, 163
Barbosa, Duarte 129
Barcelos, Pedro de 63, 143, 149, 158
Barcia, González 183
Bardi, Francisco de *tree* 452, 450
Bardi, Giacomo de 450
Barreda, Rodrigo de 217
Barreto, Afonso Telles 354
Barreto, Aldora Afonso 491
Barreto, Beatriz 491
Barreto, Brites Pereira 379
Barreto, Gonçalo Nunes 354–5, 379
Barreto, Isabel Moniz *tree* 356, 185
Barreto, Leonor Telles *tree* 366, 379
Barreto, Vasco Moniz (II) *tree* 452, 355
Barrientos, Hernando de 441
Barros, João de *quoted* 52, 133, 139, 185, 228
Basque nationality and CC 111
Bautista, Juan 139
Beatriz, Marchioness of Moya 414
Beatriz de Braganza, Duchess of Beja (wife of Prince Fernando) *tree* 344, *tree* 365, *tree* 444, 327, 351, 381, 392, 416, 489
Behaim, Martin 118–19, 122, 124, 138, 169, 172, 254, 449
Beja, Abraâo de 68
Beja (Portugal) 37, 104, 313, 327–8, 356–7, 371, 381, 413
Belém *map* 341, 332, 394
Belmont, Jean de 7
belt (baldric), in heraldry *ill* 459, 457, 460–1
Benedict XI Pope 39
Benedict XIV, Pope 20
Benin 60
Bermudas 58
Bernáldez, Andrés 82, 185, 198
Bernard, St 4, 20, 35–6, 37–8, 42, 505
Bertaldi, Nicolo 47
Béziers 6, 13
Bianca of Navarre 419
Biancasa, Gracioso 125
Bianco Map ill 172–3, 125, 137, 153, 155, 171–3
Bissipat, George 100
Bizagudo, Pedro (Cunha, Pedro Vaz) 60–1
Boa Vista (Cape Verde Islands 136, 332
Bobadilla, Beatriz de *tree* 411, 416
Bobadilla, Inés de *tree* 411, 414
Bocanegra, Don Carlos de Córdoba y 220
Bocanegra, Mécia de *tree* 362
Boniface VIII, Pope 39, 371
Book of Marco Polo ill 249, 248
Bordeille, Viscount Andrew of 169

Borgonha, Princess Maria de 345
Borromeu, Count Giovanni 179
Bourbon, Duchess of (Anne de Beaujeu) 328–9
Bovarello (son-in-law of Domenico) *tree* 190, 192
Branco, Cape 136
Brandâo, Friar Francisco 384
Brasil (Antilles) *map* 341, 332
Brazil (Azores) 169, 171
Brazil (South America) 72, 141, 159
 discovery of 57–62, 158, 160–1, 241
Brendan, St, myth of 168–71
Brito, Afonso Vaz de 374
Brito, Artur de *tree* 366, 446
Brito, Constança de 497
Brito, Estevâo de *tree* 366, 374, 420
Brito, Guiomar de *tree* 425, 425
Brito, Isabel de *tree* 411, *tree* 425, 420, 424
Brito, Margarida de *tree* 366, 374, 420, 488
Brito, Martinho de 420
Buena Vista (Cuba) 332

cabala *ill* 307, *ill* 309, *ill* 312, *ill* 317, 260, 262–81
 alchemic 248, 272–3, 287
 dextratio *ill* 277, 277–8
 gematric method *ill* 27, 244, 273, 301
 mirror method *ill* 303, *ill* 305– *ill* 306, *ill* 310, 242, 278, 284, 286
 repetition *ill* 305, 304–5
 sephiroth *ill* 264–*ill* 266, *ill* 287, *ill* 310, *ill* 313, 262–3, 272, 312, 317
 temurah *ill* 279, *ills* 286–7, 251, 273, 279–81, 283–99, 304
Cabral, Diogo *tree* 445, 353, 355, 385
Cabral, Fernão Álvares *tree* 362, 385
Cabral, Gonçalo Velho 57
Cabral, Isabel 355
Cabral, Pedro Álvares *tree* 362, 60, 158–9, 360, 385
Cabrera, Andrés de, 1st Marquis of Moya *tree* 411, 408, 413, 416
Cabrera, Juan de 414
Cabrera, Juan de, 2nd Marquis of Moya *tree* 411
Cadamosto, Luigi 53, 55, 57, 128, 136–8
Cadaval Castle 363
Caetano, Francisco 126
Caetano de Lima, Luiz *quoted* 374, 23
Calheta do Arco (Madeira) 241
Calixtus II, Pope 50
Câmara family *tree* 411, *tree* 445, *tree* 493, *tree* 497, 241, 257, 314–15, 331, 353, 364, 413, 498–504
Câmara, Beatriz da *tree* 362, *tree* 445, 386

Câmara, Cristóvão Esmeraldo Atouguia da 352
Câmara, Elena Gonçalves da (aunt) tree 445, tree 493, tree 496, 497
Câmara, Inês Dias da tree 362, tree 493, 361, 492
Câmara, Isabel Gonçalves (Sciarra) da (mother) tree 362, tree 365–tree 367, tree 411, tree 444–tree 445, tree 493, tree 497, 315, 331, 349–52, 381–2, 385, 420, 443, 489, 492, 500
Câmara João Gonçalves da (uncle, 2nd Captain of Madeira) tree 366–tree 367, tree 411, tree 445, tree 447, 351, 364, 392–3, 401, 437, 443, 448
Câmara, João Roiz (Rodrigues) da tree 445, tree 452, 381, 448
Câmara, Maria Gonçalves da 491
Câmara, Rodrigo da (seventh Captain of São Miguel) 353, 496
Câmara, Rui Rodrigues da 381
Caminha, Pedro (pero) Vaz de 61
Camões, João Vaz de 360
Camões, Lopo de tree 362, tree 445, tree 493, 361, 492
Camões, Luís Vaz de tree 362, 360–1, 479
Canada 63, 149 see also Newfoundland
Canary Islands map 73, map 76, 48–51, 54–5, 66, 169, 171, 392, 397, 399–400
Cano, Sebastian del 71, 129
Cantino Map 126, 158
Câo family 461–2
coat-of-arms ill 463, 464
Câo, Diogo 118, 133–4, 499
Cape Cross (SW Africa) 141
Cape Horn (north) 147
Cape Non 105
Cape of Good Hope 129, 133–4, 134, 227, 230, 247, 394
Cape Verde Islands map 76, 29, 67, 131
discovery of 55, 57, 137–8
Capitulaciones 184–6, 215–17, 222, 414, 416, 484
Cardona, Bereguela of 380
Cardona, Maria de (grandaughter) 219
Caribs 69, 161
Carmo, Monastery of 354–6
Carrillo, Maria Afonso de 443
Cartularios 223–5
Carvajal, Fernandino de, Bishop of Cartagena 70
Carvajal, Garcia do 72
Carvalhal, João Nunes de 420, 424
Carvalho, João Lopes de 133, 369
Cascais 363
Casenove, Guillaume de 100

Castanheira, João 390–2, 392–3
Castela, Branca de 7
Castel-Branco, Martinho de tree 445, 385
Castel-Branco, Nuno Vaz de 385
Castel-Branco, Simâo de tree 445, 385
Castelnau, Pierre de 6
Castile 10, 13, 117, 166 (see also Spain)
Castilian language 87, 99, 198, 503
cosmography in 68, 126
Castro family tree 402
coat-of-arms 419
Castro, Álvaro de tree 402, 385, 402
Castro, Álvaro Pires de 419
Castro, Brites de tree 366, 374, 419
Castro, Catarina de 361, 386
Castro, Constança de tree 362, tree 453, 369, 486
Castro, Fernando de tree 365, tree 402, 386, 402
Castro, Fernando de (II) 386
Castro, Francisco de 421
Castro, Guiomar de tree 362, tree 366, 419, 432
Castro, Inês de 415
Castro Isabel de tree 362, tree 365, tree 367, tree 402, tree 411, 360, 363, 369, 385, 402, 408
Castro, João de tree 365, tree 402, 351
Castro, Pedro de tree 362, tree 365–tree 366, tree 453, 374
Castro, Pedro de, Admiral 369, 385–7
Castro, Pedro Rodrigues de 385
Castro, Rodrigo de tree 402, 402
Castro, Vaz de 418
Castro e Albuquerque, Constance de 354
Castro-Marim 41–2
Catalá, Eurógio 210
Catalan language 87, 90
Catalonia see under Spain
Cataño, Francisco 224
Catarina, Princess (sister of Alphonse IV) 329, 408
Cathars 3–10, 13, 35, 41–2, 47–8, 180–1, 189, 272, 284–5
surnames 8
Cazzano, Francisco de 136
Cazzano, Lucas de 135
Celestine V, Pope 371
Centurione, Ludovico see Escoto, Luis Centurión
Ceuta 22, 49, 424
chain (in heraldry) 460
Chanca, Dr 198, 417
Chanca, Garcia 417
Charles I, King of Spain (Emperor Charles V) tree 444, 216–17, 440, 443, 447
Charles 'the Bold' Duke of Burgundy tree 444, 329–30, 343

Charles V, Emperor *see* Charles I, King of Spain
Charles VIII, King of France 329
Christ, Monastery of (Tomar) 260
Christ, Order of 10–11, 15, 24, 42–4, 152, 241, 256, 274, 314, 351, 356–7, 504
  CC's membership of 11, 69, 112, 347
  and Portuguese exploration 43, 49, 51–3, 118, 120, 169
  route to India 50–1, 69
  secrecy rule 55, 122, 151
Christian I, King of Denmark and Norway 62
Cistercian Order 5, 9, 37, 42
Clement IV, People 41
Clement V, Pope 20, 41, 67
Clement VI, Pope 20, 48–9, 171
Clermont, Council of 35
Cod Island *see* Newfoundland
Coelho, Gonçalo 62
  Land of 124
Coelho, Joâo 55, 60, 62, 130
Coelho, Nicolau 460
Coelho da Sylva, Martim 418
Coimbra, Cortes of 31
Colóm, Cristóvam *see under* Columbus, Christopher (names of)
Colóm, Diogo *see* Columbus, Diogo (brother)
Colombo, Amigheto (cousin) *tree* 190, 190, 192
Colombo, Antonio (uncle) *tree* 190, 189, 221
Colombo, Baldasário 199, 220–2
Colombo, Benedicto (cousin) 190
Colombo, Bianchinetta (sister) *tree* 190, 192
Colombo, Doménico (father) *tree* 190, 189, 192, 197, 199, 221–2, 326
Colombo, Giovanni Canajole de (grandfather) 189
Colombo, Giovanni (cousin) *tree* 190, 189–95
Colombo, Mateu (cousin) *tree* 190, 190, 192
Colombus, Diogo (son) 200–1
Colón, Bartolomeu (brother) *see under* Columbus, Bartolomeu
Colón, Christo-Ferens *see under* Columbus, Christopher (names of)
Colón, Cristóbal (bookseller) 198
Colón, Cristóbal (Cristóvam) *see under* Columbus, Christopher (names of)
Colón, Diogo (brother) *see under* Columbus, Diogo
Colón, Diogo (son) *tree* 190, *tree* 444, *tree* 452, 83, 101, 158, 192, 338, 416, 427, 435, 443

will of 218–19, 354
Colón Fernando de (son) *quoted* 100, *tree* 190, *tree* 444, 82–4, 89, 101–2, 104, 106, 135, 143, 182, 194, 196, 204, 216–17, 221, 227, 275, 325, 345, 395–6, 399, 414, 427, 434–5, 436, 439–40, 441–3, 446–7, 472, 482–3, 501
  biography of CC 84
  birthplace of 218–19
  on birthplace of CC 186–7
  'Columbian Library' of 83–4
  and culture of CC 23
  and Harana family 429–30, 434–6
  and Henry 'the Navigator', Prince 54
  on name of CC 183–4
  and Thule 146
Colón, Fernando (son) 206
Colón, Filipa (great-granddaughter) 219
Colón, Isabel de (grand-daughter) 220, 378
Colón, Juán 98
Colón, Juana (grand-daughter) 220
Colón, Luis (grandson) 84, 183, 218–19, 219
Colón, Maria (great-granddaughter) 219
colon 238
  in CC's signature *ill* 317, 285, 316–19
Colonna family 498–504
Colonna, Agapito, Archbishop of Lisbon 375, 434, 500
Colonna, as name of CC 187–8
Colonna, Cecilia *tree* 362, 498–500
Colonna, Giácomo 499
Colonna, Pietro 499
Colonna, Sciarra 499–500
Colonna, Stephan 499
Colonna, Vittoria 499
Colonna family 498–504
  arms of 375
  and name of CC 187–8
Colonus *see under* Columbus, Christopher (names of)
Colón-Zarco *see under* Columbus, Christopher (names of)
Columbus, Antonio (uncle) 210
Columbus, Baptistinna (aunt) 210
Columbus, Bartolomeu (brother) 107, 207
  as Colombo, Bartolomeu 107, 199, 326
  as Colón, Bartolomeu *tree* 190, *tree* 493, 215–16, 227, 247, 325, 327–30, 347, 360, 380, 488, 496
  as Joâo Afonso de Aguiar 375–84, 488
Columbus, Christopher *see also under* John II, King of Portugal
  ancestry and family *tree* 190, *tree* 362, *tree* 365–*tree* 367, *tree* 402, *tree* 411, *tree* 444–*tree* 445,

tree 452, *tree* 454, *tree* 493, *tree* 497, 393–7, 413–17; and admirals 384–7; and Aguiars 495; and Colonnas 500–2; father *see* Colombo, Doménico; Columbus, Doménico; Fernando, Prince; Henry 'the Navigator'; Jewish ancestry 120, 188, 211; marriage of 185, 255–6, 347 (*see also* Moniz Perestrelo, Filipa); mistress of *see* Henriques, Beatriz; mother *see* Fontanarossa, Susanna; Gonçalves da Câmara, Isabel; noble origin of 112, 184–5, 187, 328–9; plebeian origin of 181–2; royal origin of 473, 503; and Vasco da Gama 361
birth and nationality 81–5, 85, 97–112, 214–18, 360, 504 (*see also* Corsica; Genoa; Italy; Pontevedra; Portugal; Spain; Switzerland) birthdate of 195–7
descendants 489–90; sons of *see* Colombus, Diogo; Colón, Diogo; Colón, Fernando; grandchildren *see* Colón, Luis; Toledo, Cristóbal de; great-grandson *see* Toledo, Diego de; and house of Braganza 363
documents: *Capitulaciones* 184–6; diaries 84; *Imago Mundi* note 227–8; letters 105, 140, 223–7, 328–9, 397, 399 (*Litera rarissima* 196–7; *Libro de las Profecias* 235–6; *Memorando of Debts* 207–12; *Memorial* 205–7, 222; *Memorial* to Ferdinand and Isabella (1494) 214; *Provisión* of 1493, 184; *Royal Deed* 205; wills 199–207, 214, 217–19, 221, 427)
heraldry, siglum and signature: coat-of-arms *ill* 466, 93, 184, 202, 225, 465–7; monogram *ills* 319, 319; siglum *ill* 284, *ill* 303, 202–3, 204, 221, 238, 241–5, 252, 283–314, 380, 500–1, 504 ('Colon' in *ill* 238; Fernandez in 311, 313; Salvador in 304; Zarco in 239); signature 81, 85, 202, 207, 247
language of *see* Hebrew language, Latin language *and under* Italy, Spain, Portugal
mission of 68–9, 141, 504–6
names assumed: Christoferens 180; Christóforo Colombo 179; Christophorus 180; Colóm, Cristóvam 186, 229, 320, 326, 347, 383, 440; Colombo 82, 184, 188–9, 201; Colombo, Christóforo *tree* 190, 81, 191–5, 199; Colombo, Jacopus

de 191–2; Colón, Christo-Ferens 103–4, 110; Colón, Cristóbal (Cristóvam) 69, 81, 85, 98–9, 106, 139, 179–80, 183–4, 184–5, 187–8, 200, 205, 238, 440; Colonus 180–1, 183–4, 326; Colón-Zarco 81, 83–4, 106, 120, 161, 237–41, 316; Columbus 183; Columbus 3, 180; Coullon, Jean 102–3, 104, 110, 151; Culá 440–1; Fernandez Zarco, Salvador 239, 247, 304, 311, 313; Scolvus, Joannes 110
personal life of 486–92; childhood 381–2; death 207, 252, 298; education 81, 91, 93, 95, 236–7, 326–7, 383–4, 503; finances 206–12; and Holy Ghost 11; imprisonment 222; living in Lisbon 101, 211; physical appearance 111, 472–4; and Templars 71
seamanship and cartography of 236–7; cosmography of 119, 161; knowledge of America before 1492 130–1; *maps* of 158, 162, 164, 197; measurements used by 126–7, 128
and Spain: arrival in Castile 195–7, 449; mission to 69, 123; offers services to 105–6; receives title of Admiral 102
voyages: before 1492 129–30, 133–7; first American voyage 69–70, 124, 127, 129, 163, 388–93, 495; third American voyage 131, 330, 428–9; fourth American voyage 131, 399, 414, 481–2; Luso–Danish expedition of 1477 63, 134, 151; sails with Dias 134, 247; shipwreck 100–7, 130, 196
Columbus, Diogo (brother) 94, 107
as Afonso de Aguiar 375–84
as Colón, Diogo 217, 347
Columbus, Diogo (son) 107, 204, 211, 219, 223–4, 226
CC's memorial to 205–7
lawsuit of 214–18
will of 218–19
Columbus, Doménico de (father) 191, 209–10
columns *ill* 266, 251, 459–60, 498
compasses (in heraldry) *ill* 459, 268–9, 456
Concepción, in Portugal 333
Concepción island *map* 341, 333, 482
Consolation Point 160
Constance, Queen of Léon and Castile 9
Conti, Patricio 52

Contreiras, Miguel 30
Córdoba y Bocanegra, Don Carlos
  de 220
Corrêa, Isabel 354
Correia, Germano 168
Corsica, as birthplace of CC 98
Corte-Real family *tree* 362
Corte-Real, Fernão Vaz 360, 496
Corte-Real, Gaspar Vaz 61, 63, 149, 157,
  360
Corte-Real, João Álvares 149–51, 153, 157
Corte-Real, Miguel Vaz 63, 157–8, 360
Corte-Real, Vasqueanes 157
Corvo (Azores) *see* Santa Cruz
Costa, Fontoura da 249
Costa, Isabel da *tree* 452
Coullon, Jean *see under* Columbus,
  Christopher (names of)
Coutinho family *tree* 402, 370–1, 498
Coutinho, António de Azevedo 440–2
Coutinho, Beatriz 363
Coutinho, Catarina 414
Coutinho, Diogo (brother of 1st Count of
  Marialva) 420
Coutinho, Fernando *tree* 402, *tree* 445,
  29, 361, 402
Coutinho, Filipa 361
Coutinho, Francisco, Marshal, 4th Count
  of Marialva *tree* 365, *tree* 411, 416
Coutinho, Francisco (brother of 1st Count
  of Marialva) 420
Coutinho, Gonçalo, Marshall, 3rd Count
  of Marialva *tree* 365, *tree* 411,
  tree 453, 361, 370, 414, 416
Coutinho, Gonçalo, 2nd Count of
  Marialva 378
Coutinho, Isabel 416
Coutinho, Joana *tree* 366, 420
Coutinho, João *tree* 453, 361
Coutinho, Manuel de Sousa *quoted* 13
Coutinho, Maria *tree* 402, 402
Coutinho, Vasco de Menezes, Count of
  Borba *tree* 402, 401, 446
Coutinho, Vasco Fernandes, 1st Count of
  Marialva *tree* 366, 370, 401, 420
Coutinho, Vasco (grandfather of 1st
  Count of Marialva) *tree* 36, *tree* 402,
  361, 370, 420
Coutinho de Albuquerque, Francisco
  tree 445, 385
Covadonga, battle of 36
Covichã, Pedro da 28, 51, 68, 123, 167
Cresques, Abraham 52, 56
Cresques, Jehuda 92, 120
Cresques, Samuel 92, 120
Crusades *table* 34, 8–9, 21–2, 34–8, 46,
  371

cuba, word *ill* 332, 331
  in siglum *ill* 317, 317–18
Cuba (island) *map* 75, *map* 77,
  map 341, 69, 329, 331, 483
Cuba (Portugal) *map* 340, *map* 451, 104,
  318, 331, 351, 361, 363–4
  as birthplace of CC 380, 504
Cuña, Catarina de 433
Cunha, Álvaro da *tree* 453, 385, 395
Cunha, Filipa da *tree* 402, 401
Cunha, Isabel da *tree* 402, *tree* 453, 386
Cunha, Luiz Caetano da *quoted* 374, 23
Cunha, Pedro (Pero) Corrêa da (brother-
  in-law) 134
Cunha, Pedro (Pero) Vaz da ('the
  Bizagudo') 60–1
Cunha, Rui de Mello da, Admiral, 385
Cypangu *see* Japan

Dacuña, Cristóbal Vasquez 441–2
D'Albano, Henry 5
Damán, Alvaro *see* Domião de Aguiar,
  Alvaro
Damião (Damán) de Aguiar, Alvaro
  tree 493, 394–5, 492–5
Daniel, Filipe 23
Darce, Antónia Dias 138
Dayala, Pedro 72
D'Eça *see* Deza
Delgado, Aldonça 256, 347
De Navigatione Columbi 197–8
Dennis, King of Portugal *tree* 409, 10–11,
  23, 41–2, 47, 53, 86, 168, 371, 408,
  422, 436, 472
Deza, Alonso Goméz, Admiral 415
Deza, Antonio de *tree* 411, 416
Deza, Fernão *tree* 411, 416
Deza, Friar Diego de, Archbishop of Sevill
  tree 411, 416
Deza (d'Eza), Fernando 415–16
Deza (d'Eza), Joana 414–15
Deza (d'Eza), Maria 414–15
Djaber-ben-Afflah (astronomer) 253
Dias, Bartolomeu 117, 129, 133–4, 227–8,
  230, 247, 489
  of 'São Cristóvâo' 394, 495
Dias, Bernal 70
Dias, Diogo 395
Dias, Rui 448
Dias, Vicente 57, 135–7
Dighton, rock of *ill* 152, 63, 151–2, 158
Diogo, Duke of Viseu and Beja (half-
  brother) *tree* 344, 43, 166, 230, 351,
  478, 486, 504
Dominguez, Isabel *tree* 365, *tree* 367,
  363
Dominican Order 13, 29–30, 84

Dória, Francisco 224
Dornelas, Afonso 238
D'Ornelas family
    arms of *ill* 316, 315
dove (in heraldry) 7–8, 183, 275
Duarte, King of Portugal *tree* 344,
    tree 362, *tree* 365–*tree* 367,
    tree 409, *tree* 444, *tree* 454,
    tree 493, 26–7, 87, 256, 316, 349
Duarte, Prince 86
Dulmo (de Ulm), Fernando 60, 138–9

eagle (in heraldry) *ill* 250, 457–8, 461
Edward IV, King of England 329
Ega (Portugal) 41
Egidio, Father, of Bologna 52
Elvas (Portugal) *map* 451, 37, 380
England 47, 143–4, 148
Ennoxedo, Pedro (notary) 207
Enriques, Leonor *tree* 366
Esbarraya, Leonardo 428
Escabedo, Rodrigo de 203
Escobar, Pedro 57
Escoto, Luis, Centurión 208–10, 212
Esmeraldo family
    coat-of-arms *ill* 463
Esmeraldo, Cristóvam 342
Esmeraldo, João 342–3
    coat-of-arms of 343
Esmeraldo Atougia da Câmara,
    Cristóvam 352
Espada, Marcus de la 56
España, Luis de *tree* 409, 410
Espichel, Cape 22
Espíndola, Gaspar d' 224
Espínola, Baptista 208, 211–12
Esteves, Maria *tree* 362, *tree* 366, *tree* 367, 375
Esteves (Arco), Gonçalo *tree* 362, 498
Esteves de Aguiar, Álvaro *tree* 493, 119,
    395, 494–5
Esteves de Aguiar, António, Bishop of
    Ceuta *tree* 493, 498–9
Esteves de Aguiar, Elena *tree* 493, *tree*
    497, 494, 496
Esteves de Aguiar, Fernando, Abbot of
    Alcobaça *tree* 493, 494, 496, 498
Estreito, João Afonso do 60
Estremadura, as birthplace of CC 98
Ethiopia *see* Abyssinia
Eudes, Duke of Burgundy *tree* 9, 9
Eugene IV, Pope 43, 50, 54, 135, 355
Évora (Portugal) *map* 451, 25, 27, 37, 164, 166

Falcão, Branca de Sousa 392
Faleiro, Francisco 122
Faria, Manuel Severim de 384

Faria, Sebastião de 353
Felipe, Marcus 187
Ferdinand IV, King of Castile 41
Ferdinand I, King of Portugal *tree* 409,
    25, 415
Ferdinand V, King of Spain *tree* 444, 68,
    70–1, 103–4, 109, 159–60, 203, 215,
    254, 408, 432, 447
    CC's letter to 105, 140
Fernandes, Álvaro 149
Fernandes, João 42, 371
Fernandes, Valentim (printer) 246–9,
    247, 312, 343
    signature of *ill* 248, 247–8
Fernandez, Beatriz *tree* 410, 410, 412
Fernandez, Garcia 94
Fernandez Zarco, Salvador *see under*
    Columbus, Christopher (names of)
Fernandina (island) *map* 75, 330–1
Fernando, 2nd Duke of Braganza
    tree 365, 351, 381
Fernando, Prince, Duke of Viseu and Beja
    (father) *tree* 344, *tree* 362, *tree* 365–
    tree 367, *tree* 411, *tree* 444–
    tree 445, *tree*, 454, *tree* 493,
    tree 497, 55, 136, 327, 331, 351, 356,
    379, 382, 420
    attempted move to Naples 380–1
    as father 244, 283, 315, 349, 472, 504
    life of 486
    and Order of Christ 43
    in siglum 313–14, 318
Fernando, Prince ('La Cerda') 410
Fernando, Prince (the Martyr of Fez)
    tree 344, 69, 475
Fernando, 3rd Duke of Braganza (plotter)
    tree 365, 166, 351, 357, 361, 478
Ferrare, Ricobaldo 51
Ferreira, Santos 285–6
Fiesco, Bartolomé 210
Filipe of Castile, Lord Cabrera 413
Flanders 47, 134
Flora, Joachim de 10
Florence 47
Flores (Azores) 163, 320
Florida 61, 168
Fogaça, João 415
Fogaça de Albergaria, Diogo *tree* 411, 414
Fonseca, Juan de, Bishop 198, 237
Fontanarossa, Susanna (mother) *tree* 190,
    191, 473
Forjaz, Elvira *tree* 497
Fornari, Baliano 183, 187–8
France 102, 328–9
    as birthplace 103
France, Lançarote de 49
Francisca de Toledo, Ana 220
Frederick III, Emperor *tree* 444, 357, 387

Freire de Andrade, Rui, Admiral
  tree 362, 385
Frio, Cape 159
Frisland *see* Thule
Fróis, Estêvão 62, 130
Furtado de Mendoça family
  coat-of-arms *ill* 463

Galeazzo, Giovanni 255
Galego, Vasco 129
Galera, Ponta de *map* 341, 333
Galicia, as birthplace of CC 111
Galician language 87, 90–1, 99
Gallego, Alonso 320
Gallo, Antonio 197–8
Gama family *see also* Roiz da Gama
Gama, Estêvão da *tree* 362, 104, 361
Gama, Guiomar Vaz da 361
Gama, Sancho Roiz da 417
Gama, Vasco da *tree* 362, 50, 104, 131, 133, 361
Garcia, Urraca 433
Gaston, Count of Medinaceli 410
Gaunt, John of, Duke of Lancaster
  tree 344, *tree* 454, 285, 472
Gelasius, Pope 20
Genoa 4–5, 13, 215, 222, 470 (*see also* St Lawrence, Church of)
  Bank of St George 69, 86, 89, 97, 182, 194, 197–8, 201, 204, 206, 212, 221, 224–7
  as birthplace of CC *tree* 190, 69, 81–2, 84, 86, 90, 93–4, 97, 99, 101–4, 105–7, 110–11, 134–7, 140, 179, 181–2, 186, 189–95, 195, 197–9, 200–2, 204, 209, 217–18, 219, 221–2, 325, 343, 415, 432, 450, 491, 501
  in CC's siglum 311
  Ebro, island of (Genoa), as birthplace 109–10
Genoa (island), as birthplace of CC 109–10
Germaine (de Foix), Queen of Spain (2nd wife of Ferdinand V) 215
Ghedaliah (Portuguese Jew) 25
Ghedélia-aben-Judáh, Dom 23
Gil, João 353
Giocondo, Friar Giovanni 159
Giustiniani, Agostino, Bishop of Nebbia 181
Gogoreto, as birthplace of CC 98
Góis, Branca de 496
Góis, Brites (Beatriz) de *tree* 362, *tree* 452, *tree* 497, 355, 452, 496
Góis, Damião de *tree* 497, 51, 492–3, 495–6
Góis, Judith de *tree* 497, 360, 496
Góis, Luís Afonso de 129

Góis, Rui Dias de 497
Gomera (Canaries) 414
Gomes, Diego 57
Gomes, Diogo 137
Gomes, Estêvão 129
Gomes, Maria *tree* 409, 422, 436
Gomes, Marina *tree* 409, 422
Gonçalves, Brites (Beatriz) de 350, 353, 385
Gonçalves, Diogo 494
Gonçalves, Gabriel 139
Gonçalves, Helena 139
Gonçalves, Nuno, polyptych of *ill* 477, 472, 474–6, 544
Gonçalves, Pedro 139
Gonzáles, Tomas 199
Gorrício, Friar Gaspar 83, 95, 217, 226, 235, 320
Grã, Francisca da 493
Graciosa islands *map* 341, 334
Granada 38, 68–9, 141
Greenland *map* 147, 62, 112, 144, 150, 156–7
  CC's voyage to 123
  as Thule 143, 148
Gregory I ('the Great'), Pope 20, 35
Gregory VII, Pope 5
Gregory IX, Pope 20, 22–3, 500
Grimaldi, Giovani Antonio 182
Guadalupe (Antilles) *map* 77, *map* 341, 334, 483
Guadalupe (Portugal) *map* 451, 389
Guedáliah, Master (astrologer) 27, 256
Guinchos (Antilles) *map* 341, 334
Guinea 135–6, 392 *see also* Africa
Gusmão, Álvaro de 408
Guzmán, Afonso Henriquez de *tree* 409, 413, 437
Guzmán, Beatriz de *tree* 367, *tree* 410, *tree* 447, 412
Guzmán, Carlos Henriquez de *tree* 447, 413, 437, 446
Guzmán, Diego Henriquez de *tree* 367, 446
Guzmán, Francisco de 420
Guzman, Henriquez de, Duke of Medina Sidonia *tree* 409, 412, 420
Guzmán, Juan de, Duke of Medina Sidónia *tree* 409, 410, 412
Guzman, Leonor Nunes de *tree* 409, *tree* 411, 407, 410

Haagen, Willem van der 139
Haiti 58, 69
Harana family *see* Arana
Harana, Rodrigo Enriquez de 427–8
Hargen, Joan of 492
Hassan-ben-Haiten (astronomer) 253

Hebrew 92, 120
Heisterbach, Cesar of 6
Henriques, Afonso 412
Henriques, Beatriz tree 366–tree 367, tree 425, tree 444, tree 447, 374, 421–2, 425, 427–38, 432, 437, 489, 504
Henriques, Carlos tree 366, 443
Henriques, Diego tree 366, tree 477, 413
Henriques, Diogo 412
Henriques, Fernando tree 366–tree 367, tree 425, tree 447, tree 453, 364, 374, 378, 443, 445
Henriques, Henrique tree 445, tree 447, 374, 412
Henriques, João 364, 448
Henriques, Mécia tree 366, 487–8
Henriques, Pedro 37
Henriques, Violante tree 366, tree 424, 367
Henriques family tree 367, tree 424, tree 447
coat-of-arms ill 468, 467
Henriques (Miranda), Briolanja 424
Henriques (Pereira), Beatrice 423
Henriquez, Afonso tree 410, 412
Henriquez, Alonso 414–15
Henriquez, Beatriz tree 190, tree 445, 206, 431, 437–8, 442
Henriquez, Beatriz (daughter of above) 437
Henriquez, Diego tree 367, tree 410, 412
Henriquez, Fernando tree 410, 413
Henriquez, Fradique, Admiral tree 411, 414–5
Henriquez, Garcia 437
Henriquez, Henrique 354
Henriquez, Juan tree 447, 413
Henriquez de Guzmán, Afonso 437
Henriquez de Guzmán, Carlos tree 447, 413, 437, 446
Henriquez de Guzman, Diego tree 367, 446
Henriquez de Sotomayor, Fernando tree 447, 437
Henry, Count of Burgundy tree 9, 8–9, 22, 36
Henry II, King of Castile tree 362, tree 410, 407–8, 413–14
Henry IV, King of Castile 419–20, 432
Henry 'the Navigator', Prince tree 344, 43, 49–51, 53–6, 67, 313–14, 356, 504
death of 52
as father of CC 244
Henry VII, King of England 329
heraldry 456–67 (see also under Columbus, Christopher)
Herculano, Alexandre quoted 14–15

Hercules, Duke of 126
Hermnanus Secundus 253
Herrera (chronicler) 127
Hidgen, Ranulpho 51
Hispaniola map 75, map 77, 82, 204, 483
Hojeda, Afonso de 158, 237, 417
Holy Ghost (Paraclete), cult of 3, 7–8, 11, 39, 54, 183, 284–5, 478
and name Columbus 85
Holy Grail 9–10, 163
Homem, Álvaro Martins 62–3, 149–1
Hospitallers, Order of the 36, 41
Huerter, João 118
Hugo, St, Abbot of Cluny 8

Iceland 62, 105, 122, 129, 147–8, 150, 162–3
as Thule 143–4, 146
Idanha-a-Nova 41
Ielling rock ill 156
Ilha Longa 331
Imago Mundi note 227–8
India 49–50, 50–1, 60, 70, 122–3, 131, 251–2
Indian Ocean 71
Innocent II, Pope 20
Innocent III, Pope 6–7
Inquisition, Holy Office of 13–17, 29, 352–3, 416–17, 494
Inter Caetera, Bull 70, 180, 203
Ireland 145, 163–4
Isabel, Duchess of Braganza (wife of Fernando 3rd) (aunt) tree 365, 351, 357, 486–7
Isabel, Duchess of Burgundy (wife of Charles 'the Bold') tree 444, 329–30, 343, 495
Isabel, Duchess of Burgundry (wife of Philip 'the Good') tree 444, 343
Isabel, Duchess of Coimbra (wife of Prince Pedro) tree 344, tree 444
Isabel, Empress (wife of Emperor Charles V) tree 444, 345, 442, 446–7
Isabel, Queen of Castile (wife of Charles II) tree 444
Isabel, Queen of Portugal (the 'Holy Queen', wife of Dennis) tree 409, 10–11
Isabel, Queen of Portugal (wife of Alphonse V) tree 444, 164, 344, 357, 433
Isabel de Braganza, Princess (sister of King Duarte) 345
Isabel de Braganza (wife of Prince João) 327
Isabella, Queen of Spain ('the Catholic', wife of Ferdinand V) tree 444, 69,

93, 140, 330–1, 345, 368–9, 412,
  414–16, 432, 447
 death of 215
Isabella (island) *map* 75, 214, 331
Isabella (Puerto Rico) 331
Italián family 224
Italy
 as birthplace of CC 82, 97, 185–6
 cosmography and navigation 107,
  122–3, 126
 Italian as language of CC 86, 88, 90,
  93, 503
 navigation in 107

Jacob-ibn-Josef, 'Almansor' 37
Jácomo of Bruges 138–9
Jadeu, Dom Judas 26
Jahia-aben-Jaisch 21
Jaime, 4th Duke of Braganza *tree* 365,
  363, 373
Jamaica *map* 75, *map* 77, 218
James I, King of Aragon *tree* 409, 410
James II, King of Aragon 41
Jane, 'the Fool', Queen of France (wife of
  Philip the Fair) *tree* 444
Jan Mayen island 146
Japan 69–70, 120, 122–3, 167, 173, 238,
  254, 321
Jehuda 'the Coneso' 253
Jerusalem 8, 204
Jews 13–17, 19–32, 239, 352–3, 487
 CC as Jewish 99, 111, 188
 CC's mission for 506
 compulsory conversions of 28–30
 contribution of 55, 118, 254–5, 257
 Jew in CC's *Memorandum* 208, 210
 names of 7–8, 352–4, 502; Zarco
  239–40
 and Order of Christ 43
Joana, Duchess of Braganza (wife of
  Fernando, 2nd Duke) 351, 363, 381
Joana, 'Holy Princess' *tree* 344, 329, 345,
  357
Joana, Queen of Castile (wife of Henry
  IV) 419–20
João, Duke of Viseu and Beja (half-
  brother) *tree* 344, 43, 486
João, Martim 22
João, 'Master' 60
João, Pedro 22
João, Prince (son of John I) *tree* 365, 327
João Coelho, Land of 55, 61
John I, King of Aragon 433
John I, King of Portugal *tree* 344,
  tree 362, *tree* 365, *tree* 409,
  tree 444, *tree* 454, 20, 25–6, 31, 253,
  353, 369, 373, 419

John II, King of Portugal and Jews 27,
  tree 344, *tree* 444, 29, 60, 105–6,
  228–30, 263, 268, 329, 347, 357, 448,
  472
 and CC 129, 316, 368; letter to
  CC 103, 185, 228–30, 244; 1488
  meeting with 72, 140, 412; 1493
  meeting with 69–70, 393–7, 399;
  mission given to CC
  68–9, 141
 conspiracy against 164–6, 229, 370
 and navigation 118–19, 123, 134;
  secrecy obligations 43–4, 125
John III, King of Portugal 371, 442
John XXI, Pope 10, 37–8
John XXII, Pope 10, 20, 42, 257
Jorge, Duke of Coimbra 345, 357, 371
José, Master 29, 118
Joseph-aben-Jahia 21
Juan, Prince (son of Ferdinand and
  Isabella) 199–200, 204, 215, 331
Juana, Princess (daughter of Ferdinand
  and Isabella) 200, 215, 236, 414, 447
Juana, Princess ('la Beltrâneja') 420
Juana (island) 331
Juda fi-de-Mosé 253
Judáh, Rabbi 23
Judâh-aben-Mosséh Navarro, dom 25
Julião, Cardinal Pedro 37–8
Juromeña 37
Jusaf-aben-Abasis 25

Labrador 62–3, 103, 143, 158
Labrador (Lavrador), João Fernandes 63,
  143, 149, 158
La Cerda family *tree* 409, 423
La Cerda, Fernando de *tree* 409, 49
La Cerda, Isabel 412
La Cerda, Isabel de *tree* 409, 410
La Cerda, Juan de 422
La Cerda, Juan de, 2nd Duke of
  Medinaceli *tree* 409, 408, 410
La Cerda, Luis de 48–9, 171
La Cerda, Luis de, 1st Duke of Medinaceli
  tree 409, 410, 412
La Cerda, Manuel de 412
La Cerda, Margarida de *tree* 409, 413
La Cerda, Maria de *tree* 409, 410
La Cerda, Martim Gonçalves de *tree* 425,
  422, 436
La Cerda, Nuno Pereira de *tree* 425, 379,
  421–2, 425–6, 436
La Cueva, Beltrán de 419–20
La Cueva, Carlos de 220
La Cueva, Juan de 220
La Cueva, Juana de 219
La Frontera, Pedro (Pero) Vasques

de 58, 164, 320, 338, 441
Lamego, José de 28
Lancerote, island of 54
Lanuza, Juan de 237
La Peña, Juan de 25
Lara, Pedro Nunes de 419
La Rábida *map* 451, 104
La Rábida, monastery of 94, 367
  and friars Marchena 364, 367–8, 370
Las Casas, Friar Bartolomeu de
  quoted 82–3, 91–3, 94, 102, 109, 117,
  124, 129, 131, 185, 207, 217, 228, 236,
  325–6, 328, 368, 388, 416, 441, 472–3
Las Casas, Guillherme 54
La Torre, Juana de 226
La Vega, Elvira de *tree* 362, 410
Leão, Dom Nunes de 385
Ledesma, Fernando de 408
Ledesma, first Marquis of *see* La Cueva,
  Beltrán de
Leiria, Cortes of 30
Leitão, Belchior de Andrade 421, 423
Leme, António 62
Leon, Ponce de 61
Leonor, Empress (wife of Frederick III)
  tree 444, 387
Leonor, Lady of Duénas 445
Leonor, Princess (daughter of John
  II) 357
Leonor, Queen of Portugal (3rd wife of
  Manuel) 487
Leonor, Queen of Portugal (wife of John
  II) *tree* 344, *tree*, 444, *tree* 454,
  30–1, 119, 246–7, 263, 270, 351, 357,
  396–7, 420, 486
  arms of *ill* 270
  half-sister of CC 316
Lepe, Diogo de 135
Lima, Francisco de *tree* 402, 401
Lima, Leonel de *tree* 402, 401
Lima, Luís Caetano de *quoted* 374, 23, 384
Limi, Álvaro Gomes de 405
Linscholten, Jan Huygen van 168
lion (in heraldry) 458–9, 467
Lisboa, Friar António de 67–8
Lisbon *table* 34, 4, 25, 29–30, 46
  CC living in 101, 211
  'House of the Twenty-Four' 487
  Perestrelo family in 347
  University of 53–5
Livery Collar, Order of the 285
Lobo, Diogo Lopes 379
Lobo, Pedro, Lord of Alvito 408
Lobo, Rui Dias 363
Lopes, Fernão 419
Lorgues, Count Roselly de 191
Lorris, Guillaume de 259
Lorris, treaty of 7

Louis VIII, King of France 7, 20
Louis IX, (St Louis), King of France 7,
  410
Louis XI, King of France *tree* 409, 329,
  357
Louis XII, King of France 159
Lucena, Vasco Fernandes de 134
Lugo, Alfonso de 414
Lully, Raymond 54, 67, 108, 125, 164,
  166, 259

Macedo, Joana de 118
Machim, Legend of 55–6
Madeira *map* 76, 29, 87, 103, 128, 130,
  135, 139, 169, 171, 314, 338
  discovery of 55–6
Magalhâes Teixeira, Fernão de 496
Magellan, Ferdinand 71, 122, 129, 496
magnetic variation 126
Majorca 108
Majorca, Jácomo de 28, 52–3
Malacello, Lanzarote 48
Malaga 93
Manrique, Inés 434
Manuel, Dom Nuno 355
Manuel, Friar Dom João 355–6, 412
Manuel de Portugal, Mencia *tree* 409, 412
Manuel de Portugal, Pedro 220
Manuel do Nascimento, Francisco 15
Manuel I, King of Portugal *quoted* 402,
  tree 344, *tree* 365, *tree* 444, 133–4,
  157–9, 363, 434, 487
  and armillary sphere 71, 238
  half-brother of CC 316
  and Jews 14, 16, 29–31
  meeting with CC 396–7
  and Order of Christ 43
Marcham, Christóvam de 367
Marcham, Pêro de 364, 367–8
Marchena, Friar Antonio de 104, 364–70,
  416, 449
Marchena, Friar Juan Peres de 104, 108,
  368–70, 383, 408, 416, 449
Marchena, Juana Ponce de *tree* 409, 408
Marchena, Maria Ponce de *tree* 411, 369
Marchena, Maria (II) Ponce de 369
Marchena, Rui Gonçalves de 367, 369–70
Marchena, Rui Ponce de 410
Marchena, Urraca de *tree* 409
Madresio, Doctor Fracesco 182
margarita, significance of 476–8
Margarita (island) 478
Marguerite (de Provence), Queen of
  France *tree* 409
Maria, Queen of Portugal (2nd wife of
  Manuel I) *tree* 344, *tree* 365,
  tree 444, 487
Maria Afonso, Princess 408

INDEX   565

Marinho family   171
  arms of   315
Marrón, Catalina (sister-in-law in cuevas
  account) *tree*   190
Martel, Charles   35
Martines, Ecija Hernando   26
Martins, Fernando, Canon   166
Martins, Gil, Friar   42
Martin V, Pope   10, 20, 43, 135, 434,
  498–9
Mary of Burgundy, Empress (wife of
  Maximilian I) *tree*   444, 330, 343
Mascarenhas, Martim Vaz de *tree*   447,
  353–4
Mascarenhas, Nuno *tree*   366, 367
Mathematicians, Junta of   118
Matinimo (island)   485
Matosinhos   370–1
Mauro, Fra   125
Maximilian I, Emperor *ill*   474, *tree* 444,
  119, 247, 330, 343, 345
Maximo, Bishop of Merida   162–3
Mayoradgo deed   199–200, 202, 216–17
Mayorazgo deed   200–5, 225
Mazagan   169
Medici, Cosimo de   120
Medinaceli family   423
Medinaceli, Duke of   103, 180, 196, 408,
  416, 432
Medina-Sidónia, Duke of   408, 413, 416,
  432–3
Mello family *tree*   453, 378–82 (*see also*
  Afonso de Mello, Martins de Mello,
  Pereira de Mello, Soares de Mello)
  coat-of-arms *ill*   462
Mello, Branca de *tree*   366–*tree* 367,
  tree 447, *tree* 453, 378–9, 445
Mello, Briolanja de   379
Mello, Brites de *tree* 365, *tree* 411,
  tree 447, *tree* 453, *tree* 497, 378–9,
  385
Mello, Constança de *tree*   453, 379
Mello, da Cunha, Rui de, Admiral   385
Mello, Diogo de *tree*   453, 378
Mello, Estêvão de *tree*   402, *tree* 453, 385
Mello, Estêvão Soares de   418
Mello, Filipa de *tree*   365, *tree* 453, 379
Mello, Francisco de, 3rd Count of
  Tentúgal *tree*   365
Mello, Garcia de *tree*   447, *tree* 453, 379
Mello, Guiomar de   418
Mello, Inês de *tree*   445, *tree* 452, 448
Mello, Isabel de *tree* 365, *tree* 367,
  tree 453, *tree*, 493, 361, 363, 378–9,
  381, 385, 401
Mello, Joana de *tree*   366, 420, 488
Mello, João de *tree*   366, 378
Mello, Jorge de   379

Mello, Leonor de *tree*   453, 379, 488
Mello, Maria de *tree*   452, *tree* 453, 401,
  418
Mello, Martim Afonso de *tree*   365,
  tree 367, *tree* 402, *tree* 425, 363,
  378–9, 401, 412
Mello, Martim de (II)   378
Mello, Nuno Álvares Pereira de *tree*
  II 365, 363
Mello, Rodrigo de *tree*   365, *tree* 453, 363
Mena, Juan   87
Mendez, Diego   159
Mendez, Diogo   210
Mendoça family
  coat-of-arms *ill*   463
Mendoça, Afonso Furtado de   285
Mendoça, Ana de *tree*   411, 345, 357
Mendoça, Catarina Furtado de (Brites)
  tree 362, *tree* 493, 256
Mendoça e Vasconcelos, Francisca de
  tree 493, 491, 495
Mendoza, Diego Hurtado de *tree*   411,
  414
Mendoza, Elvira Hurtado de   434
Mendoza, Isabel de *tree*   411, 415
Mendoza, Juan Hurtado de   434
Mendoza, Juana de *tree*   411, 415
Mendoza, Lope de, Archbishop of
  Santiago   99
Mendoza, Pedro de   415
Mendoza, Pedro de, Cardinal   180, 196,
  412
Mendoza y Bobadilla, Beatriz de   413
Menéndez, Diego   217
Menezes, Afonso Telles de   491
Menezes, Beatriz de *tree*   365, *tree* 402,
  385, 420
Menezes, Duarte de *tree*   366–*tree* 367,
  tree 402, *tree* 411, *tree* 453, 363–4,
  367, 369–70, 385, 401, 408, 413, 423–4
Menezes, Garcia de, Bishop of Évora   87,
  446
Menezes, Henrique de   400
Menezes, Inês de   354
Menezes, João de, Prior of Crato
  tree 411, *tree* 367, *tree* 402,
  tree 411, 364, 395, 400–1, 413
Menezes, Leonor de *tree*   362, *tree* 365,
  357, 386
mermaid (in heraldry)   314–15, 457
Mértola   37, 371
Mesa, Don Cristóbal de   430–1
Milan   13
Mina (Elmina, West Africa) *map*   59, 58,
  60, 105, 112, 128, 130, 135, 227, 394,
  396, 424, 489
Minorita, Friar Paolino de   51
Miranda, Aires de *tree*   447, 424

566 INDEX

Miranda, Francisco de *tree* 424
Miranda, Gomes de *tree* 425, 424
Miranda, Margarida de 361, 385
Miranda, Violante de 420, 424
Miranda Henriques, Francisco de 424
mirror, symbolism of *ill* 458, 457
Mithra, rite of 271–3
Moeller, Camile Johan 449
Molay, Jacques de 40–1
Moleto, Giuseppe (publisher) 183
Moliarte, Miguel *tree* 356, *tree* 362, *tree* 452, 225, 347–8, 448–50
Moniz family *tree* 452, *tree* 497
Monetarius, Ieronimo 118
Moniz, Diogo *tree* 402, *tree* 452, *tree* 453, 346–7
Moniz, Duarte Martins *tree* 355
Moniz, Egas 22, 353
Moniz, Henrique Martins *tree* 355, 354
Moniz, Isabel *tree* 362, *tree* 366, *tree* 452, *tree* 497, 256, 355–6, 422 (*see also* Perestrelo)
Moniz, Maria 353
Moniz, Moninho 22
Moniz, Vasco Martins *tree* 452, 354–5
Moniz Barreto, Isabel *tree* 356, 185
Moniz Barreto, Vasco (II) 355
Moniz Perestrelo, Catarina (sister-in-law) *tree* 356, 346–8, 384
Moniz Perestrelo, Filipa (wife) *tree* 190, *tree* 356, *tree* 362, *tree* 366, *tree* 402, *tree* 444, *tree* 445, *tree* 452, *tree* 497, 101, 185, 206, 256, 338, 345–8, 353, 355–6, 371, 383, 399, 439, 489, 491, 496
  coat-of-arms 255–6
  death of 210–11
Moniz Perestrelo, Isabel (sister-in-law) *tree* 402, 345–7, 347–8, 385–6, 401
Moniz Perestrelo, Violante (sister-in-law) *tree* 356, *tree* 452, 221, 224–5, 347–8, 450, 448
Moniz Telles, Henrique 353
Monsanto, Count of 27
Montalboddo, Francesco 181
Montarrojo, Pedro de 67–8
Montecorvino, João de 49
Montfort, Simon de 7
Moors 13, 20–1, 21, 23, 46, 48 (*see also* Arzila, Tangier)
  and Canary Isles 169, 171
  reconquista 4–5, 66
Mosquera, Maria de 219
Mosséh-aben-Navarro 26
Mossêh Chavirel, Dom 25
Moura 37
Moura, Maria de 425
Moura, Pedro de 379

Moyseh, Rabbi 24
Mozambique 68
Mozarabs 46
Mueller, Camil Johan 119
Munzer, Jéronimo *quoted* 52
Muratori (author), document of 197–8
Muet, battle of 7
Muslims *see* Moors

Nahum, Dom Moysés 26
Nahum, Judas 26
Negro, David 26
Negro, Pablo de 208–10, 212
net (in heraldry) *ill* 459, 258, 268, 270, 456
Neto, Pedro Gonçalves 134
Newfoundland (Island of Cod) *map* 154, 58, 62–3, 103, 149, 172
Newport (Rhode Island) 153
Nicaragua *map* 77, 160
Nicholas, Lord of Alemy 495
Nicholas IV, Pope 23, 371
Nicholas V, Pope 43, 50, 135, 387
'Niña' 388, 390–2, 392–3, 395, 417
Nogaret, Guillaume de 41
Nogueira, Afonso 345
Nogueira, Madre Violante 345
Nogueira, Violante 375
Nola 128
Nola, Antonio de 136, 138
Noroña family *tree* 411 (*see also* Noronha)
Noroña, Fernando de, 2nd Count of Vila Real *tree* 411
Noroña, Isabel de *tree* 411
Noroña, João de 437
Noroña, Maria de *tree* 411, *tree* 445, 437
Noronha family 407 (*see also* Noroña)
Noronha, Beatriz de *tree* 447, 397, 413
Noronha, Constança de *tree* 447, 386
Noronha, Duarte Abreu de 418
Noronha, Fernando de *tree* 362, *tree* 402, 354, 385–6, 397
Noronha, Fernâo de 58, 159
Noronha, Filipa de (aunt) *tree* 445, *tree* 447, 351, 354, 374, 401, 412
Noronha, Isabel de *tree* 402, *tree* 447, 401
Noronha, Leonor de 353–4
Noronha, Manuel de *tree* 445, 448
Noronha, Maria de *tree* 366–*tree* 367, *tree* 402, *tree* 445, *tree* 447, 351, 354, 385, 401, 443, 448
Noronha, Martinho de 395–7
Noronha, Mécia de *tree* 445, 385
Noronha, Pedro de, Archbishop of Lisbon *tree* 356, *tree* 362, *tree* 402, *tree* 452, 345, 348, 386, 397, 401
Noronha, Sancho de 381

Nossa Senhora de las Nieves 395
Nova, João da 58
Nova Scotia map 154
Nunes, Pedro 28, 31, 122
Nuñez, Ana 427, 429, 431, 435, 437
Nuñez, Anton 430
Nuñez, Lucia 431
Nuñez, Velasco 430

Oderigo, Nicolo 95, 98–9, 223–5
Oliva de la Frontera, as birthplace of CC 98
Oliveira, Heitor de 420
Ombesto, Beringel 47
Omega, Cape map 341, 332
Orejón, Catarina de tree 409, 410
Orleans, Council of 18
Orta, Garcua de 31
Orta, Samuel d' 482
Osório, Jéronimo 15
Ourém, Count of 25
Ourique, battle of 36
Our Lady of the Assumption (channel, West Indies) 396
Oviedo, Gonzalo Fernández de (chronicler) 82, 135, 185, 473

Pacheco Pereira, Duarte quoted 61, 246–7, 342
Pais, Gualdim 36–7
Paiva, Afonso de 28, 51, 68
Palencia, Synod of 19
Palha, Simão 110, 354
Pallastelli, David Gabriel 255
Pallastrelli (Palastrelo) Filipe 256
Palmela 37, 371
Panama 160, 218
Pantoja de Góis, Luís 496
Paraclete see Holy Ghost
Paris, as birthplace of CC 98
  Treaty of (1229) 7
Passagno, Admiral 49
Patarenes see Cathars
Pavia, University of 89, 503
Payen, Hugues de 38
Pedro, Prince, Duke of Coimbra tree 344, tree 444, 10, 49, 87, 120, 433
Pedro of Castile, Dom 419
Pelagius of Cantabria 36
pelican (in heraldry) 263, 267–8
Pellegrino, Christofforus 191
Pellegrino, Johannis 191
Pepin, 'the Taciturn', King of France 35
Peraza, Gullém 414
Peraza, Inés de 414
Peraza de Avala, Hernán 414
Pereira family tree 425
Pereira, Álvaro Gonçalves, Prior 354

Pereira, Beatriz tree 411, 354, 421
Pereira, Brites tree 453, 385
Pereira, Diogo tree 366, 374
Pereira, Diogo Nunes 425
Pereira, Duarte Pacheco quoted 61, 246–7, 342
Pereira, Fernando Afonso 374
Pereira, Gonçalo, Archbishop of Braga 180, 354
Pereira, Grimaneza tree 452, 354
Pereira, Lenor Dias 434
Pereira, Maria Fernandes 434
Pereira, Nune, tree 366–tree 367, tree 447, 374, 421, 423–4
Pereira, Payo Gonçalves 354
Pereira, Ruy Dias 379
Pereira, Violante 422
Pereira Barreto, Brites 379
Pereira de La Cerda, Nuno see La Cerda, Nuno
Pereira de Serpa, Rui 381
Peres, Fernán 415
Peres, Juan 368
Perestrelo family tree 493, 255–6, 353, 422, (see also Moniz Perestrelo)
  coat-of-arms ill 462
Perestrelo, Bartolomeu (father-in-law) tree 356, tree 362, tree 366, tree 445, tree 452, tree 493, tree 497, 56, 134, 256, 347, 355, 361, 422
Perestrelo, Bartolomeu (son of above) tree 356, 256, 362
Perestrelo, Bartolomeu III 256
Perestrelo, Catarina (sister-in-law) see Moniz Perestrelo
Perestrelo, Filipa (wife) see Moniz Perestrelo
Perestrelo, Isabel (Branca) Moniz tree 362, tree 452, 360
Perestrelo, Rafael 256
Perestrelo, Violante (sister-in-law) see Moniz Perestrelo
Pessanha, Carlos 363
Pessanha, Carlos, Admiral 384–5, 385
Pessanha, Genebra 363
Pessanha, Lanzarote, Admiral 385
Peter II, King of Aragon 6
Peter IV, King of Aragon 433
Peter I, King of Castile tree 454, 410, 415
Peter I, King of Portugal tree 409, 24
Peutinger, Konrad 247, 249, 252, 312, 343
Peypertuse, fall of 7
Pezagno, Manoel, Admiral 363, 384
Pezagno family see Pessanha
Philip II, King of France 7
Philip IV, 'the Fair', King of France tree 444, 39

Philip 'the Good', Duke of Burgundy
   tree 444, 330, 343
phoenix (in heraldry) *ill* 249–*ill* 250,
   248, 258, 268, 457
Piacenza, as birthplace of CC 98, 187–8,
   501
Pierre, Abbot of Cluny 20
pigeon, symbolism of 258
Pina, Rui de 70, 186, 395–6
'*Pinta*' 117, 129, 388–9, 393
Pinzón, Arias Perez 126
Pinzón, Juan Martín 218
Pinzón, Martin Alonso 129, 320, 388
Pinzón, Vicente 160
Pires, Tomé 31
Pisano, Matheus de 87
Pius II, Pope 120, 125
Pizón, as Cathar name 8
Pizzigano, Zuana, *map* of *ill* 154,
   ill 170, 152–3, 156, 169, 171–2
'*Pleyto Successorio*' 219–21, 237, 320
Pó, Fernando 57
Poitiers, Alphonse of 7
Poitiers, battle of 35
Poland, as birthplace of CC 111
Polo, Marco 49, 54, 124
Pombal 41
Ponce de Leon, Pedro *tree* 411, 360
Pontevedra (Galicia), as birthplace of
   CC 98
Porrás, Diego 216
Porte, Geronimo de 191
Porte, Pedro Balascio de 191
Portocarerro, Guiomar de 385
Portocarrero, Aldonza de 220
Porto Rico *map* 75, *map* 77
Porto Santo (Madeira) 87, 139, 169,
   346–7, 383
Portucalense 8
Portugal 8, 15, 86–7, 246
   as birthplace of CC 82, 112, 120, 129,
      139–40, 198, 217–18, 442–3, 503
   CC in service of 196
   cosmography in 71, 151
   exploration 47, 70, 72, 133–41, 150;
      restrictions of foreign ships 134, 138
   and Inquisition 13, 15
   Jews in 19–32
   navigation in 117–31; geographic
      measurements 122–3, 127–8, 128,
      137, 144, 147; secrecy policy 43–4,
      60, 125, 135, 167–8, 338
   political estates of 31
   Portuguese as language of CC 88–9,
      91–4, 99
   shipbuilding 117–18, 135, 168
Portugal, Álvaro de *tree* 365, *tree* 453,
   220, 378
Portugal, Isabel de (grand-daughter) 219
Portugal, Jacob de 219–20
Portugal, Jorge Alberto de 219–20, 378
Portugal, Friar Laurentino de 51
Portugal, Leonor de 220
Portugal, Nuno de 220–1
Portugal, Pedro de 220
'Prester John' 50–1, 67, 105
Principe (island) 247
'Prophecy of the Western Sibyl' 247–52,
   312
Puerto, Geronimo de 208–9
Puerto Santo (Cuba) *map* 341, 334

Rabiçag-aben-Cayut 253
Radulf (monk) 20
Ragel, Aben (alchemist) 92, 120
Raimundo, Soeiro 22
Raymond, Count of Barcelona 4–5, 109
Raymond (VI), Count of Toulouse
   table 35, 6–7
Raymond (VII), Count of Toulouse 7
Recho, Nicolo da 48–9
Redinha 41
'Religion of St John' 10
Remón, Friar Alonso (quoted) 368
Resende, Garcia de 394
Reykjannes, Cape 147
Ribeiro, Gonçalo 495
Riverol, Francisco de 223–4
Rodrigues, Afonso 434
Rodrigues, Justa 355
Rodrigues, Master 118
Rodrigues de Mafra, João 125, 129
Rodriguez, Martin (notary) 199, 217
Rojo, Cape *map* 341, 334
rope (in heraldry) *ill* 270, 268, 460
Roriz, Fernando (Fernando Martins) 120
Rosa, Constanza 219
rose (in heraldry) *ill* 267, 258, 267–8, 268,
   460
Roupinho, Dom 47
Roupinho, Fuas (Admiral) 22
Ruiva (village) 327 (*see also* Vila Ruiva)
Ruiz, Juan, Canon of Cordoba 435

Sá family, coat-of-arms *ill* 463
Sá, Brites de 498
Sá, Constança Roiz (Rodrigues) de (wife
   of João Gonçalves Zarco) *tree* 362,
   tree 365–*tree* 367, *tree* 445,
   tree, 454, *tree*, 493, *tree* 497, 244,
   498–9, 500, 502
Sá, João Rodrigues de *tree* 453, 363,
   498–9
Sá, Rodrigo Anes de *tree* 362, 498, 500

Sacavém   395
Safi   424
Sagres *map*   451
'Sagres School'   49, 51–3, 68–9, 108, 356–7
Saint Giles, Council of   6
Saliniério, Julio (historian)   195
Salvador, as name of CC   183
Salvaterra-do-Extremo   41
Samba, Isabel   219
Sampayo, Diogo de   379
San Antonio *map*   341, 334–5
San Bartolome *map*   340, 335
Sancha Espada, Navarro   41
Sanches, Rui   70
Sanchez, Afonso   130, 149, 198
Sanchez, Gabriel *quoted*   417
Sánchez, Maestro Juan   428
Sanchez, Rodrigo   417
Sancho I, King of Portugal   21, 37
Sancho II, King of Portugal *tree*   409, 22–3, 37
Sancti Spiritus *map*   340, 335, 341
San Domingo   329
San Jorge *map*   341, 335
San Juan, Cape of *map*   340, *map*   341, 335
San Luis (Cuba) *map*   341, 335
San Luis (Portugal) *map*   340, 335
San Miguel (island) *map*   340, *map* 341, 335, 389–90, 391, 393
San Nicholas, channel of *map*   341, 335
San Salvador (Bahamas) *map*   75, 69, 482
Santa Catarina (Canary Islands)   399
Santa Catarina (Curaçao) *map*   341, 335–6
Santa Catarina (Portugal)   336
Santa Clara (Cuba) *map*   341, 336
Santa Clara (Madeira)   336
Santa Clara (Portugal) *map*   340, 341, 336, 389
Santa Cruz (Azores)   57
Santa Cruz (Cuba) *map*   341
Santa Cruz (Portugal) *map*   340, 336
Santa Lucia (Antilles) *map*   341, 336
Santa Lucia (Portugal) *map*   340, 336
'Santa Maria'   117, 129, 393, 417, 428
Santa Maria (Azores)   390–1, 392–3
Santa Maria das Nieves   395
Santa Maria de Souto (Portugal)   389
Sant'Ana, Friar Gaspar de   354
Sant'Ana (Guadelupe) *map*   341, 336
Sant'Ana (Portugal) *map*   340, 336
Santángel, Luis de   226, 412–13, 417
Santarém, Joâo de   57, 125
Santarém (channel, West Indies)   336–7, 396
Santarém (Portugal)   10, 21, 36–7

Santiago (Cape Verde islands)   136, 138, 337
Santiago (Portugal) *map*   340, 337
Santo Domingo (Guinea)   337
Santo Domingo (Hispaniola) *map*   341, 337
Santos, Convent of   343, 345
Santos-o-Novo, Convent of   371
Sanuto, Marino, map of   125
San Vicente *map*   341, 337–8
'São Cristovão'   394
Saona (Italy), as birthplace of CC   97
São Thomar and Principe, archipelago of   57
São Tomé   29
Sargasso Sea   68, 112, 125, 149, 153, 155, 320
   CC's knowledge of   236
   pre-Columbian discovery of   58
Sarmiento, Diego Perez *tree*   410, 445
Sarmiento, Leonor *tree*   367, *tree* 410–*tree*   411, *tree* 447, 445
Sarmiento de Villamayor, Diego Perez   445
Sarria, Juan Lorenzo de   418
Satanazes *map*   154
Saucedo, Juan Garcia de   431
Savonna   191
   documents of   191–2
Savoy, Count of   329
Sciarra Colonna   499, 500–2
Sebastian, King of Portugal *table*   34, 364
Segovia, Bishop of   419
Sequeira, Maria   375
Serpa (Portugal) *map*   340, *map*   451, 37, 335, 381, 412
Serpa, Rui Pereira de   381
Serpent (in heraldry)   285
Serrão, José   129
set-square(in heraldry) *ill*   459, 456
Setúbal (Portugal) *map*   340, *map*   451, 363
Seven Cities, Island of *map*   147, 58, 60, 120, 162–73
   as Iceland   162–3
Seven Cities, lake of   162
Sforgia, Sibila   433
Signorio, Giovani de   191
Silves, Diogo de   57
Sines, fortress of   104
Siurdi of Stephani, map of *ill*   157
Skolp, Jan   110, 150
Soeiro, Dom, Bishop of Lisbon   22–3
Soligo, Cristóforo   125
Soto, Fernando de   437
Soto, Mencia de *tree*   447, 437
Sotomayor, Maria Vargas de *tree*   366, *tree*   367, *tree*   447, 446

Sotomayor, Fernando de 413
Soure 41
Sousa, Branca de *tree* 410, *tree* 425, 412
Sousa, Briolanja de *tree* 365, *tree* 425, *tree* 453, 378
Sousa, João de 72
Sousa, Lopo de 364
Sousa, Lopo Dias de 370
Sousa, Maria de *tree* 366, *tree* 402, *tree* 445, 370
Sousa, Mécia de *tree* 453, 379
Sousa, Pedro de, Lord of Beringel *tree* 366, 364
Sousa, Pedro de, 1st Count of Prado 374, 420, 487
Sousa, Ruy de *tree* 366, 72
Spain 158
  as birthplace of CC 82, 98
  Catalonia 4, 13, 109; as birthplace of CC 100–4, 107, 140, 259, 501; navigation in 107–8
  cosmographic knowledge in 71
  division of exploration rights with Portugal 70
  shipbuilding 117–18
  Spanish as language of CC 86, 89–90, 93–5
  and western route to India 122, 141
Spinola, Antonio 211
Spinola, Baptista 211
St Christopher Island (Cape Verde) 136
St George, Bank of (Genoa) *see under* Genoa
St Helena 58
St James, Order of 42, 360, 371–2
St John, Order of *see* Hospitallers, Order of the
St John Island 159
St Lawrence, Church of 5
St Lucia *map* 77
St Matthew, islands of 58
St Nicholas (West Indies) 396
St Paul (island) 57–8
St Peter (island) 57–8
strap, leather (in heraldry) 258
St Salvador (island) 330–1
St Salvador (village) 104
Switzerland, as birthplace of CC 111
Sylva, Catarina da *tree* 366, 401
Sylva, Filipa da *tree* 453, 379
Sylveira family *tree* 366, 380
Sylveira, Fernando da 384
Sylveira, Fernão da *tree* 366, *tree* 447, 379
Sylveira, Isabel da *tree* 366, *tree* 367, *tree* 453, 378–9
Sylveira, João da *tree* 366, 363, 368, 387
Sylveira, João Lobo da, 3rd baron of Alvito 368, 370
Sylveira, Maria Lobo da 363
Sylveira, Mécia da 421
Sylveira, Nuno da 378

Tables of the Declination of the Sun *ill* 121, 119–20, 481
Talavera, Fernando de, Friar 413
Tangier 27, 256, 363, 423–4, 436
Tavera, Inés, de *tree* 411, 416
Tavira 37, 57, 364
Teggia, Angiolino del 48
Teive, Diogo de 57–8, 62, 126, 139, 149, 163–4, 320
Teixeira family *tree* 452, *tree* 497, 495–6
Teixeira, Fernão de Magalhães (Magellan), *tree* 452, 496
Teixeira, Francisca de 496
Teixeira, Guiomar 256
Teixeira, Henrique Moniz 353
Teixeira, Joana *tree* 362, *tree* 366, *tree* 452, *tree* 497, 355, 496
Teixeira, Lançarote 496
Teixeira, Tristão Vaz (I) *tree* 452, 56, 457, 495
Teixeira, Tristão Vaz (II) *tree* 452, 459
Teixeira, Violante *tree* 445, *tree* 452, 496
Telles, Henrique Moniz 353
Telles, Leonor 25
Telles, Maria 415
Templars 8–10, 15, 26–7, 39, 41, 152, 257, 272, 348–9
  customs of *ill* 40, 38–40, 257, 272, 284–5
  extinction of 24, 38–43, 353
  secrecy rule 55, 70
Terceira (Azores) 135, 138–9, 149, 389–90
Teresa, Countess of Portugal 9
Terracossa di Monoconesi, as birthplace of CC 98
Terra, Roja, as birthplace of CC 110–11
Terrarossa, as birthplace of CC 98, 326
Terra Rubra 325–7, 360, 380
Terra Ruiva (village) 104
Thule *map* 147, 62, 143–8, 163
  CC's positioning of 143–4, 145–6
  as Greenland 143, 148
  as Iceland 143–4, 146
Toledo *table* 34, 36
  Council of 19
  Law of 216
  Tables of 253
  Treaty of *map* 74, 66–7, 68, 166, 206, 228, 230, 392, 396
Toledo, Archbishop of 420, 432
Toledo, Cristóbal de (grandson) 219
Toledo, Diego de (geat-grandson) 219
Toledo, Francisca de 220–1

Toledo, Maria de (wife of Diogo Colón)
 tree 444, 219
Toledo, Pero de 93
toleta de marteloio 118–19
Tomar 17, 37, 43, 241
 derivation of 273
Tomar, Monastery of Christ in
 and tower in Newport (RI) 153
Tordesillas, Treaty of 61–2, 66, 72, 203,
 206, 216, 230, 237, 307, 383, 400, 504
Toro, battle of 414–15, 420, 432, 443,
 446–7
Torquemada, Gonzalo de 433
Torquemada, Isabel de 502
Torquemada, Juan de 435
Torquemada, Mosén Don Juan de 433
Torquemada, Pedro de 427, 432, 435
Torquemada, Thomas de, Friar (Grand
 Inquisitor) 433, 502
Torre, Juana de la 196
Torres, António 212
Torres, Francisco de 375
Torres, Luis de 417
Tortosa 4–5, 110
 as birthplace of CC 109
Toscanelli, Pablo del Pozzo 88–9, 101,
 120, 164, 192, 196, 326
 map, of *ill* 165, *ill* 166, 120, 122, 124,
 129, 167, 172, 236, 254
Toulouse 4–5
 Counts of see Raymond (VI & VII)
Trencavel, Viscount of *table* 35, 6
Trinidad, island of *map* 77, *map* 131,
 map 340, *map* 341, 337
 pre-Columban discovery of 61
Trujillo 37
Tunis 105

Ughelli, Abbot Ferdinand 4–5
Ulloa, Afonso 84
Ulloa, Isabel de *tree* 411, 416
Ulloa, Juan de *tree* 411, 416
Ulloa, Luis de 139, 224–5, 416
Ulloa, Maria de *tree* 411, 416
Ulm, Fernando de see Dulmo (de Ulm),
 Fernando
Urban II, Pope 35
Urries y Ruiz de Arana, Juan de 434
Usodomar, Antonio 137

Vagos (Portugal) 156
Vale do Paraíso 338, 395–6
Valentin, Basile (alchemist) 259
Valladolid 207
Valsequa, Gabriel 125
Vasconcelos family *tree* 493, 492
Vasconcelos, Aldonça de 491
Vasconcelos, Diogo de 493

Vasconcelos, Isabel de *tree* 452,
 tree 497, 496
Vasconcelos, Joane Mendes de *tree* 497,
 491
Vasconcelos, João Mendes de 421
Vasconcelos, Martim Mendes de
 tree 445, *tree* 493, *tree* 497, 491
Vasques, Antão 26
Vaz, Estêvão 72
Vazo, Antonio 208, 210
Velho Cabral, Gonçalo 57
Venezuela *map* 75, *map* 77, 131, 160
Vera Cruz 60
Verágua (Panama) *map* 77, 218, 329
Verzelino (jurist) 97
Vespucci, Amerigo 158–60, 160, 181
Vicente, Catarina *tree* 362, 256
Vicente, Gil 14
Vicentolo, Bernardina de 220
Vieira, Father António 15
Vikings *ill* 155–*ill* 156, 62, 153
Vila Alva 361
Vila de Alegrete 424
Vila Ruiva 328, 351, 360–3, 380
Vila Viçosa *map* 451, 351, 361, 373, 381
Vilhena, Filipa de 378
Vilhena, Joana de *tree* 367, *tree* 411, 400
Vilhena, Leonor de *tree* 411, *tree*, 445,
 tree 453, 413
Vilhena, Margarida de *tree* 365–*tree* 366,
 tree 453, 363, 378
Vilhena, Marquis of 420, 432
Villabon, Dr 330
Vinland 153
Violante, Queen of Castile (wife of
 Alphonse X) *tree* 409, 410, 422
Visinho, José 135
Vivaldi, Ugolino 105
Vivaldi, Vadino 105

West Indies see Antilles, *and individual*
 islands
wheel (in heraldry) 460
Windward Islands 61
Wynendale, Bailio de 139

Xara family 502
Xara, Juan Garcia de 502

Zacuto, Abraão 20, 118–19
Zacuto, Abraham-ben-Samuel 108
Zacuto, José 28
Zacuto, Marino 67
Zacuto, Rodrigo 28
Zacuto, Samuel ben 119
Zafra, Pedro de 434
Zarco family 239–40 see also
 Colón-Zarco (under Columbus)

Zarco, Isabel (Sciarra) Gonçalves
(mother)   364, 382, *(see also* Câmara, Isabel)
Zarco, João Gonçalves Zarco (grandfather)
tree 362, *tree* 365–*tree* 367,
tree 445, *tree* 454, *tree* 493,
tree 497, 11, 17, 43, 56, 130, 174, 239, 241, 256, 314, 350, 361, 370–1, 384–5, 412, 414, 498–9, 504
takes name Câmara 239
coat-of-arms 331
Zuñiga, Isabel de 415